ACOUSTICAL HOLOGRAPHY

RECENT ADVANCES IN ULTRASONIC VISUALIZATION

Volume 7

ACOUSTICAL HOLOGRAPHY

RECENT ADVANCES IN ULTRASONIC VISUALIZATION

A Continuation Order Plan is available for this series. A continuation order will bring delivery of each new volume immediately upon publication. Volumes are billed only upon actual shipment. For further information please contact the publisher.

ACOUSTICAL HOLOGRAPHY

RECENT ADVANCES IN ULTRASONIC VISUALIZATION

Volume 7

Edited by

Lawrence W. Kessler

Sonoscan, Inc.
Bensenville, Illinois

SPRINGER SCIENCE+BUSINESS MEDIA, LLC

The Library of Congress cataloged the first volume of this title as follows:

International Symposium on Acoustical Holography.

    Acoustical holography; proceedings. v. 1-
New York, Plenum Press, 1967-

      v. illus. (part col.), ports. 24 cm.

    Editors: 1967-   A. F. Metherell and L. Larmore (1967 with H. M. A. el-Sum)
    Symposiums for 1967-   held at the Douglas Advanced Research Laboratories, Huntington Beach, Calif.

    1. Acoustic holography—Congresses—Collected works. I. Metherell. Alexander A., ed. II. Larmore, Lewis, ed. III. el-Sum, Hussein Mohammed Amin, ed. IV. Douglas Advanced Research Laboratories. v. Title.

QC244.5.I 5   69-12533

Library of Congress Catalog Card Number 69-12533
ISBN 978-1-4757-0655-0   ISBN 978-1-4757-0653-6 (eBook)
DOI 10.1007/978-1-4757-0653-6

Proceedings of the Seventh International Symposium on Acoustical
Holography and Imaging held in Chicago, Illinois,
August 30-September 1, 1976

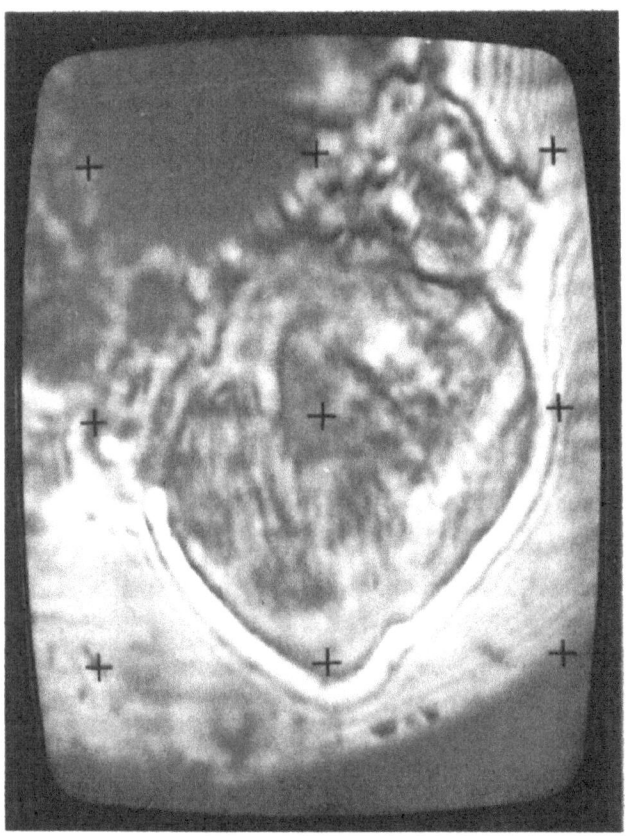

Acoustic micrograph at 100 MHz of beating mouse embryo heart. Due to opacity, the internal structure revealed here cannot be seen with an optical microscope. The cross marks are placed 1 mm apart.

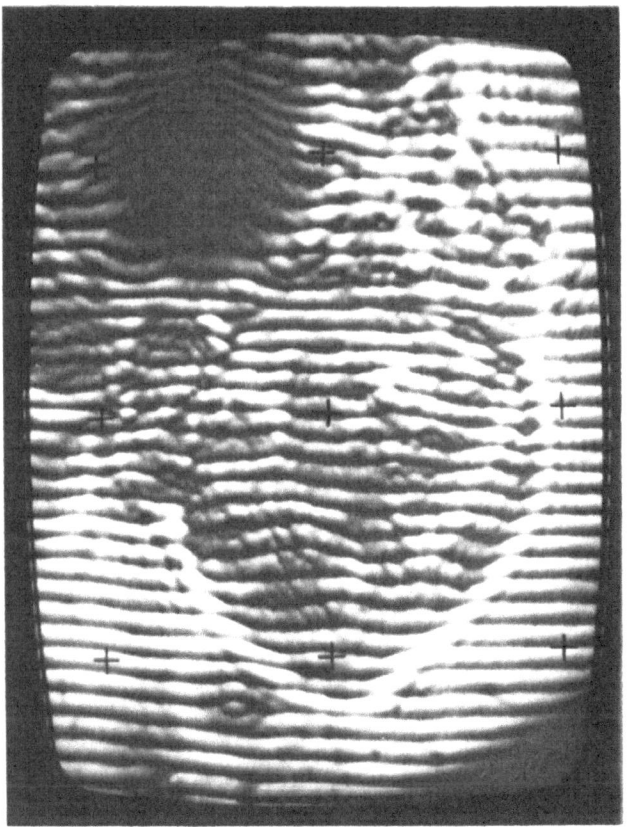

Acoustic interferogram of same from which the variations in acoustic index of refraction are revealed everywhere by the vertical displacements of the horizontal fringes. Quantitative data are obtained graphically.

APPLICATION OF ACOUSTIC MICROSCOPY TO CARDIOVASCULAR RESEARCH

Both images are still frames from a video tape recording of the cardiac contraction. The study is being carried out by R.C. Eggleton and F.S. Vinson on their Sonomicroscope 100, Scanning Laser Acoustic Microscope which is manufactured by Sonoscan, Inc. See Chapter 2.

PREFACE

This volume contains the papers presented at the Seventh
International Symposium on Acoustical Imaging and Holography.
The meeting was held in Chicago, Illinois U.S.A. on August 30 -
September 1, 1976. Since 1967 this series of conferences has
served as an international forum for bringing together the many
different facets of the ultrasonic visualization field. Medical
diagnosis, nondestructive testing, underwater viewing, seismic
mapping, and acoustic analytical microscopy are examples of the
broad range applications of this technology. Throughly broad in
scope, the common denominator that binds this volume together is
the application of engineering and physics disciplines to the
creation of images by means of ultrasonic waves.

Efforts on the part of the program committe and their institutions
are greatly appreciated. They are acknowledged below:

Newell Booth	Navelex
	Arlington, Va.
Byron B. Brendon	Holosonics, Inc.
	Richland, Wa.
Reginald C. Eggleton	ICFAR-IU Medical School
	Indianapolis, In.
Philip S. Green	Stanford Research Institute
	Menlo Park, Ca.
Henry Karplus	Argonne National Labs
	Argonne, Il.
J.D. Meindl	Stanford University
	Stanford, Ca.
Alexander F. Metherell	University of California
	Irvine, Ca.
R.K. Mueller	University of Minnesota
	Stillwater, Mn.
Frederick L. Thurstone	Duke University
	Durham, N.C.

Glen Wade   University of California
                             Santa Barbara, Ca.
Keith Wang   University of Houston
                             Houston, Tx.

The meeting was co-sponsored by Sonoscan, Inc. and the Office of
Naval Research under contract # N00014-76-C-0749, in cooperation
with:
        IEEE Group on Sonics and Ultrasonics
        The Acoustical Society of America

The next, i.e. EIGHTH INTERNATIONAL SYMPOSIUM ON ACOUSTICAL
IMAGING AND HOLOGRAPHY, is planned for Key Biscayne, Florida
during the spring of 1978 and will be organized by Dr. Alex F.
Metherell.

The editor would like to express his appreciation to Jean
Kazmerski, Gloria Kessler, Lois Michaels, Joyce Parney, and Helen
Pasik for their assistance in making the chairman task always
bearable and sometimes enjoyable. Special thanks are extended to
Jean Kazmerski for her efforts in compiling this volume.

L.W. Kessler

CONTENTS

THE USE OF ULTRASOUND METHODS TO DETECT CHANGES IN BREAST TISSUE WHICH PRECEDE THE FORMATION OF A MALIGNANT TUMOR*

E. Kelly Fry, Indiana University School of Medicine and

Indianapolis Center for Advanced Research, Indianapolis,

Indiana 46202 U.S.A.

INTRODUCTION

The approach taken, to date, by most investigators studying the use of ultrasound visualization for breast examination has been traditional, namely: scan the region of the palpable mass; compare the imaging pattern of the mass, as it appears on the echogram, to known features of benign and malignant tumors, such as wall contours; attempt to identify the physical character of the mass by its effect on reflection and transmission of the ultrasound beam; and finally, strive to detect smaller and smaller masses using this approach. In view of the fact that there is a large population of women presently harboring an undetected malignant mass within their breasts and in consideration of the statistical data on life span of the breast cancer patient in relation to size of the tumor within the breast at the time of initial treatment, this approach is valid.[1-3] However, in consideration of the known systemic nature of breast cancer, that removal of the breast is still the primary treatment even under the circumstances of identification of minimal malignant masses, and in view of the potential of ultrasound to detect pre-malignant changes in breast tissue, this limited approach is not sufficient.

Although earlier investigators recognized the continuing lethal effect of breast cancer over the lifetime of the patient, and thus the systemic nature of the disease, it is only recently that this systemic aspect has been emphasized.[2-5] The average patient who undergoes surgical treatment for a malignant mass within the breast, including the classical radical mastectomy procedure, will have a decreased life span as a result of recurrence of some aspect of the disease associated with the systemic factors.[3] The recurrence may

* Supported by Showalter Residuary Trust, Indianapolis, Indiana.

1

take place decades after the initial surgical procedure. Apparently, the disease is already systemic for most of the United States
women who currently receive a diagnosis of breast carcinoma and
undergo surgical treatment as a result of that diagnosis.[3] Clearly,
there is a need for a diagnostic technique which will recognize the
presence of breast carcinoma prior to the formation of an overt
mass.

Before discussing possible approaches to this problem, it is
important to emphasize here that recognition of the systemic nature
of breast carcinoma should not, in any way, decrease efforts to increase the efficiency of detecting small malignant masses in the
breast since both earlier and current clinical studies clearly indicate that there is a significant relation between the life span of
the patient, the size of the tumor, and extent of nodal involvement
at the time of initial treatment.[1-3,6-8] There are many factors involved in this data, (including the histological character of the
malignant mass) which cannot be discussed here since the approach of
detecting small masses in the breast is not the primary subject of
this paper. However, it is clear from existing statistical data that
efforts to identify small malignant breast masses prior to nodal me-
tastasis should continue. At the present time, knowledge of the
basic character of breast carcinoma is extremely limited. It is
known, however, that (a) there is a large population of generally
older women (well over the 90,000 United States women whose tumors
will be detected in the coming year) who now harbor a malignant mass
within their breast; and, (b) there is a population of younger women
whose breasts are already undergoing the changes that precede the
formation of a malignant mass. Therefore, two separate instrumentation approaches should be taken to the breast carcinoma detection of
problem, namely: (1) improvement of all presently existing examination techniques with a goal of accomplishing reliable detection of
malignant masses less than 0.5 cm in largest diameter and (2) development of instrumentation with the capability of detecting those
breast tissue changes which precede the formation of a malignant
mass.

Since a number of authors have found a relationship between
breast carcinoma and ductal hyperplasia and some have suggested that
detection of enlarged ducts by mammographic techniques can be used
to increase the efficiency of breast carcinoma detection, ultrasonic
detection of disturbed ductal structures may prove to be a valid
approach to early detection of malignant changes within breast
tissue.[9-17] This paper discusses the results of the present author's
and associates' earlier investigations on detection, by ultrasound
visualization techniques, of enlarged ductal structures in excised
breasts. Also included are the preliminary findings of current
studies on detection of ductal structures of in vivo subjects by an
ultrasound B-mode visualization system.

BACKGROUND

The normal procedure in the case of a patient undergoing a biopsy to determine whether neoplastic tissue is present in a breast
mass is surgical excision of the mass, and frozen section examination
of the excised tissue. If carcinoma is diagnosed, a mastectomy is
carried out and both gross and microscopic examination of the breast
and the tumor follow. Unfortunately, however, sectioning and histologic staining, in most cases, is confined to the tumor and selected
areas within the breast. The major mass of breast tissue remains
unexamined except for gross inspection. It is not possible, therefore, to correlate pre-operative diagnostic examination of the breast
by mammography, xeromammography, or ultrasound with the possible
overall tissue changes in the pathological breast or to a sufficient
degree, with the tissue changes in the area surrounding the tumor and
in the tumor itself. Adequate correlative information on the generalized changes in the breast and in the immediate region of the tumor
requires whole breast sectioning and histologic staining.

Gallager and Martin's approach to the problem of devising a
method for correlating pre-operative examination data with the tissue changes that are presumably present in breasts containing a
malignant mass is to have a mastectomy performed without any surgical intervention in the area of the tumor and to study the breast
tissues by whole breast sectioning and histologic staining.[9-12]
Such a procedure is only carried out in the case of patients whose
breast carcinoma can be diagnosed with certainty by x-ray and needle
biopsy techniques. A partial summary of the early surgical specimen
preparation technique used in the Gallager and Martin studies is
provided in the following excerpt:[9]

    "To prepare a radical mastectomy specimen for whole
organ sectioning, the axilla is removed, and the breast
(with its attached pectoral muscles) is divided into 3 or
4 parallel blocks, each 6 to 8 cm thick. The plane of
sectioning is selected by reference to the mammograms.
Each block is molded into a shape conforming as nearly
as possible to that shown in the mammogram and is fixed
in this position, dehydrated, infiltrated and embedded
in paraffin. Sections 12 to 20 microns thick are cut at
intervals of 0.5 to 1.0 mm. Thus, a series of slides is
prepared which represents the entire sectioned area 3-
dimensionally. In the earliest specimens studied, the
entire breast was sectioned. However, it was learned
that in older patients the extreme lateral and medial
parts of the breast are composed almost entirely of
adipose tissue. Accordingly, in subsequent specimens
only the center blocks have been cut, the outer ones
being embedded and reserved for later sectioning if
necessary".

"Assembled sections and mammograms are examined
by a radiologist and a pathologist working together.
Immediate comparisons and correlations are made and
recorded."

A number of experimental goals were included in these histo-
pathologic/mammographic studies but the initial, primary aim was improvement in the detection of small, early breast tumors by mammographic techniques. The results obtained in these investigations on
the histological structure of overt tumor masses are significant to
the problem of detecting such masses by ultrasonic techniques. However, since detection of small breast masses is not the subject of
this paper, these specific results will not be discussed in this
paper.

In more than three quarters of the originally examined breast
specimens containing invasive carcinoma, Gallager and Martin found
areas of intraductal carcinoma or intraductal hyperplasia with epithelial atypism.[9] Such involved ducts are dilated or surrounded by
increased amounts of hyalin connective tissue. The subsequent hyperplasia of the ductal structures is sufficient to allow detection by
mammogram inspection. Although the affected ducts may be located
close to the malignant mass, this is not necessarily the case. They
may be widely distributed to remote sections of the breast.[9-10] The
data of Gallager and Martin, as well as the findings of previously
cited authors, seems to indicate that hyperplasia of ductal structures and carcinoma are causally related.

In an effort to correlate the echogram patterns of malignant
breast tumors with the structural and histological features of such
tumors, a cooperative ultrasound/histopathologic investigation was
undertaken with the previously referenced H. S. Gallager and the
present authors and associates.[18] The fundamental approach consisted of detailed ultrasound visualization of the previously described
formalin-fixed breast specimens and comparison of the echograms with
the histological sections. This early research was not directed at
a specific study of the ductal structures. Rather, its primary aim
was to determine the relationship between the structure of various
classes of malignant tumors and the attenuating or non-attenuating
character of such structures in respect to the applied ultrasonic
beam. These aspects are not detailed in this paper. Only the observations made on enlarged ductal structures of one of the examined specimens is discussed and related to the subsequent attempts to
detect ductal structures of normal subjects.

METHODS AND PROCEDURES

A. Ultrasound Visualization of Formalin-Fixed, Excised Breasts Containing an Intact Malignant Tumor

More than one type of ultrasound transducer was used to study the excised breast tissue but the data presented in this document was obtained with a computer-controlled, B-mode visualization system, used in conjunction with a 5 cm diameter, 10 cm focal length lead zirconate titanate transceiver with a frequency of 2.5 MHz. Earlier publications discuss the specific details of the acoustic, electronic aspects of the system.[19,20]

The transducer motion consisted of a sector sweep with the transducer pivoting about its axis for an angular motion of $\pm$ 15 degrees. After each single scan, the transducer was moved in appropriate linear steps (2 to 5 mm) and another scan was recorded. This method of operation was continued until, essentially, the complete specimen of tissue was scanned. For a sagittal (vertical) scan, the transducer was positioned so that it swept the tissue along the superior-inferior axis. For transverse (horizontal) scanning, the transducer was turned 90 degrees from the position set for sagittal scanning. Polaroid photographs of the information presented on a CRT were used to record the breast scan information.

Essentially a search technique for the purpose of (1) identifying on the echogram the normal components of the tissue under study, and (2) recognizing any pathological structures within the normal framework was used in the study of these excised breasts. In order to relate data recorded on the echograms to specific areas of the tissue, an anatomical landmark on the tissue (such as the center of the nipple or the center of a prominent skin discoloration over the area of pathology) was selected, and a highly reflective and attenuating acoustic target was placed in this landmark. A scan was made with the target in place and the two linear coordinates (X and Y) and the depth coordinate (Z) of the transducer position recorded. The echogram of the breast tissue with the target in place showed a distinct, easily identified reflection of the target. Therefore, subsequent scans of the tissue with target removed could be directly related to the chosen anatomical landmark. The data on the X, Y and Z coordinates was automatically printed on each echogram with the aid of the computer.

The response signals of the transducer were fed into an attenuator calibrated in decibels from 1 to 80 db. Prior to carrying out the search procedures, appropriate ranges of attenuator settings were determined, as well as the best location in the tissue for the focus of the ultrasound beam. In addition to the method outlined above for obtaining echograms with sufficient information for the

identification of normal structures and the detection of pathological growths, another more complex method was used which can yield
echograms with a multiple focus and with a variable db setting for
each focal region. This was carried out with the aid of the computer and gating circuitry for the display system.[19] This multiple
focus technique indicated that sufficient information could be obtained with the single focus to allow detailed scans by this method.

B. Ultrasound Visualization of Breasts of In Vivo Subjects

The instrumentation applied for the examination of breasts of
in vivo subjects was a linear scan, B-mode visualization system designed to obtain information useful for clinical application but not
to fulfill the function of a routine clinical instrument.[21] The general instrumentation system included: pulsers with PRF of 500 to
1000/sec. and rise times varying from 20 nsec. to 0.1μsec. (for echograms displayed in this paper, PRF 1000/sec., rise time of 0.1μsec.),
a zero to 80 db attenuator preceding a logarithmic amplifier, a display system consisting of a PEP 400 scan converter, a 14" GBC monitor, a 9.5 × 7.5 cm dedicated monitor and a Polaroid camera. Various
amplifiers have been designed and tested in an attempt to obtain
adequate amplification, over a wide frequency spectrum, of the signals received from fine structures of the breast (present units
under test have 10μv sensitivity and a bandwidth of 600 KHz to 13
MHz; for echograms displayed in this paper, the amplifier had a
sensitivity of 40μv, frequency response of D.C. to 5 MHz, and a
dynamic range of 80 db). A series of relatively small diameter
(1.9 cm), concave disc-type transducers with center frequencies from
2.25 to 10 MHz were specifically designed for the breast visualization study. A lead metaniobate, 1.9 cm diameter, 7.5 cm focal
length transducer with a midband frequency response of 4.4 MHz and
a bandwidth of 3 MHz just above the noise and 1 MHz at the 3 db
point was used to obtain the echograms displayed in this paper (for
in vivo subjects with a direct water coupling between transducer
and breast). For the two echograms shown in this paper for in vivo
subjects with a water bag coupling between transducer and breast, a
lead metaniobate, 5 cm diameter, 10 cm focal length with a 3.0 MHz
frequency response and a bandwidth of 3 MHz just above the noise and
0.5 MHz at the 3 db point was used.

The overall instrumentation system, but in particular the transducers and amplifiers, were initially tested for probable adequacy
of display of structures such as ductal structures by using fine
grain natural sponges as test targets. In addition to using the fine
grain of the sponges as breast phantoms, a variety of types of solid
structures, varying in diameter from 0.25 mm to 5.0 mm, were either
embedded in the sponges or used on the surface as test targets of
known dimensions and locations. These structures included nylon
threads, 1 mm plastic hollow tubing, steel wires and tungsten aral-
dite attenuating spheres.

RESULTS

Prior to discussing the results obtained in these studies, it is
of value to consider the primary features of the ductal structure of
the female breast. In the lactating breast, milk is conducted to the
nipple by means of excretory ducts. These ducts are part of the
glandular system, which, on a macroscopic scale, consists of fifteen
to twenty lobes arranged radially around the nipple, with each lobe
having a single excretory duct. From the nipple surface, the ducts
run dorsally and essentially parallel and at the base of the nipple
diverge, first forming a wide reservoir (ampulla or sinus lactiferus)
and then proceed in a complex pattern of fine sub-divisions to the
peripheral regions of the breast. Each of the terminal branches of
a duct lead into small structures called lobules, containing the
basic secretory tissue (acini or alveoli). There is varying information on the size of the primary ducts. Martin and Gallager indicate diameters of 0.5 to 0.8 mm in the nipple and subareolar
region.[12] Earlier authors have indicated diameters as large as 1.5
to 2.5 mm; for the ampulla region, generally located just below the
areola, diameters as wide as 5 to 9 mm have been reported.[22,23]
Figure 1 is an artist's view of the primary breast structures in the
field of view of a single, transverse ultrasound scan across the nipple. The fine acini structures have been exaggerated in size in
order to demonstrate the glandular terminations of the ducts.

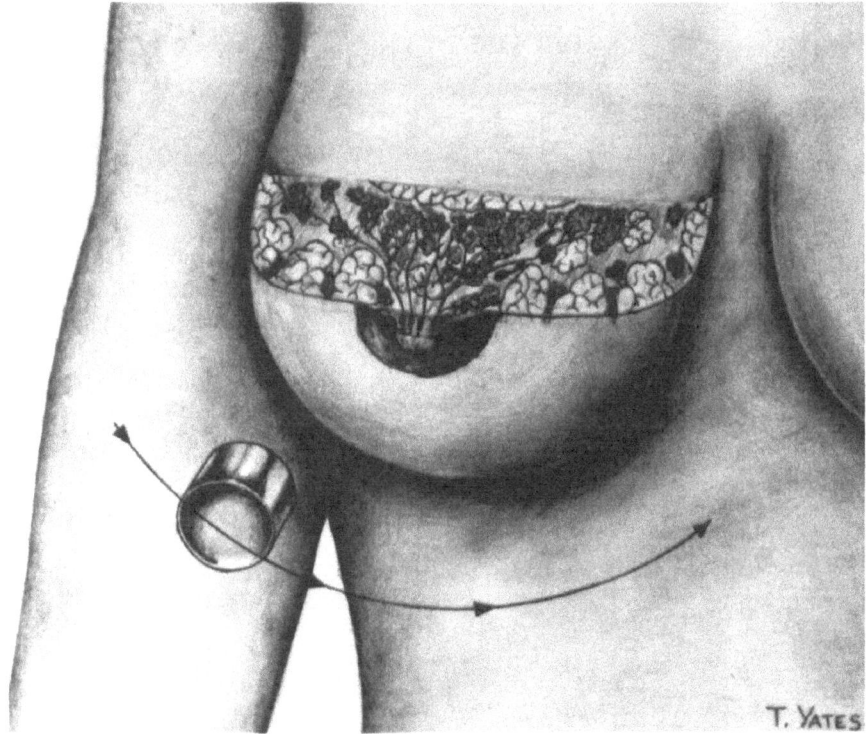

Figure 1

The echogram patterns obtained from scans of the formalin-
fixed whole breast of a 39-year-old subject with a confirmed diagnosis of breast carcinoma indicated the presence of discrete, vertical, tract-like structures in the general area beneath the nipple
and areola, with the tracts apparently traversing the areas from the
deep regions of the breast to the nipple surface. These tissue
tracts can be observed in the echograms shown in Figure 2, which resulted from a sagittal scan of the excised breast; i.e., a scan at
right angles to the plane shown in Figure 1. Although the sagittal
scans appeared to demonstrate the tract-like structures with greater
clarity, they were also visible on transverse scans, as demonstrated
in Figures 3a and 3b. (The echogram shown in Figure 3b was obtained by scanning the same tissue plane as that of Figure 3a, but at a
5 db less attenuator setting).

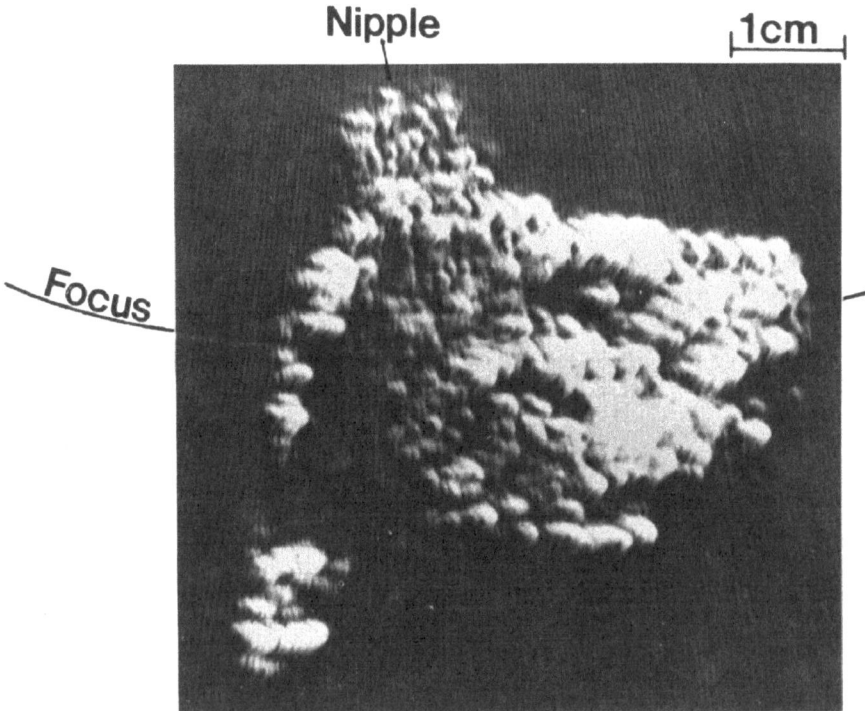

Figure 2

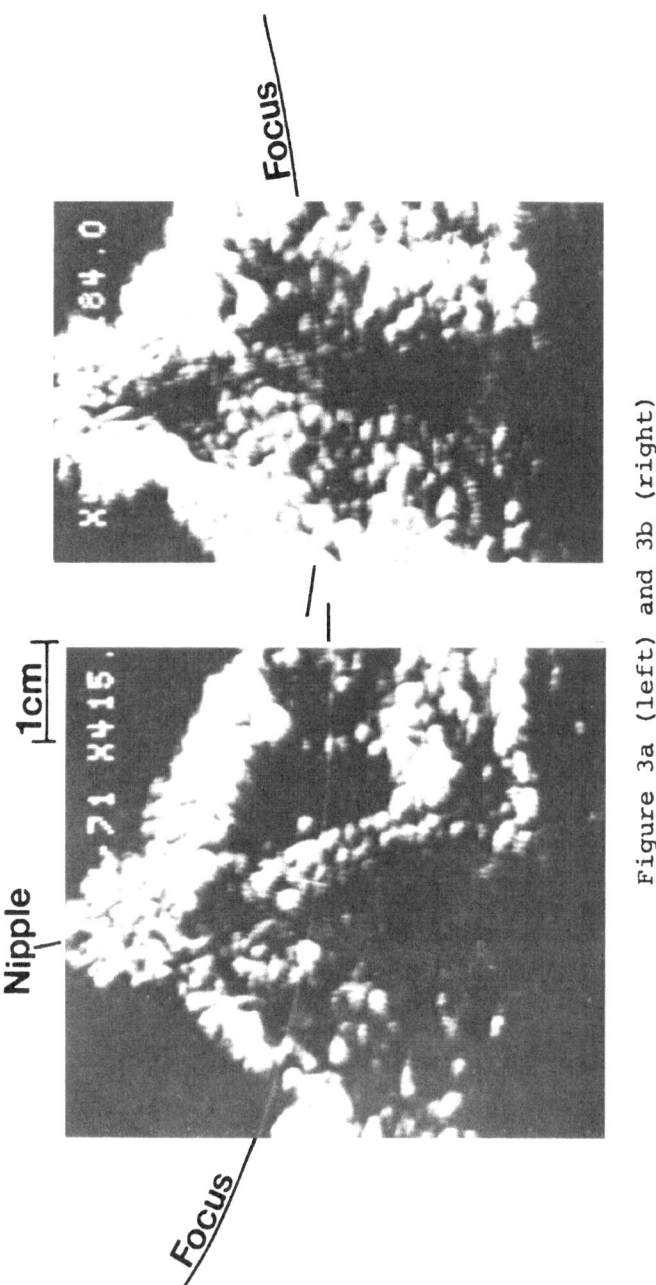

Figure 3a (left) and 3b (right)

The general character and location of the echoes located beneath the nipple in Figure 2 would lead to the tentative conclusion that ductal structures are being visualized. This assumption was confirmed by the histological sectioning and staining. Figure 4 is a duplication of just one of the histological sections for the region below the nipple. Some of the enlarged ductal structures are quite obvious in this cross-section.

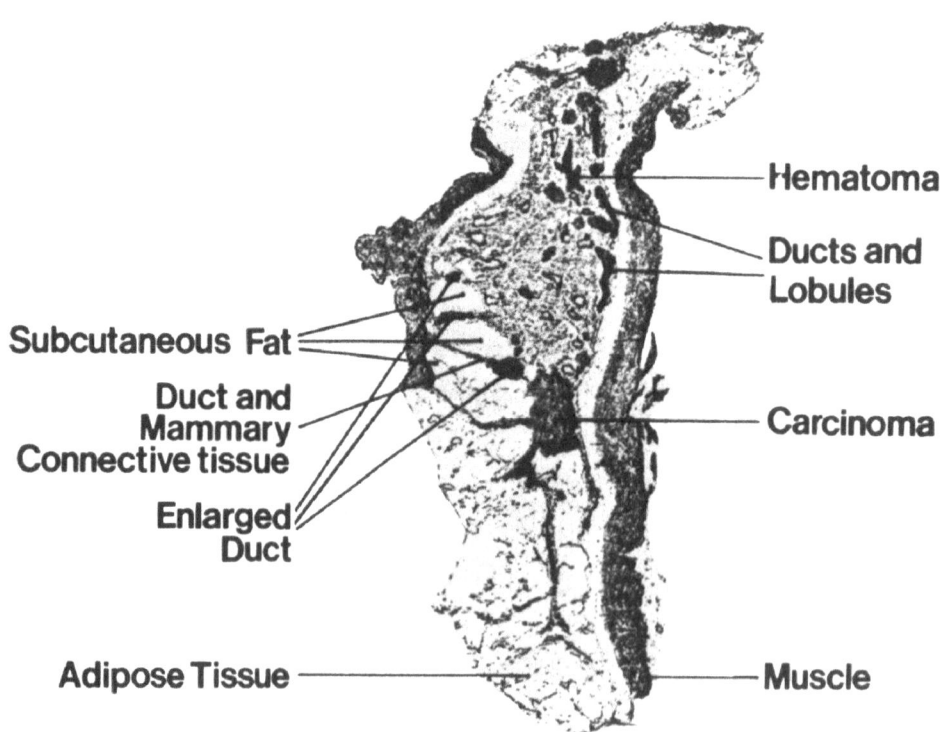

Figure 4

After considering the need for a non-ionizing radiation method
for examination of the ductal structures of the breast, and after
reviewing the earlier studies on excised breasts, investigations
were recently initiated on application of ultrasound visualization
techniques for detection of ductal structures of normal subjects.
At the present time, this investigation is in a preliminary stage.
One practical initial approach to the problem of ultrasonically detecting the fine tube-like ductal structures is to use relatively
high frequency, focused transducers to obtain adequate resolution
and to confine the search to the region directly beneath the nipple
and areola, a region where the ducts have their largest diameter and
are greatest in number per unit area. Until such time that the more
advanced transducers (such as phased array or synthetic aperture)
are readily available for ultrasound breast examination, simple
focused transducers, either of the lens or disc type, can be used in
clinical programs to obtain much-needed data on breast structure.
Small diameter, concave disc units are particularly useful because
the three basic design parameters of such transducers---1/2 diameter,
wavelength and radius of curvature of the disc---can be formulated
in varying combinations and test units can be fabricated with relative ease. Thus, the importance of factors which probably have a
specific influence on successful detection of discrete structures in
the breast (such as attenuation, beamwidth, transverse (D_t) and
axial (D_a) focal region diameters, pressure amplitudes in regions on
either side of the focus) can be experimentally determined.[24-27]
Further, although small diameter transducers have decreased sensitivity in comparison to the larger diameter discs, they have distinct
advantages in regard to designing relatively simple clinical instruments for examination of the breast. The units mentioned in this
paper were originally designed for a hand-held, rapid scanner.

Earlier studies of the author and associates demonstrated the
enhanced value of direct water coupling between breast and transducer in comparison to the more traditional water bag technique.[20,28]
Nonetheless, in view of some of the practical aspects of the water
bag coupling technique, a clinical-research study on the possible
advantages and limitations of a modified water bag technique, when
used in conjunction with wide aperture focused transducers, was undertaken.[21] Figures 5 and 6 (3 MHz transducer) resulted from this
study rather than the present specific efforts to detect normal duct
structures. They are shown here to illustrate that it is possible
to detect fine structural features, including ducts, while using the
water bag method but, in comparison to direct water coupling, this
method is less effective. As can be clearly seen in Figure 5, some
of the ductal structures below the nipple are evident but they cannot readily be traced to their exit point at the nipple. In detecting discrete breast structures, the placement of the focus of the
transducer is significant. One aspect of this is illustrated in
Figure 6, which represents the case of the focus just above the

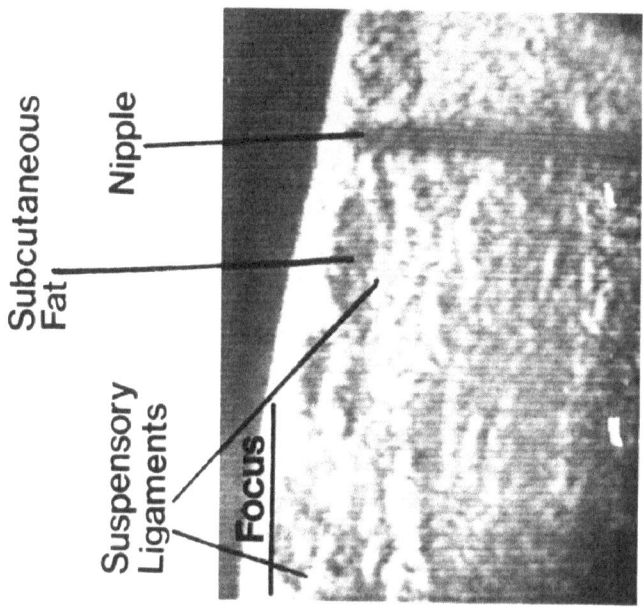

Figure 6

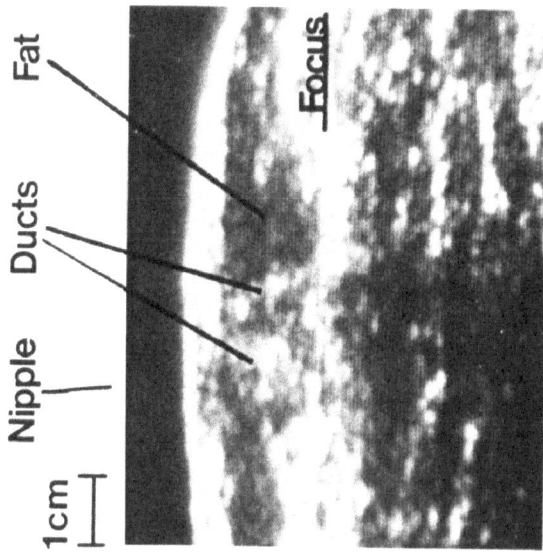

Figure 5

nipple surface. It is apparent that although the fine structures of the subcutaneous fat and suspensory ligaments are well visualized, no tissue structures can be viewed in the shadow resulting from the attenuation of the nipple. This effect was demonstrated with all ages of subjects. If, in fact, an attenuating target, which is approximately the same size as the nipple, is placed on any non-nipple/ non-areola surface of the breast, a simulation of effect of nipple structure in relation to the placement of the focus can be clearly demonstrated.

Although the water bag coupling technique is not recommended, there is one application of this method which may prove of value in studying the ducts of the breast, namely, use of the water bag for regions other than nipple and areola. If the subject is lying partly on her side, so that the outer breast surface is essentially horizontal, the water bag compresses the tissues but does not cause the type of distortion observed in the nipple. The compression of the tissue is an advantage from the viewpoint of attenuation. It may be of value, therefore, to consider the water bag technique in attempting to locate the ducts in the outlying areas of the breast.

Figures 7a and 7b are echograms obtained under the circumstance of direct water coupling between transducer and skin surface. The coupling was accomplished by an adaptation of the water bags used for the previously mentioned study (the bags are designed to contour the top and side surfaces of the breast reasonably well). A circular opening (as large as 10 cm in diameter) was cut in the mylar bag and the bag then placed on the breast surface with the nipple in the center of the opening, and the cut edges carefully formed to the skin surface. If the cut edges of the bag are properly contoured to the skin surface, and care is taken in the initial pouring of the water, it will be found that the combined effect of the upward flotation of the breast and the downward pressure of the water forms a good seal without the aid of any oil or glue. This technique worked well on both young and older subjects with average and large size breasts.

Figures 7a and 7b are echograms of the same subject (different time periods) obtained while using this coupling technique and the 4.4 MHz transducer described under Methods and Procedures. Figure 7a was obtained by means of a sagittal scan mode, while 7b resulted from the more conventional transverse scan. The ductal structures are clearly apparent in both echograms but, as in the case of the excised breast previously described, the ducts are better visualized in the sagittal scan. However, further studies on a large number of subjects need to be carried out before any conclusion is made in this regard.

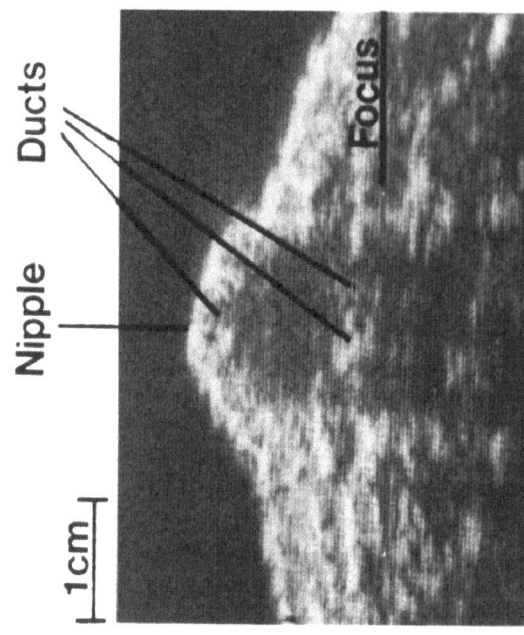

Figure 7b

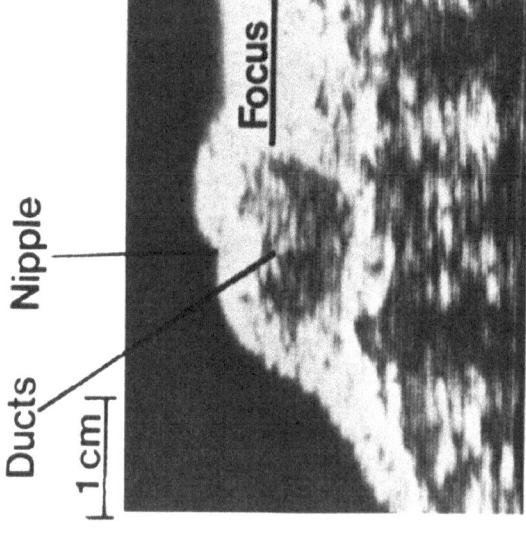

Figure 7a

It is clear from the results obtained in these preliminary studies that normal ductal structures can be visualized. Improvements in the quality of the imaging is dependent on further instrumentation development.

DISCUSSION

As pointed out in the Introduction, there is a serious need for methods that are capable of detecting breast tissue changes which may precede the formation of a malignant mass. The mammographic and histological data of the previously cited authors on ductal changes accompanying breast carcinoma are significant in this regard. However, at the present stage of knowledge, the presence of ductal epithelial hyperplasia is not a proof of malignancy. The exact nature of the causal relationship between ductal changes and breast carcinoma are not known. Gallager and Martin indicate that epithelial hyperplasia is premalignant in a nonobligate sense, that is, although epithelial hyperplasia is one of the changes that must take place during the carcinoma process, hyperplasia per se may be due to causes other than a malignancy.[9-10]

Although the histological and mammographic studies carried out, to date, on ductal changes detected in patients with breast cancer, are significant, they are limited by the nature of the examination technique. Obviously, histological studies can only be carried out on excised breasts. X-ray examination of the breast has usually been confined to special categories of subjects and may be further restricted in the future. If, however, sufficient improvements are made in ultrasound instrumentation, the ultrasound method could be used to examine large populations of women of all ages, both normal and those with a breast pathology. Thus a large amount of data, particularly on the younger women, could be collected which might, in conjunction with histological and mammogram information, lead to further understanding of the causal relationship between ductal changes and breast carcinoma.

The author would like to indicate that it is not being proposed in this paper that ductal structures be investigated without regard to the other tissues in the breast. Since this paper is concerned with the specific topic of detection of ductal structures by ultrasound, only the ductal changes which accompany a breast malignancy have been discussed. However, epithelial changes may also be found in the lobules. Further, there may be more generalized changes in breast tissue when a carcinoma is present. Gallager and Martin not only found changes in the supportive connective tissue and, in some breast carcinoma cases, an increased density of the breast due to proliferation of collagen in the mammary tissue, but consider these types of changes to be diagnostically significant.[9-10] It is important to ultrasound detection that this change in the mammary

tissue is not confined to the region of the tumor mass but may be a
widespread density effect. Wolfe has proposed that the pattern of
distributed breast tissues, as seen on xeromammography, can be used
as a predictive diagnostic tool.[29] In that regard he has developed
a classification system for xeromammographic images that is predictive in regard to women who have a high risk of breast cancer. The
tissue characteristics that are significant to this classification
are broad categories, such as fatty, dense, cystic, and prominent
ducts. Myron Moscowitz's and associates' recent studies, correlating
breast biopsy data, and Wolfe's classification scheme, is additional
evidence of the value of this approach and also lends further credence
to the concept that epithelial hyperplasia is a malignant precursor
(the University of Cincinnati College of Medicine).[30]

The generalized tissue changes (such as increased density) or
characteristics (such as fatty, cystic, etc.) briefly mentioned
above can be detected ultrasonically. The present limitations on
the extent to which such components of generalized tissue can be de-
lineated and quantified cannot be discussed in adequate detail here
because there are many factors involved in such a discussion including types of ultrasound instrumentation in relation to complexity of
breast tissue structure. However, there is little doubt, in this
author's opinion, that such identification and characterization can
be accomplished if sufficient combined clinical and research studies
are undertaken.[31-37] The primary aim of the earlier ultrasound investigations of the author and associates was echogram characterization of the tissues of the normal breast, over the age of young to
old, with the purpose of providing a means for early breast cancer
detection by recognition of slight deviations from such echogram
pattern norms.[20,21,28,31,34,35] Even in these early studies it was
shown that a breast with a clinical diagnosis of fibrosing adenosis
(a dense breast usually with hyperplasia and a proliferation of connective tissue) could be ultrasonically identified by the echogram
characterization approach.[20] It is also of interest to note the
type of tissue detail in regard to fat lobules and suspensory ligaments that can be imaged by means of relatively simple ultrasound
instrumentation, as shown in Figure 6 of this paper.

CONCLUSION

It has been shown that ductal structures located in the region
inferior to the nipple and areola can be detected by relatively
simple ultrasound visualization techniques. It is suggested that
further ultrasound investigations be carried out to improve echogram
resolution of these fine tissue structures and to increase accuracy
of detection of ducts in other regions of the breast. This approach
is recommended in order to provide a means of examining the ductal
structures of a large population of women for the purpose of adding
to present histological and x-ray data on the causal relationship
between ductal hyperplasia and breast carcinoma.

REFERENCES

1. 75 Cancer Facts and Figures; American Cancer Society, Inc.,
    New York, 1974, pp. 1-29.

2. J. W. Berg and G. F. Robbins: "Factors Influencing Short and
    Long Term Survival of Breast Cancer Patients," Surgery, Gyne-
    cology and Obstetrics, January-June, 1966, Vol. 122, pp. 1311-
    1316.

3. B. Fisher, N. Slack, D. Katrych and N. Wolmark: "Ten Year
    Follow-up Results of Patients with Carcinoma of the Breast in
    a Cooperative Clinical Trial Evaluating Surgical Adjuvant
    Chemotherapy," Surgery, Gynecology and Obstetrics, Vol. 140,
    1975, pp. 528-534.

4. I. Macdonald: "The Breast," in, Management of the Patient With
    Cancer, ed., T. F. Nealon, Jr.; W. B. Saunders Co., Philadel-
    phia/London, 1965, pp. 435-469.

5. D. C. Tormey: "Combined Chemotherapy and Surgery in Breast
    Cancer: A Review," Cancer, Vol. 36, 1975, pp. 881-892.

6. S. J. Cutler and R. R. Connelly: "Mammary Cancer Trends,"
    Cancer, Vol. 23, No. 4, April, 1969, pp. 767-771.

7. J. A. Urban: "Biopsy of the Normal Breast in Treating Breast
    Cancer," Surgical Clinics of North America, Vol. 49, No. 2,
    April, 1969, pp. 291-301.

8. N. H. Slack, L. E. Blumenson and I. D. J. Bross: "Therapeutic
    Implications From a Mathematical Model Characterizing the
    Course of Breast Cancer," Cancer, Vol. 24, 1969, pp. 960-971.

9. H. S. Gallager and J. E. Martin: "The Study of Mammary Carcin-
    oma By Mammography and Whole Organ Sectioning," Cancer, Vol. 23,
    No. 4, April, 1969, pp. 855-873.

10. H. S. Gallager and J. E. Martin: "Early Phases in the Develop-
    ment of Breast Cancer," Cancer, Vol. 24, No. 6, Dec., 1969,
    pp. 1170-1178.

11. H. S. Gallager and J. E. Martin: "The Pathology of Early
    Breast Cancer," in, Breast Cancer: Early and Late, Year Book
    Medical Publishers, Inc., Chicago, 1968, pp. 37-50.

12. J. E. Martin and H. S. Gallager: "Reflections on Benign Dis-
    ease: A Radiographic-Histologic Correlation," in, Early Breast
    Cancer, Detection and Treatment, ed. H. S. Gallager; John Wiley
    & Sons, New York, 1976, pp. 177-181.

13. R. E. Qualheim and E. A. Gall: "Breast Carcinoma with Multiple Sites of Origin," Cancer, Vol. 10, 1957, pp. 460-468.

14. J. N. Wolfe: "Mammography: Ducts as a Sole Indicator of Breast Carcinoma," Radiology, Vol. 89, Aug., 1967, pp. 206-210.

15. J. N. Wolfe: "A Study of Breast Parenchyma by Mammography in the Normal Woman and Those with Benign and Malignant Disease," Radiology, Vol. 89, No. 2, Aug., 1967, pp. 201-205.

16. L. J. Humphrey and M. A. Swerdlow: "Large Duct Hyperplasia and Carcinoma of the Breast," Arch. Surg., Vol. 97, 1968, pp. 592-594.

17. M. M. Black, T. N. C. Barclay, S. J. Cutler, B. T. Hankey and A. J. Asire: "Association of Atypical Characteristics of Benign Breast Lesions with Subsequent Risk of Breast Cancer," Cancer, Vol. 29, Feb., 1972, pp. 338-343.

18. E. Kelly Fry, H. S. Gallager and T. D. Franklin, Jr.: "In Vivo and In Vitro Studies of Application of Ultrasonic Visualization Techniques for Detection of Breast Cancer," Proc. IEEE Ultrasonics Symposium, Miami, Dec., 1971.

19. W. J. Fry, G. H. Leichner, D. Okuyama, F. J. Fry and E. Kelly Fry: "Ultrasound Visualization System Employing New Scanning and Presentation Methods," J. Acoust. Soc. Amer., 44(5), Nov., 1965, pp. 1325-1338.

20. E. Kelly Fry, G. Kossoff and H. A. Hindman, Jr.: "The Potential of Ultrasound Visualization for Detecting the Presence of Abnormal Structures Within the Female Breast," IEEE Ultrasonics Symposium, 1972, pp. 25-30.

21. E. Kelly Fry, F. J. Fry, N. T. Sanghvi and R. F. Heimburger: "A Combined Clinical and Research Approach to the Problem of Ultrasound Visualization of Breast," in, Ultrasound in Medicine, Vol. I, ed. D. White; Plenum Pub. Corp., New York, 1975.

22. L. Testut: Traite D' Anatomie Humaine Tome Cinquieme, 8th Edition; reviewed by A. Latarjet, Gaston Doin and Cie; Paris, 1931.

23. C. M. Jackson, editor: Morris' Human Anatomy, Ninth Edition; P. Blakiston's Son & Co., Inc., Philadelphia, 1933.

24. H. T. O'Neil: "Theory of Focusing Radiators," J. Acoust. Soc. Amer., Vol. 21, No. 5, Sept., 1949, pp. 516-526.

25. W. J. Fry and F. Dunn: "Ultrasound: Analysis and Experimental Methods in Biological Research," in, Physical Techniques in Biological Research, Vol. 4; Academic Press, New York, 1962, pp. 261-394.

26. J. Reid: "A Review of Some Basic Limitations in Ultrasonic Diagnosis," in, Diagnostic Ultrasound; Plenum Press, New York, 1966, pp. 1-12.

27. J. Saneyoshi, Y. Kikuchi and O. Nomoto: Chot Onpu Gijutsu Benran (Handbook of Ultrasonic Technology) [in Japanese]; Nikkon Kogyo Press, Tokyo; 4th Edition, 1971, pp. 1399-1418.

28. E. Kelly Fry: "A Study of Ultrasonic Detection of Breast Disease," Third Quarterly Report, U.S.P.H.S. Cancer Control Program, PH-86-68-193, 1969, pp. 1-18.

29. J. Wolfe: Presentation at the 14th Annual Conference on the Early Diagnosis and Treatment of Carcinoma of the Breast, sponsored by Amer. Col. Radiol. and Natl. Can. Inst., SanJuan, Puerto Rico, March 1975.

30. Private communication.

31. E. Kelly Fry: "A Study of Ultrasonic Detection of Breast Disease," Second Quarterly Report, U.S.P.H.S. Cancer Control Program, PH-86-68-193, 1968.

32. E. Kelly Fry, G. Kossoff and L. V. Gibbons: "Characterization of the Normal Female Breast by Ultrasonic Visualization Techniques," Proc. AIUM, Winnipeg General Hospital, Manitoba, Canada, October, 1969.

33. E. Kelly Fry, L. V. Gibbons and G. Kossoff: "Characterization of Breast Tissue by Ultrasonic Visualization Methods," Proc. Acoust. Soc. Amer., San Diego, November, 1969.

34. E. Kelly Fry: "A Study of Ultrasonic Detection of Breast Disease," Progress Report, U.S.P.H.S. Cancer Control Program, PH-86-68-193, April 1, 1970.

35. E. Kelly Fry, D. Okuyama and F. J. Fry: "The Influence of Biological and Instrumentation Variables on the Characteristics of Echosonograms," in, Ultrasono Graphia Medica, ed. J. Bock and K. Ossoinig; Verlag der Wiener Medizinischen Akademie, Vienna, Austria, 1971.

36. G. Baum: "Ultrasonic Examination of the Breast," in, Fundamentals of Medical Ultrasonography, ed. G. Baum; G. P. Putnam's Sons, New York, 1975, pp. 380-402.

37. J. Jellins, G. Kossoff, T. S. Reeve and B. H. Barraclough:
     "Ultrasonic Grey Scale Visualization of Breast Disease,"
     Ultrasound in Med. Biol., Vol. 1, No. 4, March 1975, pp. 393-
     404.

HEART MODEL SUPPORTED IN ORGAN CULTURE AND ANALYZED BY ACOUSTIC MICROSCOPY*

Reginald C. Eggleton and F. Stephen Vinson

Indiana University School of Medicine, and Fortune-Fry Research Laboratories of the Indianapolis Center for Advanced Research, 1219 West Michigan Street, Indianapolis, Indiana 46202

INTRODUCTION

Various cardiac models have been described in the literature which have been used extensively in the study of the physiology, biochemistry and pharmacology of the heart. Since the original work of Langendorff in 1898[1] there have been numerous improvements in cardiac models. Bleehen and Fisher[2] simplified the preparation by adding recirculation. In the original cardiac model, no ventricular filling occurred; thus little work was done by the left ventricle, i.e., work done in contraction is a function of the initial length of the fibers. To correct this deficiency, Morgan et al.[3] added loading to the heart. Opie[4] in England and Arnold and Lochner[5] on the continent prepared an isovolumic constant volume, and later, Neely (1967)[6] and Rabitzsch (1968)[7] improved on the working heart model. Since that time various other improvements have been added.[8-10]

The cardiac model described here offers the opportunity to obtain a new spectrum of data not available previously.[11-14] In this study, isolated fetal hearts are maintained, using a continuous flow culture system, for extended periods in a stable, functioning state. The mammalian heart best suited for this application is the 15-day fetal mouse heart, which measures approximately 1 mm in length. This size of heart is suitable for analysis using the Sonomicroscope 100 acoustic microscope[15] to provide information related to morphology (external and internal), mechanical behavior, and elastic properties of the heart tissue. In addition, electrical and pressure events are also recorded. Environmental perturbations involving temperature, oxygen, and pH, as

* Work supported in part by HL-16311.

well as a variety of cardioactive drugs, may be used to study the
physiology and pharmacology of the heart.

INSTRUMENTATION

The Sonomicroscope (see Figure 1) specimens are illuminated by
100 MHz continuous wave ultrasound. The acoustic energy propagates
through the specimen and impinges on a gold-plated lucite block, as
shown in Figure 2. The acoustic pressure waves ripple the surface
of the mirror and the presence of the specimen modulates the degree
of rippling, following the contour of the acoustic wavefront. The
acoustic image appearing on the mirrored surface is scanned by a
laser beam. The angular modulation of the beam resulting from the
irregularities in the mirrored surface is converted to light intensity modulation by passing the beam over a knife edge. A photodiode is employed to convert the modulated laser beam into the video
signal displayed on the synchronized TV raster. Both magnitude and
phase information are retained. The scan raster for two-dimensional
transmission acoustic microscopy employs a TV format so that standard monitors and recorders can be utilized.

*Figure 1. The Sonomicroscope 100 acoustic microscope operates at
100 MHz and provides simultaneous acoustic and optical images of
the specimen. The interference mode records changes in transit
time of sound through the specimen.*

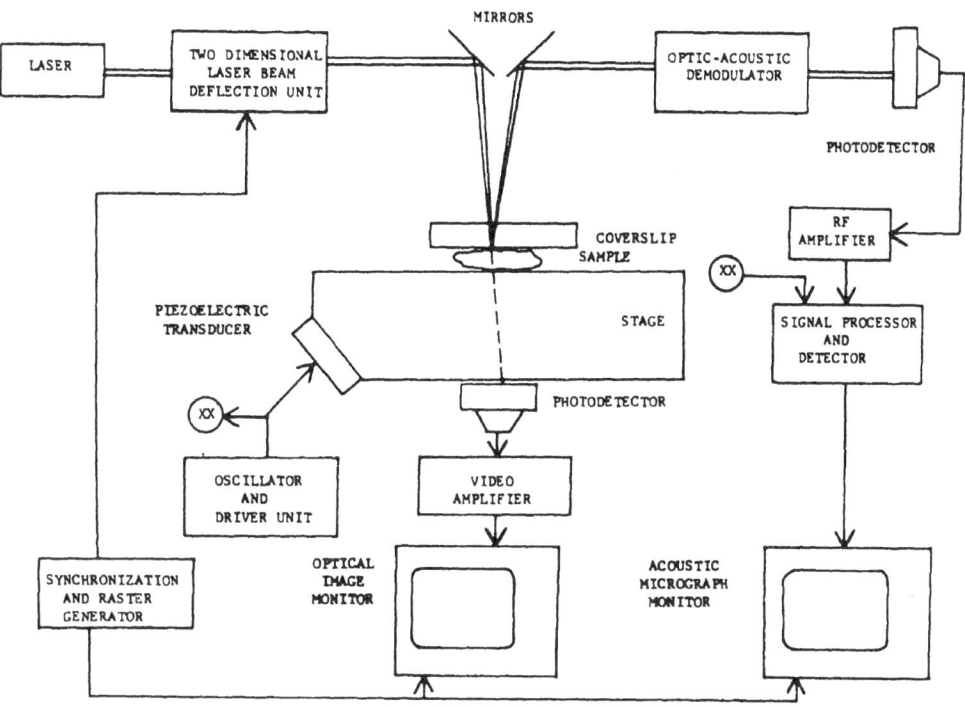

Figure 2. *A block diagram describing general operating scheme of the Sonomicroscope 100. A 100 mw C.W. gas laser output is scanned in a horizontal plane using an acousto-optical deflection system, and in the vertical plane using a mechanical deflection unit. The scan raster utilized is synchronized with the TV displays; thus there are 60 fields/second with 262½ lines/field. The laser scan raster is reflected from the mirrored coverslip, which has impressed upon it the acoustic image. The reflected beam is modulated by the angular deflection of the mirrored surface caused by the acoustic image, and is demodulated as the light passes over a knife edge from the photo detector. The amplified output of the photo detector is the input to the acoustic monitor. The acoustic image is formed by plane wave acoustic field which propagates through the stage and emerges at a 10.2° angle into water media. The intensity of the signal impinging on the mirrored coverslip is modulated by the attenuation through the specimen. The sound ripples the surface of the mirror and the degree of rippling is related to the attenuation of the specimen. Magnification is the ratio of the length of the TV scan to the length of the laser scan. The magnification is a function of the size of the monitor utilized to display the image. Magnification of 75 is achieved in the present instrument. The resolution is limited by the wave length of the sound used to illuminate the specimen (15μ at 100 MHz) and upon the size of the laser spot (15μ). The simultaneous optical image is obtained by placing a photocell below the stage collecting the transmitted laser beam as it passes through the specimen.*

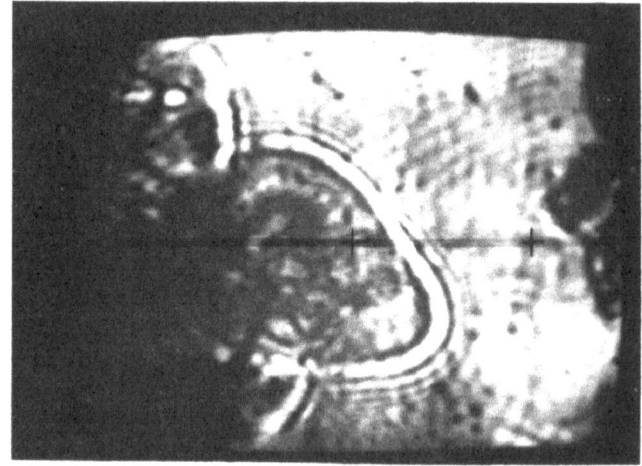

A

*Transmission acoustic
photomicrograph*

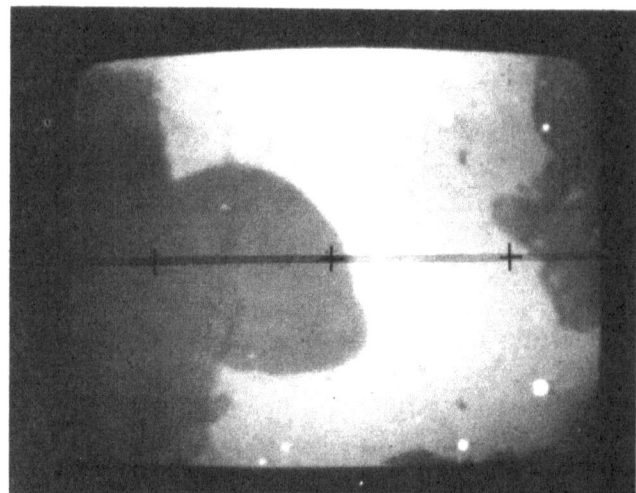

B

*Simultaneous optical
photomicrograph*

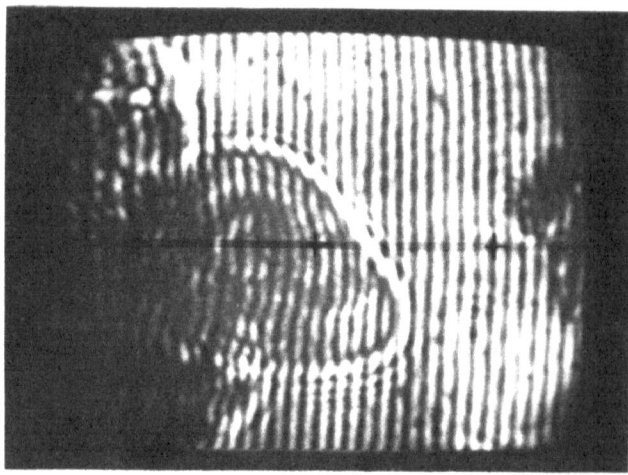

C

Interferogram

Figure 3. *Acoustic microscope images of 15-day fetal mouse heart
supported in organ culture. Taken on the 4th day of perfusion.*

IN VITRO HEART MODEL

Fetal mouse hearts are maintained with a closed circuit organ culture system shown diagramatically in Figure 4. The culture chamber can be placed on the microscope stage (see Figure 5) with a flow circuit placed adjacent to the stage. The peristaltic pump is used to increase the pressure in the supply flask, causing the culture medium (Grand Island 199, 10% bovine calf serum, 1% L-glutamine and 50 mg/ℓ insulin) to flow from the flask to the oxygenator at a rate of about 0.2 mℓ/min.

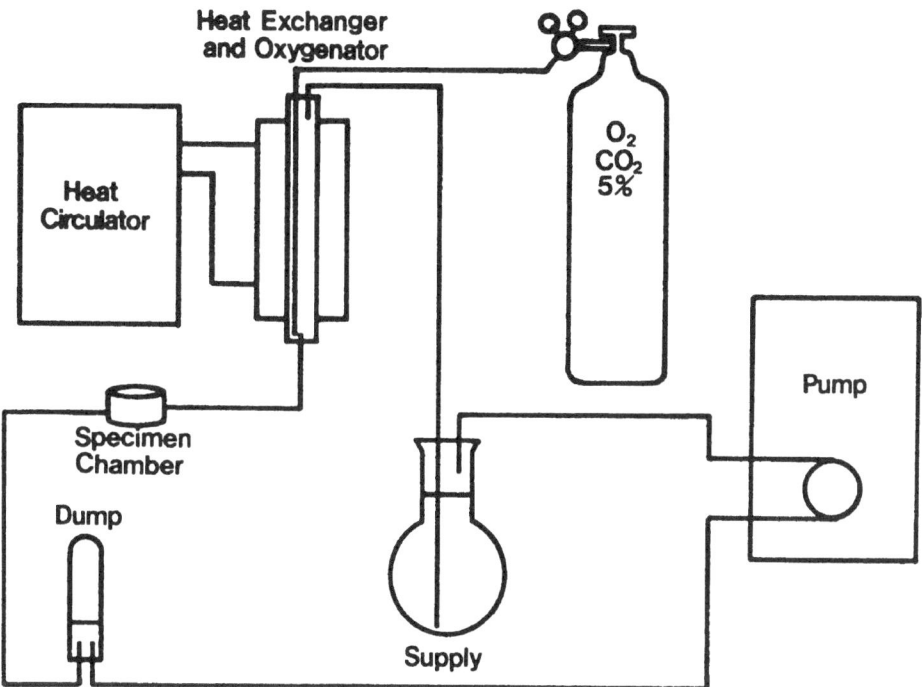

Figure 4. Schematic diagram of organ culture flow circuit. The heat exchanger and oxygenator is made from a reflux condenser. The heat circulator pumps water at 37° through the water jacket to warm the culture medium. Oxygen with 5% CO$_2$ is bubbled through the culture medium to adjust the oxygen tension. The prepared culture medium then flows through an O-ring sealed culture chamber. A mirrored coverslip forms the upper boundary of the 30$\mu\ell$ culture chamber and a thin membrane attached to the bottom of the chamber completes the seal. The culture medium flows into the chamber and exits around the edge of the mirror. A compartment above the mirror surface provides a bubble trap. The output from the specimen chamber is then returned to the pressurized supply bottle via the dump and pump. Only the oxygenator is at atmospheric pressure, and the rest of the system is closed.

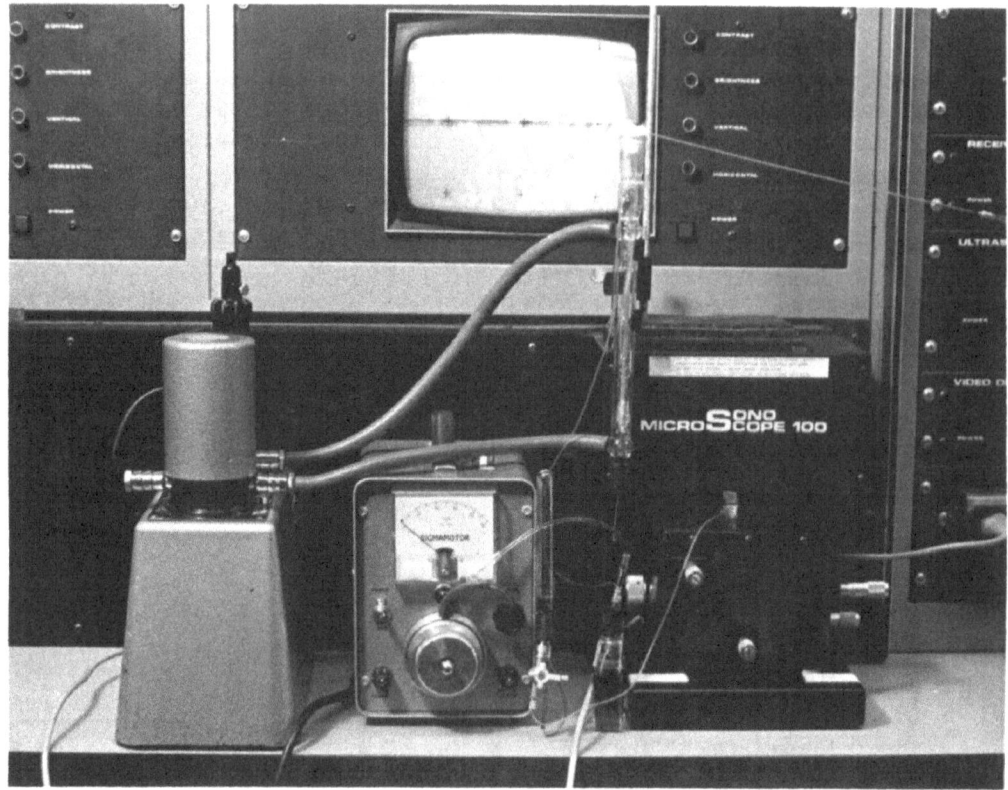

Figure 5. Organ culture perfusion system.

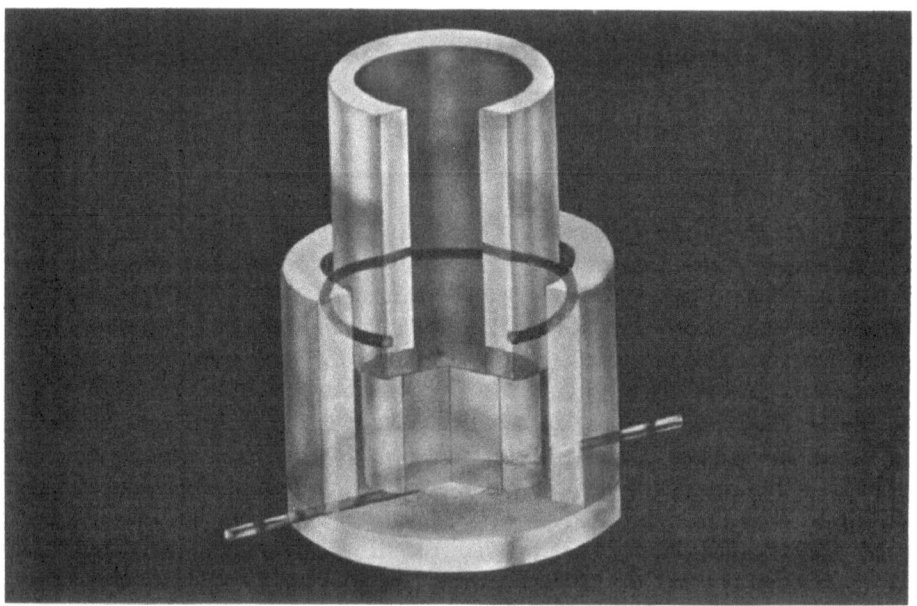

Figure 6. Sealed culture chamber and mirror system.

The oxygenator consists of a reflux condenser in which oxygen with 5% CO_2 is bubbled through the culture medium and the water jacket is used to heat the medium to 37°C. Warm, oxygenated medium flows directly to the specimen chamber (see Figure 6) on the stage, and then is returned to the supply bottle. When drugs are used, a dump prevents contamination of the supply. Dumping is achieved by inverting the dump tube. This can be accomplished without breaking the closed circuit. With this system, up to five hearts can be placed in the 30μℓ chamber on Monday morning, and they will be functional Friday evening if they are not compromised by the experimental procedure. It is interesting to note that the hearts appear to survive better in the presence of the sound field. This phenomenon may be explained by hypothesizing a higher diffusion rate for O_2, CO_2 and nutrients in the presence of the sound field. The acoustic intensity of the 100 MHz sound field is approximately 25 mw/cm^2.

MEASUREMENT METHODS

The analytical procedures used in conjunction with the acoustic microscope provide important insight into the functioning of the cardiac model. The speed of sound in the media is a function of density and adiabatic bulk modulus of the material. The microscope, operating in the interference mode, can be used to make precise measurements of changes in transit time for sound to propagate through the specimen.

The transit time for sound to propagate through 1 millimeter of tissue, assuming the speed of sound equal to 1500 meters per second, is 666 nanoseconds. The microscope configuration in Figure 7 shows the sound emerging from the stage at 10.2° into a water media. At this angle there are 32 interference lines across the width of the monitor, and each interference line corresponds to the wavefront, thus the distance between the center of two adjacent lines is 10 nanoseconds. Changes in the transit time of sound through the specimen amounting to as little as one nanosecond can be readily recognized. Thus, it is possible to measure changes in transit time to one part in 666.

Density measurements are made with relative ease by using a density graded column. This approach is most suitable for small tissue samples. The adiabatic bulk modulus is calculated

$$K_a = \rho c^2$$

where ρ is the density of the medium, and c is the velocity of sound. Observations of the hearts have revealed that the speed of sound changes during contraction and the magnitude of change depends upon the loading of the heart. The heart can be loaded by obstructing the aorta, or by cannulating this vessel. Changes in transit

Interference Mode

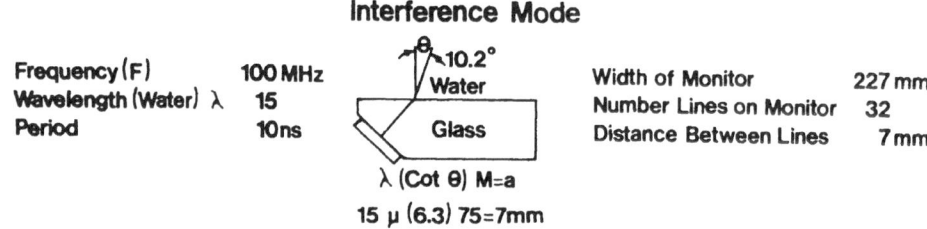

$$\lambda \ (Cot \ \theta) \ M=a$$

$$15 \ \mu \ (6.3) \ 75=7mm$$

Interference Between Examining Beam
and Reference Signal on Acoustic Image Monitor

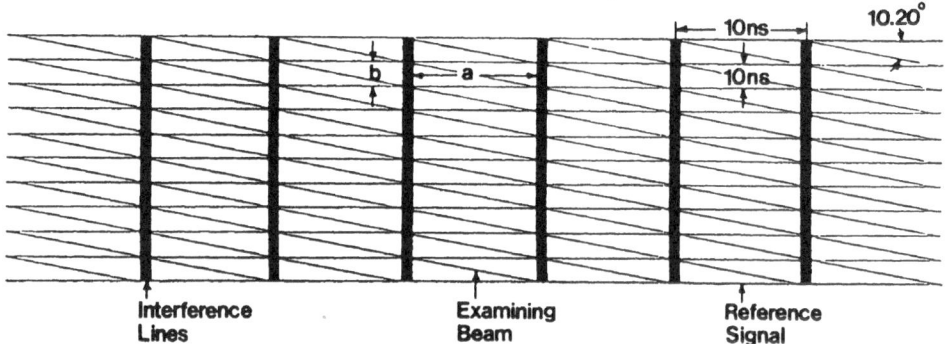

Figure 7. *The interference mode is created by injecting an electrical reference signal into the video signal, causing a series of interference fringes to appear across the image. There are 32 interference fringes spaced 7 mm apart. A calculation of the spacing distance involves multiplying the wavelength of the sound times the co-tangent of the angle of the examining beam times the magnification. This also reveals 7 mm spacing between fringes. If the speed of sound in the medium above the stage increases, then the angle, θ, increases. If the velocity of sound decreases, θ also decreases. For a medium with a velocity of 1500 meters per second, the angle of emergence is 10.2°.*

time measured for the unloaded heart were in the range of one to two nanoseconds, whereas the change in transit time for the loaded heart was observed to be in the order of 10 nanoseconds. It has long been known (Goldman and Hueter) that the speed of sound propagating parallel with the fibers is higher than normal to the fibers. However, it has not been known previously that the speed of sound in muscle varies with its state of contraction. To facilitate the study of the relationship between contraction, the speed of sound in tissue, and the elastic properties of muscle, the microscope was modified to provide what we have termed "line scan" mode of operation. This involves collapsing the ventricle sweep of the laser raster and placing 15,750 scan lines per second along a single path to observe changes in the transit time through the specimen as a function of time. The line scan technique is analogous to M mode

echo techniques. The line scan interferogram displays directly
changes in transit time, and as the path length remains constant,
it displays changes in the speed of sound. Such a line scan interferogram is shown in Figure 8, displaying a change in the speed of
sound through the heart associated with contraction.

Because of the intricate structures of the heart and the presence of ventricular spaces, it is difficult to measure tissue
thicknesses in the heart. To avoid the complexity of the heart
geometry, frog sartorius muscle was selected for initial studies
for investigation of muscle mechanics.

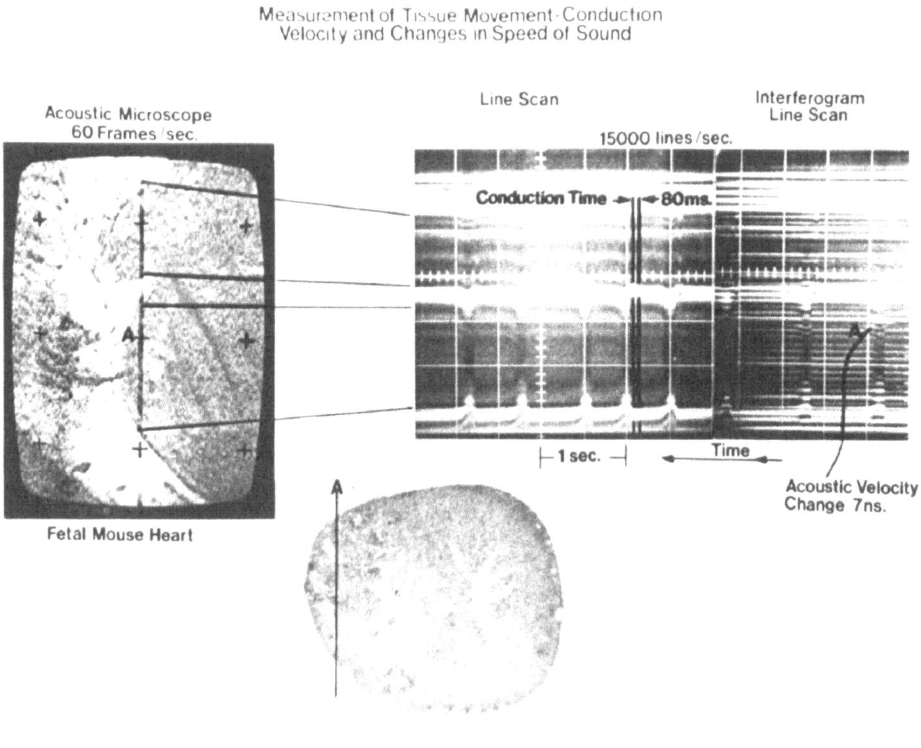

Figure 8. *Time varying events may be recorded from an acoustic
image by using a line scan technique. In this event, the horizontal laser scan is collapsed so that approximately 15,000 lines per
second are scanned across the central portion of the image, as marked on the monitor by the gray line at A. As the tissue moves, details of that movement are recorded on the line scan. For example,
the cardiac period can be measured, as well as the phase relationship between atrial and ventricular beat. The histological section
shows the propagation path of the sound through the specimen along
path A. The interferogram line scan also displays the contractile
events but, in addition, records changes in transit time of sound
through the specimen.*

RESULTS

The frog sartorius muscle is supported in an isometric frame
on the stage of the microscope (see Figure 9). Spacers are selected equal to the thickness of the muscle at rest; thus the mirrored
coverslip is in contact with the muscle and no change in thickness
is allowed to occur during contraction. In an isometric twitch,
the increase in thickness is negligible, therefore the deflection
of the lines indicates a change in the speed of sound through the
tissue. In the figure, the muscle was stimulated with a unipolar
pulse of one volt amplitude for a duration of 50 milliseconds.
(The vertical lines correspond to 16 2/3 milliseconds, and the horizontal or interference fringes correspond to a period of 10 nanoseconds).

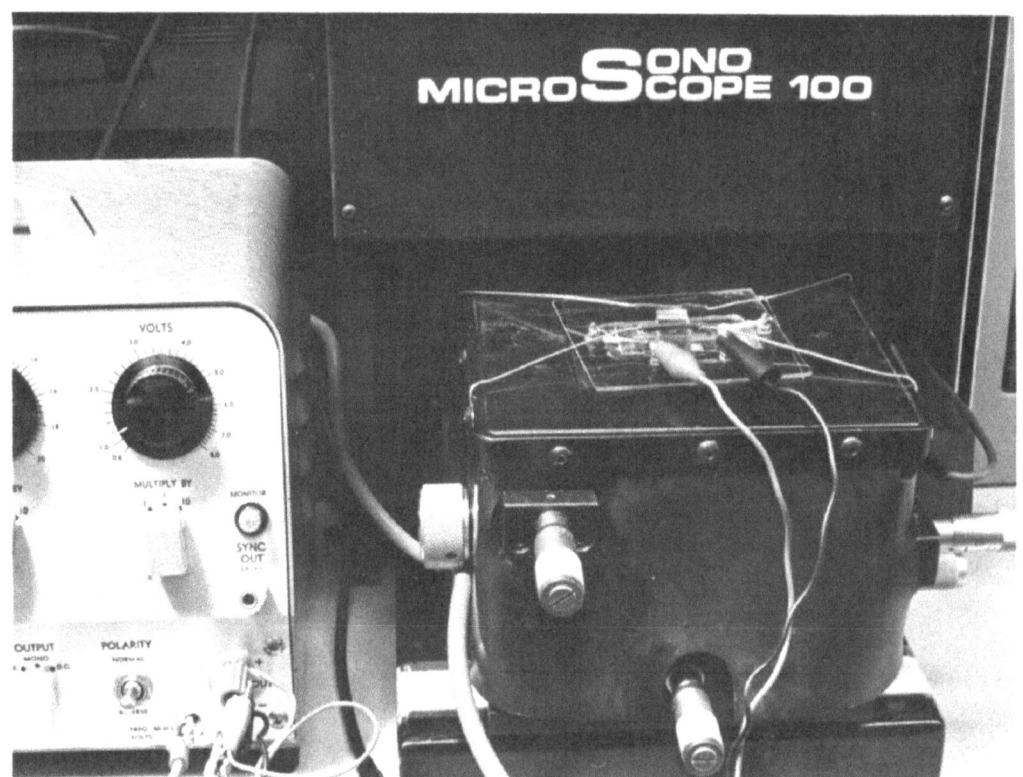

*Figure 9. Isometric frame for microscope stage showing frog sar-
torius muscle in position.*

To interpret the pattern of velocity changes shown in the interferogram, it is important to recognize that muscle fibers are
enervated periodically along their length. Because the propagation
of the exciting action potential is much faster in nerve than it is
in muscle tissue, we hypothesize that there is a motor endplate just
above the top edge of the picture, and a second one near the center
of the picture; thus explaining what appears to be a wave of contraction propagating along the muscle fiber. (See Figure 10).

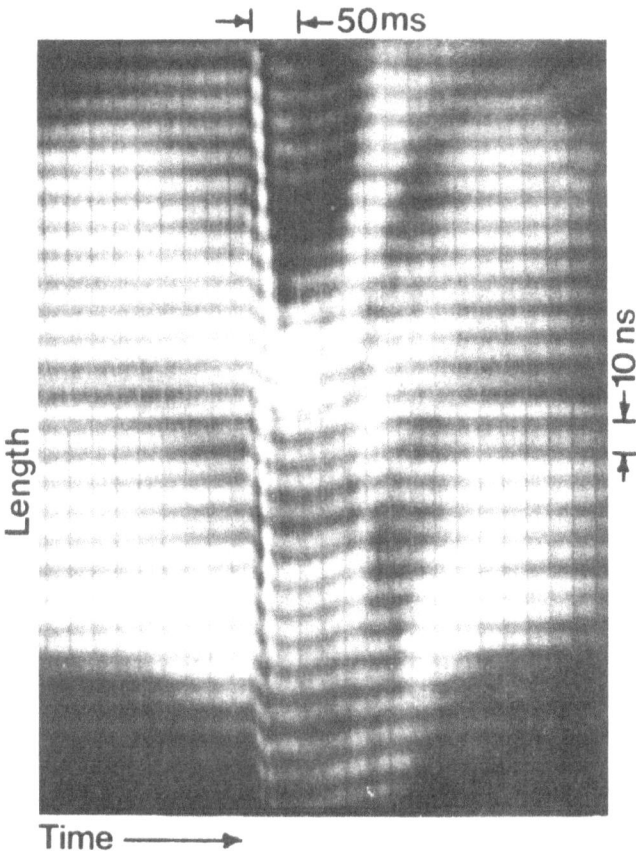

Figure 10. *Frog sartorius muscle is stimulated in isometric con-*
traction and the resulting line scan interferogram indicates a
change in the speed of sound through the specimen resulting from
contraction. The muscle is dissected with a bony attachment, and
is secured to the frame using wire attachments to the bone. Stim-
ulation is provided by a Grass stimulator providing a one-volt
monophasic pulse to gross stimulating electrodes. A spacer is used
to bring a mirrored coverslip into contact with the muscle, but
without compressing it. Thus there is a constant propagation path-
way through the muscle tissue between the microscope stage and mir-
rored coverslip and the deflection of the interference lines can be
interpreted as a change in the speed of sound through the tissue.

In addition to displaying interference fringes, this figure
also displays changes in the transmission properties of the muscle
along the scan line as it varies with time. The appearance of the
line scan interferogram ls very intriguing and will require much
additional study.

As the muscle continues to be stimulated, its threshold changes, and not all muscle fibers fatigue at the same rate. In the example shown in Figure 11, only a few fibers were contracting with a one-volt stimulation pulse, and only a portion of the length of the fiber was active. It will be noted that there is a change in the speed of sound along part of the muscle fiber, and no change along the lower portion. It seems reasonable to assume that the tension in the fiber is constant throughout its length, and that the upper portion is contracting at the expense of the stretching in the lower portion. Our tentative interpretation, assuming the A. V. Hill three-element muscle model,[16] is that we are detecting a change in the contractile element independent of the series or parallel elastic elements. If this hypothesis holds up under further investigation, then the microscope provides a means of examining independently the elements of the muscle model.

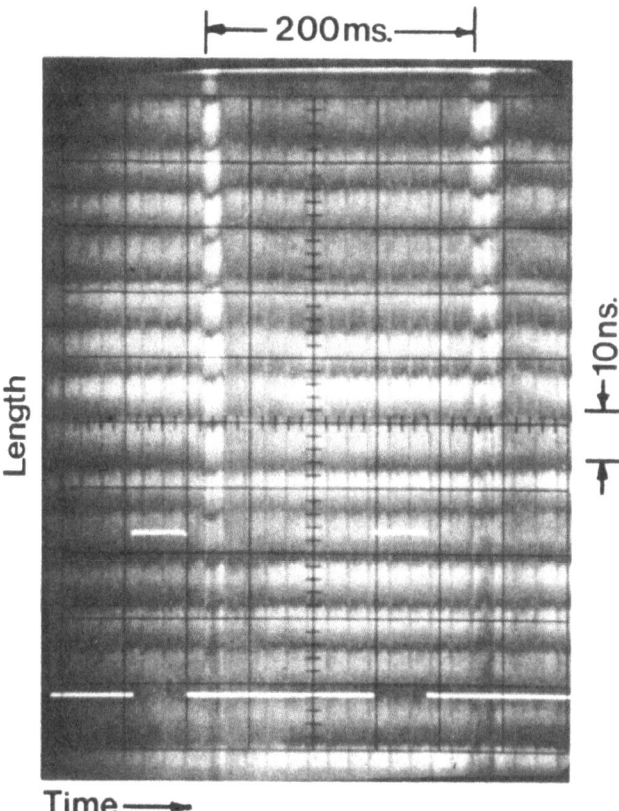

Figure 11. *Partially contracting fatigued muscle, showing changes in transit time in the upper portion of the muscle fiber, and no change in transit time for the lower portion of the specimen. This suggests that the change in the speed of sound through the tissue is associated with the contractile mechanism rather than the elastic elements, because it is reasonable to expect that the tension in the fiber is uniform throughout its length.*

DISCUSSION

Heretofore, information on muscle mechanics has been derived from force transducers attached to the ends of the muscle, and details concerning the internal behavior of the muscle could be investigated only by inference. The microscope provides an opportunity to investigate in detail internal features of muscle contraction. We are now in the process of constructing a system of manipulators and force transducers which will be necessary for a more complete analysis of the acoustic micrograph images.

We are also developing methods for cannulating the 1 mm diameter fetal mouse hearts. Figure 12 illustrates our method of accomplishing this task. A Teflon tube 0.013" in diameter is slipped into the aorta and glued in place, using Eastman 910 adhesive. This permits adjusting the ventricular pressure. It was noted previously that hearts which were pumping against an isovolumic load caused by a clot in the aorta exhibited a substantial change in the transit time of the sound through the heart, compared with a heart without such a clot.

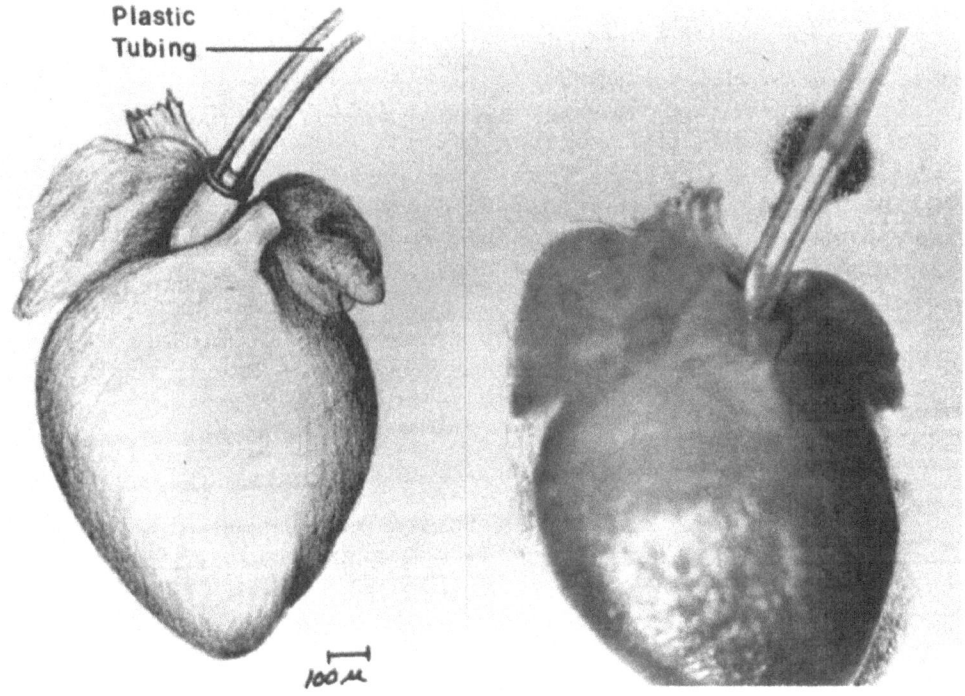

Figure 12. *Cannulation of the fetal mouse hearts is accomplished by introducing a 0.013" diameter tube into the aorta and gluing the aorta using Eastman 910 adhesive; thus providing a means of controlling the back pressure in the ventricle.*

REFERENCES

[1] Langendorff, O.: Untersuchungen am überlebenden Säugetierherzen. *Pflügers Arch. ges. Physiol., 61, 291 (1895)*.

[2] Bleehen, N. M. and Fisher, R. B.: The action of insulin on the isolated rat heart. *J. Physiol., Lond., 123, 260 (1954)*.

[3] Morgan, H. E., Henderson, M. J., Regen, D. M. and Park, C. R.: Regulation of glucose uptake in muscle. I. The effect of insulin and anoxia on glucose transport and phosphorylation in the isolated, perfused heart of normal rats. *J. biol. Chem. 236, 253, (1961)*.

[4] Opie, L. H.: Coronary flow rate and perfusion pressure as determinants of mechanical function and oxidative metabolism of isolated perfused rat heart. *J. Physiol., Lond., 180, 529 (1965)*.

[5] Arnold, G. and Lochner, W.: Die Temperaturabhängigkeit des Sauerstoffverbrauches stillgestellter, künstlich perfundierter Warmblüterherzen zwischen 34° und 4°C. *Pflügers Arch. ges. Physiol., 284, 169 (1965)*.

[6] Neely, J. R., Liebermeister, H., Battersby, E. J. and Morgan, H. E.: Effect of pressure development on oxygen consumption by isolated rat heart. *Am. J. Physiol., 212, 804 (1967)*.

[7] Rabitzsch, G.: Koronarrperfusion isolierter Warmblütterherzen mit geringen Umlaufsvolumina und Kontinuierlicher Kontractions --- und koronarflußregistrierung. *Acta biol. med. germ., 20, 33 (1968)*.

[8] Wildenthal, K.: Long-term maintenance of spontaneously beating mouse hearts in organ culture. *J. Appl. Physiol., 30(1), 153 (1971)*.

[9] Opie, L. H., Mansford, K. R. L. and Owen, P.: Effects of increased heart work on glycolysis and adenine nucleotides in the perfused heart of normal and diabetic rats. *Biochem. J., 124, 475 (1971)*.

[10] Neely, J. R., Denton, R. M., England, P. J. and Randle, P. J.: The effects of increased heart work on the tricarboxylate cycle and its interactions with glycolysis in the perfused rat heart. *Biochem. J., 128, 147 (1972)*.

[11] Eggleton, R. C. and Kessler, L. W.: Mouse embryo heart in organ culture visualized by the acoustic microscope. In *Ultrasound in Medicine, Vol. 1*, ed. D. White (Plenum Publ. Corp., New York 1975) pp. 537-542.

References (continued)

[12] Eggleton, R. C., Kessler, L. W., Vinson, F. S. and Boder, G. B.: Acoustic microscopy of live mouse embryo hearts. In *1975 Ultrasonics Symposium Proceedings (IEEE, Inc., New York 1975), pp. 57-58.*

[13] Eggleton, R. C., Kessler, L. W., Vinson, F. S. and Boder, G. B.: Effect of drugs on mouse embryo hearts in organ culture visualized by acoustic microscopy. In *Ultrasound in Medicine, Vol. 2, eds. D. White and R. Barnes (Plenum Publ. Corp., New York 1976), pp. 441-443.*

[14] Vinson, F. S., Eggleton, R. C., Boder, G. B. and Kessler, L. W.: Measurement of time related cardiac parameters by acoustic microscopy. *Presented at WFUMB '76 Scientific Sessions, San Francisco, August 1-7, 1976.*

[15] Korpel, A., Kessler, L. W. and Palermo, P. R.: Acoustic microscope operating at 100 MHz. *Nature 232, No. 5306, pp. 110-111, July (1971).* Sonomicroscope 100 manufactured by Sonoscan, Inc. 720 Foster Avenue Bensenville, Il. 60106.

[16] Fung, Y. C.: Comparison of different models of the heart muscle. *J. Biomechanics 4, 289 (1971).*

THE ROLE OF COLLAGEN IN DETERMINING ULTRASONIC PROPAGATION PROPERTIES IN TISSUE

W. D. O'Brien, Jr.

Department of Electrical Engineering

University of Illinois, Urbana, Illinois 61801

The processes responsible for affecting the propagation of an ultrasonic wave as it passes through biological tissue are poorly understood. In large part, the research into this area is at a correlation level as contrasted to a modelling level. In other words, contemporary research is studying the relationships between ultrasonic properties of tissue and other tissue properties in order to elucidate trends. With very few important exceptions, minute information has prevented any successful modelling.

This paper represents a comparison of the ultrasonic attenuation and velocity properties of tissue to the tissue properties of water, protein and, specifically, collagen contents. Table I lists the ultrasonic attenuation, velocity and water percentage in approximate order of increasing ultrasonic attenuation at 1 MHz for the indicated biological materials and tissues. Table II lists the total protein and collagen percentages of the materials detailed in Table I. In those cases where blanks exist, either the information was not available or not found. The following discussion is intended to provide additional details which could not necessarily be quantified and placed in the tables.

The ultrasonic attenuation includes not only the absorption of the ultrasonic signal which is degraded to heat but also all other mechanisms by which the energy is extracted from the propagating wave or redirected by virtue of the inhomogeneous nature of the materials. The ultrasonic velocity, and the characteristic acoustic impedance which can be determined with density information, embody within them both the inertial and restoring parameters of the particular material. Thus knowledge of the ultrasonic velocity and loss terms may, in some sense, provide a basis for developing "tissue signatures" for various tissues.

TABLE I

Ultrasonic Attenuation, Velocity and Water Content
for Various Tissues

Tissue	Attenuation At 1 MHz (cm^{-1})	Velocity (m/s)	% Water
Water (20°)	0.0003[a]	1483[b]	100
Amniotic Fluid	0.0008[c]	1510[c]	97[c]
Aqueous Humor		1497[d]	99[ef]
Vitreous Humor		1516[d]	99-99.9[ef]
CSF		1499-1515[g]	99[e]
Plasma	0.01[h]	1571[i]	90-95[ej]
Testis	0.019[k]		84[l]
Blood	0.02[h]	1571[i]	74-83[ej]
Milk	0.04[mn]	1485[n]	87[e]
Fat	0.04-0.09[iopq]	1410-1479[io]	10-19[jrs]
Spleen	0.06[p]	1520-1591[i]	76-80[rs]
Liver	0.07-0.13[ipt]	1550-1607[i]	68-78[lrs]
Kidney	0.09-0.13[i]	1558-1568[i]	76-83[lrs]
Brain	0.09-0.13[i]	1510-1565[i]	75-79[lrs]
Spinal Cord	0.09-0.12[iu]		64-80[el]
Striated Muscle	0.18-0.25[i]	1568-1603[i]	66-80[ljrv]
Against Grain	0.08-0.12[o]	1592-1603[o]	
With Grain	0.16[o]	1576-1587[o]	
Heart	0.25-0.38[iw]		77-78[r]
Tongue			62-68[r]
Against Grain	0.58[w]	1575[x]	
With Grain	0.28[w]	1585[x]	
Lens		1616[d]	63-69[fy]
Articular Capsule	0.38[oq]		
Integument	0.40[o]	1498[o]	60-72[jz]

Cartilage	0.58^{oq}	1665°	$23\text{-}34^{z}70^{aa}$
Tendon			63^{z}
Against Grain	0.54^{oq}	1750°	
With Grain	0.58^{oq}		

a) Pinkerton (1949)
b) Greenspan and Tschiegg (1957)
c) Zana and Lang (1974)
d) Begui (1954)
e) Altman and Dittmer (1961)
f) van Heyningen (1962)
g) van Venrooij (1971)
h) Carstensen et al. (1953)
i) Goldman and Hueter (1956)
j) Wolf (1976)
k) Brady et al. (1976)
l) Neufeld (1937)
m) Hueter (1958)
n) Maynard and Goss (1976)
o) Dussik and Fritch (1955, 1956)
p) Chivers and Hill (1975)
q) Dussik et al. (1958)
r) Watt and Merrill (1963)
s) Ruch and Patton (1966)
t) Pauly and Schwan (1971)
u) Dunn and Brady (1974)
v) Giese (1962)
w) Hueter (1948)
x) Ludwig (1950)
y) Davson (1972)
z) Chvapil (1967)
aa) Robb-Smith (1954)

TABLE II

Total Protein and Collagen Content for Various Tissues
(Parentheses   indicate   a calculated value)

Tissue	% Total Protein		% Collagen	
	Wet	Dry	Wet	Dry
Amniotic Fluid	0.27^a	(9)		
Aqueous Humor	$0.005-1^{bc}$	(0.5-100)		
Vitreous Humor	$0.02-0.25^{bcd}$	(2-100)	$0.014-0.067^{de}$	(1.4-67)
CSF	0.03^b	(3)		
Plasma	7^f	100^f		
Testis	(12)	72^g	Trace^h	
Milk	$3-4^b$	(23-31)		
Fat	$5-7^i$	(6-9)	yes^j	
Spleen	$17-18^i$	(71-90)	(0.5-1.2)	$2.4-4.8^k$
Liver	$20-21^i$	(66-95)	(0.1-1.3)	$0.6-3.9^k$
Kidney	$15-17^i$	(63-100)	$0.5-1.5^k$	$1.8-5.3^k$
Brain	10^i	(40-48)	(0.04-0.3)	$0.2-1.2^k$
Striated Muscle	$20-21^i$	(59-100)	(0.7-1.2)	3.5^k
Heart	17^i	(74-77)	(0.4-1.6)	$2-7^k$
Tongue	$14-17^i$	(37-53)		
Lens	$30-36^{cd}$	(81-100)		
Integrment			$7-30^k$	$60-80^{ekl}$
Cartilage	(49-63)	$74-81^k$	$10-20^e$	$46-62^k 80^m$
Tendon	35^k	(95)	32^k	$51-95^{kl}$

a)   Zana and Lung (1974)   h) Bloom and Fawcett (1968)
b)   Altman and Dittman (1961)   i) Watt and Merrill (1963)
c)   van Heyningen (1962)   j) Bradley (1972)
d)   Davson (1972)   k) Chvapil (1967)
e)   Mathews (1975)   l) Crisp (1972)
f)   Wolf (1976)   m) Robb-Smith (1954)
g)   Wolf and Leathem (1955)

There are principally three skeletal support systems in living organisms. These are the cellulosic system in plants, the collagenous system in animals and the chitunous system in both plants and animals. The basis of these three systems is found at the macromolecular level. The major fibrose elements are, respectively, the linear polysaccharide cellulose and the asymmetric protein molecule collagen in the cellulosic and collagenous systems. Chitin provides an alternative to cellulose in plants and collagen in animals. (Mathews, 1975).

Animal tissues are usually classified into one of five catagories, vi z., epithelial, connective, muscular, nervous and blood (Giese, 1962). Epithelial tissue is an aggregration of cells which cover the surfaces of organs. The epithelial cells lie on a noncellular basal membrane which is composed of collagen fibers embedded in a matrix. Epithelial tissue is found throughout the body such as the lining of the digestive tract, windpipe, lungs, mouth, esophagus, kidney tubules and urinary bladdar and the outer layer of the skin. Epithelial cells specialize to form nervous tissue during embryological development. Other epithelial cells specialize to form muscle tissue (Biology Today, 1972).

Collagen is closely associated in connective tissue of vertebrates. It turns out, in fact, that collagen is the most abundant single protein in the human body and the most common protein in the entire animal kingdom. Collagen comprises somewhere between onequarter to one-third of the total protein in the human body and therefore about six percent of the total body weight (White et al., 1968). One reason for the wide variation in collagen is that the aging process is intimately involved in intercellular changes and as aging proceeds, more and more collagen fibers are developed between cells. Injury and disease are additional causes. Thus, simply due to its abundance within the human body, it would seem logical that collagen would play an important role in determining ultrasonic properties of tissue. In addition, within recent time, more and more investigators (Fields and Dunn, 1973) have been suggesting that collagen may have an important role in determining ultrasonic propagation properties in tissue, especially in terms of echographic imaging. Therefore, this article is an attempt to compile the ultrasonic propagation properties of tissue, compare them to the tissue constituent properties such as percentages of water, total protein and collagen, and suggest some general observations.

In the early 1950's, at the time when pulse-echo diagnostic ultrasound was in its infancy, the observation was made that very few reflections were observed from breast fat alone but these reflections were much more pronounced from connective tissue sheets in lobulated fat (Wild and Reid, 1954). They (Wild and Reid, 1953)

also examined the ultrasonic reflection from an in vitro cube of
striated muscle wherein it was observed that no echos were detected
when the sound beam was directed parallel to the muscle fibers and
many echos were received when the orientation was changed by 90°.
This anisotropy in echo return was disrupted by mechanically rup-
turing the cube of muscle in order to break up the connective tissue
but the rupture procedure was not detailed.

In the mid 1950's the ultrasonic propagation properties of articular tissue were examined (Dussik and Fritch, 1955, 1956; Dussik
et al., 1958) and it was concluded that tissues with higher collagen
content exhibited higher values of ultrasonic attenuation and velocity, for the most part, as compared to soft tissue with lesser
amounts of collagen. These values have been included in Table I.

Dussik and Fritch (1955, 1956) also suggested that aging of
dense fibrose tissue is accompanied by an increase in the ultrasonic
attenuation.

Examination of the human female breast, pre- and post-
menopausal, with fibrosing adenosis exhibited an increase in the
ultrasonic reflectivity over normal breast tissue. This condition
is characterized by a proliferation of connective tissue which is
replacing the normal glandular breast tissue (Fry et al., 1972).
The post-menopausal breast has been shown to exhibit a three to
four percent lower ultrasonic velocity than that of the pre-meno-
pausal breast. This velocity difference has been attributed to the
proliferation of fatty tissue interlaced with an increased amount
of connective tissue (Kossoff et al., 1973).

Greenleaf and his colleagues (Greenleaf et al., 1975, 1976)
have shown in excised, unfixed breast specimens, that fat yields
the lowest attenuation and lowest velocity as compared to all other
surrounding tissue. Additional relative comparisons of the ultrasonic propagation properties of breast showed the following: normal parenchymal breast tissue exhibited relatively high attenuation
and medium high velocity, infiltrating medullary carcinoma exhibited
an attenuation between fat and normal breast tissue and a high velocity, and connective tissue associated with muscle boundaries of a
scirrhous carcinoma clearly exhibited the highest attenuation and
velocity.

Measurements of ultrasonic attenuation in the aqueous and vitreous humor indicate it to be from 50 to 100% greater than water at
30 MHz (Begui, 1954). While both are extremely high water content
materials, the vitreous contains vitrosin, a basement membrane-like
collagen (Mathews, 1975) which may account for the slightly higher
ultrasonic velocity.

Cerebrospinal fluid, CSF, also an extremely high water content material, is a clear, colorless liquid which contains small amounts of protein, glucose, urea, salts and leukocytes (Schneidermann and Hosek, 1976).

The early studies that examined the ultrasonic properties of blood and plasma are largely responsible for our understanding of the importance of protein in that the protein content largely determines the ultrasonic properties and that the absorption is directly proportional to the protein concentration (Carstensen et al., 1953; Carstensen, 1960). Over the frequency range from 1 to 10 MHz, the absorption of blood and plasma exhibit a frequency dependence to the 1.3 power (Carstensen, 1960). Hemoglobin, the major protein in red blood cells, in aqueous solution exhibits a similar frequency dependence over a much wider frequency range (O'Brien and Dunn, 1972).

Mammalian testicular tissue, a relatively high water content tissue, exhibits an ultrasonic absorption lower than any other intract parenchymal tissue and, additionally, appears to exhibit a frequency dependence similar to that of a single relaxation phenomenon, over the frequency range from 0.7 to 7 MHz (Brady et al., 1976)

Milk, a suspension of fat particles and hydrated casein complexes, exhibits an ultrasonic absorption which is proportional to frequency to the 1.1 power (Hueter, 1958; Maynard and Goss, 1976). Cow milk is composed of 87.4% water, 3.5% protein, 3.5% fat and 4.9% total carbohydrates. In comparison, human milk contains about one-third the concentration of protein, twice the concentration of total carbohydrates with water and fat approximately the same (Watt and Merrill, 1963).

Fat is an almost water free tissue. Total body water is dependent upon the total amount of body fat. On the average, babies have less fat than young males, and young males have less fat than young females and this is reflected in the average total body water, viz., 76%, 60% and 50%, respectively (Wolf, 1976). Fat develops in loose connective tissue and consists of numerous fat cells lying in close contact with one another (Bradley, 1972). For most fat tissue, the frequency dependence of attenuation is between the 0.93 and 1.3 power over the frequency range from 1 to 10 MHz (Goldman and Hueter, 1956; Dussik and Fritch, 1956; Chivers and Hill, 1975). The low ultrasonic velocity may be attributed to the low water content.

Dussik and Fritch (1956) indicated that the attenuation of fat tissue located in the sole of the foot yielded a consistently higher value than from other body areas such as the abdomen. It is interesting to observe that the fat in the sole serves one of the few structural and protective functions whereas most fat tissue serves primarily the function of energy storage (Windle, 1976).

Orbital fat tissue has been measured by a backscatter spectral analysis technique and has yielded an ultrasonic attenuation of 0.3/(MHz-cm) over the frequency range from 6 to 12 MHz (Lizzi and Laviola, 1975). More recent measurements indicate the attenuation is 0.17/(MHz-cm) over the frequency range from 5 to 15 MHz and, additionally, when compared with abnormal orbital fat (Graves' disease), is increased to 0.20/(MHz-cm). Histology showed that the abnormal tissue was infiltrated with connective tissue (Coleman et al., 1976).

For purposes of examining the ultrasonic propagation properties, spleen, liver and kidney may be grouped together principally because the scatter in the data does not permit extraction of trends between them. On the average, all three have approximately the same percentages of water, total protein and collagen. The spleen contains a connective tissue framework in which the lymphatic vessels are found. For its large size the liver has relatively little connective tissue. The interlobular septa, called Glisson's capsule, accounts for the major fraction of connective tissue. The kidney is a highly heterogeneous organ. On the average, the cortex has a lesser amount of collagen than the kidney as a whole (Chvapil, 1967; Bloom and Fawcett, 1968).

The frequency dependence of attenuation for spleen is approximately to the 1.5 power over the frequency range 1 to 8 MHz (Chivers and Hill, 1975). For liver, over the frequency range 200 kHz to 10 MHz, and kidney, 200 kHz to 100 MHz, the frequency dependence of attenuation is approximately a linear relationship (Goldman and Hueter, 1956; Kessler, 1973). Acoustic microscopic images of the kidney appear to yield details of connective tissue boundaries (Lemons and Quate, 1974; Kessler et al., 1974). Both cirrhotic and fibrotic livers have a greater acoustic impedance than normal livers, with an average increase of about 8 percent (Yamakawa et al., 1964).

Hueter (1958) showed that the attenuation of ultrasound in liver decreased markedly during the period from 1 hour post mortem to 2 days post mortem. At 1.5 MHz, the attenuation changed from 0.26 cm^{-1} to 0.06 cm^{-1}. Chivers and Hill (1975) measured the attenuation of ultrasound in liver tissue which was several weeks post mortem and obtained approximately the same value that Mountford and Wells (1972) determined under in vivo conditions at 1.5 MHz, viz., 0.20 cm^{-1}. Thus in relatively high water parenchymal tissue, details of the state of preservation are necessary for analysis.

Except for the membranes which surround the brain and the stroma of fibrous connective tissue associated with the main blood vessels, the brain is relatively free of connective tissue. It is assumed that the spinal cord also possesses relatively little connective tissue. It is interesting to observe that while the percentages of total protein and collagen are at least half those of spleen, liver and kidney, the ultrasonic attenuation and velocity do not appear to be different.

Water, however, appears to appreciably affect the ultrasonic propagation properties in brain tissue. Kremkau et al. (1976) observed that the ultrasonic attenuation in infant brain was approximately one-third that of adult brain. Infant brain exhibits one of the highest, if not the highest water content for intact tissue, somewhere around 90% as compared to 76-79% for adult brain (Altman and Dittmer, 1961). Also, the ultrasonic attenuation of an adult hydrocyphalic brain was slightly less than that of the infant brain (Kremkau et al., 1976). Oka and Yosioka (1976) reported that the attenuation of an edematose brain was less than that of normal adult brain. Wladimiroff et al. (1975) measured the speed of sound in infant brain from the sixteenth day of gestation to term and observed an increase with age in the velocity from 1513 to 1540 m/s. This change was attributed to the content of solids, or conversely, to the content of water.

Johnston and Dunn (1976) developed a model to describe the transmission of ultrasonic energy into the brain, through the meninges from physiological saline. The meninges consists of the three membranes which envelop the brain, viz., the outer dura mater, the intermediate arachnoid and the pia mater. The model assumed was a three transmission layer model (Kinsler and Fry, 1962). The two outer media, the brain and the physiological saline, were assumed to possess the same impedance. In order for the model to then fit the transmission data as a function of frequency, the intermediate layer was assigned a thickness of 250μm and a speed of sound of 1800 m/s, 300 m/s greater than the other two media. An examination of Table I indicates that such a high velocity would correspond to a very high collagen content material which, in fact, the meninges is.

The data on striated muscle provides some confusion in terms of the affect of attenuation as a function of grain. This would also include tongue since the deep tissue is striated muscle. For striated muscle, it was found that the attenuation coefficient was lower against the grain (Dussik and Fritch, 1955, 1956) whereas the opposite has been observed for tongue (Hueter, 1948). The same contradiction exists for the velocity. Tongue tissue consists of much more than striated muscle (Bradley, 1972). Therefore, it may not be proper to compare tongue directly with striated muscle.

The density of connective tissue is always greater in the right ventricle as compared to the left ventricle, in the range of 2 to 5% dry weight in the left ventricle and 4 to 7% in the right ventricle, and the density in the atrium is approximately twice that of the ventricles (Chavpil, 1967). On the average, heart tissue has a

slightly greater percentage of collagen than the other tissues above
it in Table II. Correspondingly, it also has the greatest attenuation coefficient.

Namery and Lele (1972) reported that the acoustic impedance of
infarcted myocardium is lower than normal myocardium. This would
indicate that the velocity decreased and may be explained in terms
of increased fluid in the area of the infarct since the ultrasonic
measurements were performed 20 to 30 minutes after ligation. A
quantitative study of the ultrasonic attenuation on normal and infarcted canine myocardium, around 2 months after the infarct, over
the frequency range from 2 to 10 MHz clearly indicates that the
attenuation is increased in the infarcted tissue (Yuhas et al., 1976;
Mimbs et al., 1976). The pathology of an infarct indicates that
within 24 hours fibroblasts and capillaries appear. Within a few
days, the fibrin is replaced by collagen and eventually, within a
couple of months, becomes dense collagenous tissue (Friedberg, 1966).
Thus, the increase in ultrasonic attenuation can be correlated with
an increase in collagen.

The ultrasonic attenuation in lenticular tissue, if extrapo-
lated to a frequency of 1 MHz by assuming a linear dependence upon
frequency, would be between 0.09 and 0.23 cm^{-1}. Begiu's (1954)
measurement at 3 MHz yielded an attenuation in the range of 0.59 to
0.69 cm^{-1} and the data of Lizzi and colleagues (Lizzi et al., 1976)
reported a value of 0.92/(MHz-cm) in enucleated human eye over the
frequency range from 10 to 17 MHz. The latter report also qualita-
tively indicated that the lenticular attenuation was greater in the
rabbit. Lenticular tissue is a high protein material with a varied
spatial distribution of water. The innermost zone of the lens
typically has less than half the water concentration of the outer-
most zone (Davson, 1972).

The balance of the tissues in Table I and II are high collagen
content materials. These include articular capsule, integrument,
cartilage and tendon which typically have collagen contents in the
range from 7 to 30 percent wet weight. This is at least a factor
of five greater than the other tissues listed in the two tables;
those values of collagen content range from 0.04 to 1.6 percent wet
weight. Yet it is unmistakable that the high collagen content
materials exhibit a greater attenuation than the low collagen content materials. Thus, it must be suggested that the scatter in the
data does not permit more than a qualitative observation on the role
which collagen plays to influence attenuation. If one were to question the accuracy of the velocity of sound in integument, and use
the velocity range from 1665 to 1750 m/s to describe high collagen
content tissues, then it could be strongly suggested that collagen
has a marked influence on velocity. At 1700 m/s, this is around a
10 percent increase in the velocity as compared to the low collagen
content tissues. This is highly supportive of the Fields and Dunn's

(1973) hypothesis that the structural collagen-containing components of tissue are largely responsible for echographic visualizability.

Bibliography

P. L. Altman and D. S. Dittmer. Blood and Other Body Fluids. Federation of American Societies for Experimental Biology, Washington, D. C. (1961).

Z. E. Begui. Acoustic Properties of the Refractive Media of the Eye. J. Acoust. Soc. Amer., 26, 365-368 (1954).

Biology Today. CRM Books, Del Mar, California (1972).

W. Bloom and D. W. Fawcett. A Textbook of Histology. W. B. Saunders, Co., Philadelphia (1968).

J. V. Bradley. Elementary Microstudies of Human Tissue. Charles C. Thomas, Springfield, Illinois (1972).

J. K. Brady, S. A. Goss, R. L. Johnston, W. D. O'Brien, Jr. and F. Dunn. Ultrasonic Propagation Properties of Mammalian Testes. J. Acoust. Soc. Amer., (in press). (1976).

E. L. Carstensen. The Mechanism of the Absorption of Ultrasound in Biological Materials. IRE Trans. Med. Electronics, ME-7, 158-162 (1960).

E. L. Carstensen, K. Li and H. P. Schwan. Determination of the Acoustic Properties of Blood and its Components. J. Acoust. Soc. Amer., 25, 286-289 (1953).

R. C. Chivers and C. R. Hill. Ultrasonic Attenuation in Human Tissue. Ultrasound Med. Biol., 2, 25-29 (1975).

M. Chvapil. Physiology of Connective Tissue. Butterworth, London (1967).

D. J. Coleman, L. A. Franzen and F. L. Lizzi. Spectral Analysis Evaluation of the Eye Involvement of Thyroid Ophthalmopathies. Presented at the First World Federation for Ultrasound in Medicine and Biology, paper no. 930, San Francisco (1976).

J. D. C. Crisp. Properties of Tendon and Skin. In Biomechanics: Its Foundations and Objectives, eds., Y. C. Fung, N. Perrone and M. Anliker, pp 141-179, Prentice-Hall, Inc., Englewood Cliffs, New Jersey (1972).

H. Davson. The Physiology of the Eye. Academic Press, New York (1972).

F. Dunn and J. K. Brady. Absorption of Ultrasound in Biological Media. Biophysics, 18, 1128-1132 (1974).

K. T. Dussik and D. J. Fritch. Determination of Sound Attenuation and Sound Velocity in the Structures Constituting the Joints, and of the Ultrasonic Field Distribution within the Joints on Living Tissues and Anatomical Preparations, both in Normal and Pathological Conditions. Progress Report, Project A-454, Public Health Service, April (1955), September (1956).

K. T. Dussik, D. J. Fritch, M. Kyriazidou and R. S. Sear. Measurements of Articular Tissues with Ultrasound. J. Phys. Med., 37, 160-165 (1958).

S. Fields and F. Dunn. Correlation of Echographic Visualizability of Tissue with Biological Composition and Physiological State. J. Acoust. Soc. Amer., 54, 809-812 (1973).

C. K. Friedberg. Diseases of the Heart, Third Edition, W. B. Saunders Co., Philadelphia (1966).

E. K. Fry, G. Kossoff, and H. A. Hindman, Jr. The Potential of Ultrasound Visualization for Detecting the Presence of Abnormal Structures within the Female Breast. In 1972 Ultrasonics Symposium Proceedings, ed. J. deKlerk, pp 25-30, IEEE Catalog No. 72 CHO 708-8SU, New York (1972).

A. C. Giese. Cell Physiology. W. B. Saunders Co., Philadelphia (1962).

D. E. Goldman and T. F. Hueter. Tabular Data of the Velocity and Absorption of High-Frequency Sound in Mammalian Tissues. J. Acoust. Soc. Amer., 28, 35-37 (1956).

J. F. Greenleaf, S. A. Johnson, R. C. Bahn, W. F. Samayoa and C. R. Hansen. Images of Acoustic Refractive Index and of Attenuation: Relationship to Tissue Types within Excised Female Breast. Presented at First World Federation for Ultrasound in Medicine and Biology, paper no. 1154, San Francisco (1976).

J. F. Greenleaf, S. A. Johnson, W. F. Samayoa and F. A. Duck. Algebraic Reconstruction of Spatial Distribution of Acoustic Velocities in Tissue from their Time-of-Flight Profites. In Acoustical Holography, vol 6, ed. N. Booth, pp 71-90, Plenum Press, New York (1975).

M. Greenspan and C. E. Tschiegg. Speed of Sound in Water by a Direct Method. J. Res. Nat'l. Bur. Std., 59, 249 (1957).

T. H. Hueter. Measurement of Ultrasonic Absorption in Animal Tissues and its Dependence on Frequency (in German). Naturwiss., 35, 285-287 (1948). Translation in Ultrasonic Biophysics, eds. F. Dunn and W. D. O'Brien, Jr., Dowden, Hutchinson and Ross, Inc., Stroudsburg, Pennsylvania (1976).

T. F. Hueter. Viscoelastic Losses in Tissue in the Ultrasonic Range. WADC Tech. Report No. 57-706 (1958).

R. L. Johnston and F. Dunn. Influence of Subarachnoid Structures on Transmenninges Ultrasonic Propagation. J. Acoust. Soc. Amer., 59, S76 (1976).

L. W. Kessler. VHF Ultrasonic Attenuation in Mammalian Tissue. J. Acoust. Soc. Amer., 53, 1759-1760 (1973).

L. W. Kessler, S. I. Fields and F. Dunn. Acoustic Microscopy of Mammalian Kidney. J. Clinical Ultrasound, 2, 317-320 (1974).

L. E. Kinsler and A. R. Frey. Fundamentals of Acoustics. Wiley and Sons, New York (1962).

G. Kossoff, E. K. Fry, and J. Jellins. Average Velocity of Ultrasound in the Human Female Breast. J. Acoust. Soc. Amer., 53, 1730-1736 (1973).

F. W. Kremkau, C. P. McGraw and R. W. Barnes. Attenuation and Velocity in Normal Brian. J. Acoust. Soc. Amer., 59, S75 (1976).

R. A. Lemons and C. F. Quate. Advances in Mechanically Scanned Acoustic Microscopy. In 1974 Ultrasonics Symposium Proceedings, ed. J. deKlerk, pp 41-44, IEEE Catalog No. 74 CHO 896-1SU, New York (1974).

F. Lizzi, L. Katz, L. St. Louis and D. J. Coleman. Applications of Spectral Analysis in Medical Ultrasonography. Utrasonics, 14, 77-80 (1976).

F. L. Lizzi and M. A. Laviola. Power Spectra Measurements of Ultrasonic Backscattering from Ocular Tissue. In 1975 Ultrasonics Symposium Proceedings, ed. J. deKlerk, pp 29-31, IEEE Catalog No. 75 CHO 944-4SU, New York (1975).

G. D. Ludwig. The Velocity of Sound Through Tissues and the Acoustic Impedance of Tissues. J. Acoust. Soc. Amer., 22, 862-866 (1950).

M. B. Mathews. Connective Tissue. Macromolecular Structure and Evolution. Springer-Verlag, New York (1975).

V. M. Maynard and S. A. Goss. Personnel Communication (1976).

J. W. Mimbs, D. E. Yuhas, J. G. Miller, A. N. Weiss and B. E. Sobel. Detection of Myocardial Infarction In Vitro Based on Altered Attenuation of Ultrasound. Submitted to Circulation Research (1976).

R. A. Mountford and P.N.T. Wells. Ultrasonic Liver Scanning: The Quantitative Analysis of the Normal A-Scan. Phys. Med. Biol., 17, 14-25, 1972.

J. Namery and P. P. Lele. Ultrasonic Detection of Myocardial Infarction in Dog. In 1972 Ultrasonics Symposium Proceedings, ed. J. deKlerk, pp 491-494, IEEE Catalog No. 72 CHO 708-8SU, New York (1972).

A. H. Neufeld. Canadian J. Research, B15, 132 (1937).

W. D. O'Brien, Jr. and F. Dunn. Ultrasonic Absorption Mechanisms in Aqueous Solutions of Bovine Hemoglobin. J. Phys. Chem., 76, 528-533 (1972).

M. Oka and K. Yosioka. Ultrasonic Absorption of Human Brain Tissue. Presented at First World Federation for Ultrasound in Medicine and Biology, paper no. 1302, San Francisco (1976).

H. Pauly and H. P. Schwan. Mechanism of Absorption of Ultrasound in Liver Tissue. J. Acoust. Soc. Amer., 50, 692-699 (1971).

J. M. M. Pinkerton. The Absorption of Ultrasonic Waves in Liquids and Its Relation to Molecular Constitutions. Proc. Phys. Soc. London, B62, 129 (1949).

A. H. T. Robb-Smith. Normal Morphology and Morphogenesis of Connective Tissue. In Connective Tissue in Health and Disease, ed. G. Asboe-Hansen, pp 15-30, Philosophical Library, Denmark (1954).

T. C. Ruch and H. D. Patton. Physiology and Biophysics. W. B. Saunders Co., Philadelphia (1966).

N. Schneiderman and R. S. Hosek. An Overview-The Nervous System. In Biological Foundations of Biomedical Engineering, ed. J. Kline, pp 415-438, Little, Brown and Co., Boston (1976).

R. van Heyningen. The Lens. In The Eye; Vegetative Physiology and Biochemistry, ed. H. Davson, pp 213-287, Academic Press, New York (1962).

G. E. P. M. van Venrooij. Measurement of Ultrasound Velocity in Human Tissue. Ultrasonics, 9, 240-242 (1971).

B. K. Watt and A. L. Merrill. Composition of Foods. Agriculture Handbook No. 8, U. S. Department of Agricultrue, Superintendent of Documents, U. S. Government Printing Office, Washington, D. C., December (1963).

A. White, P. Handler and E. L. Smith. Principles of Biochemistry. McGraw Hill Book Co., New York (1968).

J. J. Wild and J. M. Reid. The Effects of Biological Tissues on 15-mc Pulsed Ultrasound. J. Acoust. Soc. Amer., 25, 270-280 (1953).

J. J. Wild and J. M. Reid. Echographic Visualization of Lesions of the Living Intact Human Breast. Cancer Research, 14, 277-283 (1954).

W. F. Windle. Textbook on Histology. McGraw Hill, New York (1976).

J. W. Wladimiroff, I. L. Craft and D. G. Talbert. In vitro Measurements of Sound Velocity in Human Fetal Brain Tissue. Ultrasound Med. Biol., 1, 377-382 (1975).

M. B. Wolf. The Body Fluids. In Biological Foundations of Bio-medical Engineering, ed. J. Kline, pp 391-411, Little, Brown and Co., Boston (1976).

R. C. Wolf and J. H. Leathem. Hormonal and Nutritional Influences on the Biochemical Composition of the Rat Testis. Endocrinology, 57, 286-290 (1955).

K. Yamkawa, A. Yoskioka, K. Shimizu, T. Moriya, S. Higashi, T. Sakamaki, K. Sawada, K. Kuramochi and H. Okuda. Studies on the Acoustic Impedance of Experimentally Illed Rats. Jap. Med. Ultrasonics, 39-40 (1964). Reported in Field and Dunn (1973).

D. E. Yuhas, J. W. Mimbs, J. G. Miller, A. N. Weiss and B. E. Sobel. Correlation between Changes in the Frequency Dependence of Ultrasonic Attenuation and Regional CPK Depletion Associated with Myocardial Infarction. Presented at First World Federation of Ultrasound in Medicine and Biology, paper no. 1123, San Francisco (1976).

R. Zana and J. Lang. Interaction of Ultrasound and Amniotic Liquid. Ultrasound Med. Biol., 1, 253-258 (1974).

HIGH RESOLUTION, HIGH SENSITIVITY ULTRASONIC C-SCAN

IMAGING SYSTEM

R. Mezrich

RCA Laboratories
David Sarnoff Research Center
Princeton, N. J. 08540

I.   Introduction

A particular goal of much of the research in ultrasonic imaging is the development of systems that can image a volume inside the body with high resolution in all dimensions.

It is well known, both in optics and acoustics, that a necessary condition for achieving high resolution images is the utilization of a high numerical aperture system. In practical terms, this means that the target should be illuminated with waves that cover a large angular spectrum and the echoes scattered from the target must be collected over as large an angular range as possible. For a variety of reasons however, such as ease of manipulation or convenience in placement over intercostal spaces, conventional ultrasonic transducers have relatively small effective diameters, with the result that the resolution characteristics of these systems are anisotropic, having very high resolution in the axial direction and significantly lower resolution in the lateral direction.

Previous experimental efforts using large numerical aperture techniques[1-3] have been limited by the complexities of the mechanical or electrical configuration and by the relatively long times needed to form an image. In the following sections a relatively simple system is described that gives high resolution acoustic images in both lateral and axial directions. In this method a large transducer is coupled, by means of mechanical deflection techniques,

to an ultrasonic lens capable of insonifying and receiving sound
over a large angular range and over large fields. Both C-scan and
B-scan images can be shown, with the B-scan displayed in real
time and the C-scan image formed in 4-8 seconds. Wavelength
limited resolution is obtained in both dimensions.

II. Principles of Operation

The method used to generate the acoustic images depends on
two simple properties of a well corrected lens. The first is that a
plane wave incident on a lens will be focused to a spot in the focal
plane of the lens and these echoes scattered from objects at the
focal plane will be received by the lens and converted back to a
plane wave. The second property is that the position of the spot in
the focal plane is proportional to the angle of the incident plane
wave. Thus (as shown in Fig. 1) as the incident angle of the plane
wave is varied the insonified spot will be scanned over the focal
plane. At each position of the scan, echoes from the insonified
spot will be collected by the lens and, as stated above, be converted back again to plane waves inclined at the incident angle but
traveling in the opposite direction. By using short acoustic pulses,
and with the knowledge of the focal length of the lens, those echoes
coming from the focal plane can be detected and processed for display on a CRT that is scanned synchronously with the acoustic spot.
In this way, by scanning the spot in two dimensions an image of
the plane normal to the acoustic axis (i.e. a C-scan) can be developed. With large numerical aperture lens the resolution of the
image can be of the order of the acoustic wavelength.

The time needed to develop the image is limited by the acoustic
velocity. If we consider a lens with a 20 cm focal length, the time
needed to insonify and receive the echo from any spot (assuming an
acoustic velocity of 1500 m/sec) will be ~ 270 μ sec and so the time
needed to image a 128 x 128 element image will be ~ 4.4 seconds.
While this is a short and reasonable time in which to form and view
an ultrasonic image of internal structures, it is sufficiently long
that relatively simple mechanical systems can be considered for
use in scanning the acoustic wave.

Several simple mechanical systems have been developed, one
of which is shown schematically in Fig. 2, where a large (10 cm)
diameter transducer is placed above an acoustic (styrene) lens.
The transducer is held so that it can pivot (or tilt) about a diameter
and this mount is designed so the transducer can be rotated while

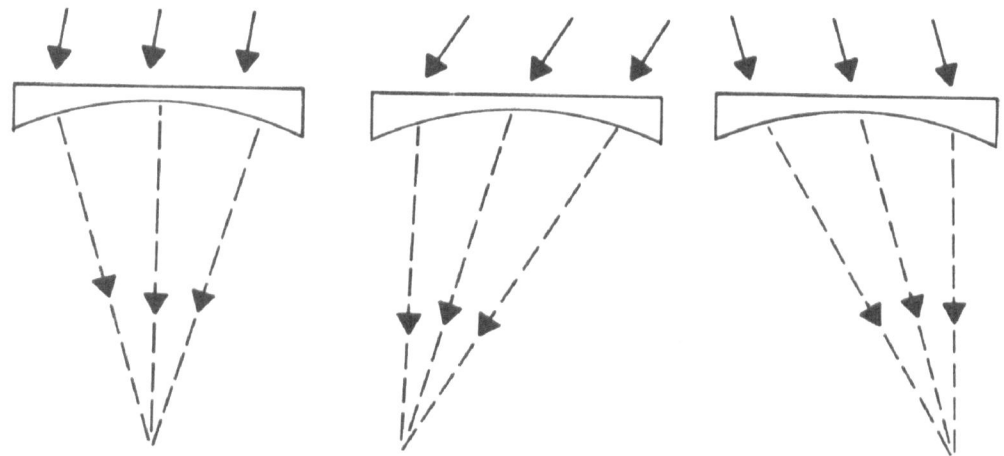

Fig. 1.   Spot Scanning by wave angle inclination.

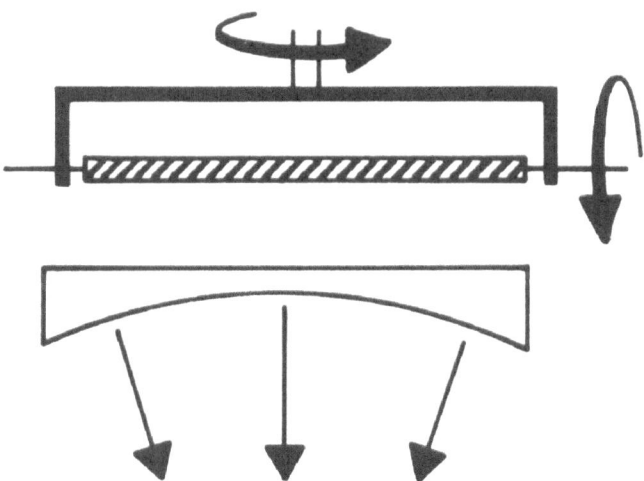

Fig. 2.   System for Spiral Scanned Acoustic Imaging.

it is tilted. The entire structure is surrounded by fluid, and short pulses of sound are emitted from the transducer through the lens and to the object at a repetition rate determined by the round trip travel time of the acoustic wave.

As the transducer tilts, the spot of sound is scanned along one axis (say the x-axis). If while tilted at some angle the transducer structure is rotated, the locus of scanned spots will describe a circle, with the radius equal to x_0 (proportional to the tilt angle of the transducer). By continuing the tilting motion as the structure is rotated a spiral pattern will be generated,[4] scanning the focal plane. Note that with this arrangement a plane is scanned, not a spherical surface.

An example of the image obtained with this arrangement is shown in Fig. 3. The object was a metal caning, with 2.5 mm small holes and 6 mm large holes and, as can be seen, a nickel was placed on the caning. The ultrasonic frequency used was 1.6 MHz and the time to form the image was ~6 seconds. The structure is clearly well resolved.

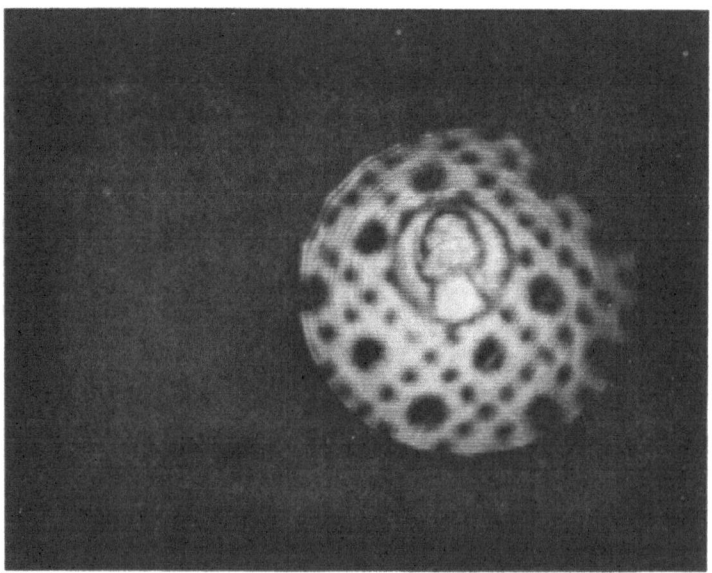

Fig. 3.   Acoustic image of metal caning with nickel. Small hole diameter is 2.5 mm, large hole diameter is 6 mm. Acoustic frequency was 1.6 MHz.

For a variety of reasons, especially the implementation of the
B-scan display to be discussed later, a raster scan system rather
than a spiral scanned system is preferred. Although x and y motion
of the transducer could be used, a simple method is to use Risley
prisms.[5,6]

In the Risley prism arrangement, as seen in Fig. 4, the combined action of two counterrotating prisms will produce a scan in
one dimension. By using two such sets of prisms, one set oriented
to produce a scan along one direction (say the x direction) and the
other set offset by 90° to produce a scan in the orthogonal direction
a raster scan will be generated.

The complete arrangement is shown in Fig. 5 and includes the
transducer, the two sets of Risley prisms, and the lens. The operation is as before, except with this arrangement the spot is scanned
back and forth along the plane in an x, y format, rather than in a
spiral. The number of lines in the image is determined by the ratio
of rotary speeds of the two sets of prisms (100 in the image to be
shown below). It should be noted that in this arrangement, as in
the previous one, the sound waves are always emitted and received
normal to the face of the transducer regardless of which point in the
scan is insonified.

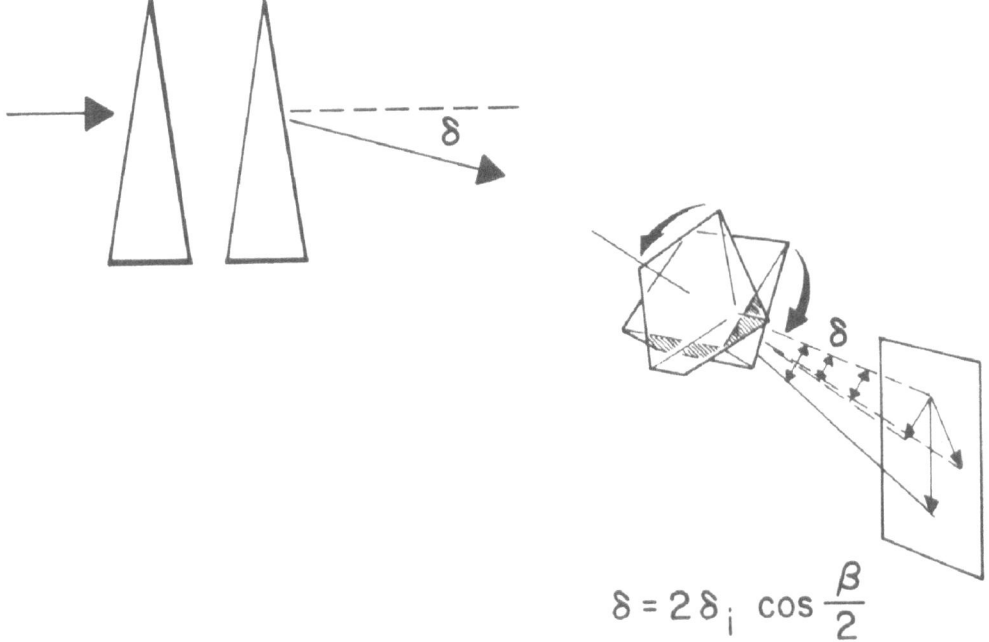

$$\delta = 2\delta_i \cos \frac{\beta}{2}$$

Fig. 4.   Risley prism arrangement for beam deflection in one plane.

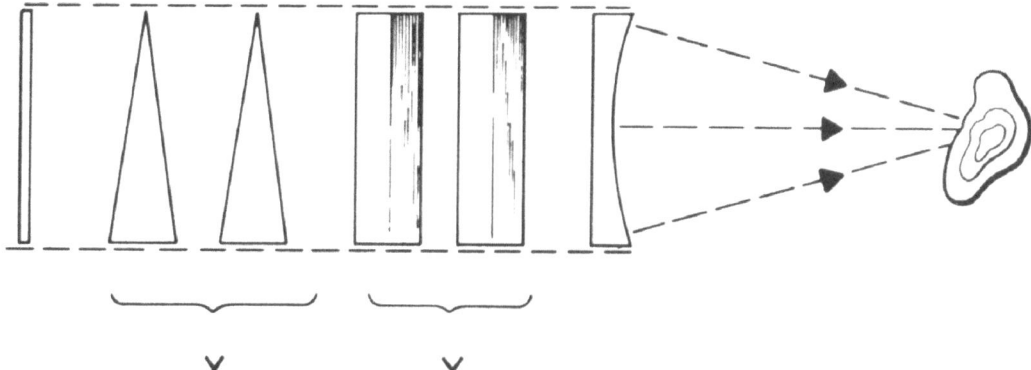

X   Y

Fig. 5. Raster scanned imaging systems showing, from left,
         transducer, prism set for x-deflection, prism set for
         y-deflection, and lens.

An example of the image with this arrangement, of the same
object used above, is shown in Fig. 6. Again 1.6 MHz waves
were used and the time to form the image was ~ 6 seconds. (The
slight loss of detail is due to an inadvertent misalignment of one of
the prism sets which is now being corrected.)

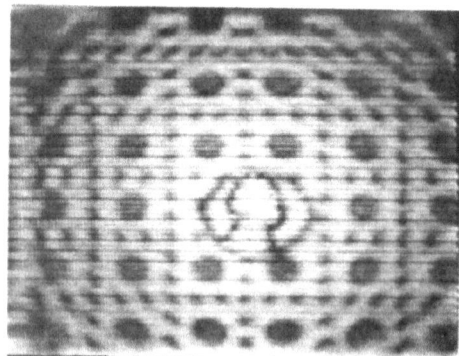

Fig. 6. Raster scanned acoustic image of caning.

III. Features

As simple as these scanning arrangements are they have some useful features with regard to resolution, angular tolerance and intensity, that are important in ultrasonic imaging applications.

The resolution of any imaging system depends critically on the manner in which the object is illuminated, for it affects the extent to which the imaging lens can capture the full angular spectrum of waves scattered by the object. The greater the angular range of the incident illumination, the greater will be the spectrum of waves entering the lens pupil, and so the greater the resolution. The optimum condition occurs, as is well known in microscopy, when the cone of illumination just fills the lens pupil. The variation in resolution with different illumination can be substantial, with, for example, the resolution in the case using the optimum cone of illumination being as much at two times larger as an alternative case where a normally incident plane wave illuminates the object. In practice this implies that a system using optimum illumination could achieve as good a resolution with longer wavelength (higher frequency) waves as a system with sub-optimal illumination could get with shorter wavelengths. This is particularly important for ultrasonic applications, where the attenuation of tissue increases with frequency, for it implies that the effective resolution possible with high frequency sound waves can be realized while only suffering the decreased attenuation of lower frequency insonification. The system described here, with the same lens used to focus and receive the waves, inherently uses the optimum illumination.

A further consequence of using a system in which waves are transmitted and received over a high angular range is the increased angular tolerance of the system. Even if the target has surfaces tilted away from the acoustic axis some of the waves scattered from the target will still be received as long as the deviation from the axis is less than the cone angle of the illumination. With cone angles of the order of 15-30°, as used in practice, this implies a considerable tolerance for imaging even specular objects tilted away from the axis.

Another consequence of a focusing system is the intensity "gain" that obtains when one compares the intensity at the focal plane with that at the lens aperture. As an example, if the spot size at the focal plane is ~1 mm while the aperture diameter is 10 cm (as used in practice) the "gain" in intensity is 10^4 (40 db).

This "gain" can offset some of the effects of attenuation and ensure a constant intensity at each overlying plane of tissue. Thus, in contrast to the case with collimated or plane wave transducers, the intensity at the surface of the patient need not be higher than that at the site being imaged.

IV. Examples

Some examples of the imaging capabilities of this system are shown in Figs. 7 and 8. Fig. 7 illustrates the resolution characteristics when imaging through overlying tissue. Fig. 7 shows the results of an experiment in which we obtain images of a metal caning lying under a 6 cm thick piece of liver. The top photograph shows the image before the liver was put in place while the lower images shows the caning through the liver at two slightly different power levels. Other experiments, with similar results, have been done with other tissues. The caning is clearly resolved even through the overlying tissue. Some of the overlying structures, for example the vasculature in the liver, cast shadows, which could be minimized by increasing the acoustic intensity. These overlying structures could themselves be imaged as the focal plane was moved (for example by "racking" the system to the plane of the structure).

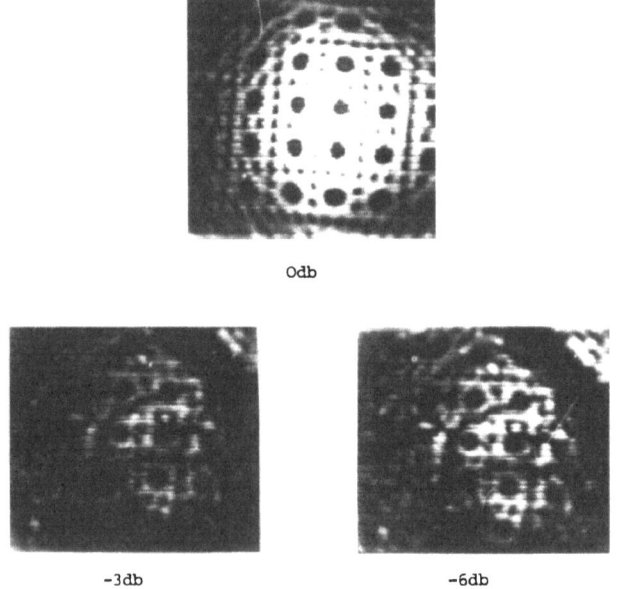

Fig. 7.   Metal caning imaged through 6 cm thick calf liver. Top shows image before liver is put in place. Lower right made with 3 db higher acoustic intensity.

A more interesting example of the imaging capabilities, ultrasonic images of a lamb kidney, is shown in Fig. 8. The upper
photograph shows the intact kidney and the lower photograph shows
a cross-section of the kidney (the kidney was sliced after the imaging experiment) in which some of the internal structures are visible.
The bottom photographs show the ultrasonic images of the same
(intact) kidney, with the two images displaying two different planes
in the kidney.

Examination of these figures, and comparison with actual kidney
cross-sections (e.g. in Gray's Anatomy[7]) indicates the high detail
possible with this ultrasonic scanning system.

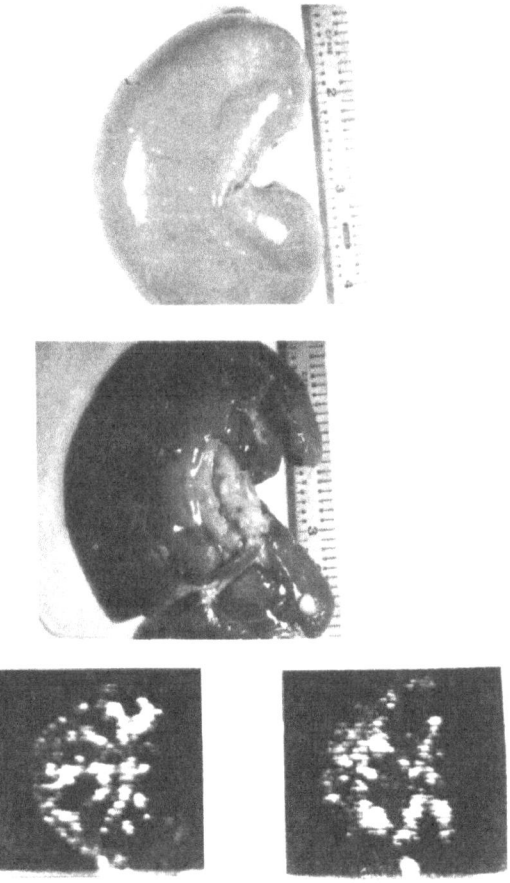

Lamb Kidney

Fig. 8. Acoustic C-scan image of lamb kidney. Upper photograph
shows intact kidney. Middle photograph shows kidney
cross-section from slice made after imaging experiment.
Images of two different planes shown at bottom.

V. Depth of Field and B-Scan

To this point the scanning system has been described with the implication that the lens focuses the wave precisely to a single plane. Practical lenses do not behave this way; rather there is a range around the focal plane over which the wave is well focused. The actual situation is much like that schematically illustrated in Fig. 9. If D is the spot size at the focal plane, then there is a range about the focal plane (δ) over which the spot size does not vary appreciably from this value D. The extent of this depth of field is determined by the lens geometry and is given by

$$\delta = 4\lambda \left(\frac{F}{A}\right)^2$$

and through this depth various planes can be displayed without moving the imaging system. Any desired plane may be viewed by simply using an electronic gate to select only those echoes from a particular depth. The resolution in depth (the finesse with which any plane can be selected) is given by the characteristics of the transducer (i.e. how well it is damped) and is of the order of a wavelength.

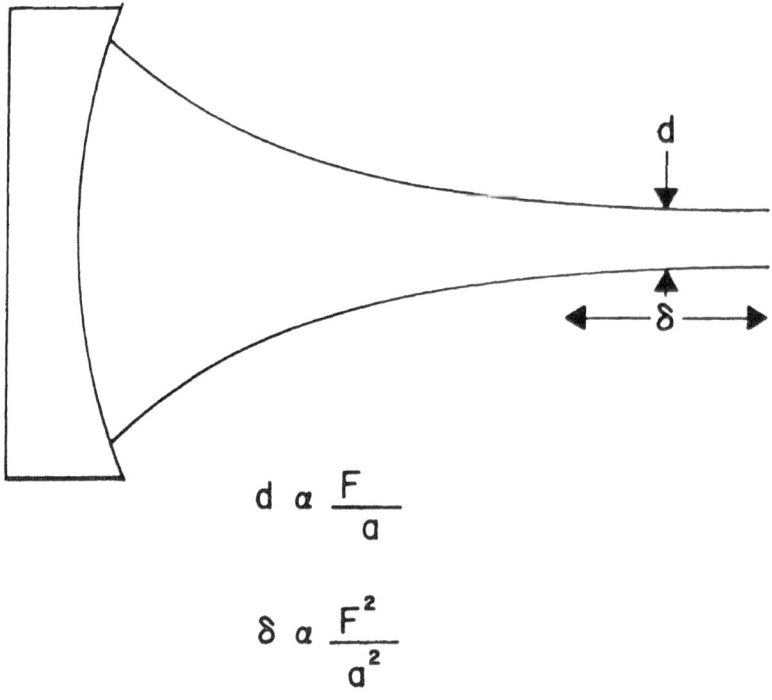

$$d \; \alpha \; \frac{F}{a}$$

$$\delta \; \alpha \; \frac{F^2}{a^2}$$

Fig. 9. Beam shape near focal plane.

A good example of this depth resolution is shown in Fig. 10. The top figure shows the nickel lying on the metal caning imaged as before. In the lower right the electronic gate has been set to receive echoes only from the plane of the nickel and so the caning is not seen. In the left photograph the gate has been set at the plane of the caning and so the nickel is not seen. From this example we see that the depth resolution is approximately 2 mm.

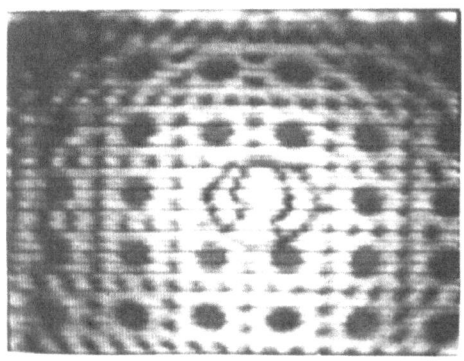

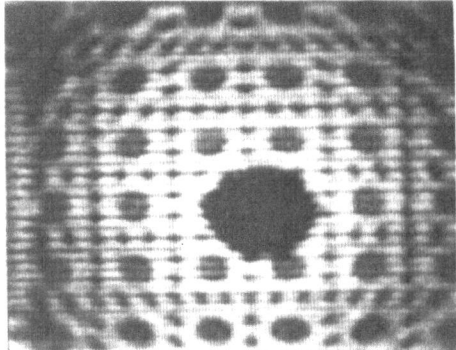
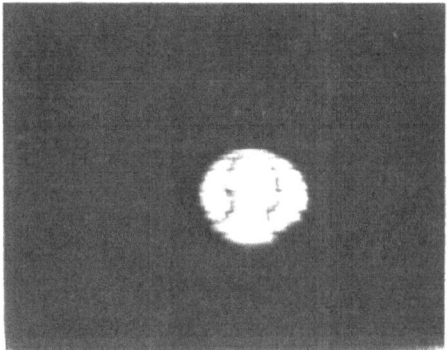

Fig. 10. Illustration of depth resolution. Top shows image of caning with nickel. Lower right shows nickel, lower left shows caning. Images made by changing detection gate delay.

The extent of the depth of field is considerable. To illustrate this, images of the caning were taken at different distances, with the imaging system fixed and only the delay of the electronic gate varied to select the echo. The results are shown in Fig. 11 (where the depth dimension is in inches). As can be seen there is excellent resolution for at least one inch around the nominal 10-inch deep focal plane and tolerable resolution over at least a two-inch range.

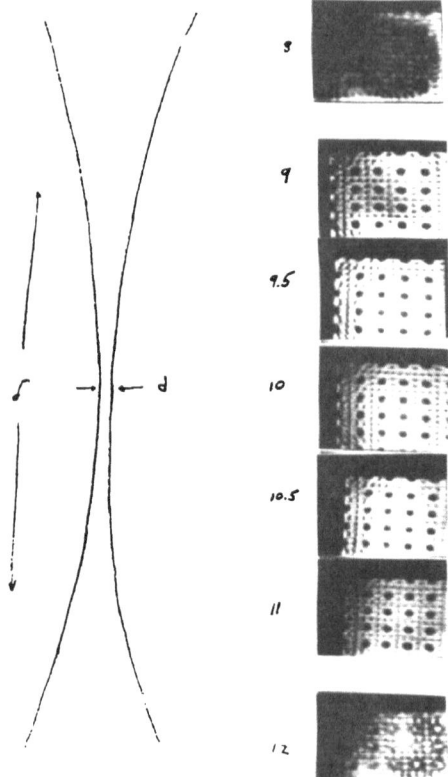

Fig. 11. Extent of depth of field images shown by moving caning to
various depths.

The depth of field may be used to advantage in two ways. The
first, as has been illustrated above in the case of the kidney image,
is to simply vary the gate delay to form detailed C-scan images of
different planes. The second is to use this depth of field to display
high resolution, essentially real time, B-scan images. We remember
that the beam is scanned in raster fashion, and we further note that
one axis (say the x-axis) is scanned quickly (15-20/sec). Instead
of using a gate to select the return echoes from one plane, all the
echoes emanating through the depth of field can be detected and
processed to show a B-scan image of a slice along the plane of the
fast axis scan. One set, the slow set, of the Risley prism can be
stopped to allow careful examination of a particular plane, or, with
both prism sets rotating, various planes can be examined as the
beam is scanned. The depth of the B-scan plane is equal to the
depth of field, which can be varied by changing the aperture of the
imaging system. This, of course, also affects the lateral resolution,
but since the resolution is inversely proportional to the aperture,

while the depth of field is inversely proportional to the square of
the aperture, the compromise between depth and resolution is not
severe. As an example, if the resolution is ~2 mm and the depth
of field is ~1-2 inches (as seen in Fig. 11), a decrease in resolution to ~3 mm will increase the depth of the field to ~2-4 inches.

An example that illustrates the B-scan and C-scan capabilities
is shown in Fig. 12. The object was a machinist's "hold-down"
clamp with many steps along the angled surface, placed next to the
caning. The C-scan image shows the caning and only those steps
at the plane of the caning. The B-scan image shows the orthogonal
cross-section of the caning and the sequence of steps. Both images
can be displayed simultaneously, and both show high resolution.

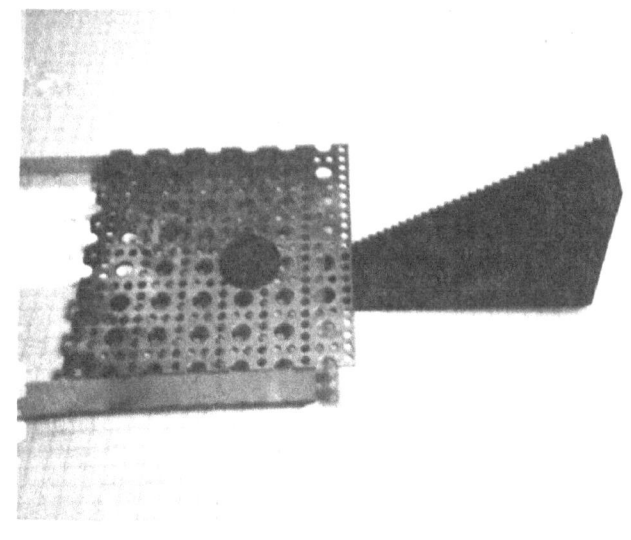

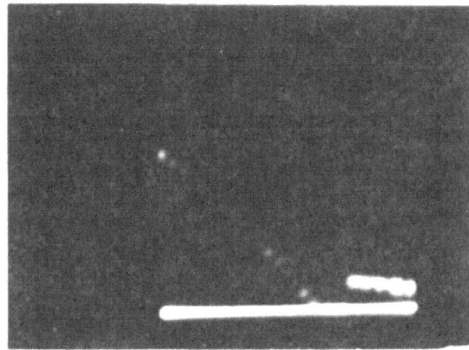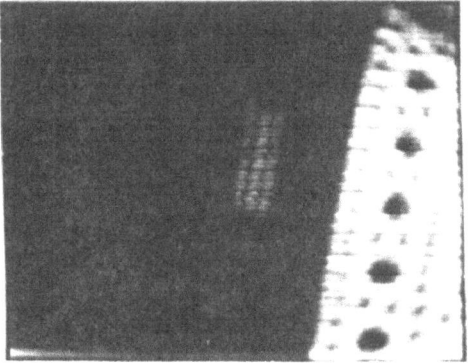

Fig. 12. B-scan and C-scan images of machinists clamp next to
caning.

Summary

Simple, high resolution methods for obtaining C-scan and B-scan ultrasonic images have been described which offer advantages in resolution, angular tolerance and sensitivity. C-scan images are formed in 4-8 seconds and orthogonal B-scan images are displayed simultaneously in near real-time.

References

1. Brown, T. G., "Visualization of soft tissues in two and three dimension-limitations and developments," Ultrasonics 5, p. 118 (1967).

2. Wells, P. N. T., "Physical Principles of Ultrasound Diagnosis," Chapt. 4.7, Academic Press (1969).

3. Green, P. S., Schaefer, L. F., and Macovski, A., "Considerations for Diagnostic Ultrasonic Imaging," Acoustical Holography, Vol. 4, p. 97 (1972).

4. McCready, V. P. and Hill, C. R., "Constant Depth Ultrasonic Scanner," Brit. J. of Rad. 44, p. 747 (1971). (A method for spiral scan is described, but with the focal plane being a spherical surface.)

5. Jenkins, F. A. and White, H. E., "Fundamentals of Optics," p. 23, McGraw-Hill (1957).

6. Green, P. S., Schaeffer, L. F., Jones, E. D., and Suarez, J. R., "A New High-Performance Ultrasonic Camera," Acoustical Holography, Vol. 5, p. 493 (1974).

7. Gray, H., "Anatomy of the Human Body," edited by H. Goss, p. 1282, Lea & Febiger, Philadelphia (1966).

A LARGE APERTURE REAL-TIME EQUIPMENT FOR VASCULAR IMAGING*

Anant K. Nigam
NYIT Science and Technology Research Center
8000 No. Ocean Drive, Dania, Florida 33004
and
Charles P. Olinger
Univ. of Cincinnati, Cincinnati, Ohio 45267

ABSTRACT

A reflective scanning pulse-echo technique has been utilized to provide real-time imaging in conjunction with large aperture transducers. The larger apertures have been employed to provide sub-millimeter resolutions in all three space directions. Use of reflecting scanning mechanism provides real-time operation without sacrificing portability. To improve equipment repeatibility, a novel technique termed the Interactive Gain Compensation (IGC) is used to preprocess the RF signals. This also enhances equipment detectability together with simplification to the conventional TGC arrangements. The input dynamic range is also favorably affected by the new IGC processing. Preliminary utilization of this equipment has been in the in-vivo imaging of the human carotid artery with planned subsequent applications for the coronaries. Atherosclerotic plaques less than 1mm in size have been detected in the carotid. These present less than 5% occlusion in the artery and may not be readily identified by X-ray angiography. Because of greater gray scale repeatability and higher resolutions, it has also been possible to visualize the generally complex plaque topography and the relative "hardness" of the plaque in a consistent and repeatible manner. The accompanying presentation includes typical real-time video-taped data of images obtained in the carotid arteries of normals and patients with known atherosclerotic lesions.

*This work is supported in part under Contract N01-HV-5-2964 from the National Heart Lung and Blood Institute, N.I.H.

A.K. NIGAM AND C.P. OLINGER

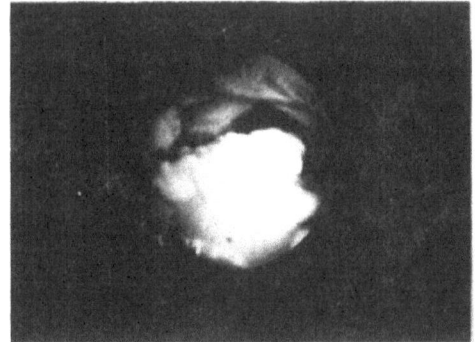 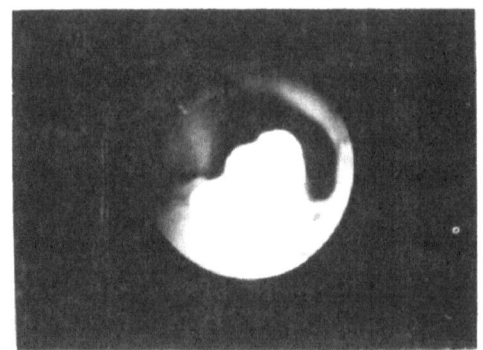

FIG. 1   ENDOSCOPIC VIEWS OF TWO DIFFERENT PLAQUE TYPES IN
THE HUMAN CAROTID ARTERY (See also Ref. 2)

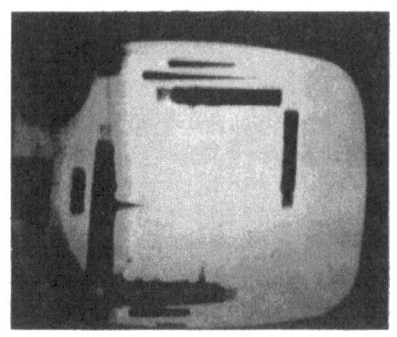 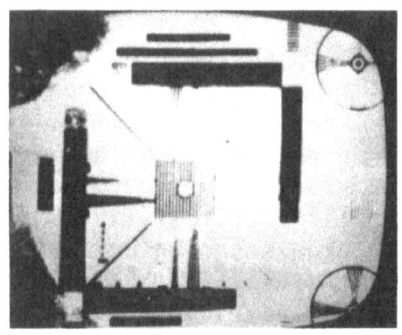

50 lines resolution
No gray levels

200 lines resolution
No gray levels

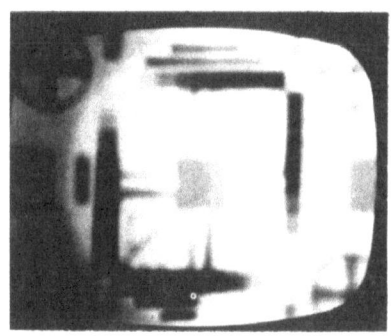 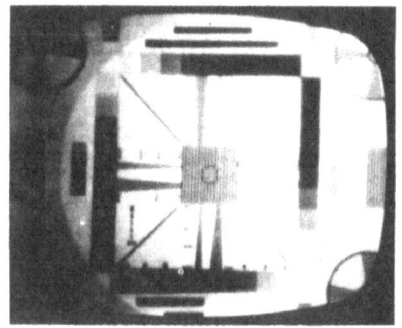

50 lines resolution
10 gray levels

200 lines resolution
10 gray levels

FIG. 2   OPTICAL IMAGES SHOWING EFFECTS ON IMAGE QUALITY
WITH HIGHER RESOLUTION AND GRAY SCALE

INTRODUCTION

Our goals in ultrasonic imaging and analysis of the vascular system, with particular emphasis for the coronary and carotid artery in humans, is for the early detection and quantization of athero-sclerosis. Ultrasound is further suited for this diagnostic role insofar as it being a non-invasive and non-traumatic technique, it may be readily used to monitor and assess the role of diets, drugs and surgical procedures in the possible reversibility of atherosclerosis in humans and clinical animals.

Atherosclerosis is characterized by a thickening of the intimal wall of arteries and/or the formation of lesions and plaques within the lumen[1]. The physical nature of the athero-sclerotic plaque is typified by a complex surface topography as evidenced by pathologically obtained endoscopic views[2] of a few different lesion types shown in Fig. 1. The mechanical nature; i. e., the texture of the lesion also varies widely depending possibly on the causative mechanism. In terms of ultrasound imag-ing requirements, therefore, both high resolution and gray scale appear to be important in the satisfactory detection of athero-sclerosis. The role of these two parameters on the resulting image quality may be demonstrated by simulated comparative optical images as shown in Fig. 2.

This paper outlines a description of some of the important considerations in the design of a B-scan pulse-echo imaging equip-ment and its utilizations in the preliminary imaging of the human carotid artery. Most of the preliminary clinical data acquired by the equipment exists in videotape format and only a few typical results are included here. The accompanying talk includes presentation of a 5-minute selected portion of the real-time videotaped data.

EQUIPMENT DESIGN CONSIDERATIONS

Clinical conditions and the anticipated utilization dictate equipment portability, simplicity of operation and repeatability. Equipment portability is provided by maintaining minimum physical size together with a hand-held, self-scanning portable probe. The overall view of the equipment is shown in Fig. 3a. The probe, shown in Fig. 3b is hand-held and the membrane end is made to contact the patient's skin at the imaging location. It contains a large aperture transducer and an electromechanical scanning mechanism for 30 frames/second scanning. The membrane, transducer and scanning mechanism contain a small volume of water which serves several purposes. First, because of the water path, it is possible to collect meaningful image data right from the skin surface of

(b)

(a)

FIG. 3   OVERALL VIEW OF THE ULTRASONIC EQUIPMENT.   (a) OVERALL VIEW OF THE ULTRASONIC EQUIPMENT.   (a) OVERALL CHASSIS CONTAINING THE ELECTRONICS AND THE TV DISPLAY AND RECORDING DEVICES.   (b) THE PORTABLE PROBE CONTAINING TRANSDUCER, SCANNER AND MEMBRANE.   THE PROBE ATTACHES AT THE END OF CABLE SHOWN WITH EQUIPMENT.

the patient. This is of importance for imaging the sub-surface
vessels. These studies may be necessary, especially in the initial
clinical validation of the equipment. Secondly, the use of water
path coupling enables the use at present of the more cost-
effective electromechanical scanning. Third, it aids in providing
a uniform and repeatable contact surface between transducer and
the patient. Further comments on repeatability are contained in a
later section describing the signal processing.

The nature and size of the atherosclerotic plaque as well as
the overall imaging problems have suggested equipment performance
specifications. A large aperture imaging system is proposed to
specifically aid in the visualization of arteries. It is recognized that the artery cross-section is circular and the plaque
surface is irregular. The path of the artery also follows a complicated three-dimensional trajectory over the heart and in the
neck. Thus the structures to be imaged present a wide range of
inclinations to the ultrasound beam -- from normal to parallel
incidence and the larger the numerical aperture of the system the
more sensitive it will be to the scattered echoes leading to more
favorable imaging of the entire lumen.

The need for very high resolutions is also apparent. A resolution of 0.3 - 0.4 mm in both transverse directions and 0.15 -
0.2 mm in the depth direction provides approximately 15-20 resolution elements in imaging a 1 mm size plaque. This leads to less
than ± 5% inaccuracy in quantification of plaque size. In contrast,
with 0.5 mm resolutions, the image contains only four resolution
elements with about 30% inaccuracies in quantification. With the
higher resolutions mentioned above, atherosclerotic plaques as
small as 0.3 mm cover two resolution elements and should be just
detectable.

The mechanical hardness of the plaque varies primarily in
relation to its degree of calcification and cholesterol content,
and depending on this the plaque may cast very strong or weak
echoes. It is therefore necessary to provide for a large dynamic
range during ultrasound imaging. With careful design of the equipment, it is possible to achieve over 50 dB dynamic range for the
received echoes. With non-linear compression of the signal, this
dynamic range may be displayed continuously over a 10-to-15 shade
video gray scale. Controlled signal pre-compression is also provided by utilizing the conventional Time Gain Compensation[3] (TGC)
together with a novel Interactive Gain Compensation (IGC) described
in the next section.

With the large dynamic range for the processing circuits, it
is also necessary to maintain a very large dynamic range at the
transducer. Thermal noise, both in the receiver electronics and

in the medium (Brownian Motion noise) is not generally a limiting
factor, instead the coherent "noise" caused by transducer side-
lobes and multiple ringing is generally dominant. Sidelobe
suppression greater than 60-80 dB achieved by utilizing apodized
lenses as described in an earlier paper[4]. Multiple ringing
artifacts may be substantially minimized by careful design of the
probe and transducer.

In addition to large dynamic range and low noise, two other
important considerations in the design of the electronic processing circuitry may be mentioned here. One adds significantly to
equipment operation and the other is related to equipment repeatability and enhanced detectability.

For providing greater ease of operation, the incoming RF
signal is processed to make it compatible with CCTV format so
that (1) it may be directly recorded on standard CCTV peripheral
equipment such as videotape recorders, etc., without the need of
an intervening vidicon or frame-store; (2) low-cost high-quality
displays developed primarily for the TV industry can be utilized
and (3) standard image enhancement and image processing techniques, also available for TV image processing can be used.

For providing greater detectability and higher degree of
equipment repeatability, a novel signal processing called the
Interactive Gain Compensation, is utilized. This is described in
the following section.

INTERACTIVE GAIN COMPENSATION

In ultrasound pulse-echo imaging, the returning echoes from
increasing depths undergo a higher degree of attenuation. Because
increasing depth is directly proportional to increasing time, the
attenuating effect is conventionally offset by a Time-Gain Compensation (TGC) circuit which monotonically increases receiver gain as
a function of time. The ultrasound absorption properties of most
imaging media can be approximated by an exponential expression.

$$I = I_i e^{-\alpha z} = I_i e^{-\alpha c t} \tag{1}$$

where I is the ultrasound intensity, z the propagation path-length,
α and c the attenuation constant and ultrasound velocity in the
medium, and t is the arrival time. A TGC circuitry which only
provides an exponential time gain $\exp(+\alpha c t)$, however, is not
sufficient to offset the attenuation effects, because apart from
absorption there are various other attenuation mechanisms. Chief
among these are the geometrical damping mechanisms associated with
beam spreading and reflections.

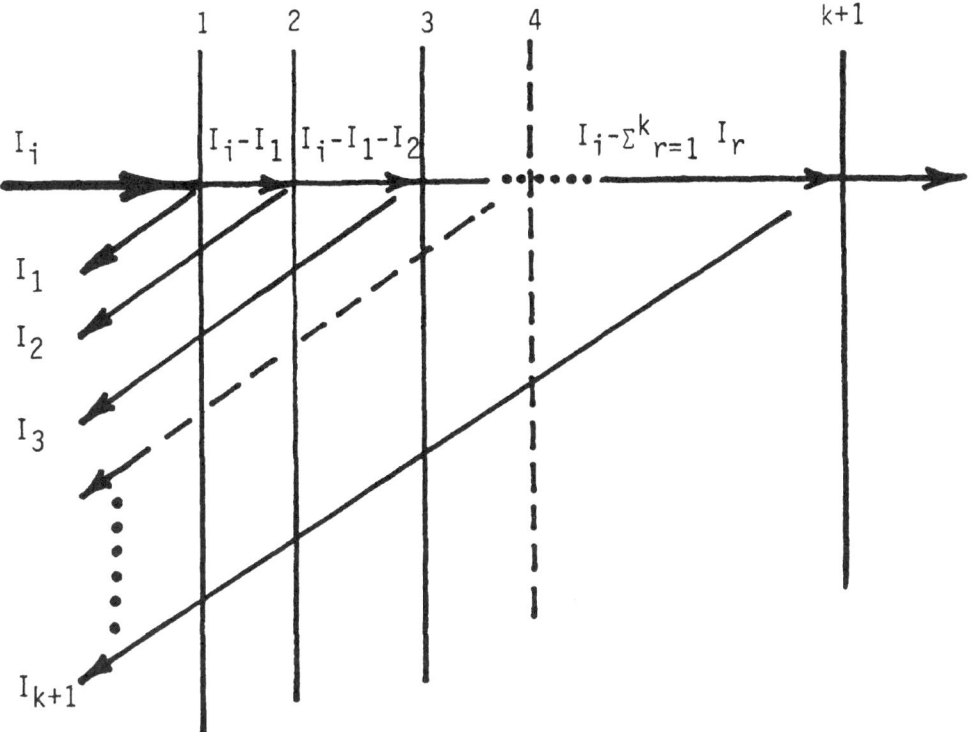

FIG. 4a   SCHEMATIC REPRESENTATION OF THE REFLECTIVE LOSSES
ASSOCIATED IN IMAGING THE DEEPER LYING INTERFACES.

To take into account these other effects, TGC circuits are
conventionally modified to allow for arbitrarily-shaped time-gain
characteristics. This large degree of flexibility gives rise to
decreased image repeatability. It, however, appears possible to
compensate for the other geometric damping terms in a less arbitrary manner as described below.

Geometrical damping due to reflections is caused primarily
because as the ultrasound pulse propagates across an interface, a
portion of it is reflected. The reflected and transmitted
intensities, I_r and I_T respectively, are algebraically related:

$$I_T = I_i - I_r \qquad\qquad (2)$$

where I_i is the incident ultrasound intensity. For imaging of
subsequent interfaces, therefore, only a smaller portion of the

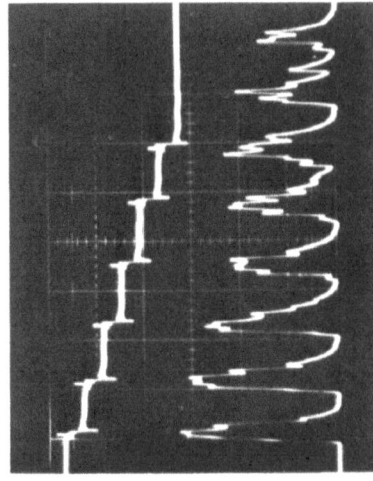

FIG. 4c   ECHOES FROM 8 SUCCESSIVE REFLECTORS
WITH USUAL (TOP) AND IGC PROCESSING
(BOTTOM). TOP TRACES SHOW THE
INTERACTIVELY COMPUTED IGC FUNCTION.

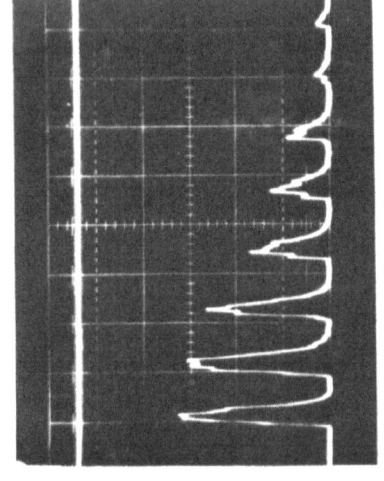

FIG. 4b   SCHEMATIC BLOCK DIAGRAM SHOWING A
POSSIBLE IMPLEMENTATION OF THE NEW
INTERACTIVE GAIN COMPENSATION: IGC
SIGNAL PROCESSING.

energy, I_T, is available (see Fig. 4a). To simulate uniform intensity imaging, therefore, the receiver must provide an additional gain proportional to

$$I_i/I_T = I_i/(I_i - \Sigma_{r=1}^{k} I_r) \tag{3}$$

subsequent to each reflection. This can be accomplished because knowledge of both I_i and I_r is available at the receiver. If the approximation $V\alpha\sqrt{I}$ is made, where V is the associated voltage amplitude, then the expression in Eq(3) may be rewritten in terms of the voltage gain as follows:

$$1/(1 - \Sigma_{r=1}^{k} V_r^2/V_i^2)^{\frac{1}{2}} \tag{4}$$

A straightforward implementation of the necessary compensating gain in equation 4 may be described as follows. As returning echoes V_r are received, the energy associated with each echo is instantaneously and continuously computed (such as with a square-law detector). The computed energy is cumulatively stored, with the storage register being continually updated with each echo arrival. The storage register level is continuously sensed by a control circuit which produces a signal proportional to the algebraic difference between this and a pre-stored reference level (which is proportional to I_i). Another control signal proportional to the inverse of this signal is then generated to produce a proportional gain at the receiver. In other words, as each echo is received the circuit instantaneously and automatically computes and provides an electronic gain to all subsequent echoes in the prescribed manner. For convenience, we will call this circuit as the Interactive Gain Compensation, or IGC. Its function may simplistically be viewed as one that attempts to simulate uniform illuminating beam intensity at all interfaces.

The entire IGC circuit is functionally illustrated in the block diagram of Fig. 4b. This may be fabricated on a single 8"x4" PC board. Fig. 4c shows the effect of this circuit on the received echoes from a target comprising of eight plastic blocks. The top picture is the reference and the bottom picture with the IGC operational. The top traces in each picture is the output at test point B (see Fig. 4b, lower level implies higher gain) and the bottom traces show the resulting video signal. For the testing, the value of the reference level (see Fig. 4b) was set to the value of the measured voltage at test point A observed when an ideal reflector is placed in front of the transducer. Once set, this value need not be changed unless transducer output changes.

It is important to realize that a full implementation of the IGC processing would require some further processing of the returning echoes prior to their being applied at the IGC circuit.

This further processing would include corrections for finite capture angle of the transducer as well as internal reverberation
effects in the imaged object which cause less than complete capture
of all the reflected energy. One effect is minimized by using very
large aperture transducers and the other by preprogramming a
slightly different proportionality coefficient in the gain computation by the IGC, which may be done with the aid of test targets.

   With the desired use of large aperture highly focussed transducers an additional geometric damping term is introduced which
is caused by beam spreading/focussing effects. The axial beam
pattern of the transducer is experimentally measured and a time-
gain-compensation which is inversely proportional to this measured
lens gain as a function of depth is additionally provided. This
function is distinguished by calling it the Lens Gain Compensation
or LGC and is relatively straightforward to implement.

   The overall effects of the IGC circuit on the ultrasound image
may be assessed as follows. First, the IGC leads to greater detectability. This is because it pre-compresses the RF signal to
reduce the input dynamic range. Consider the imaging situation
where a weakly reflecting plaque is situated behind a strongly
reflecting artery wall. If the overall gain in the equipment is
increased in an attempt to enhance plaque echoes, the signals from
the artery wall are also amplified causing blooming effects which
may cause spatial smearing of the wall-image across the extent of
the plaque. With the IGC circuit, the increased gain is automatically (and interactively) applied only after the artery wall
location so that the weaker plaque echoes are enhanced without
blooming of the earlier stronger echoes from the artery wall. This
leads to greater dectectability.

   Second, the IGC processing leads to greater image repeatability
in terms of its observed gray scale. This is of singular importance
because certain diagnoses, such as between cystic and solid tumors,
cirrhotic and normal liver and even between hard and soft plaques,
are increasingly being based on the observed gray scale of the
structure. With the IGC, the effect of intervening tissue layers
is minimized and greater repeatability in the image gray scale is
attained not only from one patient to the next, but also from one
site to another for the same patient.

   Third, the IGC processing is expected to simplify the arbitrary
assignments of TGC waveforms. The IGC together with the compensation for beam spreading/focussing effects (or LGC-lens gain compensation) mentioned above simplifies the TGC compensation to one
which is the familiar exponential function in equation (1). This
is not only simpler to implement electronically but also requires
the operator to make a single adjustment, that for the value of the

selected absorption coefficient α, thereby providing greater
repeatability in overall equipment performance for subsequent
testing and cross-correlative studies.

CAROTID ARTERY IMAGING

The technique described above has been utilized in imaging of
the human carotid artery. The carotid artery in both normal subjects and those with known atherosclerotic disease was examined
and in cases where angiographic results were also available,
favorable correlation was observed between the ultrasound image and
the angiographically obtained data. A detailed clinical study is
planned in the near future. This would include clinical correlative
studies for the atherosclerotic lesions in the carotid as well as
the coronary arteries.

The accompanying talk includes a brief videotaped presentation
of typical data which highlights both the potential for ultrasound
as well as the technique and signal processing employed here for
this application. A few images are also presented here in Fig. 5,
6 and 7. Fig. 5 shows a hard and a soft plaque. The hard plaque,
distinguished by its brighter image is about 1.5 mm thick and 3 mm
long, whereas the softer plaque is distinguished by a fainter image
and is 2 mm thick and 4 mm long. Both these plaques are well resolved to the extent that the surface topography is visualized in
great detail. These plaques correspond to less than 10% occlusion

   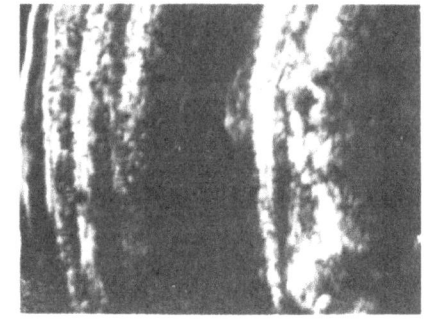

(a)   (b)

FIG. 5 LONGITUDINAL-SECTION ULTRASONOGRAM OF THE HUMAN NECK,
        SHOWING THE CAROTID ARTERY WITH (a) "HARD" PLAQUE ABOUT
        1 mm THICK, 3 mm LONG AND (b) "SOFT" PLAQUE ABOUT 2 mm
        THICK, 5 mm LONG. THE SKIN SURFACE IS TO THE LEFT AND THE
        PATIENT'S HEAD TO THE TOP, LUMEN DIAMETER IS ABOUT 6-7 mm.

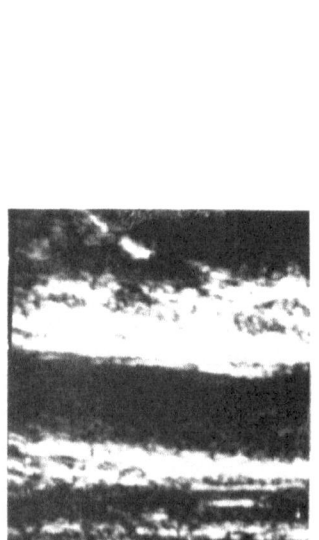

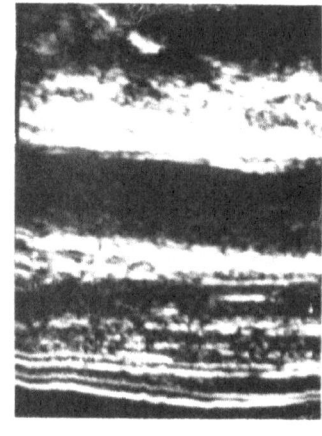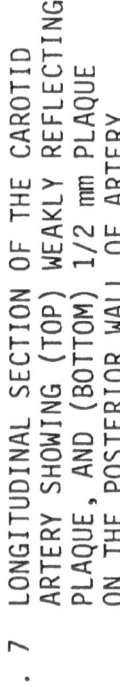

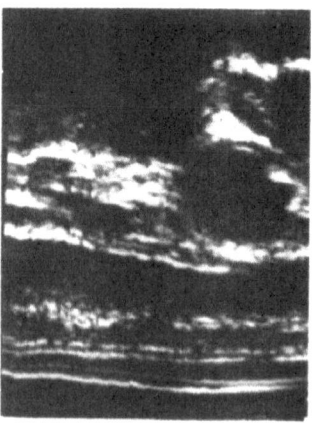

FIG. 7   LONGITUDINAL SECTION OF THE CAROTID
ARTERY SHOWING (TOP) WEAKLY REFLECTING
PLAQUE, AND (BOTTOM) 1/2 mm PLAQUE
ON THE POSTERIOR WALL OF ARTERY.

FIG. 6   CAROTID ARTERY IN LONGITUDINAL VIEW
(TOP) WITH PLAQUE, IN CROSS-SECTION
VIEW (BOTTOM) SHOWING PLAQUE IN LOWER
RIGHT-HAND QUADRANT OF THE LUMEN.

of the lumen. Although not unambigiously seen, some internal structure may also be identifiable in the "soft" plaque image of Fig. 5. Data of this nature, if observed consistently and reliably, would provide important information for plaque growth and buildup. On the extreme right end of the images the vertebral column is visualized. In the soft plaque image of Fig. 5, two vertebral ridges are seen. With proper orientation the vertebral artery may also be imaged. However, in the preliminary studies this has been of secondary concern so far.

Fig. 6 shows a "hard" plaque in longitudinal-section and cross-section views of the neck. It is noticed from the cross-section view that the plaque occupies the entire lower right quadrant of the artery wall. Similar views at different sections permit accurate volumetric computations of plaque sizes to be made. Fig. 7, top, shows an extremely weakly reflecting plaque on the posterior wall of the common carotid. The plaque is of complex geometry and almost 10 mm long. Fig. 7, bottom, is an image of the smallest observed plaque with the limited clinical experience so far. This plaque is approximately 0.5 mm thick and about 8 mm long and corresponds to less than 5% occlusion of the artery lumen.

ACKNOWLEDGMENTS

The authors wish to thank Mr. J. G. Loukas and Mr. J. W. Marcinka for their extensive support in the optimization of the overall equipment performance, and to Mr. K. R. Solomon for overall program management.

REFERENCES

1. See for example, J. R. A. Mitchell and C. J. Schwartz, Arterial Disease, F. A. Davis Co., Philadelphia (1965).

2. C. P. Olinger, to be published in Surgical Neurology.

3. See for example, P. N. T. Wells, Physical Principles of Ultrasonic Diagnosis, Academic Press (1969) Chapter 4.

4. Anant K. Nigam, "Standard Phantom Object for Measurements of Gray Scale and Dynamic Range of Ultrasonic Equipment." in Acoustical Holography, Vol. 6, (N. Booth, Ed.,) Plenum Press, 689 - 710 (1975) See Appendix A.

ULTRAFAST ECHOTOMOGRAPHIC SYSTEM USING OPTICAL PROCESSING OF ULTRASONIC SIGNALS [*]

R. TORGUET, C. BRUNEEL, E. BRIDOUX, J.M. ROUVAEN, B. NONGAILLARD

Laboratoire d'Opto-Acousto-Electronique, Centre Universitaire de Valenciennes - 59326 - VALENCIENNES CEDEX - FRANCE

E.R.A. C.N.R.S. n° 593

INTRODUCTION

Since a few years considerable progress have been done in the acoustical imaging techniques for biomedical applications. The superiority of such a technique comes from its non invasive character, ease of use and lack of hazard, even for long and redundant insonification periods. The commercially available apparatuses have however a limited resolution and the interpretation of the acoustic images is therefore difficult. Ultrafast systems are furthermore needed in the cardiologic field in order to get slow motion pictures of the very fast movments of the cardiac valves. An ultrafast apparatus of good resolution has been devised in our laboratory.

The basic principle is the image reconstruction using an optical processing of the ultrasound echo (1 - 2). This enables us to reach a resolution of about 1.5 mm and an image rate higher than 2000 per second, the apparatus being merely intended for cardiological applications.

BIOMEDICAL IMAGING

The most currently used acoustical imaging technique is the mode B - echography. One axis of the image is given by the round-trip time of the ultrasonic waves. A single linear array of piezoelectric transducers is therefore needed to get an image.

(*) This work was supported by the D.G.R.S.T., Paris, France.

The construction of mode B imaging devices is then more simple
than that of mode C apparatuses, which use a bidimensionnal
mosaïc array of piezoelectric transducers. The actual apparatuses
have a diffraction limited resolution. Two solutions have been
studied in order to circumvent this problem :

- acoustic beam focusing, the more appropriate tool being a
  Fresnel zone plate which gives an appreciable depth of focus
  (3)

- true image reconstruction, using for example as here an optical
  processing of the ultrasonic echo.

PRINCIPLE OF THE METHOD

     An acoustic wave induces, inside the medium where it propa-
gates, periodic compression and dilatation zones, responsible
for a periodic optical index modulation. The analog of an optical
grating is thus produced and incident light may be diffracted by
the acoustic wave. Theoretical calculations (4 - 5) show that, in
the small interaction regime, the diffracted plus and minus one
light orders are phase and amplitude modulated by the acoustic
wave. Owing to the Huyghens principle, a complete information
transfer from the acoustic beam to the light one may be accomplished. Optical lenses are then used to get a true image of the
insonified object. The basic principle of this transfer is illustrated in the Fig. 1, the system giving the image of a single

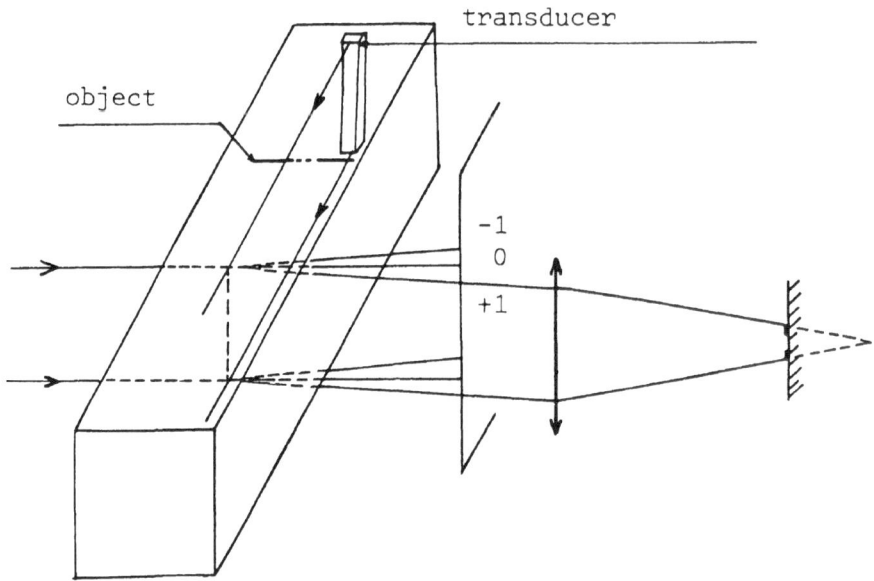

Fig. 1 : basic principle

line normal to both ultrasound and incident light propagation
directions. The second image dimension is given by the echographic technique. The ultrasonic waves are pulse modulated and the
image is shaped in the reflection mode, the round trip time of
the ultrasonic echo being used for discrimination between lines
lying at different depths. A rotating galvanometer mirror converts
the time modulation of the light beam into a spatial information
and restitutes therefore the second dimension of the image
(see Fig. 2). The mirror movments are synchronous with the ultrasonic pulse generator.

In mode B echography, the depth of the visualized lines
lies between a few centimeters and fifteen centimeters and the
total round trip time is about 200 microseconds. A fast zoom
effect is therefore needed in order to get a well in focus image
over all the previous depth. A conical optical lens set behind
the rotating mirror is used for this purpose. The focal length
variation is so designed in order to compensate for the varying
object distance.

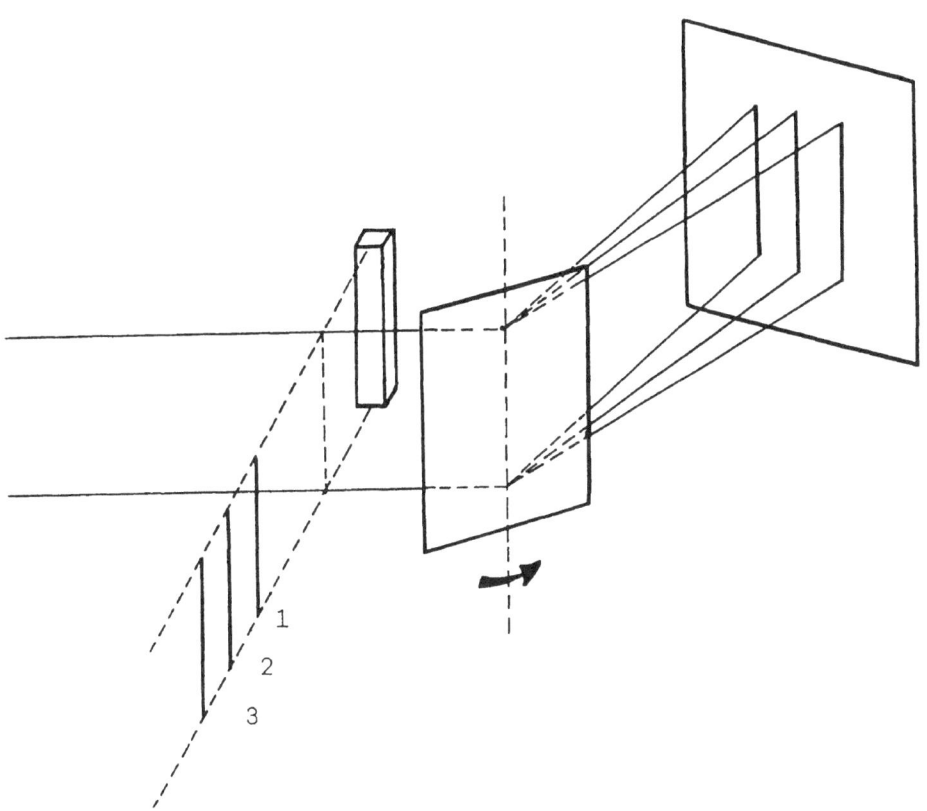

Fig. 2 : galvanometric mirror

BIOMEDICAL APPLICATIONS

A direct acousto-optic interaction system is not well
matched to biomedical applications, since it has a very low sensitivity and the patient must lie in a water bath. The informations
about the phase and amplitude of the acoustic field along a single
line are needed in the application of the Huyghens principle.
These informations may be sampled by a linear array of piezoelectric transducers and converted into electrical signals.
After convenient amplification, the electrical signals are
fed to a second linear array of piezoelectric transducers and the
initial acoustic field may thus be reconstructed. The optical
processing unit is therefore completely separated from the
patient's body (see Fig. 3) on which is only applied a probe
consisting of a single linear piezoelectric array. The amplification is performed for all electrical signals in a parallel fashion,

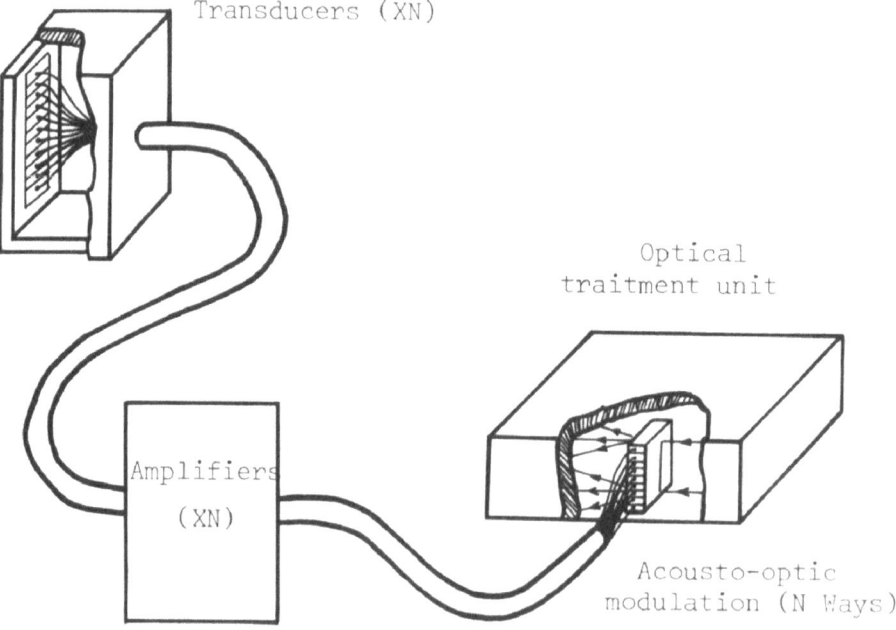

Fig. 3 : diagram of the set-up

so that the image rating time is only given by the ultrasonic
round trip time, that is about 200 to 300 microseconds.

 The actual prototype comprises 20 ways, the exploration
field being approximatively equal to 5 x 15 cm^2 and the resolution being about 2.5 mm. The images are picked using a vidicon
tube, thus use is made of the excellent grey scale capability of
the T.V. monitors but the image rate is reduced to 50 per second.
A greater image rate may be obtained by using either a movie-
camera or a photodiode array. The Fig. 4 and 5 show typical
results. The image of the tomography of 1.5 mm diameter rods
lying on the peripheral of a diameter circle is given in Fig. 4.
The image of the tomography of an human heart is shown in Fig. 5,
in which the aortic valves are clearly resolved.

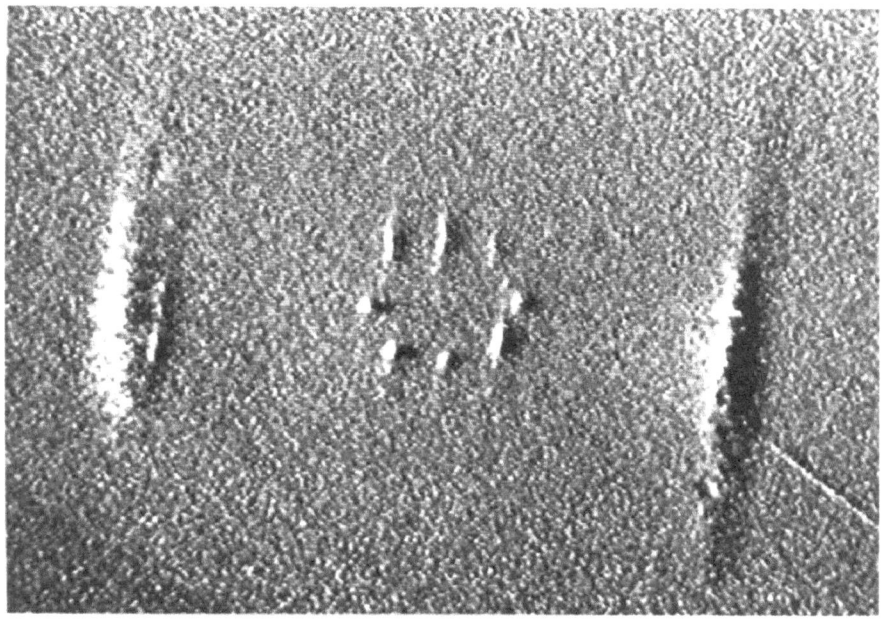

Fig. 4 : image of rods

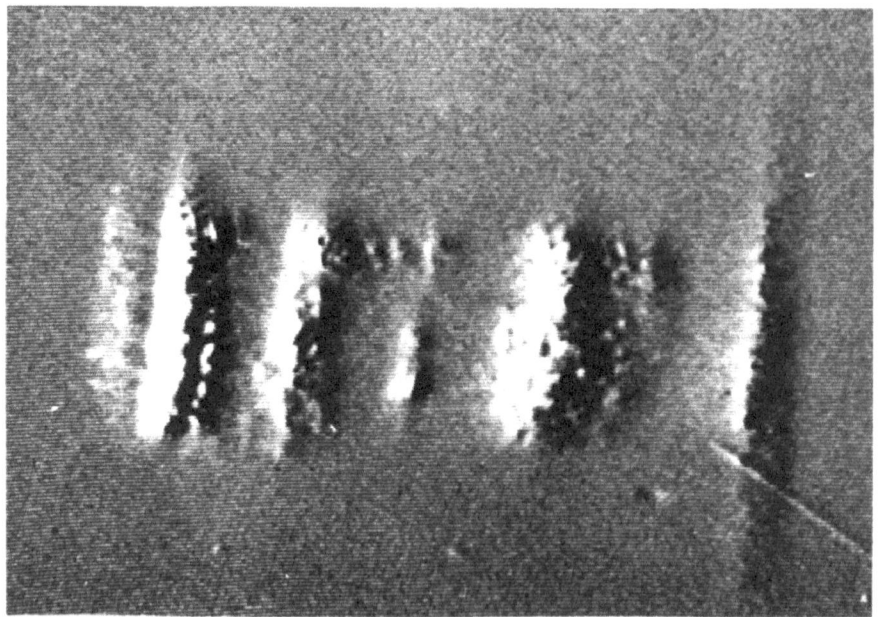

Fig. 5 : image of the heart

CONCLUSION

The first results are very encouraging. The resolution is better than that of actual apparatuses. The rising up of the way number will allow an observation field of up to 10 x 15 cm^2 with about a 1.5 mm resolution. The high motion movie camera will allow the decomposition of the aortic and mitral valves movments, since the slow motion pictures taken from our actual images taken at a 50 per second rate show that these valves appear blurred during their displacements.

REFERENCES

(1) John LANDRY, Roy SMITH and Glen WADE : "Optical heterodyne detection in Bragg Imaging". Acoustical Holography, Vol. 3 Eds A.F. Metherell, Plenum Press 1971.

(2) Byron B. FRENDON : "A comparison of acoustical holography methods" Acoustical Holography Vol. 1, Plenum Press 1969.

(3) P. ALAIS Fink colloque Biocapt 75 et Acoustical Holography
    Vol. 7.

(4) R. TORGUET : Doctorat d'Etat 1973 unpublished

(5) C. BRUNEEL, R. TORGUET, J.M. ROUVAEN, E. BRIDOUX :
    "Ultrafast Echotomographic system using optical processing
    of ultrasonic signals" to be published in Acoustical Imaging
    and Holography, publisher Crane, Russak, INC, N.Y.

AN IMPROVED SYSTEM FOR VISUALIZING AND MEASURING

ULTRASONIC WAVEFRONTS

D. Vilkomerson; R. Mezrich; K-F Etzold

RCA Laboratories
David Sarnoff Research Center
Princeton, N. J. 08540

I.   Introduction

At the previous Symposium we described an instrument to visualize and measure ultrasonic wavefronts [1]. In that system, a
very thin, flexible, optically reflective membrane, called a pellicle,
is immersed in a tank of water. The pellicle is so thin and so-well-
coupled to the water that an ultrasonic wave moves the pellicle at
every point with almost exactly (>99.95%) the same motion as the
water. This motion is detected by a scanning Michelson interferometer, and the changes in interference so caused are converted to an
electronic signal by a photodiode. As indicated schematically in
Fig. 1, this signal is used to modulate the intensity of a spot on a
CRT that is raster scanned synchronously with the laser beam raster-
scanning the pellicle; in this way a map of the displacement amplitude on the pellicle is formed on the CRT. The electronic signal
also is displayed for exact measurement of the displacement of a
particular point or of a scan line on the pellicle. As shown in Fig. 1,
acoustical lenses may be used to form an acoustic image on the
pellicle that is made visible on the CRT.

The difficult part of the interferometric detection of motion is
negating the effects of the large (in terms of the displacements to
be measured) vibrations in the components of the interferometer and
the phase shifts caused by air currents in the optical paths. These
ambient disturbances, while large, are of lower frequency than the
ultrasonic displacements we wish to measure, so there should be

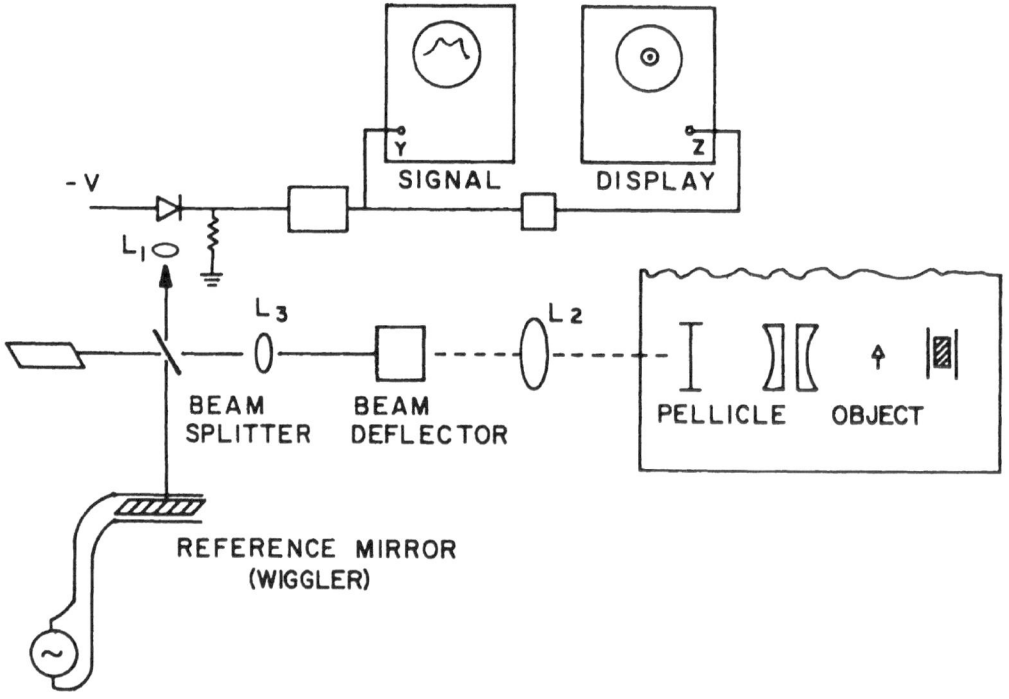

Fig. 1.   Schematic representation of the previously reported
          system taken from Ref. 1.

ways of discriminating against the environmental noise.

In the system we previously described we defeated the ambient disturbances with the "wiggler" shown in Fig. 1. As described in detail in Section II, by overwhelming the ambient disturbances by a known motion at a frequency below the signal frequencies we could ensure reproducible measurements. This technique was completely successful for ultrasound pulses lasting 10 μs or more. However, for short pulses, such as those from medical diagnostic transducers, the technique could not work.

We have changed our means of eliminating the effects of the ambient so that even very short pulses can be measured. The new system measures the temporal as well as the spatial variations in the ultrasound wavefront.

We will discuss in this paper the details of the new stabilization technique that replaces the wiggler method formerly used. The

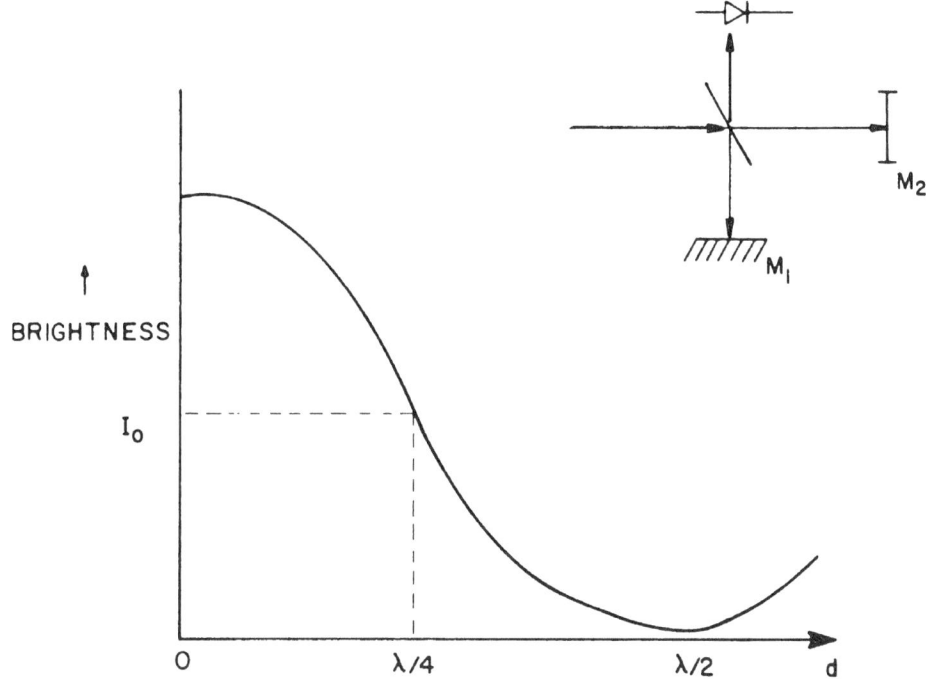

Fig. 2.   The brightness, B, at the photodiode as a function of the
          difference, d, in length of the optical paths traveled by the
          two beams. Note that if M_2 moves by δ, d changes by 2δ.

rest of the system remains unchanged; we refer the interested
reader to references 1 and 2 that discuss the other parts of the
system in detail, and we will not treat them here.

II.   Theory and Implementation of the New Stabilizing System

Fig. 2 shows the relation between the output brightness, B, of
a Michelson interferometer and the optical-path-difference, d, the
difference between the optical distances travelled by the two beams
in the interferometer. A small motion in the movable mirror, M_2,
causes a change in brightness, B. The amount of change in brightness depends upon the slope of the curve in Fig. 2, i.e. the derivative of the function B vs d. Fig. 2 shows

$$B = I_0 + I_0 \cos 2\pi \frac{d}{\lambda} \qquad (1)$$

where I_0 is the sum of the intensities in each optical path. (We
have assumed here equal intensities and perfect interference.) The
sensitivity to motion of one of the mirrors, i.e. the change in

brightness with a small change in optical path d of Δd, is

$$\Delta B/\Delta d = (2\pi I_0/\lambda) \sin (2\pi d/\lambda) \qquad (2)$$

This sensitivity depends on I_0 and d, the optical path difference. When d is an integer multiple of $\lambda/2$, the sensitivity is zero.

The optical path difference is sensitive to ambient disturbances; air currents and vibrations easily cause a change of a few thousand angstroms which varies $\sin 2\pi d/\lambda$ over its range of +1 to -1. Those who have set up a Michelson interferometer know how the fringes dance.

In the systems we reported on previously, the sensitivity was effectively stabilized by forcing d to change a few thousand angstroms every 10 microseconds or so. We did this by vibrating the reference mirror, mirror M_1 of Fig. 2, at 50 KHz with an excursion of $\sim\pm 1600$Å (which causes a change in d of ± 3200Å or $\sim\pm\lambda/2$). This "wiggler" changed d much faster than the ambient so we could be sure that at least once during every 10 microsecond the sensitivity passed through its maximum, that is $\sin (2\pi d/\lambda)$ equalled one. The ultrasonic signal of $\delta \sin wt$ vibrating the pellicle would cause a signal from the photodiode, after high-pass filtering, of

$$S = (2 \delta \sin wt) (2\pi CI_0/\lambda) \sin (2\pi d/\lambda) , \qquad (3)$$

where C is the conversion coefficient of the photodiode from light to electric current. If that signal is passed through a peak-detector [3], the value of S at its maximum, i.e. when $\sin wt = 1$ and $\sin 2\pi d/\lambda = 1$, is retained. This value is equivalent to the signal that would be obtained from an interferometer operating at its most sensitive operating point.

If, however, the ultrasonic signal appears for only a few microseconds (as would a pulse from a diagnostic transducer), the detected signal would depend on the value of d of the interferometer at the particular microsecond the pulse occurs. The peak-detected signal is as likely to be at zero as at the correct value of (2δ) $(2\pi CI_0/\lambda)$.

For short pulses of ultrasound, an interferometer whose sensitivity is constant in time is needed. The way we have achieved this is by exploiting the two independent orthogonal polarizations possible in a beam of light; in effect, we set up two interferometers,

one for each polarization, that use the same beams. The optical-
path-difference d of the two interferometers are offset so that zeroes
in the sensitivity of one interferometer occur at maximal sensitivity
of the other.

Fig. 3 illustrates one example of interferometers, which we
call quadrature-dual-interferometers [4]. The linearly-polarized
light from the laser can be thought of as having two equal amplitude, temporally in-phase polarization components, one at $+45°$
(in space) and the other at $-45°$ to the linearly polarized light from
the laser. The $\lambda/8$-plate shown is made from a material such as
mica that has an index of refraction different for one polarization
than for the orthogonal one [5]. It is called a $\lambda/8$-plate because
its thickness is such that one polarization emerges retarded by $\lambda/8$
compared to the other polarization, i.e. the optical path for one
polarization is longer by $\lambda/8$. By placing a properly oriented $\lambda/8$-
plate in the reference arm of the interferometer, as shown in Fig. 3,
one polarization advances $\lambda/4$ ($\lambda/8$ on each pass to and from the
mirror) compared to the other. When the two polarizations are separated by a polarizing beam-splitter, the optical path difference of
one polarization will be different from that of the other polarization
by $\lambda/4$. The resulting brightness versus optical-path-difference
curves are shown in the right of Fig. 3.

Note that the zeroes of sensitivity, i.e. where the slope is
zero, of B_1 are at the points of maximum sensitivity of B_2. There
are no longer points of total zero sensitivity for this interferometer.

Moreover, we can achieve <u>constant</u> sensitivity by utilizing the
trigonometric identity

$$\sin^2 x + \cos^2 x = 1. \tag{5}$$

Consider the high-pass filtered signals from each photodetector in
Fig. 3, assuming an ultrasonic vibration in the measuring mirror of
$\delta \sin wt$:

$$S_1 = (2\,\delta\,\sin wt)\,(\pi\,CI_0/\lambda)\,\sin\,(2\pi\,d/\lambda) \tag{6a}$$

and

$$S_2 = (2\,\delta\,\sin wt)\,(\pi\,CI_0/\lambda)\,\cos\,(2\pi\,d/\lambda) \tag{6b}$$

Note that the $\lambda/4$ shift in the optical path difference d changes the
"sin" of eq. 3 and eq. 6a to the "cos" of eq. 6b. (The two signals

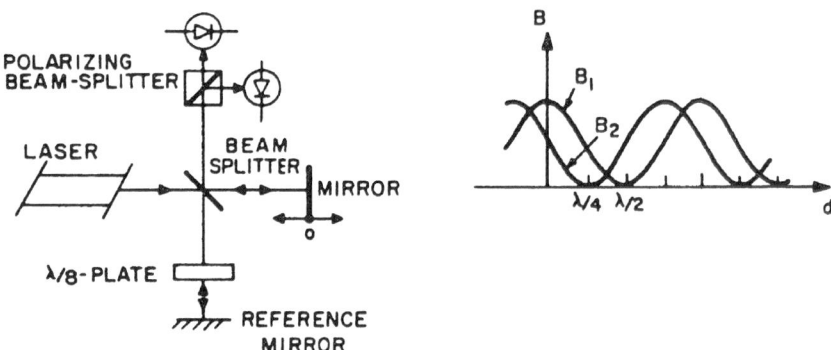

Fig. 3. A quadrature-dual-interferometer and its associated brightness vs optical-path-difference functions. The brightness
B_2 shows the effect of a $\lambda/4$ relative retardation caused by
the $\lambda/8$-plate.

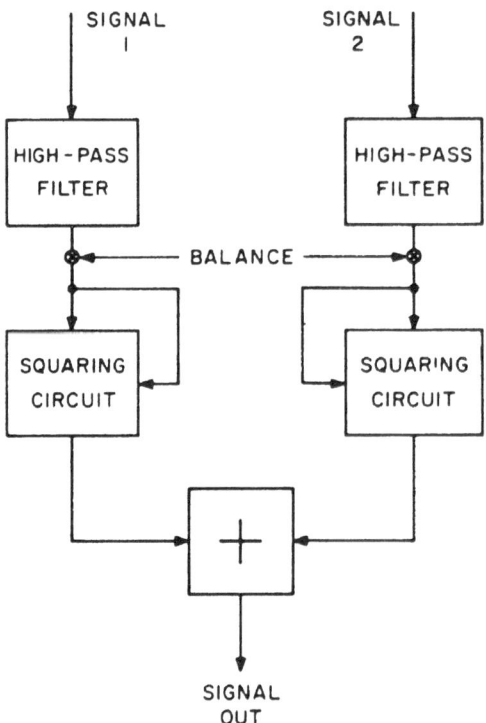

Fig. 4. The signal-processing for producing a stable output signal
from the two signals of a quadrature-dual-interferometer.

are said to be in <u>quadrature</u> with respect to d. Now remembering eq. 5, if the two signals are squared and summed, as indicated schematically in Fig. 4, the result is a single output signal of

$$S = S_1^2 + S_2^2 = (2 \delta \sin wt)^2 (\pi CI_o/\lambda)^2 \qquad (7)$$

This signal is proportional to the square of the ultrasonic displacement, i.e. proportional to the ultrasonic intensity and, most important, <u>independent of optical-path-difference d</u>! Such a system achieves the temporally constant sensitivity that we require for measuring short pulses.

For eq. 7 to be valid, two conditions must be achieved: (1) the two signals must differ in dependence on d by 90° and (2) the maximum sensitivity of the two channels must be equal.

Considering the first condition, if there is lack of quadrature by a small angle $\mathcal{e}$, i.e.

$$S_1 = (2 \delta \sin wt) (\pi CI_o/\lambda) \sin (2\pi d/\lambda)$$

and   (8)
$$S_2 = (2 \delta \sin wt) (\pi CI_o/\lambda) \cos (2\pi d/\lambda + \mathcal{e})$$

then the output will be, to first order,

$$S_{out} = (2\delta \sin wt)^2 (\pi CI_o/\lambda)^2 [\, 1 + 2\,\mathcal{e} \sin(2\pi d/\lambda)\cos (2\pi d/\lambda)\,] \quad (9)$$

The output signal will vary $\sim \pm\mathcal{e}$ as d changes; a 5° error, for instance due to an incorrectly cut phase plate, results in a $\pm 8.7\%$ variation in sensitivity.

Considering the second condition, if the maximum sensitivity of one channel is normalized to 1 and the other is $(1 + \propto)$, the output signal is, to first order,

$$S = (2\delta \sin wt)^2 (\pi CI_o/\lambda)^2 [\, 1 + 2\propto \cos^2 (2\pi d/\lambda)\,] \qquad (10)$$

An unbalance of 5% will result in a total variation of 10% in sensitivity as d varies.

We incorporated a dual-quadrature-interferometer into the previous system; the resulting arrangement is shown in Fig. 5. The reference and pellicle components of the combined beam are orthogonally polarized. In the former system a polarizer was oriented on the axis 45° to each beam's polarization; the common components of each beam interfered, and, by means of a photo-diode, produced the output signal. As shown in Fig. 5, the detection process uses two polarizers and photodiodes, with one half of the output beam split off by a nonpolarizing beamsplitter and being sent through a $\lambda/4$-plate before reaching its polarizer. The $\lambda/4$-plate is oriented so as to shift the phase of the reference beam polarization by 90° relative to the pellicle beam. The resulting signals are in quadrature, and processed as in Fig. 4.

One particular advantage of this arrangement for a quadrature-dual-interferometer is that it does not change the optical arrangements of the previous system. The benefits of that system, described in more detail in ref. 2, of high light efficiency and isolation of the laser from back-reflections are retained. Also, the new system is easily retrofitted to older systems.

In operation the described optical system is easily adjusted to satisfy the two requirements for stable operation discussed above. The 90° phase shift is obtained by first rotating the polarizer to obtain a null in the reference beam (with the pellicle beam blocked) and inserting the $\lambda/4$-plate and rotating it until the null reappears. This ensures that the axes of the $\lambda/4$-plate are aligned with the axes of polarization and a $\lambda/4$ retardation is obtained (if the $\lambda/4$-plate is accurately made). Balance between the two signals is obtained by adjusting the angle of the polarizers in the output beam; as the polarization components mixed in each channel depends upon the angle of the polarizer, changing the angle of the polarizer in the more sensitive channel away from the optimum reduces the signal from that channel so that balance can be achieved.

Fig. 6 shows how well such a system can work. Successive measurements of about 100 pulses on the same point of the pellicle were made; Fig.6a shows the stable response of the system, while Fig. 6b, obtained from just one of the two signals of the interferometer, shows large variations in output signal due to changing d, indicating why we require the stabilization technique.

III. Examples of System Operation

The new system can operate with "long" pulses of ultrasound (greater than 10 μs) like the previous system. We have published reports of using that system for imaging and measuring tissues (6),

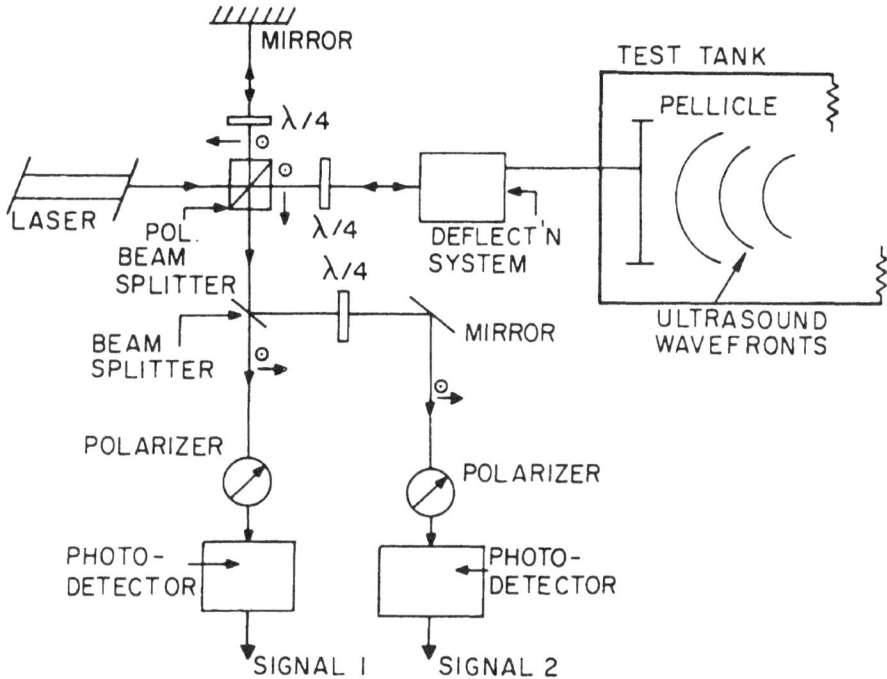

Fig. 5. The optical arrangement of the improved system. The system is the same as described in reference 2 up to the point
where the non-polarizing beam splitter sends half the output beam through the $\lambda/4$-plate. The $\lambda/4$-plate is aligned
so as to retard the reference beam by $\lambda/4$. The polarizers
pass a polarization component common to both the reference
and pellicle beams so that interference between them occurs.
The change in brightness is detected by the photodiodes.
(The polarization of each beam is indicated by the dot or
arrow at the left of the beam path, looking in the direction
of propagation.)

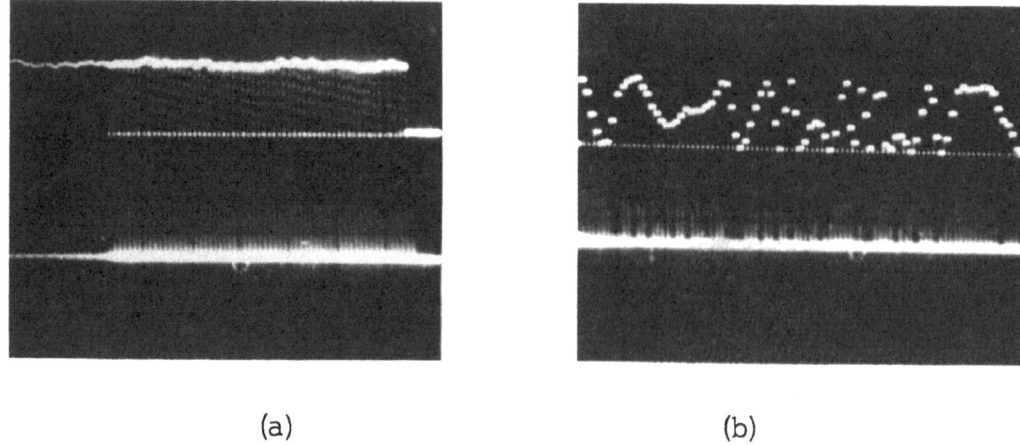

(a)   (b)

Fig. 6.   (a) The output signal for a sequence of 100 identical
               pulses at one spot on the pellicle. The lower trace is the
               intensity signal, the upper the energy signal derived by
               integrating the intensity. The variation is $< \pm .4$dB.

               (b) The output signals for a similar sequence of pulses
               with one of the photodiodes of Fig. 5 blocked.

examining the condition of transducers (7), and measuring changes
in ultrasound velocity in materials by phase-contrast imaging (8).
We also can examine scattering characteristics of materials by
using a lens to produce the spatial power spectrum of the material
on the pellicle (9).

    The new system can perform all these functions, and for these
long pulse applications has an advantage over the old: greater sensitivity. In place of the sample-and-hold circuit of the previous
system we use an integrate-and-hold circuit, wherein the signal is
integrated for the length of the pulse. The integrated value is held
until the next pulse appears, and it is this integrated intensity, the
energy of the pulse, that is displayed. The peak-detection scheme
required a bandwidth approximately equal to the operating frequency;
however, the bandwidth of the integrated signal is approximately
the inverse of the pulse length, so the bandwidth is reduced by the
number of cycles in the pulse. As the noise power is proportional
to the bandwidth, the integrated signal has a greater signal-noise
ratio than the peak-detected signal used in the previous system.

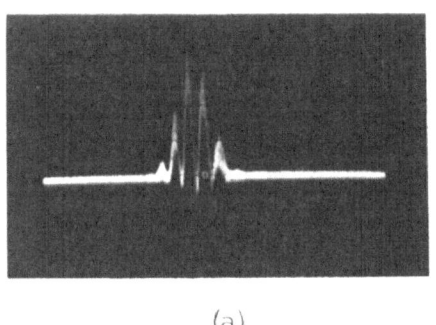
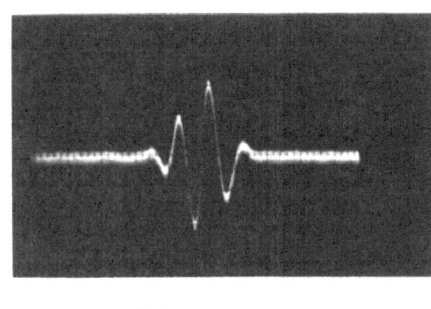

(a)   (b)

Fig. 7.   (a) The intensity of a function of time of a pulse from a
           2.25 MHz NDT transducer. The zero-crossings are 200
           microseconds apart, and the peak corresponds to 75mW/cm^2.

           (b) A signal obtained by the transducer receiving the pulse
           shown in (a) from a reflector.

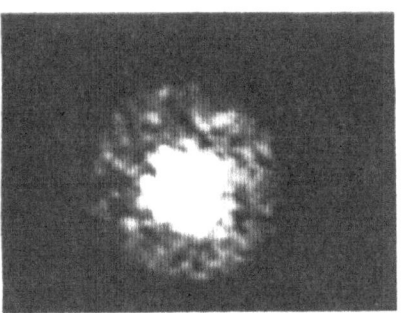

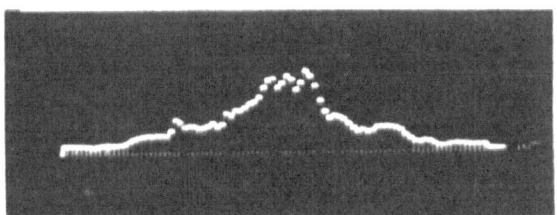

Fig. 8.   (Top) The map of the energy in the near field of the trans-
           ducer.

           (Bottom) The energy distribution along a line through the
           center of the field pattern.

The main advantage of the new system is that it can measure short pulses. In Fig. 7a is the output for an ultrasonic pulse from an NDT transducer. The intensity vs time behavior of the pulse is clearly shown, and, as the system is calibrated, the peak intensity of 75 mW/cm^2 can be read off the screen of the oscilloscope. For comparison, Fig. 7b shows the reflected pulse received by that transducer under the same drive. As our system displays the intensity, which is proportional to the square of the pressure, the peaks of Fig. 7a are positive and in the ratio of the square of the relative amplitudes shown in Fig. 7b.

Fig. 8 is the map of the integrated intensity, i.e. the energy, in the near field of the transducer. The top figure is the map of energy as a function of position. Below the energy map is the energy signal for a single line in the map. The variation in energy with position is easily measured. The capability for measuring both the spatial and temporal characteristics of wavefronts produced by transducers is crucial for the design of diagnostic instruments.

The basic interferometric measurement range of acoustic displacement is limited on the low end by the shot noise in the laser beam [1] and at the high end by nonlinearities when δ approaches λ/4. For our system this corresponds for frequencies in the diagnostic range of intensities from a few nanowatts/cm^2 to tens of watts/cm^2. This range of 10^{10} or so far exceeds the range of our electronics. One easy way of adjusting the sensitivity of the system while retaining calibration is by inserting neutral density filters of known value into the beam as it comes from the laser. For example, a neutral density filter of 2.0 (meaning it attenuates by a factor of 100) reduces the sensitivity by a factor of 10^4 because the sensitivity is proportional to I_0^2 (as shown in eq. 7).

The short pulses that can be used with the new system allow discrimination among pulses that arrive at different times at the pellicle by going through different velocity materials. In Fig. 9, the transmission image of a 15 mm thick block of styrene partially inserted into a uniform field present in the test tank as shown. (The edges scatter the sound and cause the dark bands.) For the top image the receive gate of the system was set to be wide enough to accept the range of pulse arrival times that occur (the velocity through styrene is 1.57 times as fast as through water). In the image below left the gate was set to accept only those pulses propagating through the styrene; on the right is the image with the receive gate set for the pulses traveling through water. Fig. 10

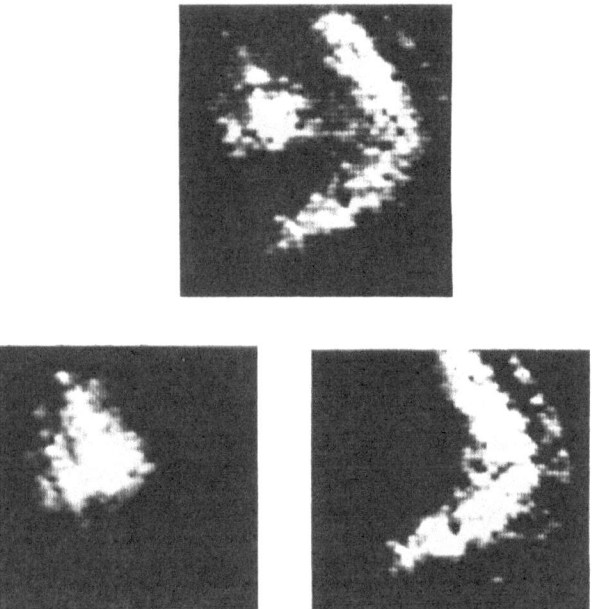

Fig. 9. Top: The image of a 15 mm thick block of styrene in a uniform ultrasound field; the black bands are due to scattering
from the edges of the block. Below left: the receive gate
adjusted to admit only early-arriving pulses. Below right:
the receiving gate adjusted to receive only late-arriving
pulses.

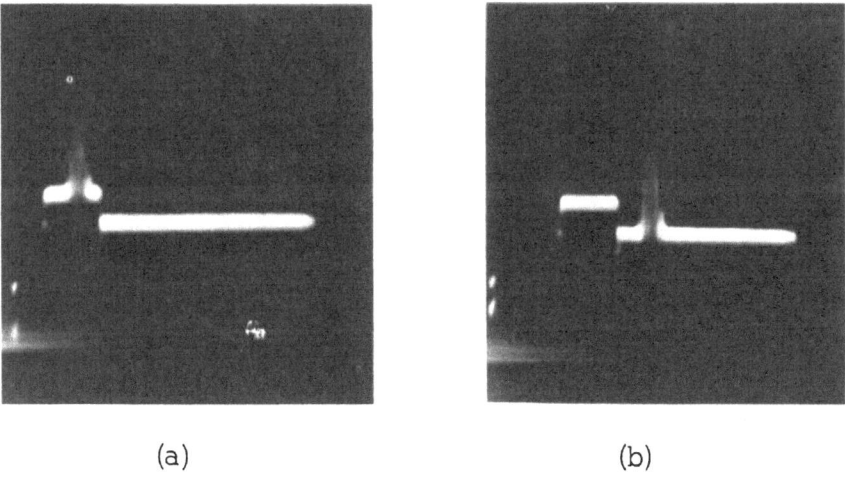

(a)   (b)

Fig. 10. (a) The receive gate and arriving pulse from a point on the
left of the "below left" image of Fig. 9; (b) the receive gate
and pulse for a point on the right of the same image. The
shift in pulse position is 3.6 μs, corresponding to the
velocity and thickness of the styrene plate.

shows the difference in arrival times can be measured, from which the differences in velocity can be calculated. This technique complements the phase-contrast imaging in that it can be used for delays of a wave period or more, while phase-contrast imaging produces unambiguous measurements of delays of less than a half a wave period. By varying the receive gate a map of equi-delay contours can be generated, like those shown as "time-delay spectroscopy" [10].

With the improved instrument we can investigate "pulse imaging", images formed by short pulses, a kind of imaging almost impossible with optics. The "side lobes" or higher-order diffraction patterns that result from constructive interference between wavefronts that have traveled paths that differ in length by an integer number of wavelengths cannot exist if the pulse is only one wave long.

Fig. 11 shows one such experiment. An ultrasonic plane wave pulse 20 wavelengths long is used to insonify a lens placed a focal length away from the pellicle. The well-known Airy pattern of a focused lens is visible in the intensity-position histogram through the focal spot shown in Fig. 11a. When the ultrasonic pulse was shortened to one and a half waves, the first Airy ring disappeared, as can be seen in Fig. 11b. (Theory would predict it should be lowered to one-ninth of its value, as there would be some overlap between the pulses, however, the reduction by one-ninth puts the Airy ring into the noise.)

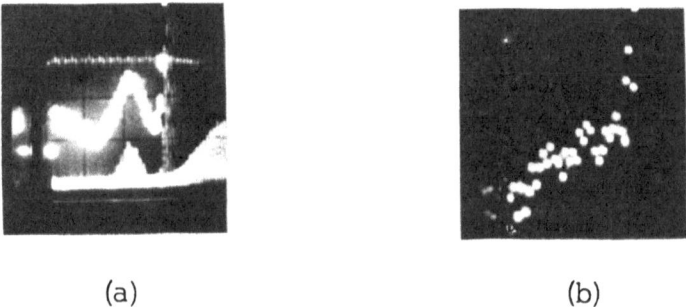

(a)   (b)

Fig. 11. (a) The energy distribution along a line through the focus
         of a lens insonified with 20 wave burst of 2.25 MHz ultra-
         sound. Only the region to the left of the focus is shown,
         and the main focus intensity is off the top of the picture.
         (b) The same as (a) except that the pulses are 660 nS long
         (1.5 waves).

These examples were chosen to show some of the capabilities of this system. The ability to measure both the temporal and spatial characteristics of an ultrasonic wavefront make this system an excellent general purpose research and design instrument.

References

1. R. S. Mezrich, K. F. Etzold, D. H. R. Vilkomerson, "System for Visualizing and Measuring Ultrasonic Wavefronts," Acoustical Holography, Vol. 6, N. Booth, Ed., Plenum Press, N. Y. (1975).

2. R. Mezrich, D. Vilkomerson, K. Etzold, "Ultrasonic waves: their interferometric measurement and display," Appl. Optics 15, 1499 (1976).

3. A peak-detector can consist of a diode in series with a capacitor; the highest positive voltage charges up the capacitor to that value and it remains at the peak value until reset.

4. D. Vilkomerson, "Measuring Pulsed Picometer-Displacement Vibrations by Optical Interferometry," Appl. Phys. Lett. 29, 183 (1976).

5. M. Born and E Wolf, "Principles of Optics," 3rd Ed. Pergamon Press, N. Y. (1965), page 691.

6. C. Calderon, D. Vilkomerson, R. Mezrich, K. F. Etzold, B. Kingsley, H. Haskin, "Differences in the Attenuation of Ultrasound by Normal, Benign, and Malignant Breast Tissue," Journal Clinical Ultrasound 4, 249 (1976).

7. See reference 1, pages 168-170.

8. R. Mezrich and D. Vilkomerson, "Ultrasonic Phase-Contrast Imaging," Appl. Phys. Lett. 27, 177 (1975).

9. To be published. See J. W. Goodman, "Introduction to Fourier Optics," J. Wiley & Sons, N. Y. (1968), Chapter 5, for a discussion of the Fourier transforming properties of lenses.

10. R. Heyser and D. LeCroissette, "Transmission Ultrasonography," Proc. 1973 IEEE Ultrasonics Symposium, IEEE Press, N. Y. (1973), page 7.

REAL TIME IMAGING WITH SHEAR WAVES AND SURFACE WAVES

T. M. Waugh and G. S. Kino

Stanford University

Stanford, California 94305

ABSTRACT

A chirp focused phased array B-scan imaging system has been employed to obtain cross sectional images in metals. Good quality images with a field of view of 5 - 20 cm x 5 cm of arrays of holes have been observed using shear waves, Rayleigh waves, Lamb waves, and flexural waves in thin sheets of metal. With 2.5 MHz Rayleigh waves, we have detected and located surface flaws as small as 250 μm deep and 1000 μm diameter.

I. INTRODUCTION

We have used the chirp focused phased array B-scan imaging system reported on at last year's conference to make reflection mode images in aluminum and stainless steel. The basic system was originally designed to operate at frequencies of the order of 2.5 MHz in water, and would not be expected to have sufficient resolution using longitudinal waves in metals, for which the acoustic velocity is approximately 4 times that in water, to give useful images. We therefore decided to excite and receive the scanned focused acoustic beam in water, and mode convert to shear waves in the metal, for which the acoustic wave velocity is approximately twice that in water. We were able to demonstrate that the beam remains focused and can be scanned, and we obtained good two-dimensional cross sectional images of arrays of holes in the metal. Later we were able to demonstrate that good imaging could be obtained using Rayleigh waves on the surface of the metal, or Lamb waves in thin sheets of metal. The

latter result was encouraging because it involved the use of highly
dispersive waves. As the focusing system behaves like a very rapidly moving lens scanning a focused beam along a line parallel to
the array at a velocity comparable to the velocity of acoustic
waves, considerable deterioration in the focusing might have been
expected in such media. This was not, in fact, the case. As far
as we are aware, this is the first real time scanned focused images
that have been obtained with Rayleigh waves and Lamb waves in thin
metal sheets.

II. PRINCIPLE OF OPERATION

The basic principles of operation of this device and the detailed theory of chirp focusing has been given in other papers.[1-4]
So we will only give a short outline of how the device operates,
and describe how focused shear waves or other waves are excited in
the material of interest.

A linear FM chirp signal of frequency $\omega = \omega_1 + \mu t$ is fed
into a tapped acoustic surface wave delay line, there being one
tap for each element of a linear acoustic transducer array. This
gives rise to an acoustic wave propagating along the delay line
with a parabolic variation of phase:

$$\phi_n = \omega_1(t - x_n/v) + (\mu/2)(t - x_n/v)^2$$

where v is the surface wave velocity and x_n is the coordinate
of the n-th tap.

The signals from each tap are mixed with a constant frequency
signal of frequency ω_2 so as to excite rf signals with a center
frequency $\omega_s = 2\pi f_s = \omega_1 - \omega_2$ and the same parabolic variation of
phase. These signals, after amplification, are inserted into the
corresponding elements on the transducer array. They give rise to
an acoustic beam with a parabolic phase variation along the transducer array; this beam is scanned rapidly along the array at a
velocity v determined by the acoustic surface wave velocity.
This is just the type of beam which would be emitted by a moving
lens moving at a velocity v parallel to the array and excited
by a plane wave, as illustrated in Fig. 1. As we have shown
earlier, the equivalent focal length of this lens is

$$z = \omega_s v^2/zv_w \quad .$$

In operation the system is used to focus on a line a distance
z from the array, and scan along this line. Then after a time delay $T = 2z/v_w$ where v_w is the velocity in the medium, the system

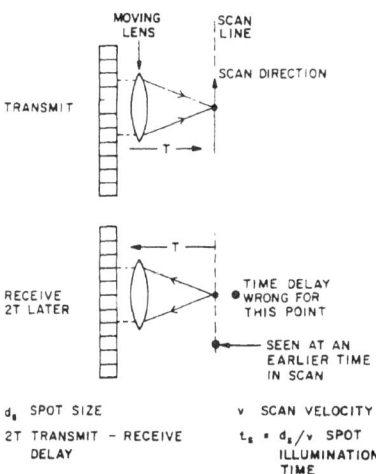

FIG. 1--A schematic of the principles of the B-scan system in
        which the device is regarded transmitting or receiving
        a moving focused beam with a lens moving parallel to
        the array.

is operated as a receiver with the same type of chirp focusing and
scanning. The system is then refocused on another line at a different distance from the array, the time delay between transmit
and receive is changed correspondingly, and a complete raster is
scanned in this way as shown in Fig. 2.

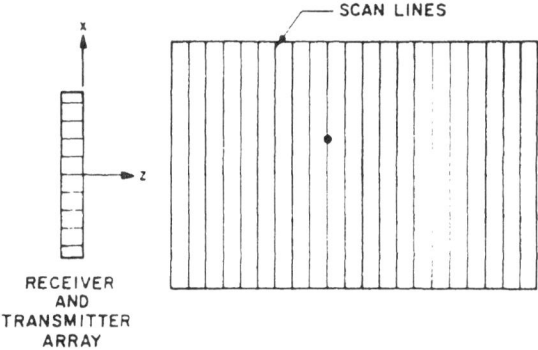

B SCAN - SCAN LINES PARALLEL TO ARRAY

FIG. 2--A schematic of the raster scanned with phased array B-scan
        system.

The calculated definitions in the x and z directions are

$$d_x = z\lambda/D$$

$$d_z = v_w z\lambda/Dv$$

respectively, where $\lambda = v_w/f_s$ and D is the width of the lens; i.e., the spatial length of the chirp along the array.

In order to prevent aberrations there is a restriction on the lens aperture to a maximum length D_{max} , where

$$D_{max} \cong 2(z^2\lambda v_w/v)^{1/3} \quad .$$

Ideally, to obtain the same definitions in the x and z directions, the scan velocity should be equal to the acoustic wave velocity in the medium. In this case

$$d_x = d_z = z\lambda/D \quad .$$

Suppose now we use the array system in water and allow the center of the beam (the ray normal to the array) to be incident at an angle θ_i on the surface of a metal, as illustrated in Fig. 3. Then it follows from Snell's law that the center of the beam leaves the surface of the metal at an angle θ_T where

$$\frac{\sin \theta_T}{\sin \theta_i} = \frac{v_M}{v_w} \quad .$$

v_M is the wave velocity in the metal and v_w is now the wave velocity in water.

Now consider a ray emitted from the array at an angle ϕ_i to the normal (the y-z plane). Let it enter the metal at an angle ϕ_T to the y-z plane as illustrated in Fig. 3. Then the propagation constant of the waves in the x direction on both sides of the metal surface must match. So that

$$\frac{\sin \phi_T}{\sin \phi_i} = \frac{v_M}{v_w} \quad .$$

We note that for paraxial focusing ($\sin \theta_i \approx \theta_i$; $\sin \theta_T \approx \theta_T$) , θ_T is still proportional to θ_i and the beam still remains focused within the metal at a point which can be calculated, but is not too critical because of the relatively large depth of focus of this

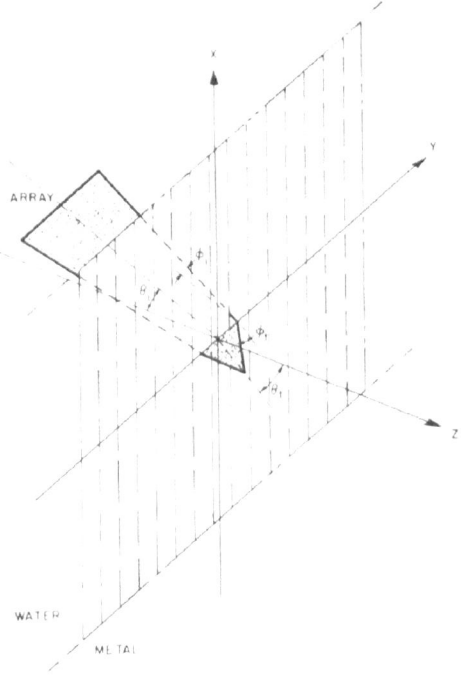

FIG. 3--Illustration of the refraction of the focused acoustic
    beam at a water metal interface. The length of the
    array is assumed parallel to the metal.

system, for the range resolution is determined by time delay.
However, it should be noted that the beam will still scan along
a line parallel to the array at a distance determined by the total
delay time to this line and back. We can estimate the required
time delays fairly easily, especially, if the array is placed quite
close to the metal surface.

It will be noted that by tilting the array sufficiently, the
angle θ_i can be made large enough so that the longitudinal waves
in the metal are cut off, and only shear waves can be excited.
When the angles are made still larger, the shear waves are cut off
and only Rayleigh waves are excited and $\theta_T = 90^\circ$. At intermediate angles, Lamb waves of various kinds can be excited, but all
these waves can be focused and scanned by the chirp focusing system.

II. EXPERIMENTAL SYSTEM

The systems which we have constructed have been operated with
scan velocities between 1 and 2.5 mm/μsec. A block diagram of the
system is shown in Fig. 4. It uses a BGO surface acoustic wave
delay line with a center frequency of 50 MHz. This line has a
total of 60 taps spaced 1 μsec apart along the line, although we
are presently only utilizing the first 28 elements of the line due
in part to defects in the line and in part to the fact that only
30 channels of transmitter and receiver electronics have been constructed. A number of different transducer arrays all using PZT-5
on a tungsten epoxy backing have been used. The array elements are
1 cm long and are on either 0.5 mm centers or 0.64 mm centers depending upon the particular array used. The elements are connected in pairs to decrease the first grating side lobe.[5] The center frequency of these arrays is in the range of 1.2 to 2.7 MHz.
Both the transmitter and the receiver are electronically apodized
to reduce the side lobe levels encountered. The measured side
lobe level is in the range of 25 dB to 30 dB down from the main
lobe. The peak output power of the system is approximately 0.25
watts into each element of the array.

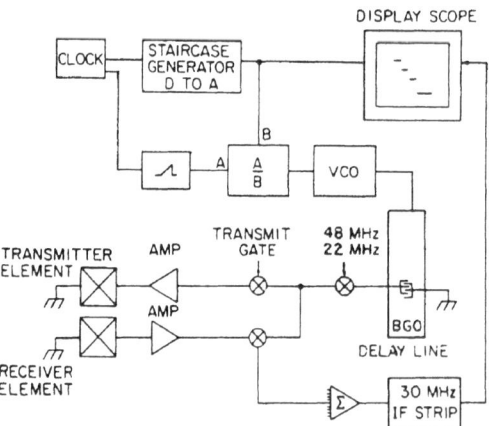

FIG. 4--Block diagram of transceiver system used for shear,
        Rayleigh, and Lamb wave ultrasonic imaging.

III. REFLECTION MODE IMAGING

Using our focused and scanned system in a reflection mode, we have been able to produce a number of B-mode images using shear, Rayleigh, and Lamb waves. As we have pointed out, for optimum resolution in both directions, we would like to have $v = v_M$ where v_M is the acoustic wave velocity in the material of interest. Thus we have attempted to match the scan velocity v to the velocity of the medium v_M . For imaging in aluminum with shear waves or Rayleigh waves, or Lamb waves, we require a scan velocity of approximately 3 mm/μsec for optimum resolution; we are actually using a scan velocity of 2 mm/μsec.

We can produce shear wave B-mode images in aluminum by tilting the array past the critical angle (13.8°) for longitudinal waves, then a longitudinal wave coming from water will be mode converted at the surface of the block to a shear wave. By cutting the end of a target block so that the shear waves propagate straight down the block, as shown in Fig. 5, we have been able to image arrays of holes drilled in the block. As illustrated in Fig. 6, we have carried out shear wave imaging at a number of different frequencies and with different scan velocities and have found that when we use a target of approximately 1 wavelength in diameter, the theoretical 3 dB transverse resolution is 2.5 mm, with a measured transverse resolution of 2.6 mm; the theoretical 3 dB range resolution is 3.9 mm, and the measured range resolution is 3.5 mm. These measurements were made at a center frequency of 2.75 MHz on targets which were 150 mm from the array, with a scan velocity of 2 mm/μsec.

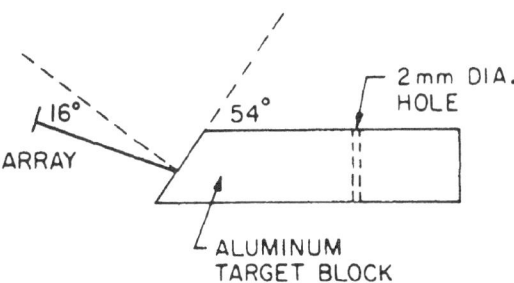

FIG. 5--A schematic of the mode conversion scheme used to produce shear waves.

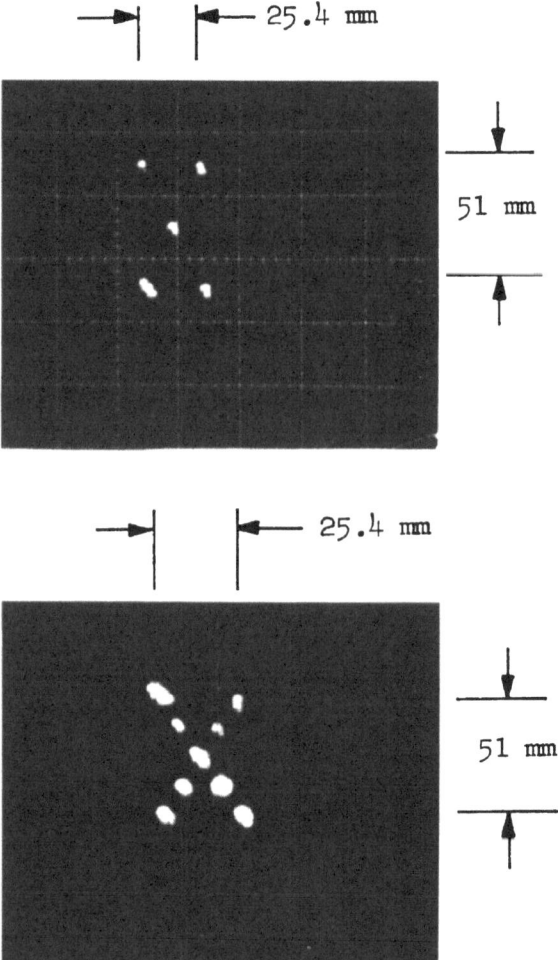

FIG. 6--B-mode shear wave images of 2 mm diameter holes in aluminum block.

(a) f_o = 1.7 MHz; scan velocity = 2.54 mm/μsec

(b) f_o = 2.75 MHz; scan velocity = 2.0 mm/μsec; central
hole is 150 mm from input end of target block.

By changing the experimental setup slightly, so that we now
have the array tilted to the critical angle for Rayleigh waves,
as illustrated in Fig. 7, we have been able to image the same
targets using Rayleigh waves as shown in Fig. 8. Because the
velocities for Rayleigh waves and shear waves are approximately
the same in aluminum (3.16 mm/μsec for shear waves and 2.96 mm/μsec
for surface waves), the expected resolutions are also very similar.
For 2.75 MHz surface waves incident on a 1 mm target located 165 mm
away from the array with a scan velocity of 2 mm/μsec, we obtain a
theoretical transverse resolution of 2.5 mm and measured resolution
of 3 mm with a theoretical range resolution of 3.7 mm and a measured
range resolution of 4 mm. Thus, we have now been able to image the
same set of targets with both shear waves and Rayleigh waves, and
are able to obtain comparable images with both of these methods.
We have also looked at very small shallow holes with Rayleigh waves
and can easily pick out a hole which is 1000 μm in diameter and
250 μm deep at a distance of 15 cm on an aluminum target block.

Since we were able to launch both shear waves and Rayleigh
waves on the thick aluminum target blocks, it was felt that the
next logical step was to attempt to image holes in thin plates
using Lamb waves. We have been able to launch at least two different Lamb waves on a 2 mm stainless steel plate. One of these
modes has a launch angle approximately that for a surface wave
(31°) and the other appears to be the Lamb wave in the plate with
an incident angle of 11°. On a 6 mm thick aluminum plate, we were
able to make B-scan images using the Lamé wave in the plate. This
is a Lamb wave in which only the shear wave is present, propagating
at 45° from the surface and reflecting off of the boundaries.
Figure 9 shows images taken of 6 mm holes, 12 mm apart at a distance of 165 mm in a 6 mm thick aluminum plate with 1.5 MHz acoustic
waves and a scan velocity of 2.5 mm/μsec. Figure 10 shows B-mode
images of 6 mm holes, 20 mm apart in 2 mm thick stainless steel
plate, using a 1.5 MHz array with a scan velocity of 2.5 mm/μsec.

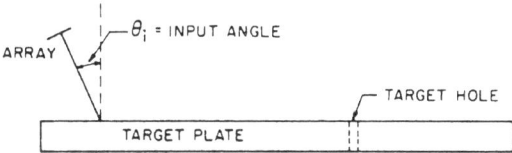

FIG. 7--A schematic of mode conversion scheme used to produce
        Rayleigh waves and Lamb waves.

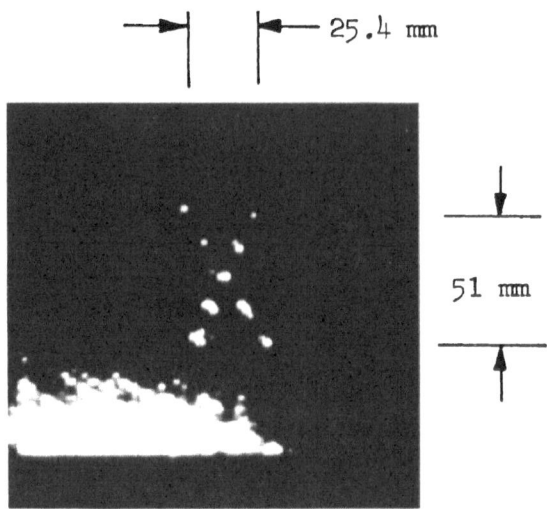

FIG. 8--B-mode surface wave image of 2 mm diameter holes shown in
    Fig. 6.
        f_o = 1.3 MHz; scan velocity = 2.54 mm/μsec.

We have also looked at very small defects with our surface
wave imaging system. We have been able to get good images at
2.5 MHz of a 1000 μm hole which was only 250 μm deep and located
approximately 150 mm away from the array on an aluminum target
plate. Thus we see that this system is rather sensitive to small
defects in the surface as would be expected for surface waves.

CONCLUSION

We have demonstrated a real time B-mode imaging system for
use in metals which uses either shear waves, Rayleigh waves or
Lamb waves. This imaging system is capable of scanning large
areas 5 cm x 20 cm at frame rates of up to 16 frames/second and
200 lines/frame with resolutions on the order of 3 mm in both
directions. This device has been demonstrated to give good images
of small drilled holes in aluminum and stainless steel plates, and
give their location, range and transverse positions accurately.

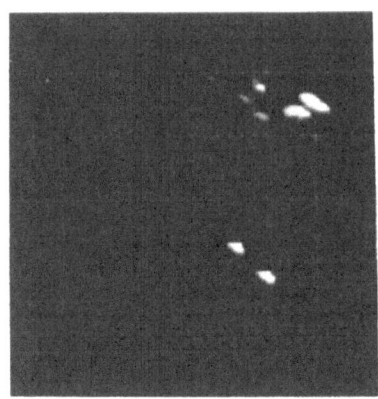

$\theta_i = 28^o$

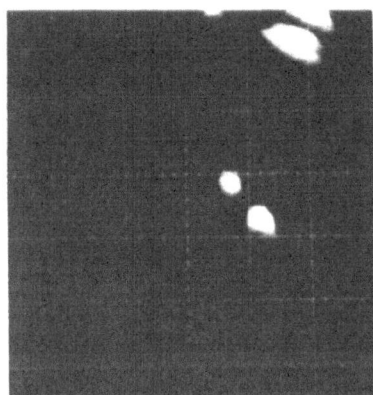

$\theta_i = 20^o$

FIG. 9--B-scan images in 6 mm aluminum plate with 6 mm holes
        12 mm apart at 1.5 MHz. Array is 165 mm from nearest
        hole. Upper photo is a surface wave mode, lower photo
        is Lamé wave image.

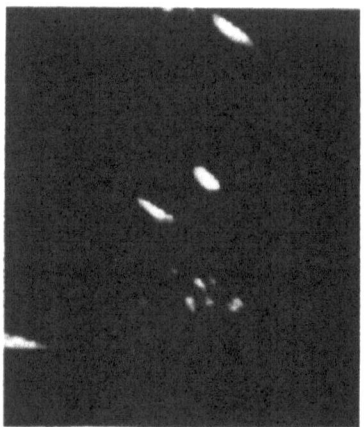

FIG. 10--B-scan Lamb wave images. A 2 mm stainless steel plate
           with 6 mm holes 20 mm apart at 1.5 MHz. Array is
           165 mm away from nearest hole. Angle of incidence is
           35° in upper photo, 11° in lower photo.

ACKNOWLEDGEMENTS

This research was sponsored by the Center for Advanced NDE operated by the Science Center, Rockwell International for the Advanced Research Projects Agency, and the Air Force Materials Laboratory under Contract F33615-75-C-5180. The authors would also like to thank C. S. DeSilets and J. Fraser who designed and constructed the arrays in the course of their research.

REFERENCES

1.   G. S. Kino, C. DeSilets, J. Fraser, and T. Waugh, "New Acoustic Imaging Systems for Non-Destructive Testing," IEEE Ultrasonics Symposium Proceedings, 1975, pp. 94-101.

2.   J. Fraser, J. Havlice, G. S. Kino, W. Leung, H. J. Shaw, K. Toda, T. Waugh, D. K. Winslow, and L. T. Zitelli, "A Two Dimensional Electronically Focused Imaging System," IEEE Ultrasonics Symposium Proceedings, 1974, pp. 19-23.

3.   J. F. Havlice, G. S. Kino, and C. F. Quate, "A New Acoustic Imaging Device," IEEE Ultrasonics Symposium Proceedings, 1973, pp. 13-17.

4.   T. M. Waugh, G. S. Kino, C. DeSilets, and J. Fraser, "Acoustic Imaging Techniques for Nondestructive Testing," IEEE Transactions on Sonics and Ultrasonics, to be published (Sept. 1976).

5.   J. Souquet, G. S. Kino, and T. Waugh, "Chirp Focused Transmitter Theory," to be published in Acoustical Holography (1976).

ACOUSTIC IMAGE CONVERTER FOR THREE-STEP ACOUSTIC HOLOGRAPHY

Amin Hanafy and Mauro Zambuto

Hewlett Packard and New Jersey Institute of Technology

Waltham, MA   02154 and Newark, NJ   07102

Abstract

A unique Acoustic Image Converter with microhorn structure in the output stage is developed. The converter constitutes a fundamental link in the three step method of acoustic holography. Using real time optical holographic interferometry for the detection of the acoustical hologram. The sensitivity, already enhanced by the step bias method, is further increased by the improved Acoustic impedance match of the converter and by the microhorn structure. The most important function of the latter, however, is the minimization of lateral cross coupling in the transfer of acoustic interference patterns from water to air media. Recent results of acoustic holography method incorporating the new converter are presented.

I.   Introduction

Visual data acquisition, which is so important in our life is possible because a light wave, after being diffracted by an object acquires characteristics that contain all the information about the optical properties of the object. In the final analysis this is due to the wave nature of the light.

A sound wave, being also characterized by the wave equation, carries, after diffraction, similar information about the diffracting object.

Seeing with sound, rather than with light can open perspective not available with conventional optical viewing, even when utilizing x-rays. At the same time, if properly processed, the information could be presented in the familiar visual way, permitting the observer to utilize directly his experience in interpreting optical images.

117

The realization of the full advantages of the acoustic imaging process is impeded, however, by finding a suitable detector which provides an optical transparency corresponding to the acoustic holographic information.

Detection surfaces have been classically sub-divided into two broad categories. The first category is the linear detectors which sense both phase and amplitude of the acoustic pressure at the recording plane. The second category is the square law detectors which measure the time average acoustic intensity at the recording plane. In this paper a novel holographic acoustic image converter of the second type is discussed and recent results of the three-step acoustic holography method incorporating the new converter are also presented.

II.   Design Criteria for the Acoustic Image Converter

The present holographic acoustic image converters consist of the acoustic image converter which transfer the acoustic diffraction patterns from surface bounded by water to surface bounded by air, by partially or completely matching the two media together, and the real time holographic interferometer which provides an optical transparency corresponding to the transferred acoustic diffraction patterns. Unlike previous acoustic image converter the present converter exhibits the ability to simultaneously detect a quantity related to the vibration amplitude at each point of the acoustical diffraction pattern, as well as the advantages of frequency selectivity due to the use of mechanical resonance and the amplification achieved by the microhorn structure.

The present acoustic image converter consists of two sections each of which is half wave resonant at the operating frequency. The first half wave section consists of two quarter wave matching plates of Beryllium and Epoxy, while the second section is a half-wave microhorn structure which performs as a high gain mechanical velocity transformer. Figure (1) shows the construction of the acoustic image converter.

Consider the first two quarter wave section in the converter and assume a plane sound wave propagating in water and having a maximum displacement amplitude A_i is normally incident on the Beryllium plate, of this incident energy a part is reflected back into the water and a part is transmitted through to the Epoxy plate and then to the air.

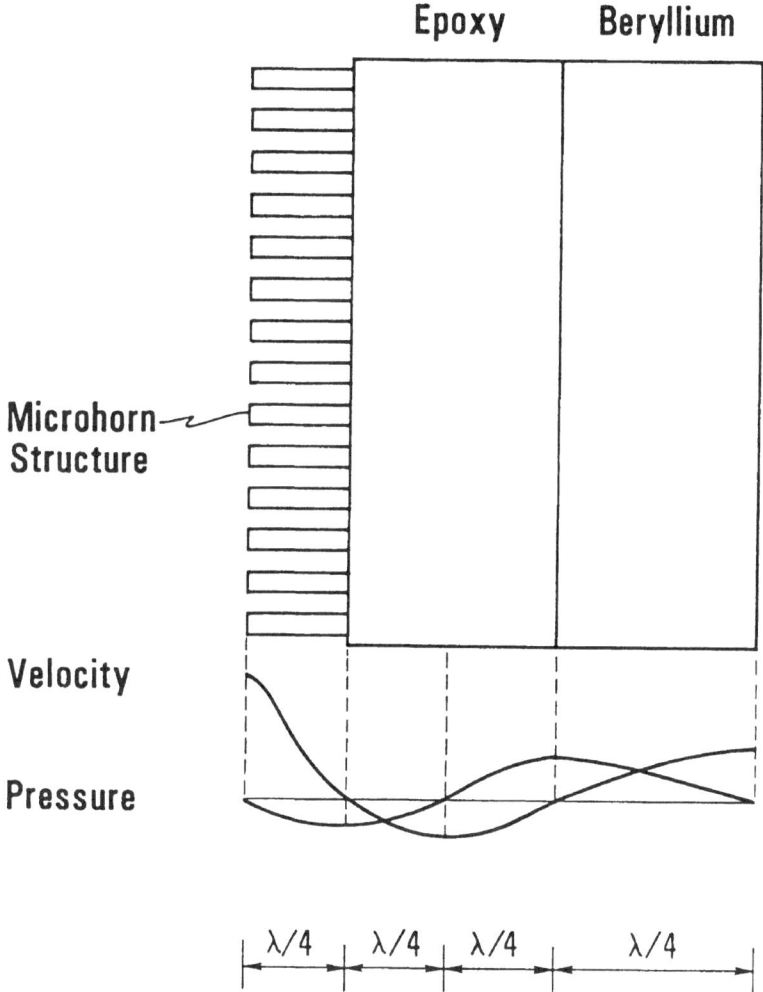

Figure (1) Acoustical Image Converter

The displacement amplitude transmission coefficient t_{d_1} (the gain) is defined as

$$t_{d_1} = A_o/A_i \qquad (1)$$

where

A_o = maximum vibration amplitude of the transmitted wave.

A_i = maximum vibration amplitude of the incident wave

From the theory of multilayer films[1] each individual plate can be described by a unique transfer metrix, which is given by

$$M_i = \begin{vmatrix} m_i & m_2 \\ m_3 & m_4 \end{vmatrix} = \begin{vmatrix} \cos\gamma_i & JZ_i \sin\gamma_i \\ \dfrac{J}{Z_i} \sin\gamma_i & \cos\gamma_i \end{vmatrix} \qquad (2)$$

where

$m_1, m_2, \ldots, m_4$ = matrix element

Z_i = acoustic characteristic impedance of the medium

γ_i = complex phase angle of the medium

$= \dfrac{2\pi}{\lambda} d - J\alpha d$

and

d = medium thickness

λ = wave length of sound wave in medium

α = attenuation coefficient in medium

For multilayer system the overall equivalent transfer matrix M_{eq} is the product of all the individual transfer matrices, i.e.

$$M_{eq} = M_1 \ M_2 \ \cdots \ M_n$$

$$= \prod_{i=1}^{i=n} M_i = \begin{vmatrix} M_1 & M_2 \\ M_3 & M_4 \end{vmatrix} \qquad (3)$$

Once M_{eq} has been determined the system gain is given by[2]

$$t_d = \frac{A_o}{A_i} = \frac{2Z_w}{m_1 Z_a + m_2 + m_3 Z_w Z_a + m_4 Z_w}$$

(4)

where m_1, m_2, . . ., equivalent matrix elements

Z_a = acoustic characteristic impedance of air medium

Z_w = acoustic characteristic impedance of water medium

The two quarter wave acoustic transformer has been studied by several investigators,[2,3,4] and most of the previous converter design suffered from severe angle sensitivity, which means that the incident sound wave from water must impinge perpendicularly to the surface of the first matching plate. An off-set by approximately $5°$ from normal incidence will decrease the converter gain by 50%, this is mainly due to mode coupling in the materials used or inter-element cross coupling in diced systems. Also the image converter gain was limited by the acoustic losses in the matching plates.

In the present design by using Beryllium to receive the acoustic diffraction patterns from the water medium, mode coupling was minimized by the unique acoustic properties of Beryllium. The Poisson's ratio of Beryllium is unusually low value of 0.02 compared with values of 0.3 to 0.4 for other metals. This ability to keep the different modes of propagation separated is of greater importance to our application since it will reduce the microhorn inter-element coupling and therefore improve the contrast of the output image. Also since mode coupling minimized by using Beryllium and by dicing the microhorn structure to greater depth, the angle sensitivity of the system was reduced as shown in Figure (2), an off-set by $15°$ from normal incidence was found experimentally to reduce the gain of the image converter by only 20%.

The second stage of the acoustic image converter is mechanical velocity transformer of microhorn structure. A 24 x 32 stepped horn array was fabricated to operate at 410 KHz. For a stepped horn the gain is given by[5]

$$t_{d_2} = \frac{\xi_o}{\xi_i} = \frac{S_i}{S_o}$$

(5)

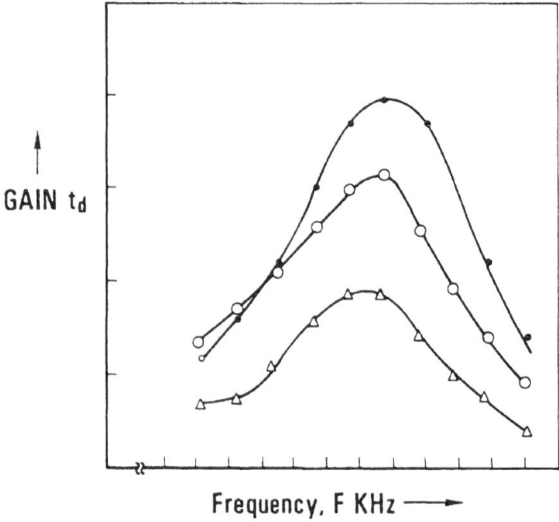

Figure (2)   Frequency   Response of a Tuned   Acoustic   Image
             Converter for a)   Normal Incidence
                           b)   15° off-set from normal in-
                                cidence
                           c)   20° off-set from normal in-
                                cidence

where

ξ_i = maximum amplitude of vibration at the input cross section

ξ_o = maximum amplitude of vibration at the output cross section

S_i = Cross section area of the input section

S_o = Cross section area of the output section

Therefore, the total gain of the image converter t_{d_t} is

$$t_{d_t} = t_{d_1}, t_{d_2}, \ldots, t_{d_n}$$

$$= \prod_{i=1}^{i=N} t_{d_i} \tag{6}$$

Theoretically the maximum achievable gain is

$$t_{d_{max}} = \frac{S_i}{S_o} \cdot \sqrt{\frac{Z_{water}}{Z_{air}}}$$

Figure (2) shows the gain frequency response of the image converter for normal incidence and for 15° off-set from normal incidence.

III. Construction and Tuning Procedure

Starting with the quarter wave Beryllium plate grinded to the calculated thickness, we used the pulse-echo immersion technique and tuned the plate as a half wave resonance at twice the operating frequency by using the sweep generator.

The process is repeated and the plate is grinded until maximum transmission is achieved at 820 KC/s by monitoring the echo display on a CRT. Figures (3), (4) shows the echo display of an untuned and tuned plate.

After tuning the Beryllium plate the second quarter wave plate of Epoxy is bonded to the Beryllium and the converter is tuned as a half wave resonant at the operating frequency 410 KC/s. The microhorn structure is then added to the converter and similarly tuned as a full wave a 410 KC/s. The final tuning of the acoustic image converter is done by using two different optical techniques;

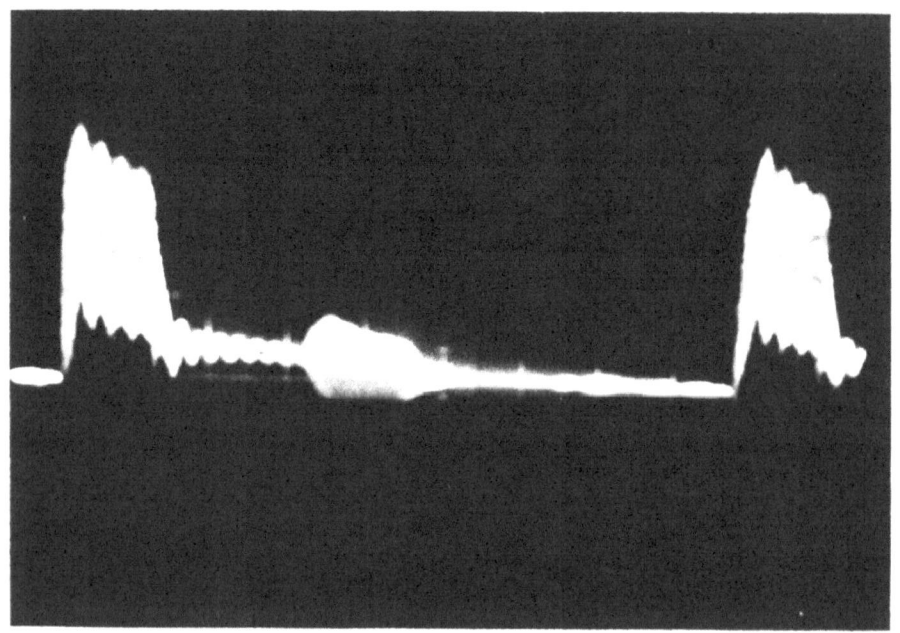

Figure (3) Echo Display of Untuned Plate

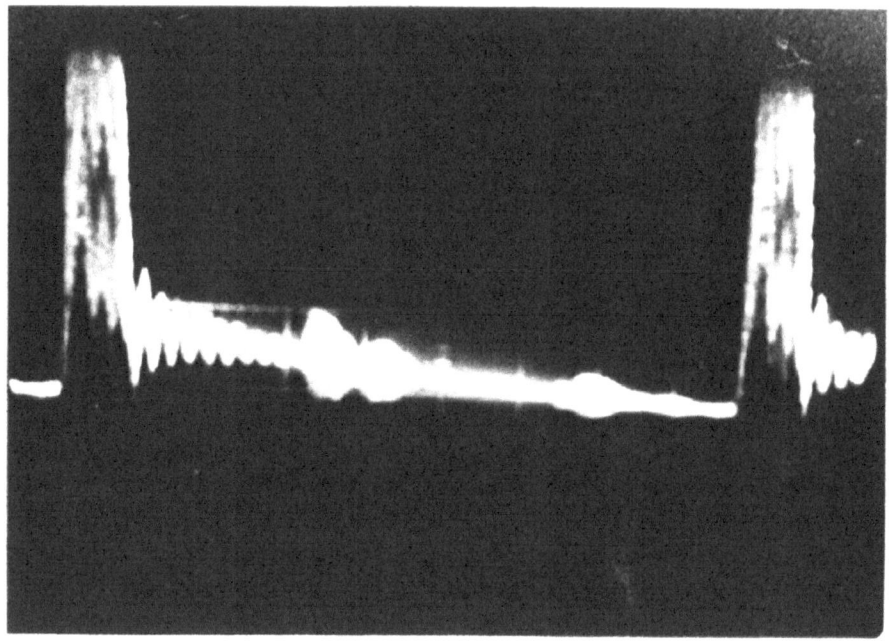

Figure (4) Echo Display of Tuned Plate

as shown in Figure (5). At this stage the image converter is
mounted in the operating position, and we excite it acoustically
from the water medium and detect the output vibration patterns
which correspond to the incident acoustic energy. In the first
detection method, the reflected modulated light beam was inter-
cepted by knife edge and then received by a photo multiplier tube,
from which the amplified output is displayed on a CRT.

The second optical method we used for tuning, was the step
bias real time holographic interferometer which gives a quantitative information about the vibration patterns on the output
stage. Figure (6) shows the set up used for evaluating the image
converter performance.

IV.   The Step Bias Holographic Interferometer

To record the output vibration patterns from the image
converter, a technique of real time holographic interferometry
yielding a one order of magnitude increase in sensitivity over conventional techniques was developed.[6,7] The technique based on
the superposition of a calibrated quarter wave step motion on the
motion to be detected. The sensitivity improvement are due to the
reversal of brightness distribution, so that vibrating points
appear lighter, nodal points darker.

Figures (7,8) show the brightness distribution in the reconstruction of a vibrating point in the conventional technique and
in the step bias techniques. Using the new technique an amplitude
of vibration of 20A° can be detected. In order to implement the
new technique in our system, the image converter rested on three
piezo electric crystal that are bonded to the back plate which is
bolted to the water tank. The aperture was sealed by RTV to the
back plate, therefore, the converter was free to move as a piston
by applying D.C. voltages on the PZT Crystals as shown in Figure
(9). In order to obtain exact quarter wave step motion, a real
time hologram is taken for the converter without any excitation,
then its image is viewed through the hologram, which will appear
bright. We then apply the D.C. voltages to the three Crystals and
we vary the voltages till the image is completely dark, which
indicates exact quarter wave step motion is applied to the converter, now by adding the excitation from the water medium the
acoustic hologram of the object with reversed brightness distribution will appear on the aperture.

V.   Three-Step Acoustic Holography

The technique consist of three steps as illustrated in Figure
(10). These steps are as follows:

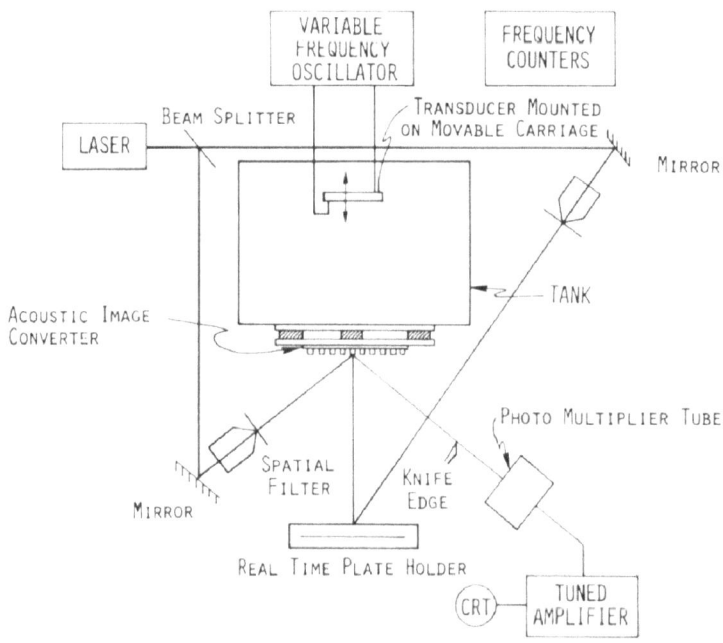

Figure (5)   Arrangement Used to Test the Performance of
the Acoustic Image Converter

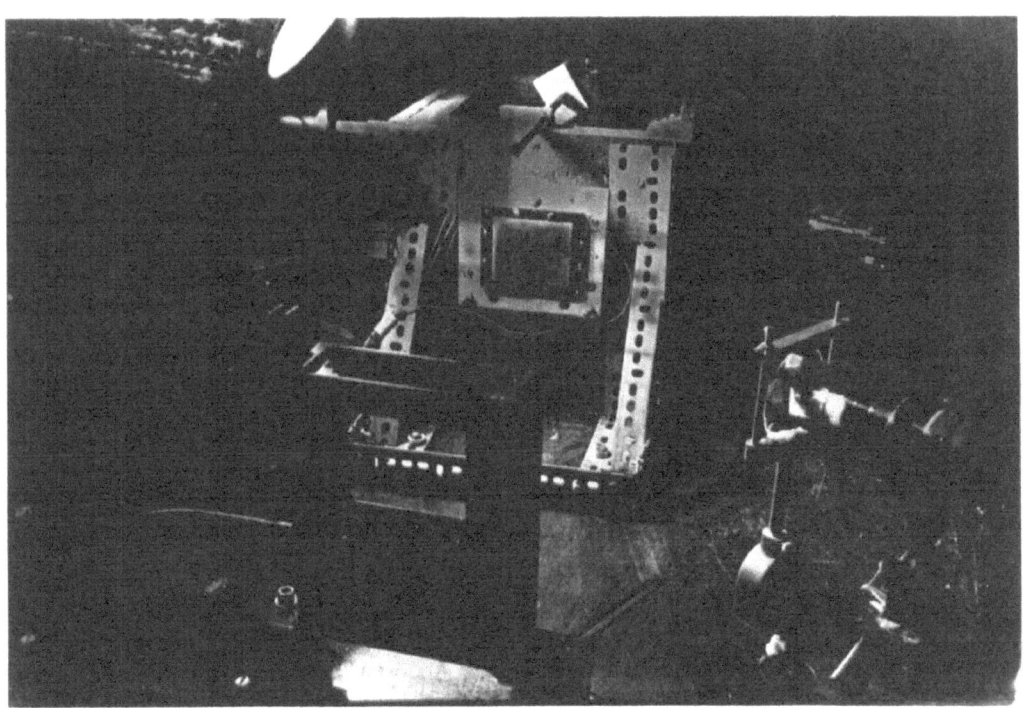

Figure (6) Setup Used for Evaluating the Image Converter Performance

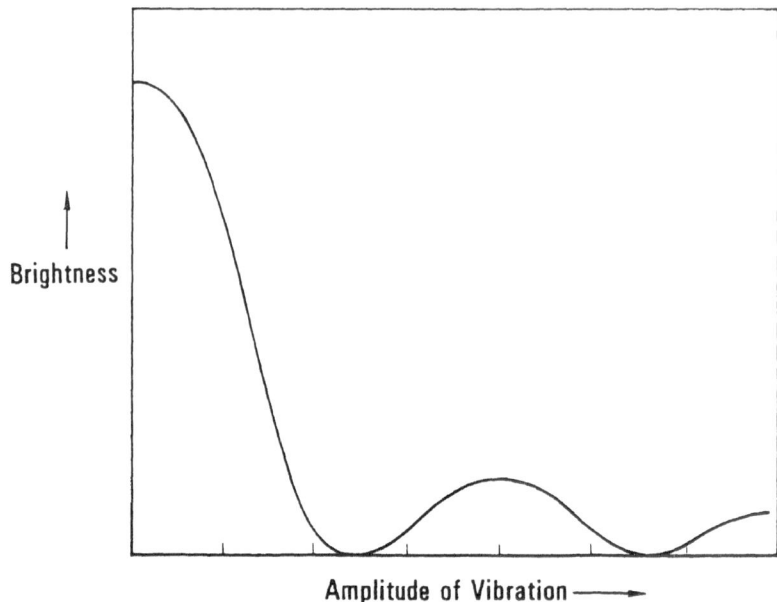

Figure (7) Conventional Brightness Distribution in the
Reconstruction of Sino Soidal Vibrating Points

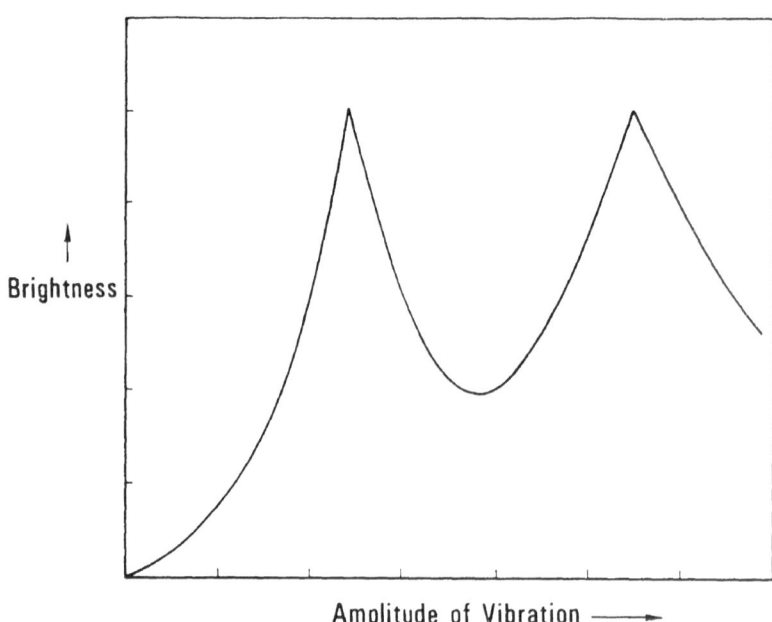

Figure (8) Step Biased Method Brightness Distribution in the
Reconstructing of Sinosoidal Vibrating Points

Figure (9)   Acoustic Image Converter Construction Including
             Biasing Technique

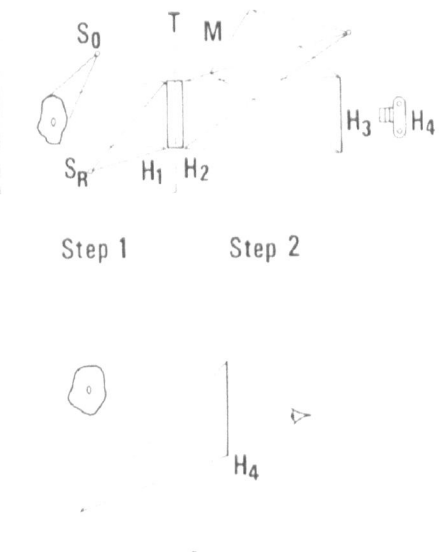

Figure (10)   Three Step Acoustic Imaging System

Step 1. Acoustic Hologram Formation and Transfer.

From Figure (10), insonification of object 0 by supersonic source S_O generates an acoustic object beam B_O. Supersonic source S_R generate an acoustic reference beam B_R which interferes with B_O, at the first surface of image Converter T, forming hologram H_1. This pattern of vibration propagates through the Converter T and generates a second acoustic hologram H_2 on the microhorn structure surface. H_2 is a faithful reproduction of H_1, but the amplitude of the vibration pattern is greatly amplified. Step one then produces H_2, the amplified acoustic hologram.

Step 2. Converted Acoustic Hologram Formation and Recording.

Hologram H_3 is a hologram of the output surface of the Converter T, recorded once and for all before the acoustic hologram formation and while the surface was therefore completely stationary. Reconstructing this hologram by means of mirror M, while at the same time observing T, now in vibration, the reconstructed static virtual image is superimposed in space on the real surface of T. The interference between these two optical patterns generates in the camera the converted acoustic hologram H_4 in one single step.

In accordance with the principles of the step bias holographic interferometry, H_4 is an optical reproduction of H_2 in which vibrating points (peaks of the interference pattern) appear bright on a dark background of non-vibrating points (nodes of the interference pattern). H_4 therefore is an optical representation of the pattern of mechanical vibration that constitutes the acoustic hologram. During the recording of H_4 a scale change is produced by the camera to partially compensate for the discrepancy between the acoustic wave length used to form the acoustic hologram and the optical wave length used for recording the optical hologram H_3.

Step 3. Image Reconstruction

Converted acoustic hologram H_4 is reconstructed by conventional optical holographic techniques. The result is a three dimensional image of the originally insonified object 0.

Scaling process is added here to completely compensate for the wave length discrepancy. By this three step technique the time delay between acoustic hologram formation and final image reconstruction can be brought down to less than a second by computer processing or by kalmar type technique.

VI. Experimental Results

The object chosen was a letter A of aluminum 5 cm height and 4 cm base and was insonified with lead titanate zirconate disc transducer (2 1/2 inch diameter and 0.2 inch thickness).

Figure (11) shows the converted acoustic hologram obtained using conventional interferometry technique, with threshold intensity of 370 mw/cm^2 and Figure (12) shows the image obtained by step bias interferometry method with threshold intensity of 47 mw/cm^2.

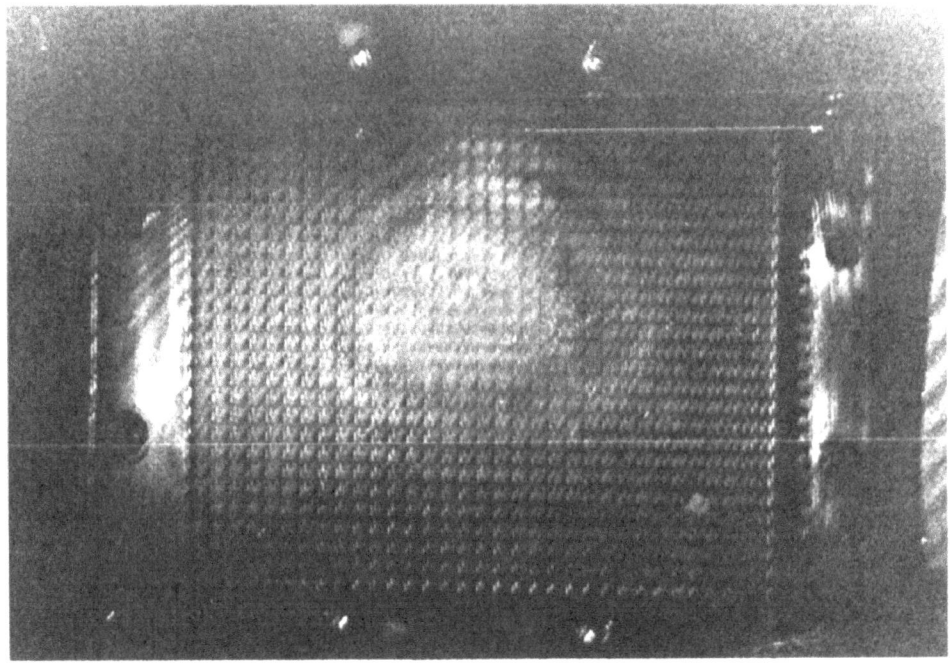

Figure (11) Converted Acoustic Hologram H_4 obtained using
Conventional Interferometry Method and with Threshold Intensity
of 370 mw/cm^2.

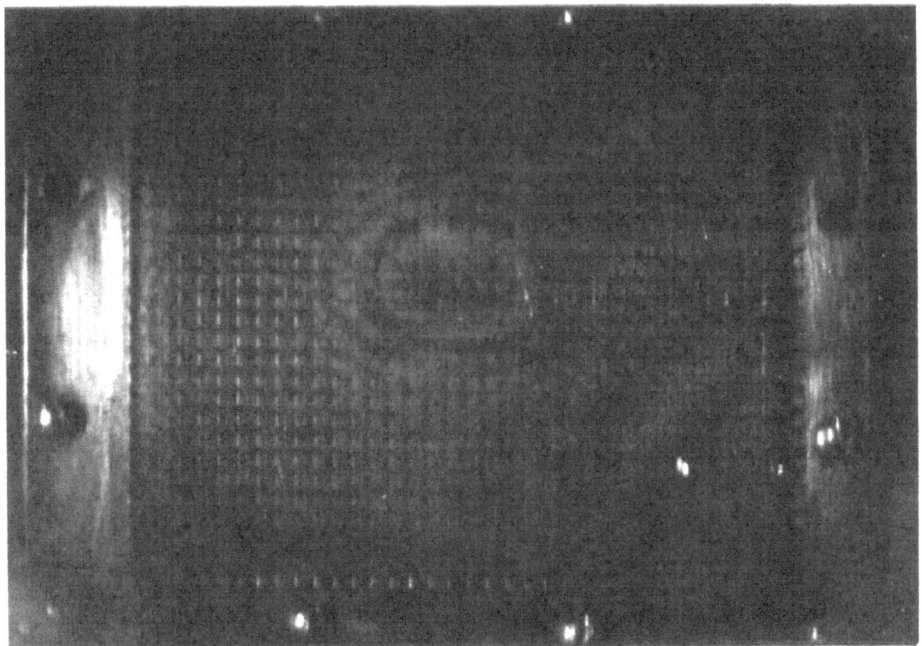

Figure (12) Converted Acoustic Hologram H_4 obtained Using Step-
Bias Interferometry, with Threshold intensity of 47 mw/cm^2

References

1. Kinsler, L.E. and Fry, A.R.,"Fundamental of Acoustics,"
    J. Willey Inc., NY, 1952.
2. Fischer, W. K., "Methods for Acoustical Holography and
    Acoustical Measurements," Doctoral Dissertation, NCE, 1972.
3. Ahmed, M., Whitman, R.L., and Korpel, A., "Response of an
    Isotropic Imaging Faceplate," Sonic and Ultrasonics, 20,323,
    1973.
4. Harris, L.A., "Element Directivity in Ultrasonic Imaging
    Systems,", Sonic and Ultrasonics, 22,336, 1975.
5. Ensminger, D. "Ultrasonic, the Low and High Intensity
    Applications," M. Dekker, NY, 1973.
6. Hanafy, A. and Zambuto, M, "Step Bias Holographic Interferometry,"
    Paper presented at Annual Meeting, OSA, Boston, MA, 1975.
7. Zambuto, M. and Hanafy, A., "Sensitivity Improvement by Step-
    Biasing in Holographic Interferometry," Optical Engineering, 14,
    372, 1975.
8. White, D.W., and Burke, J.E., "The Metal Beryllium," The
    American Society of Metals, 1955.

LENSES AND ULTRASONIC IMAGING

H.W. Jones and C.J. Williams

Acoustic Group, University of Calgary

Calgary, Alberta   T2N 1N4

ABSTRACT

The processes of image formation by acoustic lenses are
discussed. First, the process of refraction in lens materials, which
are relatively thin compared to the wavelength of sound, is commented
upon. Some experimental results for lucite wedges are presented,
and it is shown that spatial modulation of the refracted beam occurs
because of thickness-related transmission effects.

The progress in theoretical aspects of lens design is discussed
by reference to work relating to single refracting surfaces, Luneberg
lenses, two and multiple layer lenses and multiple element lenses.
The performance of various types of lenses are compared.

Some commentary is given on insertion losses and diffraction
related effects. One possible effect of turbulence is discussed.
Mention is made of aperture shading techniques.

INTRODUCTION

The uses of lenses (or mirrors) for the formation of images was
essential in conventional optics because no other method of obtaining the required processing of the wavefronts was a practical possibility. The purpose which is most usually served by a lens in imaging
is the production of a converging spherical wave from a divergent
one. Further, the energy of the wavefront is summed at a suitable
point and detected by a sensor. In recent years, the same effect
has been achieved by other methods (particularly in acoustics) as
has been demonstrated for example, by Quate's use of electronic

focussing.[1] To the extent that it is possible to produce the focussing effect by alternative methods, the lens has ceased to be an essential artifact. It still remains, however, an elegant and economical device for the production of images. The acoustic version of lenses are much less developed than their optical counterparts. Difficulties exist in the discussion of the physical theory of such lenses and conflicting opinions are expressed in regard to some topics concerning their design. It is the purpose of this paper to discuss the development of acoustic lenses and to make comment on the theory and the state of the art as it appears at this time.

REFRACTION AND ASSOCIATED PROCESSES

The refraction process in acoustics is only approximately similar to that of light. It is, in fact, much more complicated because of mode conversion effects. The classical theory of acoustic waves requires continuity conditions of displacement and velocity to exist at the boundary between two substances. These conditions lead to complex solutions which give rise to the generation of sound waves in the second medium which can be extremely complex in character. If we write:

$$v = \text{grad } \phi + \text{curl } \psi \tag{1}$$

when v is the particle velocity caused by the wave and the potentials ϕ and ψ are related by the wave equations:

$$\nabla^2 \phi = \frac{1}{c^2} \cdot \frac{\partial^2 \phi}{\partial t^2} \quad , \tag{2a}$$

$$\text{and} \quad \nabla^2 \psi = \frac{1}{b^2} \cdot \frac{\partial^2 \psi}{\partial t^2} \quad . \tag{2b}$$

where c and b are connected to the Lamé constants λ and μ and the density ρ by:

$$c = \sqrt{\frac{\lambda + 2\mu}{\rho}} \tag{3a}$$

$$\text{and} \quad b = \sqrt{\frac{\mu}{\rho}} \tag{3b}$$

At the interface the physical conditions require the continuity of stress and displacement so that (see fig. 1):

$$\lambda_1 \nabla^2 \phi_1 = \lambda_2 \nabla^2 \phi_2 + 2\mu \left(\frac{\partial^2 \phi_1}{\partial z^2} + \frac{\partial^2 \psi_1}{\partial x \partial z} \right) \quad , \tag{4}$$

$$2\frac{\partial_2\phi_1}{\partial x\partial z} + \frac{\partial_2\psi_2}{\partial x^2} - \frac{\partial_2\psi_2}{\partial x^2} = 0 \quad , \tag{5}$$

and $\quad \dfrac{\partial\phi_1}{\partial z} = \dfrac{\partial\phi_2}{\partial z} + \dfrac{\partial\psi_2}{\partial x}$ \hfill (6)

If the substances through which the sound is travelling are noniso-topic, then the solution of the equations becomes extremely difficult. Particular solutions have been obtained by Havlice[2] and Ahmed[3]. There has been discussion on how far mode conversion effects are important both from the point of view of interfacial energy losses and from the transfer of energy from one part of the wavefront to another. Folds[4] gives the opinion that the energy transfer is relatively unimportant for "narrow angle lenses" and recommends the use of normal computer optical programs for the design of such lenses. Other researchers have reported only slight differences[5,6] from that which would have been expected if the presence of shear waves, caused by mode conversion, had been neglected. We do not know how far this approximation is valid; it is attractive because it considerably simplifies computational problems.

The variation of sound transmission with lens thickness is a fairly complex problem. Brekhovskikh[7] provides a solution to the variation of transmission with thickness for layers with parallel sides. Figure 2 shows a typical result. Tarnoczy[8] suggested that a lens which has surfaces of large radii of curvature can be treated as if it consisted of a series of slabs with parallel sides (see fig. 3). The variations of transmission can then be treated by the approximation[9]:

$$T = \left[\frac{1}{1 + 1/4(\gamma + {}^1/\gamma)^2 \sin^2 (2\pi t/\lambda)}\right] 1/2 \tag{7}$$

Where $\gamma = Z_1/Z_2$, Z_1 being the characteristic impedance of the fluid, Z_2 that of the plate which has the thickness t. He showed that the

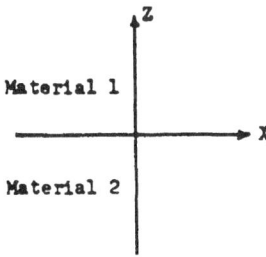

Fig. 1 Coordinates for an interface separating two media.

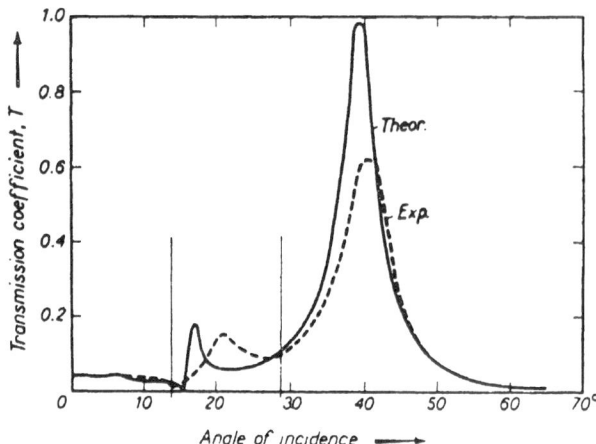

Fig. 2   Transmission coefficient at 4.02×10^5 Hz for a 2.5 mm.
thick aluminum plate in water (after Brekhovskikh, ref. 7)

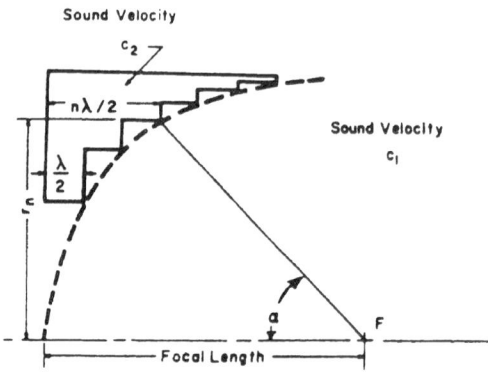

Fig. 3   Zone lens geometry (after Golis, ref. 10)

effect of this variation of the transmission could lead to a lens
which acted as a zone plate. He pointed out that a lens could have
the focal point associated with the zone plate properties which was
not necessarily coincident with the geometric focus. He designed
power concentrating lenses to take this factor into account.
Golis[10] designed lenses similar to those of Tarnoczy and compared
them with their unzoned counterparts. He reported that the results
obtained from the different types of lens were not particularly
different. A difficulty with this study is that zoning effects may
exist in any case in the conventional lens and its influence may
have been overlooked in assessing the results. More precisely, a
lens can be approximated by a series of wedges (or prisms) in a
suitable juxtaposition. Consequently, the transmission properties
of wedges is of much concern. A detailed analysis of this problem
presents some difficulties. Our studies show that transmission is

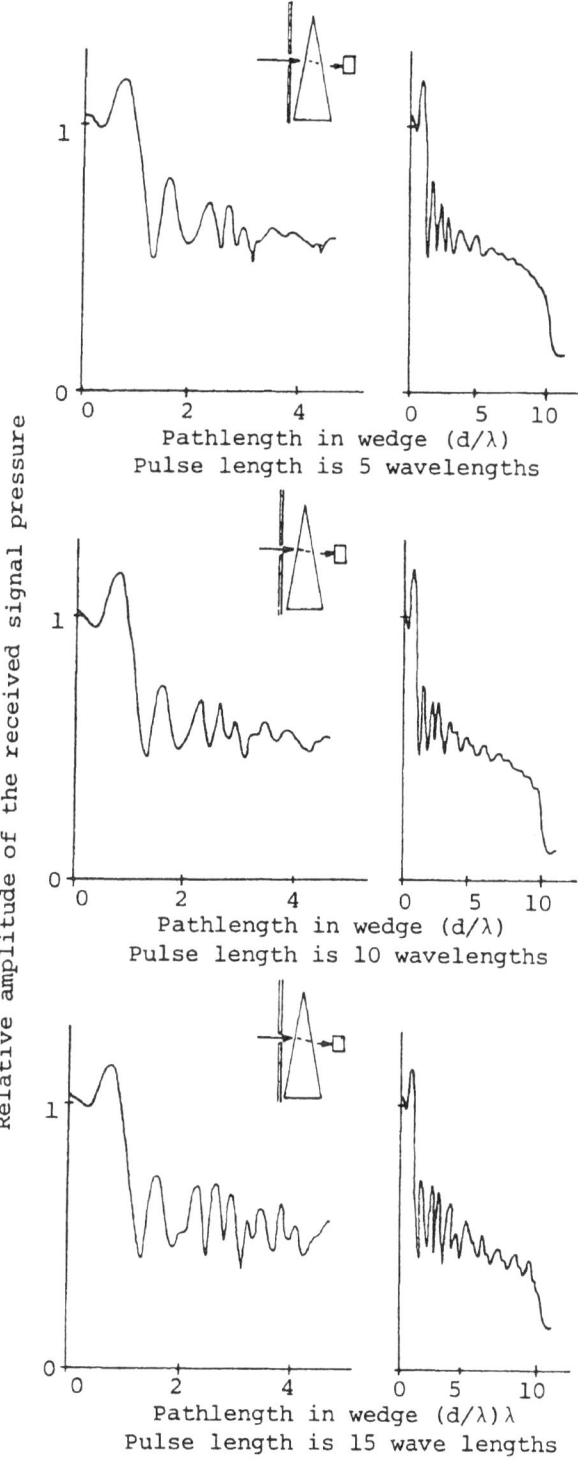

Fig. 4   Transmission through a 20° lucite wedge using ultrasound
with various pulse lengths.

related to the path length of the sound in the wedge. Figure 4 shows the variation of the transmission through a 20° lucite wedge for an angle of incidence of 10° (the left hand graphs are a detail of those shown on the right hand side). Three different pulse lengths were used to obtain these results, i.e., 5, 10, and 15 wavelengths. The plots show a periodic variation of the transmission coefficient with the path lengths in the wedge. The transmission falls to zero when the beam of sound is interrupted by the mechanism which supported the wedge. Figure 5 shows similar plots for a 20° wedge with the incoming sound at various angles of incidence. Clearly, a wavefront passing through such wedges will become spatially modulated and it is reasonable to suppose that similar effects will occur in lenses. The only direct evidence of this apart from that reported by Tarnoczy and Golis appears to be that presented by Roudebush[11]. He looked for the spatial modulation of a beam which was refracted by a liquid filled lens and failed to find any modification of the wavefront.

PRESENT POSITION OF ACOUSTIC LENS DESIGN

Much of the work on acoustic lenses has been undertaken with the beam-producing or power-focussing properties in mind. The following review is selective to the extent that the part of the literature most appropriate to image formation has been quoted.

A detailed analysis of the focussing properties of homogeneous spherical structures was developed by Boyles[12]. He obtained numerical solutions for lenses with refractive indices of 1.8, 2.25 and 2.82. The aperture diameters (d/λ) were in the range of 1 to 15. He compared his calculations with the experiments of Golis. Of particular interest was the emphasis he placed on the difference between the pressure and the intensity foci produced by ultrasonic lenses. The figures 6 and 7 show the differences which occurred. This called attention to the point that acousto-optics primarily deals with pressure and not intensity because most receivers are pressure sensitive.

Folds and Brown[13] determined the optimum refractive index for a cylindrical liquid filled lens. They sought a solution for the mean square wavefront deformation E_0 such that:

$$E_0 \leqslant \lambda^2/348 \tag{8}$$

Equation 8 arises from an acceptable focussing criteria related to diffraction theory and the Strehl criterion[14]. They showed that the equation can be satisfied if the refractive index of the lens is given by:

$$N_0 = 2 - 0.5(a/R)^2 + 0.325(a/R)^4 + \ldots \tag{9}$$

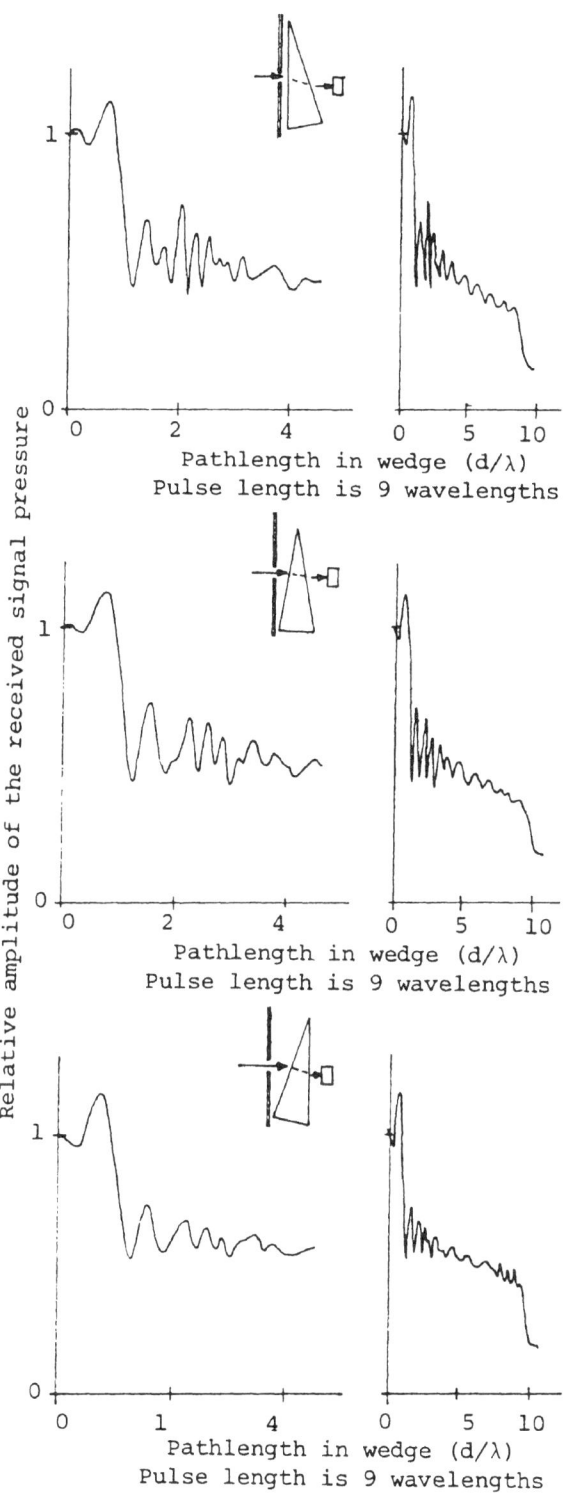

Fig. 5 Transmission through a 20° lucite wedge at various angles
        of incidence.

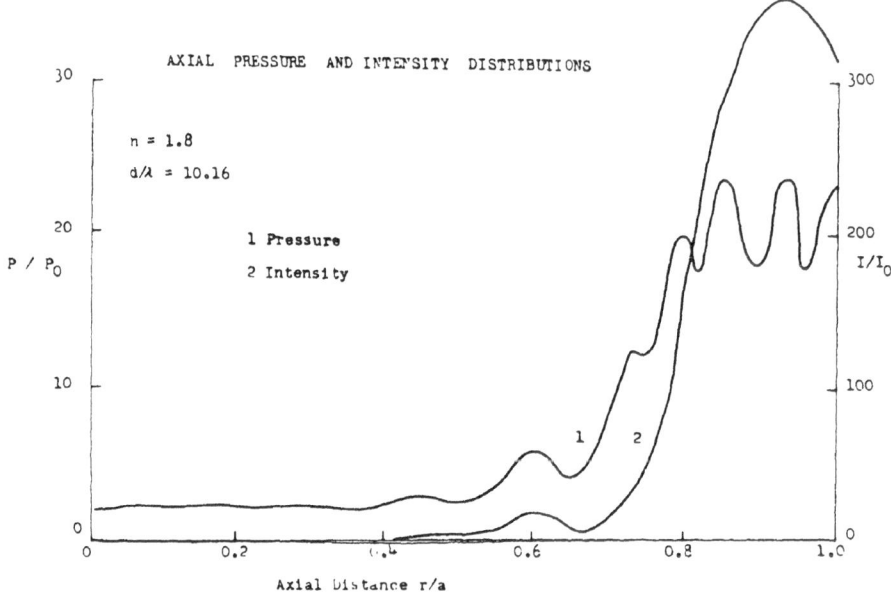

Fig. 6   after Boyles, ref. 12

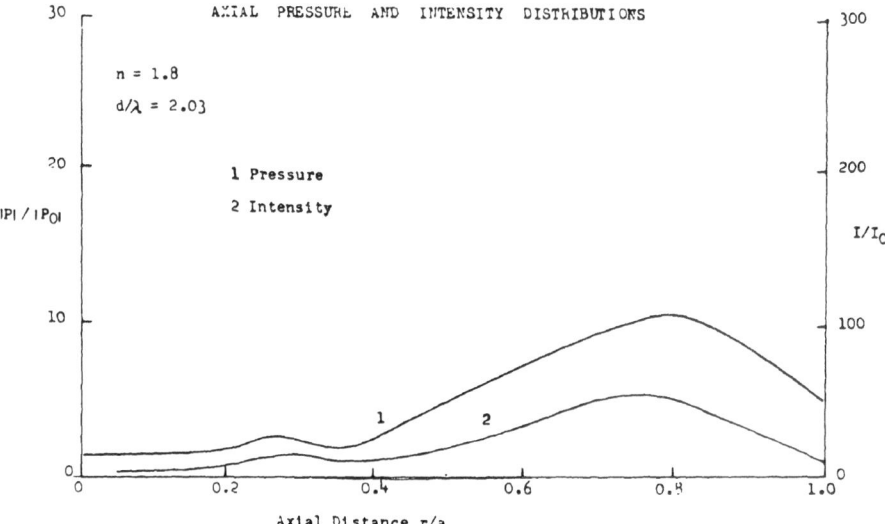

Fig. 7   after Boyles, ref. 12

where a is the radius of the aperture and R is the radius of the Gaussian reference cylinder (at the aperture). They also showed that this implied an aperture limitation given by:

$$(d/\lambda)_{max} = \frac{2.87}{(a/R)^4 - 1.5(a/R)^6 - 1.33(a/R)^8} \qquad (10)$$

Kanevskii and Surikov[15] showed that in obtaining equations (9) and (10), a physically unacceptable argument was used. As a result, those equations are only approximations. They corrected the analysis and extended its treatment to spherical lenses. Figure 8 shows their plot of E_0 versus n (refractive index) for an aperture (semi) angle of 45°. Figure 9 shows the longitudinal spherical aberration coefficient of the lens conforming to their analysis. These papers indicate the limit of obtainable focussing for liquid/liquid single surface refraction with unshaded apertures.

Luneberg[16] showed that perfect focussing occurs for spherical lens if:

$$n(r) = \sqrt{2 - r^2} \qquad\qquad (11)$$

where r is the local radius of the refracting medium.   Luneberg

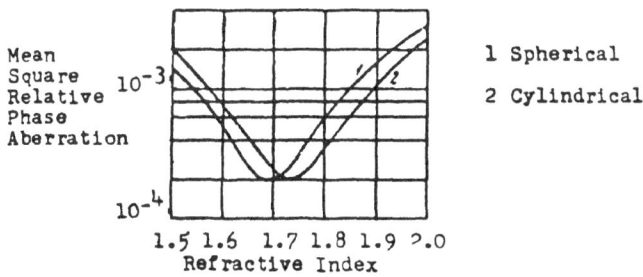

Fig. 8   The mean square relative phase aberration as a function of the refractive index (after Kanevskii and Surikov, ref. 15).

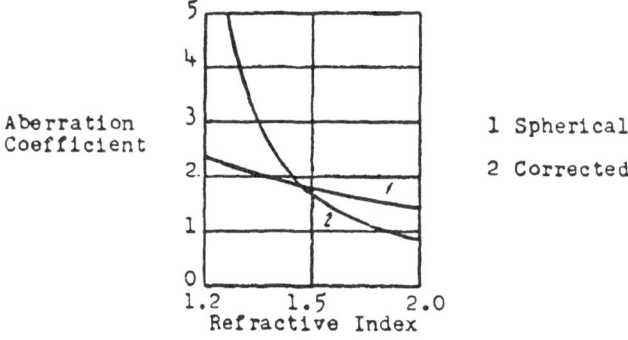

Fig. 9   The longitudinal spherical aberration coefficient as a function of the refractive index (after Kanevskii and Surikov, ref. 15).

indicated that equation 11 was not a unique solution to the
particular problem. Much attention has been paid to the Luneberg
lens; the application and elaboration of his proposal fascinated
many researchers. Gutman[17], Brown[18] and particularly Morgan[19]
directed their attention to the general problem and, in some cases,
from interest arising out of antenna theory. Gutman showed that
the variant:

$$n^2(r) = (1/r_f^2)(a^2 + r_f^2 - r^2) \qquad\qquad (12)$$

(where r_f is the radial distance to the focus) satisfied Luneberg's
requirements. Lord[20] used this relationship to provide the basis
of an analysis of a lens design in which equation (12) was obeyed
in the refracting medium. The analysis supposed that density ρ was
constant and that the sound velocity varied.

The practical achievement of equations (11) and (12) has posed
something of a problem for the experimentalist. Toulis[21] modified
the effective modulus of the liquid in a 6-foot diameter lens by
inserting compliant tubes in a suitably spaced array. This arrangement produced an approximation to the Luneberg condition, see figures
10a and 10b. Lenses in which the refracting medium consists of a
series of concentric shells approximating to the required condition
have been made. Peeler and Coleman[22] built a microwave lens with
ten concentric layers and reported their results. These authors,
together with R.E. Webster[23], showed that a fairly satisfactory
approximation is obtained "with an error of not more than 20%" as

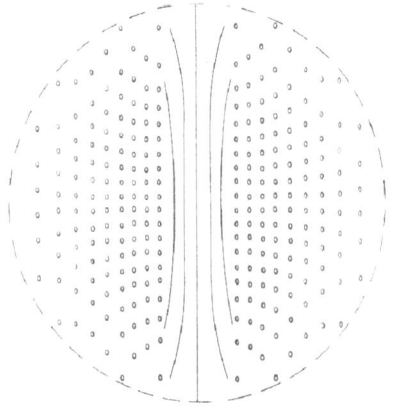 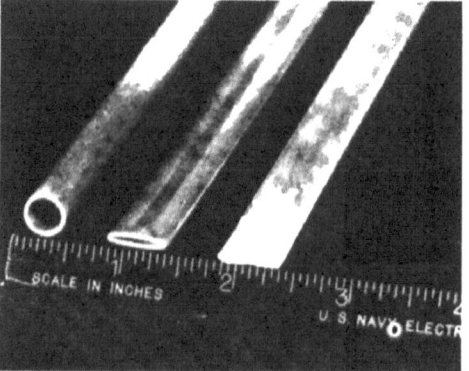

Fig. 10   a) Distribution of compliant tubes within the lens.

b) Compliant tubes (after Toulis, ref. 21).

defined by beam widths, if:

$$N = 0.3 \ d/\lambda \tag{13}$$

where N is the number of concentric layers. Unfortunately, this
reveals that 40 layers are needed to produce the quality of image
which was obtained by Folds in a lens with $d/\lambda \geq 138$. Another
approximation to a Luneberg lens can be obtained by using a two
layer lens in which the outside layer is concentric with a varying
refractive index inner sphere. Toraldo di Francia[24] investigated
this problem and showed that if the outer layer is of high refractive
index (3 < n < 4) then a good approximation can be obtained if the
inner sphere has a constant refractive index. He commented that
"for an infrared microscope objective, this lens would be practically
perfect for a numerical aperture up to values very close to unity".
Figure 11 shows his determination of the spherical aberration of such
a lens. Kanevskii and Surikov[25] applied these ideas to spherical
and cylindrical acoustical lenses and they determined:

   i) the condition for minimum aberration as a function of the
      two refractive indices (and their relationship to each
      other),

  ii) the radii of the spheres for optimum performance, and

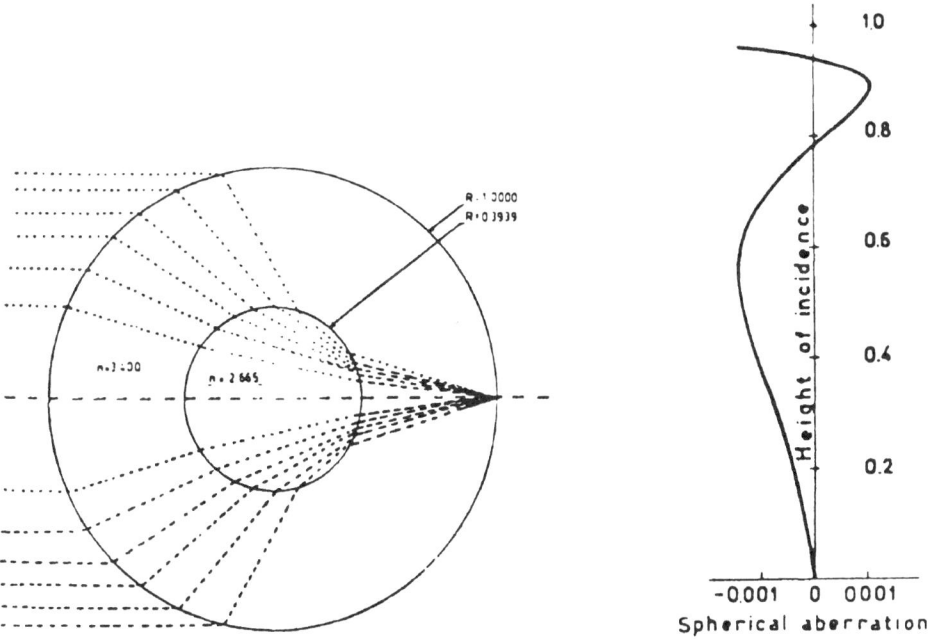

Fig. 11 Spherical aberration of a two layer lens as a function of
        the height of the incident rays (after Toraldo di Francia,
        ref. 23).

iii) the maximum aperture (d/λ) obtainable with minimum aberration.

Figure 12 summarizes their results. This work was elaborated by Metelkina and Surikov[26] who investigated the focussing of rays within the outer shell of the lens. Folds[27], apparently independently of Kanevskii et al., and at about the same time undertook a similar analysis of the two layer lens. He related his theoretical findings to experimental studies on liquid filled spherical lenses and demonstrated that his theory closely agreed with experimental findings.

Thin and multiple element solid lenses had been discussed by several authors. Folds and his co-workers[4,28] reported on this

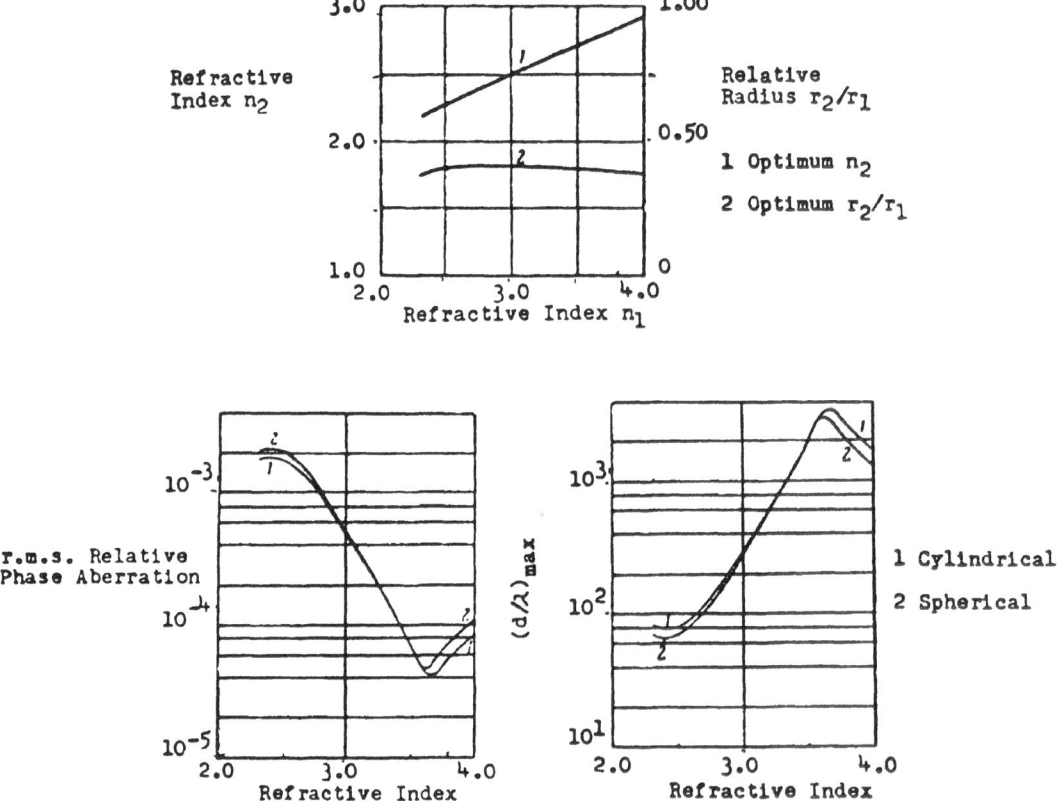

Fig. 12   Optimum relative radius and refractive index of the inner
          material as a function of the refractive index of the outer
          material with the relative phase aberration and maximum
          diameter to wavelength ratio possible as a function of the
          refractive index. (after Kanevskii and Surikov, ref. 25).

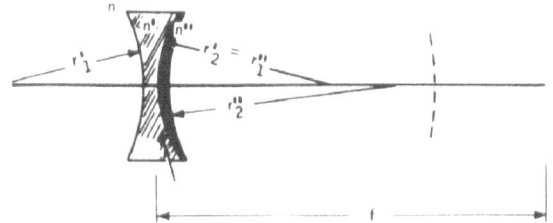

Fig. 13 Athermal doublet (after Folds, ref. 4).

topic. They established a condition for athermal lenses which had
a constant focal length over a temperature range of 30°C. The
correction process was obtained in a manner reminiscent of that
used in an achromatic doublet, see fig. 13. This work culminated
with the design and evaluation of a four element lens. Standard
modern numerical procedures were used to obtain minimum aberration.
Polystyrene and RTV 602 were used as refracting materials. The
design allowed for a series of manufacturing tolerances, both in
the material properties and the lens dimensions. It was partially
temperature corrected to give good focussing over the ranges of 10
to 25° C. It was operated at frequencies between 600 KHz to 3 MHz.
Its focal plane was curved, see fig. 14. Jones, et al.[6] designed a
single and two element lenses for underwater viewing and submitted
these to an extensive series of experimental tests. Havlice[2] designed
a microscope objective to work at 860 MHz. This work followed his
researches into the design of mirrors and lenses. He became aware
of the low losses, at high frequencies, in gallium and used this
material in combination with fused quartz. The final lens, see

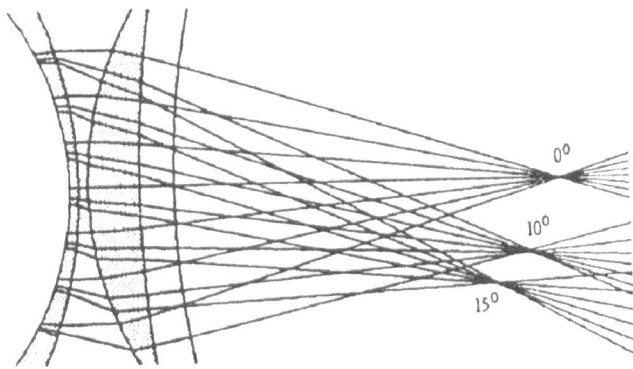

Fig. 14   Four element lens ray diagram (after Folds and Hanlin,
          ref. 28).

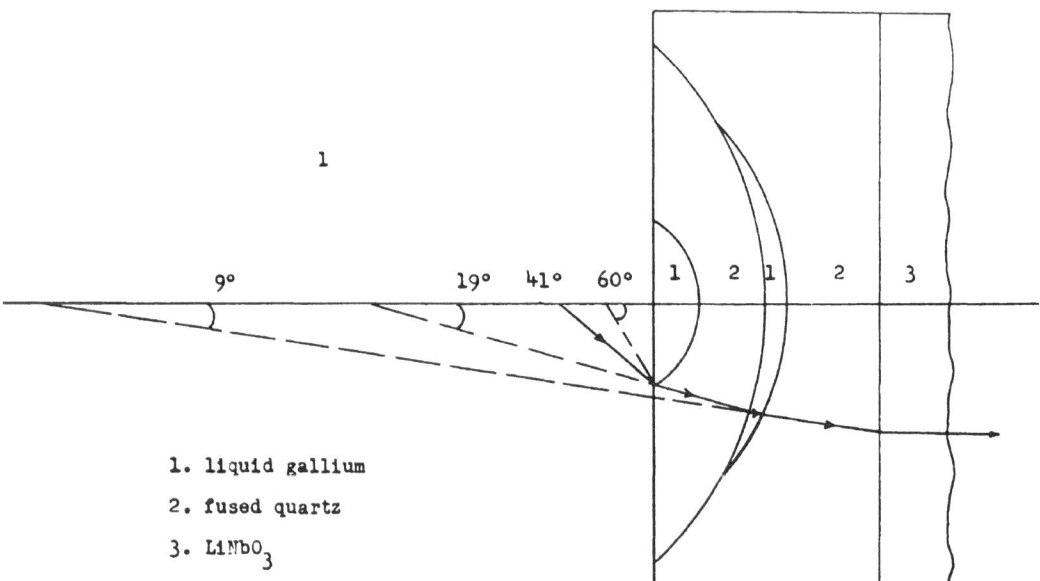

1. liquid gallium

2. fused quartz

3. LiNbO$_3$

Fig. 15 Acoustic Microscope objective lens (after Havlice, ref. 2).

fig. 15, was designed to use aplanatic points as far as this was possible so that aberration was minimized. The lens was reported to have a resolution of 5.1 microns with a phase variation of ±.36 λ over the total aperture of f/.656.

The relative performance of the various lenses which operate at the lower frequency is shown in fig. 16 where their directivities are compared to that of a plain circular piston. It should, perhaps, be pointed out that both experimental and theoretical data appear in this figure. Clearly, Boyle's spherical lens is of some interest.

INSERTION LOSSES

General comment on the insertion losses is difficult. The data on lucite wedges presented earlier in this paper indicates that a transmission loss of about 30% occurs at frequencies in the region of 1 MHz. This suggests insertion losses of about 32 dB for a 4-element lens. Folds[29] reported a 2 dB insertion loss for a f/2.8 nylon lens which was operated at 600 KHz and an 8 dB loss for a polyethylene lens working at the same frequencies. Havlice's lens had an insertion loss of 9 dB, which is remarkably low in view of the operating frequency.

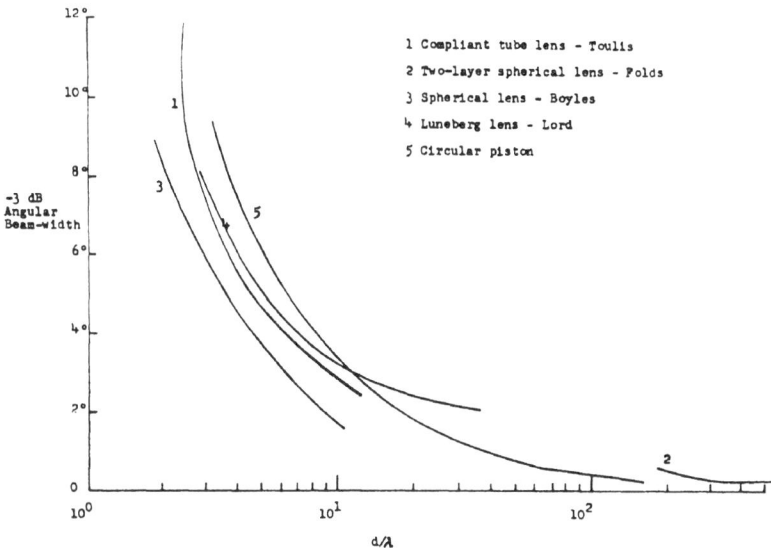

Fig. 16 Directivity pattern comparison.

DIFFRACTION AND RELATED EFFECTS

The relatively long wavelength of sound makes it necessary to consider coherence effects and aperture shading. In our numerical studies on the lens used by C.H. Jones[30] we found that the diffraction patterns predicted by computations were less well resolved than the experimental observations indicated. In the process of the calculation, the "spot patterns"[33] were obtained by plotting of some 450 "rays" per half lens. The latter calculation indicated that the image should have been better resolved than those which were actually observed. This procedure also predicted that the image quality should have been relatively better for off-axis cases. The essential difference between the two calculations should perhaps be emphasized; the first was the detailed diffraction calculation accounting for the phase and amplitude variations in the incoming wavefront. The second plotted the density of the rays, from the refracted wavefront, at the point where they intersected the image plane. In a search for a possible explanation of the differences between the two predicted and observed images, one possible solution occurred to us.

If turbulent motion affects the medium, then the phase relationship of the vector diagram (or the ϕ_n terms in equation (14) below) is disturbed so that the vector sum is continually varying. It is required that a solution should be found for these conditions. If it is assumed that:

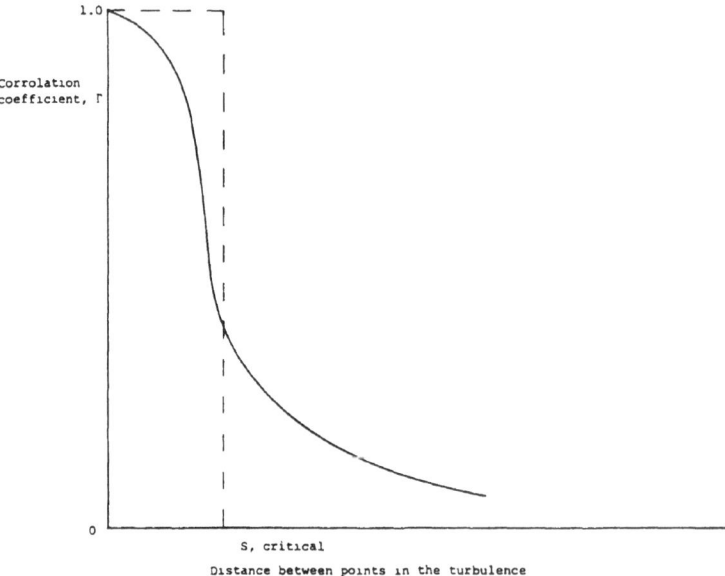

Fig. 17 Correlation coefficient.

   i) the turbulence is random

  ii) that the phase difference produced between adjacent rays
      is small

 iii) that beyond these "adjacent" rays, the phase variation
      tends rapidly, with distance, to complete randomness as
      shown in figure 17.

  iv) that the correlation coefficient referred to above can be
      approximated to, as indicated in figure 17.

   v) the turbulence is not so severe that it appreciably scat-
      ters energy from the wavefront or causes a large displace-
      ment of the rays,

the pressure of the receiver will be given by

$$\langle p_t \rangle = \left\langle \sum_{n=1}^{m} p_n \cos(\omega t + \phi_n) \right\rangle \tag{14}$$

where p_n is the contribution from element n of the original wavefront and ϕ_n is the turbulence induced phase change.

From (14) and fig. 17

$$\langle p_t \rangle = \left\langle \sum_{n=1}^{r} p_n \cos (\omega t + \phi_1) \right\rangle + \left\langle \sum_{n=r+1}^{m} p_n \cos (\omega t + \phi_n) \right\rangle.$$

$$= \left\langle \sum_{n=1}^{r} p_n \cos (\omega t + \phi_1) \right\rangle + 0 \qquad\qquad (15)$$

where r represents the value of n over which the correlation coefficient $\Gamma = 1$; in the range r + 1 to m, $\Gamma = 0$.

As an example with an adroit choice of $S_{critical}$ an approximation for the pressure developed at the image plane (or at the receiver) can be obtained by supposing that part of the wavefront which is directly converging onto this element of the receiver is the only part which makes a contribution to the output signal (see figure 18).

In the case of interest, 3 MHz sound travelled 20 feet through water. We find that extremely small turbulent velocities are sufficient to cause variations of π in this case. Perturbations of about 1 part in 10^5 in the observed sound velocity could produce the effect observed. If this is correct, the phase modulation produced in this case causes the lens to be "stopped down" and produces the sharper images.

APERTURE SHADING

As the diameter of ultrasonic lenses are small in terms of d/λ by comparison to conventional optics, the possibility of using shading techniques for the aperture is attractive. In principle, this requires the Kirchkoff integral:

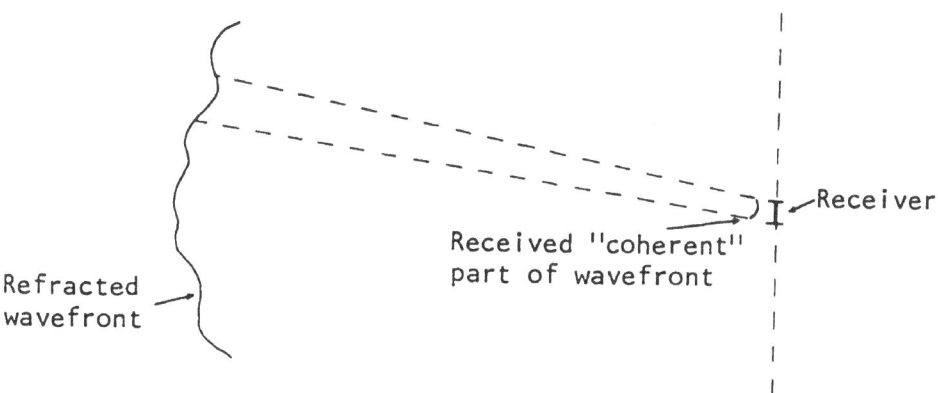

Fig. 18 Stopping effect of turbulence.

$$U(p) = -\frac{Ai}{2\lambda} \iint_A \frac{e^{ik(r+s)}}{r,s} \cdot [\cos(n_1 r) - \cos(n_1 s)] \, dS \qquad (16)$$

to be modified to include a transmission factor T such that:

$$T(x,y) = T'(x,y) + i\,T''(x,y) \qquad (17)$$

and the integral becomes:

$$U(p) = -\frac{Ai}{2\lambda} \iint_A \frac{Te^{ik(r+s)}}{r,s} \cdot [\cos(n_1 r) - \cos(n_1 s)] \, dS \qquad (18)$$

This suggestion was made initially by Luneberg. Later, Toraldo di Francia[31] showed that in principle it is possible to obtain any shape of diffraction pattern which is required simply by judicious choice of T. The latter author proposed a method which required the specification of the normalized intensity at a number of points in the diffraction pattern. He set up a system of simultaneous equations, using these values which allowed the determination of T. In practice, the solutions are difficult to obtain. Alternative methods have been suggested and particular solutions obtained. For example, Wild[32] proposed a method of obtaining an extremely sharp diffraction pattern from an annular ring. It can be supposed that many more solutions will be obtained to meet the demand of future problems. It should be possible to produce a lens with superresolution for on-axis images. A considerable literature on this topic exists in antenna theory but there has been relatively little corresponding development for the case of acoustic lenses.

CONCLUSION

An attempt has been made to discuss the whole topic of ultrasonic lenses and their application to imaging. The various effects which occur during the refraction process are discussed. It is shown that solid lenses produce spatial modulation of the refracted wave. This modulation arises directly from the variation of the transmission of the lens which is a function of its thickness. This effect appears to be absent from liquid/liquid lenses.

A fairly complete review of the present state of knowledge on the design of lens has been given. It is shown that the limits of geometrical optical performance for the majority of practical configurations has been reasonably well explored. The importance of refraction related effects has not been included in these researches even though they may be of importance. Some commentary on insertion losses has been included so that the usefulness of multiple element lenses can be assessed.

The question of turbulance in the transmitting media has been discussed. It is shown that under certain conditions, it is possible for turbulence to cause sharper images (with a lower signal to noise ratio).

A limited discussion of aperture shading has been provided.

REFERENCES

1. HAVLICE, J.F., G.S. Kino and C.F. Quate. Appl. Phys. Lett., 23, (1973), 581-583.

2. HAVLICE, J.F. "Optical Imaging of Acoustic Waves—An Acoustic Microscope". PhD Thesis, Stanford University, (1971).

3. AHMED, M. Acoustical Holography, 6, (ed. N. Booth), Plenum Press, New York, (1975), 671-687.

4. FOLDS, D.L. NSRDL-P 5216/3 (10.70), (1971).

5. JONES, C.H. J. Acoust. Soc. Am., 59, 1, (1976), 74-85.

6. JONES, C.H., H.W. Jones, J.W. Kesner, C.J. Williams. (To be published.)

7. BREKHOVSKIKH, L.M. Waves in Layered Media. (New York: Academic Press), (1960).

8. TARNOCZY, T. Ultrasonics, 3, (1965), 115-127.

9. KINSLER, L.E., A.R. Frey. Fundamentals of Acoustics, John Wiley & Sons, New York (1962).

10. GOLIS, M.J. IEEE Trans. S. and U, SU-15, (1968), 105-110.

11. ROUDEBUSH, J.L. "An experimental Biconvex Liquid-Filled Acoustic Lens", E. Eng. Thesis, Naval Postgraduate School, 1969.

12. BOYLES, C.A. J. Acoust. Soc. Am., 38, (1965), 393-405.

13. FOLDS, D.L., D.H. Brown. J. Acoust. Soc. Am., 43, (1968), 560-565.

14. BORN, M., E. Wolf. Principles of Optics. (Oxford: Pergamon Press), (1970)

15. KANEVSKII, I.N., B.S. Surikov. Sov. Phys.-Acoust., 17, (1971), 43-47.

16. LUNEBERG, R.K. Mathematical Theory of Optics. Brown University, Providence, Rhode Island, (1944).

17. GUTMAN, A.S. J. Appl. Phys., 25, (1954), 855-859.

18. BROWN, J. Wireless Eng., 30, (1953), 250.

19. MORGAN, S.P. J. Appl. Phys., 29, (1958), 1358-1368.

20. LORD, G. J. Accoust. Soc. Am., 45, (1969), 885-891.

21. TOULIS, W.J. J. Acoust. Soc. Am., 35, (1963), 286-292.

22. PEELER, G.D.M., H.P. Coleman. I.R.E. Trans. Ant. and Prop., (1958), 202-207.

23. WEBSTER, R.E. I.R.E. Trans. Ant. and Prop., (1958), 301-302.

24. TORALDO di FRANCIA, G. J. Appl. Phys., 32 (L), (1961), 205.

25. KANEVSKII, I.N. and B.S. Surikov. Sov. Phys.-Acoust., 16, (1971), 457-461.

26. METELKINA, R.V. and B. Surikov. Sov. Phys.-Acoust., 17, (1971). 138-139.

27. FOLDS, D.L. J. Acoust. Soc. Am., 49, (1971), 1591-1595.

28. FOLDS, D.L. and J. Hanlin. J. Acoust. Soc. Am., 58 (1975), 72-77.

29. FOLDS, D.L. J. Acoust. Soc. Am., 53, (1973), 826-834.

30. JONES, C.H. Private communication.

31. TORALDO Di FRANCIA, G. Del Nuovo Cimento, Supplemento Al, IX, Series IX, (1952), 426-438.

32. WILD, J.P. Proc. Roy Soc. 286 A, (1965), 499-509.

33. PALMER, J.M. Lens Aberration Data, (New York: American Elsevier, 1971.)

METHOD FOR THE INVESTIGATION OF ACOUSTIC LENSES WITH OPTICAL

INTERFEROMETRIC SYSTEM

J.K. Zieniuk

Institute of Fundamental Technological Research
Polish Academy of Sciences
00-049, Warsaw, Poland*

It is well known that the output function (image) formed by coherent waves with the use of convergent lenses is a convolution of the input function (object) and the spread function. In case of a diffreaction limited lens the spread function is just Fourier transform of the aperture. This is true not only in case of optical lenses and electromagnetic waves but also when the ultrasonic lens form the pressure distribution corresponding to the image.

In a sepcial case when the input function is a delta function the output function is simply the Fourier transform of the aperture function. In this paper we describe the preliminary investigation of ultrasonic lenses in such a case.

INTRODUCTION

More and more ultrasonic apparatus for visualization and medical diagnosis use ultrasonic lenses (1,3,4,11) to control the beam and form an image. Since there are close similarities between the Fourier optics of coherent light and acoustic imaging using lenses,

*Visiting Scientist of the Division of Electronic Products
 Bureau of Radiological Health, Rockville, Maryland 20852 U.S.A.

a Fourier optics approach can give valuable information about
acoustic systems (1,2,5,8,10). Basic factors influencing the
quality of the ultrasonic image can be investigated in this way.
In this paper we describe the investigation of the acoustic Airy
disc formed by an ultrasonic lens and factors influencing the shape
of it. The ultrasonovision, an interferometric system developed
by RCA, gives the distribution of the displacement amplitude in a
plane normal to the acoustic axis and has been used for this purpose (6,9).

APPARATUS

The ultrasonovision system has been described in detail elsewhere so we shall outline here only the basic principles. The
ultrasonic beam travelling in water passes through a very thin
opaque gold coated membrane. Since the membrane is very thin compared to the wavelength of ultrasound (4 micrometers to about 1.5
millimeters) it is transparent to the sound but at the same time,
the displacement of this membrane (pellicle) is equal to the displacement of the particles of the surrounding medium (water). This
displacement is measured by means of an optical interferometer.
The optical interferometer is of the Michelson type.

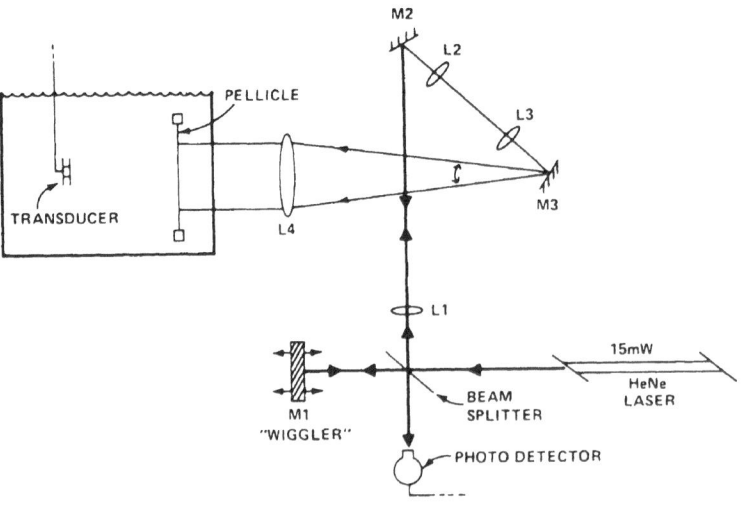

Figure 1.   Basic arrangement of the ultrasonovision

Figure 1 shows the basic optical arrangement. The beam from the HeNe laser is split into the two legs of the interferometer. In the reference leg is mirror M1. In the scanning leg are galvanometer mirrors M2 and M3 which provide the vertical and horizontal scanning, respectively. Lens L4 collimates the raster scanning beam to provide normal incidence to the thin gold coated membrane. The optimum operating condition for the ultrasonovision occurs when the relative phase difference is $\pi/2$. At that point any phase change caused by motion of pellicle will cause a maximum brightness change and the functional relationship between them will be nearly linear. To guarantee this operating condition the reference mirror M1 is "wiggled" along the optic axis at a frequency of approximately 30 kHz. The amplitude of the wiggler is such that the relative phase difference is swept through at last π radians. At the photodetector the combined effect of the pellicle motion at the acoustic frequency and the wiggler motion is a signal varying in time as the acoustic pulse and contained within an envelope having the wiggler frequency. The peak of that envelope occurs when the relative phase difference is $\pi/2$. Therefore, peak detection of the photodetector signal results in the displacement amplitude at that scan location in the acoustic field. Another advantage of this technique is to eliminate noise due to thermal drift and vibration. It has been estimated that the ultimate system sensitivity is 10^{-10} Watts per sq/cm. (acoustic) at 1.5 MHz. The system used here presently can detect displacements lower than 0.1 Angstroms corresponding to

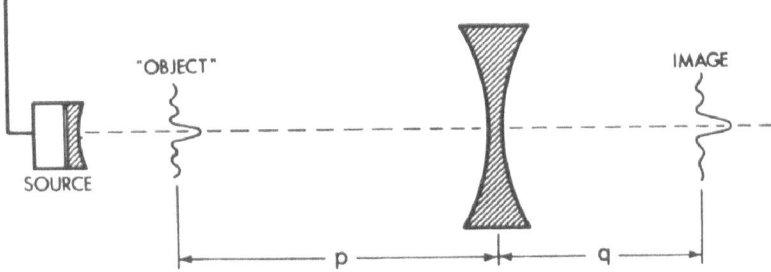

Figure 2.   Configuration of the ultrasonic system.

10^{-6} Watts per sq/cm. (acoustic) at 1.5 MHz. The system was calibrated by illuminating the photodetector with a known intensity of
light and measuring the resulted voltage after the photodetector
preamplifier. The total estimated uncertainties is on the order
of plus or minus 4%.

The system shown in Figure 2 consists of an ultrasonic focused
transducer, "pinhole", lens, and above mentioned pellicle. In
this experiment a ceramic (lead zirconate titanate) transducer with
a fundamental frequency of 1.343 MHz with a permanently attached
lens of 9 centimeter local length and of 2.54 centimeter diameter
has been used. This transducer produced ultrasonic pulses 40 microseconds in duration with a repetition rate of 3.2 kHz.

In some of the experiments an acoustically opaque screen with
a round hole of 8 millimeters in diameter, playing a similar role
as a pinhole in an optical system, was placed at the focal plane
of the transducer on the acoustic axis. This transducer and acoustical pinhole formed a secondary source of ultrasonic waves. In
these preliminary experiments, two ultrasonic lenses made of polystyrene have been used. The first lens of focal length 19.2 centimeters and 9 centimeter diameter with radii of curvature of 215
millimeters and 96.15 millimeters has been designed to have a
minimum of spherical aberration. The second lens with a focal
length of 14.7 centimeters, 7 centimeters in diameter, and radii
curvature of 32.9 centimeters and 6.3 centimeters was not specifically designed for low aberrations.

In these experiments the ultrasonic lens was located on common axis with a secondary ultrasonic source. The acoustical image
was co-planer with the pellicle such that the lens law was fulfill-
ed. In the case described here, the source was located 3 focal
lengths from the lens and the image located at the back principal
plane with the magnification of 1.5. It should be noted that the
ultrasonovision system detects only displacement amplitude.

All important elements (transducer, pinhole, and lens) were
mounted on separate holders located on an optical bench over the
water tank. This particular arrangement enables a precise movement of the elements. In some experiments an acoustical aperature
was located in the front of the lens. It was made of layer of
rubber with a circular opening of the diameter equaled to the
diameter of the lens.

EXPERIMENTS AND COMPUTATIONS

As mentioned above, in this experiment two lenses were used
and one focused transducer as the source of ultrasound.

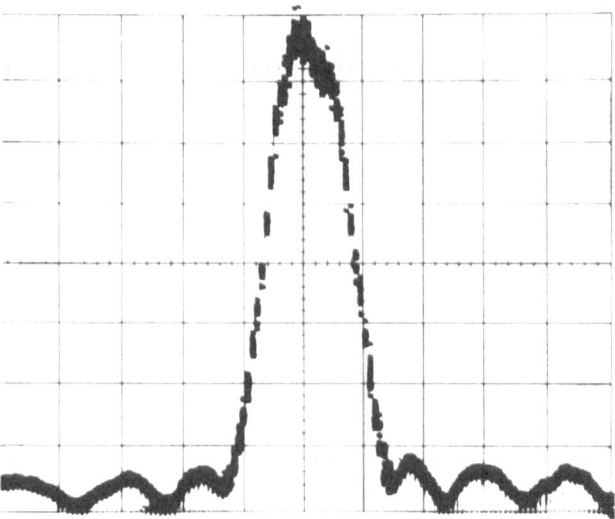

Figure 3.   Distribution of displacement amplitude across the focal
              plane of the focussed transducer.

     Figure 3 is a distribution of displacement amplitude across
the focal plane of the transducer. In this case, it is the object.
Because of technical difficulties it has not been possible to obtain
such a distribution of the output from the acoustical pinhole using
the ultrasonovision. This pinhole has been designed in such a way
that only the central part of the beam between the first zeros could
be transmitted. In this the second case, the beam cross section
was approximately Gaussian in shape.

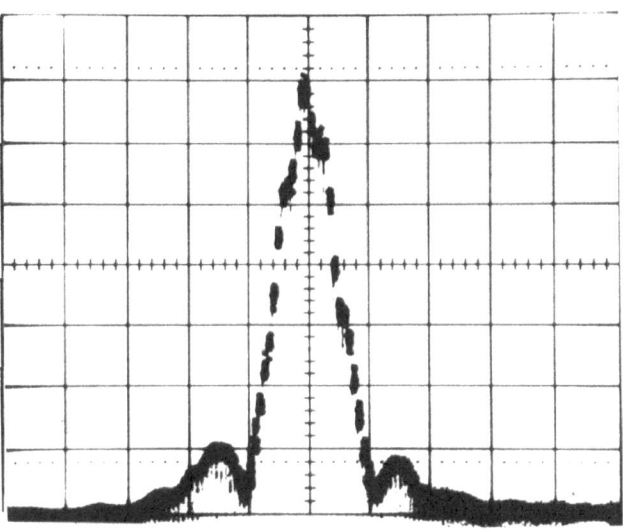

Figure 5.   An image of the acoustical source of nearly Gaussian
                distribution in case of the lens of small spherical
                aberration.

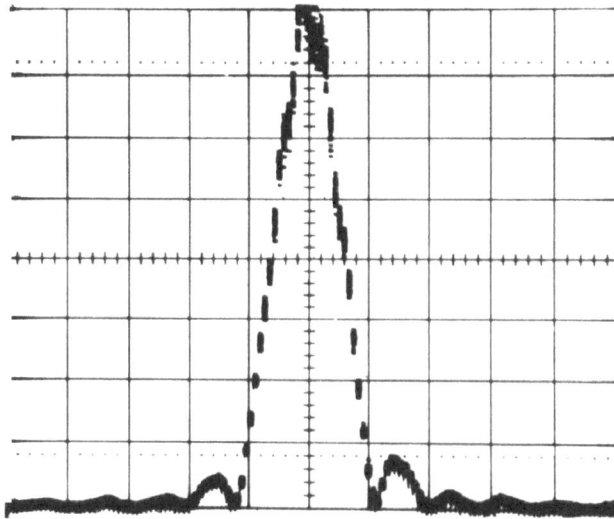

Figure 4.   An acoustical image of the distribution shown on Figure 3.

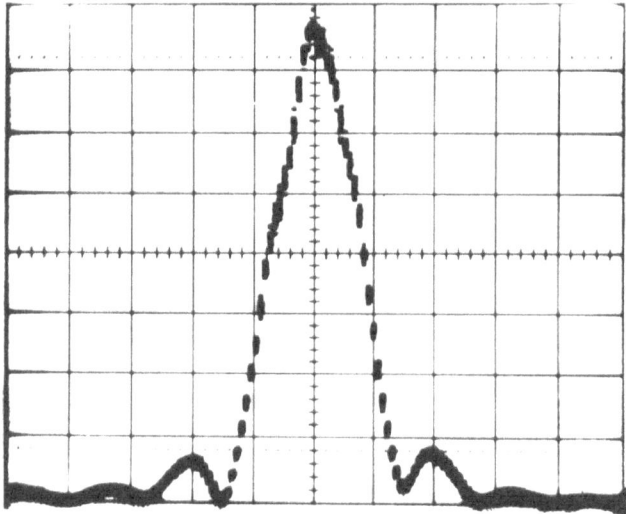

Figure 6.   Analogous image obtained with a second lens having
             greater spherical aberration.

Figure 4 is an acoustical image of Figure 3 obtained with the use
of the acoustical lens of 13.2 cm focal length. Figure 5 is an
acoustical image of a nearly Gaussian distribution obtained with
the use of the acoustical pinhole. Figure 6 is an acoustical image
of the same source but obtained with the second lens.

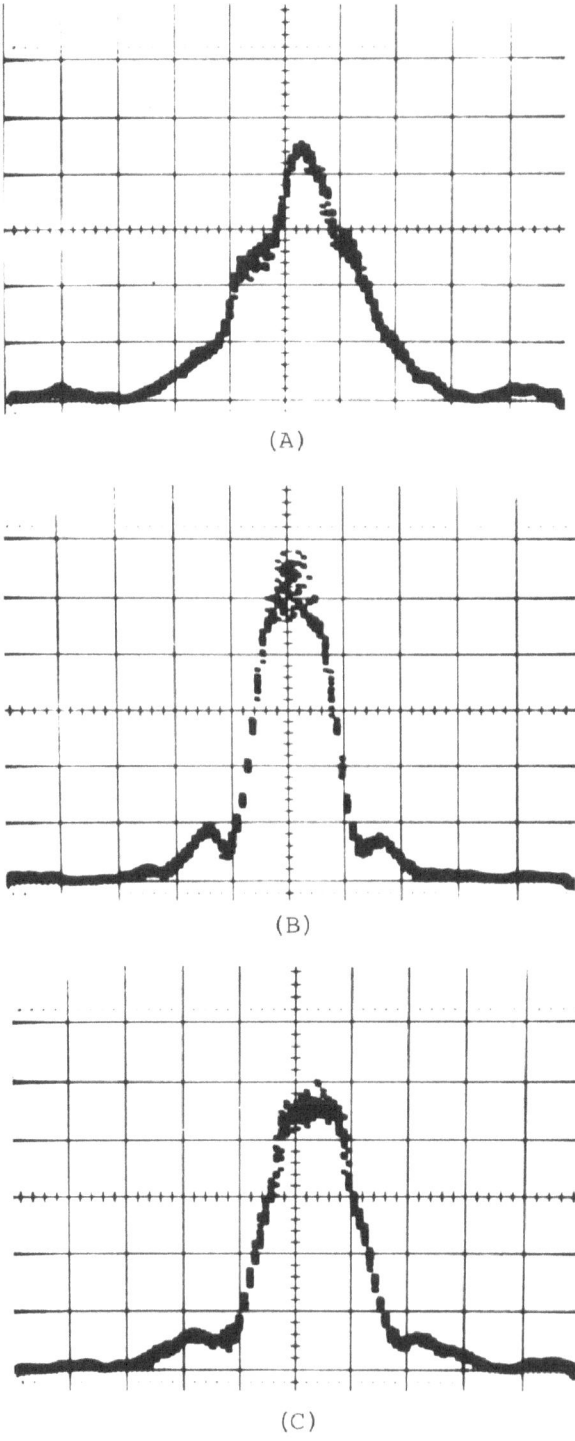

(A)

(B)

(C)

Figure 7.   (A) An acoustical image of the displacement amplitude
                distribution 2 cm before the focal plane. (B) At the
                focal plane and (C) 2 cm after the focal plane of
                focussed transducer of f=6.0 cm and the diameter
                of 2.54 cm.

Figure 7a, 7b, 7c are the distributions of the displacement amplitude normal to the acoustic axis at 2 cm before the focal plane, at the focal plane, and 2 cm after the focal plane, respectively. It can easily be seen that the width of the central peak is changing very remarkably.

The scope of this particular experiment was to compare the theoretical amplitude distribution across the acoustic Airy disc with the similar distribution in the case of real acoustic lenses having some aberrations.

To do this the acoustic displacement amplitude was measured with the ultrasonovision system. There are a hundred and twenty-eight points across each image. The numerical values are digitized into the memory of an HP9820 programmable calculator.

With the assumption that the input function is a delta function, the image at the back principle plane of the lens is a Fourier transform of the lens pupil. This Fourier transformer was calculated and compared with the experimental results. The kernel of the transform was $\pi rd/\lambda z$, where r is the radial coordinate, d is the diameter of the lens aperature, z is the lens to image plane distance, and λ is the acoustic wavelength.

The theoretical curve against which experimental points were plotted was

$$\frac{2J_1(\pi RD/\lambda z)}{\pi RD/\lambda z}$$

Figure 8 is the result of superimposing those curves for the first lens and Figure 9 for the second one. It can be easily seen that for a system arranged axially, the influence of aberration is minimal compared with the diffraction effects. There is even some suppression of the Bessel rings caused perhaps by apodization effects of the lens material.

FINAL REMARKS

Taking the Fourier approach and using the ultrasonovision system, it is easy to obtain experimentally, the distribution of the displacement amplitude on the planes perpendicular to the acoustic axis.

It can be easily seen that the shape of the Airy disc (and) subsequently the quality of the acoustical image obtained with lens) is determined mainly be diffraction effects even in the case of the second lens having greater spherical aberrations. This means that relatively simple lenses having low f-number can be used to reduce the size of the Airy disc and improve the image resolution.

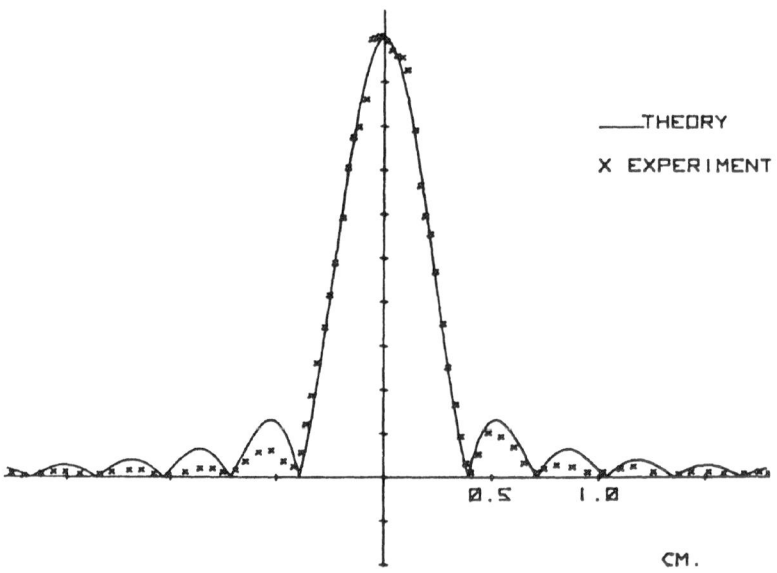

Figure 8.   Superimposition of experimental values of displacement
              amplitude with the Fourier transform of the aperture
              function in case of the lens of lower spherical
              aberration.

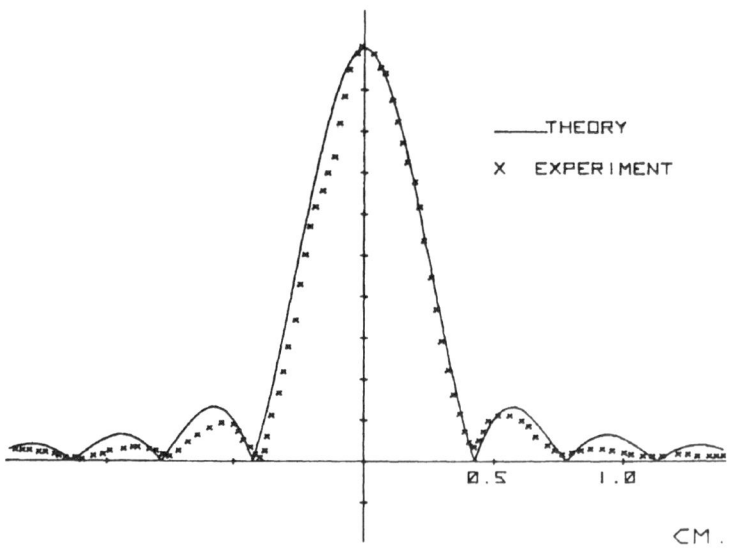

Figure 9.   As on Figure 8, but for the lens with larger amount of
              aberration.

In the near future the investigation of non-axial arrangements, where the influence of aberrations is more profound, it will be investigated.

ACKNOWLEDGEMENTS

The author wishes to acknowledge the important role of Michael Haran in preparing this paper, Dr. Harold Stewart for encouragement and support, Mr. Joseph Geary for help for the numerical computations, and the National Academy of Sciences for its grant.

BIBLIOGRAPHY

1.   E.E. Aldridge, Fourier Transform and Acoustic Diffraction.
     Ultrasonics, July 1966 pp. 131-135.

2.   R.J. Collier, C.B. Burckhardt, L.H. Lin, Optical Holography.
     Academic Press, New York, 1971.

3.   D.L. Folds, Focusing Properties of Solid Ultrasonic Cylin-
     drical Lenses. J.A.S.A. 53, 826-834 (1973).

4.   D.L. Folds, Progress Report on Ultrasonic Lens Development.
     Eight International Congress on Acoustics, Proc. of the
     Satellite Symposium on Underwater Acoustics. Aug 1974,
     Paper 2.2.

5.   J.W. Goodman, Introduction to Fourier Optics. McGraw-Hill,
     New York 1968.

6.   M.E. Haran, H.F. Stewart, Optical Interferometric Measure-
     ments of Ultrasonic Radiation and its Applications to Medi-
     cine. Proc. of Symposium Measurements for Radiation.
     NBS, March 1-4, 1976.

7.   I.Heht, A.Zajac, Optics, Addison-Wesley, 1975.

8.   T.C.Lee, D. Gossen, Generalized Fourier-Transform Holography
     and its Applications. Applied Optics, 10, 961-963 (1971).

9.   R.S. Mezrich, K.F. Etzold, D.H.R. Vilkomerson, System for
     Visualizing and Measuring Ultrasonic Wavefronts.

10. G.B. Parrent, Jr., B.J. Thomson, Physical Optics Notebook,
     Soc. of Photo-Optical Instrumentation Engineers, 1971.

11. Yasuaki Tannaka, Tsuneji Koshikawa, Solid-Liquid Compound
     Hydroacoustic lens of Low Aberration. J.A.S.A. 53, 590-595
     (1973).

A NEW ULTRASONIC IMAGING SYSTEM BY USING A ROTATING RANDOM PHASE DISK AND POWER SPECTRAL AND THIRD ORDER CORRELATION ANALYSIS

Takuso Sato, Shusou Wadaka, Junichi Ishii and
Tadasu Sunada

Faculty of Science and Engineering, Tokyo Institute of
Technology
Nagatsuda, Midori-Ku, Yokohama 227, Japan

ABSTRACT

A new ultrasonic imaging system which uses neither scanning of
a receiver nor beam forming by a phased array is presented. The
fundamental idea is the transformation of the spatial distribution
of wave intensity, which is the object to be reconstructed as the
image, into temporal information by using a rotating random phase
disk placed just behind the object and the analyses of the signal
detected by a fixed receiver placed far from the disk. It is shown
that the power spectrum of the signal is the shadow of the object
on the plane which includes the point detector and the plane
parallel to the object plane. Moreover if the 3rd order correlation
function of the signal is calculated, this value gives directly the
required reconstructed image of the object, provided that a point
source is placed near the object. A complete system with random
phase disk made of acryl and operating at 3 MHz with FFT algorithm
of a mini-computer is constructed. The theoretical discussions by
using spectral representation of the wave field and resolution
analyses as well as reconstructed images are presented. Although
most of the results have been submitted as papers in JASA, Refs. 1
and 2, comprehensive discussions are given in this article.

I. INTRODUCTION

When ultrasonic imaging systems which use transducers to detect
ultrasonic wave fields are classified from the point of view of
techniques for image formations of extended objects, conventional
methods may be divided into two main categories: i) mechanical
scanning of transmitter and/or receiver,[3] and ii) electronical beam

forming by phased array.[4] The former technique has been used as the
most popular means mainly because of the simplicity and the sureness
of the scanning as well as the one to one correspondence between the
position on the object and the scanning point, inspite of the
intrinsic limitation on the speed of the scanning. As for the latter
method, it has been developed rather recently from the requirements
to observe moving objects in real time and to make systems more
compact, although in order to increase the resolution in this system
fairly complicated electronics are required.

In this paper, another new technique which uses neither
mechanical scannings nor electronical beam formings is proposed.
The fundamental idea is based on the transformation of the spatial
distribution of the intensity of the wave field, which is the object
to be reconstructed as the image, into the information in the
temporal domain by using a moving random phase mask placed just
behind the object. The operations for the image reconstruction are
carried out as the signal analyses of the signal detected at a fixed
single point receiver located far from the mask, such as power
spectral, 2nd order or higher order correlation function analyses.
The theoretical developments by using spatial frequency
representation of the wave field show that the power spectrum of the
signal represents the shadow of the object on the plane which
includes the point receiver and the plane parallel to the object
plane. If a point reference source is placed near the object, then
the 3rd order correlation function of the signal gives directly the
object wave field itself.

A complete system is constructed by using a random phase disk
made of acryl with random surfaces and operating at 3 MHz with FFT
algorithm of an on-lined mini-computer. The experimental results
show fairly good agreement with the theoretically expected ones.
The resolution analyses and the extention of the method by adopting
a random phase mask made artificially based on M-sequences are also
presented. The details are given in Refs. 1, 2 and 5.

II. PRINCIPLE

The principle of the method is schematically illustrated in

Fig. 1. Signal transformation and information flow diagram.

Fig. 1. The spatial intensity distribution, which is the object to be imaged, is transformed into a random time series by using a moving random phase mask. Operations required for the reconstruction of the image of the object are carried out as analyses of the random time series.

Let us take the one-dimensional system shown in Fig. 2 for the mathematical simplicity and develop the mathematics. An object is illuminated by coherent ultrasonic waves and a random phase mask is moved with a uniform velocity v just behind the object. The moving random phase mask scatters the object waves and the scattered waves are detected by a fixed point receiver far from the mask. Provided that the temporal changes of the random phase shift caused by the random phase mask is much smaller than the angular velocity of the ultrasonic waves, the signal $V_s(t)$ detected by the point receiver after it is heterodyned is written as follows:

$$V_s(t) = \frac{\exp(jkD)}{j\lambda D} \int_{-\infty}^{\infty} V_o(u) \exp[(jk/2D)(x_0-u)^2 - j\theta(u,t)]du \quad . \qquad (1)$$

where $V_o(u)$ is the object wave field and $\theta(u,t)$ is the phase shift caused by the random phase mask, D is the distance between the object and the receiver planes and x_0 is the transverse coordinate of the receiver, $k=2\pi/\lambda$ is the wave number of the ultrasonic waves and λ is the wave length of the ultrasonic waves. As easily verified, the detected signal $V_s(t)$ given by Eq. (1) is a random time series due

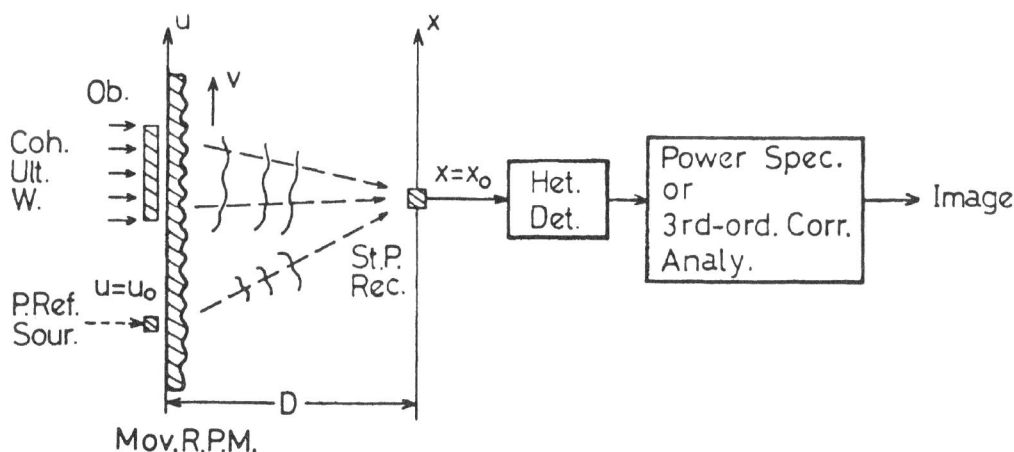

Fig. 2. Fundamental construction of system. Coh. Ult. W., coherent ultrasonic waves; P. Ref. Sour., point reference source at $u=u_0$; Ob., object; Mov. R. P. M., moving random phase mask; v, velocity of the mask; St. P. Rec., stationary point receiver at $x=x_0$.

to the randomness of the function $\theta(u,t)$. In Sec. II-A, we will show the image reconstruction method by means of the power spectrum analysis and in Sec. II-B, the method by means of the 3rd order correlation function analysis.

II-A.   IMAGE RECONSTRUCTION BY POWER SPECTRUM ANALYSIS [1,6]

The power spectrum $S(\omega)$ of the detected signal $V_s(t)$ given by Eq. (1) is defined by

$$S(\omega) = \lim_{T \to \infty} T^{-1} E\left\{ \left| \int_{-T/2}^{T/2} V_s(t) \exp(-j\omega t) dt \right|^2 \right\} , \tag{2}$$

where T is the observation time and E denotes an ensemble average. Here let us assume the following ideal condition for the random phase mask

$$E\{\exp[-j\theta(u_1,t)+j\theta(u_2,t)]\} = \delta(u_1-u_2) . \tag{3}$$

Then on substituting from Eq. (1) in Eq. (2) and by making use of the relation of Eq. (3) and also the relation that $\theta(u,t+\tau)=\theta(u-v\tau,t)$, we have

$$S(\omega) = (\lambda D)^{-2} \iint_{-\infty}^{\infty} V_0\left(u+\frac{v\tau}{2}\right) V_0^*\left(u-\frac{v\tau}{2}\right) \exp\left[-\frac{jk}{D}(x_0-u)v\tau\right] \exp(-j\omega\tau) du d\tau. \tag{4}$$

Let $\hat{V}_0(\nu)$ be the spatial frequency representation of the object wave field $V_0(u)$,

$$\hat{V}_0(\nu) = \int_{-\infty}^{\infty} V_0(u) \exp(-j\nu u) du . \tag{5}$$

Then by inserting the inverse Fourier transform of Eq. (5) in Eq. (4) we obtain

$$S(\omega) = \frac{1}{\lambda D |v|} \left| \frac{1}{2\pi} \int_{-\infty}^{\infty} \hat{V}_0(\nu) \exp\left[j\nu\left(x_0+\frac{D\omega}{kv}\right) - j\frac{D\nu^2}{2k}\right] d\nu \right|^2 . \tag{6}$$

On the other hand, the wave field $V_D(x)$ on the receiver plane when the random phase mask is removed is written as follows:

$$V_D(x) = \frac{\exp(jkD)}{j\lambda D} \int_{-\infty}^{\infty} V_0(u) \exp[(jk/2D)(x-u)^2] du , \tag{7}$$

and in the spatial frequency domain Eq. (7) yields

$$\hat{V}_D(\nu) = -j(\lambda D)^{-1/2} \exp\left(jkD + \frac{j\pi}{4}\right) \exp(-jD\nu^2/2k)\hat{V}_0(\nu) . \tag{8}$$

where $\hat{V}_D(\nu)$ is the Fourier transform of $V_D(x)$. Hence from Eqs. (6) and (8) it is easily verified that the power spectrum $S(\omega)$ of the detected signal is expressed as follows:

$$S(\omega) = |v|^{-1} |V_D(x_0 + \frac{D\omega}{kv})|^2 . \tag{9}$$

Equation (9) means that the power spectrum of the detected signal represents the shadow of the object on the receiver plane, that is, the intensity distribution on the receiver plane when the random phase mask is removed.

Now when the spectrum range of the object $V_0(u)$ is limited as follows:

$$|\nu|_{max} \leq \pi \sqrt{2/\lambda D} , \tag{10}$$

where $|\nu|_{max}$ is the maximum range of the spectrum of the object, then Eq. (6) may be approximated as follows:

$$S(\omega) = (\lambda D |v|)^{-1} |V_0(x_0 + \frac{D\omega}{kv})|^2 . \tag{11}$$

Equation (11) implies that the image of the object can be reconstructed by the power spectrum analysis, provided that the condition of Eq. (10) is satisfied. The condition of Eq. (10) means that when the structures of objects are relatively large and smooth compared to the wave length λ and the distance D between the object and the receiver planes, then the image of the objects can be reconstructed by this method.

II-B.   IMAGE RECONSTRUCTION BY 3RD ORDER CORRELATION FUNCTION ANALYSIS [2]

In the reconstruction method presented in Sec. II-A, the resolution of the system is determined by the condition of Eq. (10), that is, the information of the finer structures of the object than that determined by Eq. (10) can not be obtained. Here image reconstruction by using the 3rd order correlation function analysis is presented. This method puts no constraints on the objects. It also gives exact images of objects. Furthermore, a distinct feature of this method is also discussed. As the full development has been given in Ref. 2, here only the outline of the method is shown.

For the purpose, let us impose a point reference source $\delta(u-u_0)$ near the object as shown by the dotted line in Fig. 2 and let us use an asymmetric random phase mask. The asymmetricity of the random phase mask implies that the random phase shift caused by the mask

has skewness, that is, the utilization of the asymmetric random
phase mask results in the transformation of the spatial distribution
of the wave field of the object into a non Gaussian time series.

Let $\Gamma_3(\tau,0)$ be the 3rd order correlation function of the
detected signal $V_s(t)$, that is,

$$\Gamma_3(\tau,0) = E\{V_s^2(t)V_s(t+\tau)\} \quad . \tag{12}$$

Then we have

$$\Gamma_3(\tau,0) = \iiint\limits_{-\infty}^{\infty} V(u_1)V(u_2)V(u_3)\exp\{(jk/2D)[(x_0-u_1)^2+(x_0-u_2)^2$$
$$+(x_0-u_3)^2]\}E\{\exp[-j\theta(u_1,t+\tau)-j\theta(u_2,t)-j\theta(u_3,t)]\}du_1du_2du_3 \quad , \tag{13}$$

where $V(u)=V_0(u)+\delta(u-u_0)$ and a constant factor is omitted. Here
let us assume the following ideal δ correlated property in the 3rd
sence for the asymmetric random phase mask,

$$E\{\exp[-j\theta(u_1,t)-j\theta(u_2,t)-j\theta(u_3,t)]\} = \delta(u_1-u_2)\delta(u_1-u_3) \quad . \tag{14}$$

As easily verified, the lefthand of Eq. (14) vanishes for symmetric
random phase masks, thus in order to put the δ correlated property
as expressed in Eq. (14), an asymmetric random phase mask which has
skewness is required. By making use of the relation that $\theta(u,t+\tau)=$
$\theta(u-v\tau,t)$, from Eqs. (13) and (14) we obtain

$$\Gamma_3(\tau,0) = \int\limits_{-\infty}^{\infty} V_0(u)V_0^2(u-v\tau)\exp\{(jk/2D)[2(x_0-u+v\tau)^2+(x_0-u)^2]\}du$$
$$+ V_0(u_0+v\tau)\exp\{(jk/2D)[2(x_0-u_0)^2+(x_0-u_0-v\tau)^2]\}$$
$$+ V_0^2(u_0-v\tau)\exp\{(jk/2D)[2(x_0-u_0+v\tau)^2+(x_0-u_0)^2]\}$$
$$+ \delta(v\tau)\exp[(3jk/2D)(x_0-u_0)^2]$$
$$= \Gamma_3^{(1)}(\tau,0) + \Gamma_3^{(2)}(\tau,0) + \Gamma_3^{(3)}(\tau,0) + \Gamma_3^{(4)}(\tau,0) \quad . \tag{15}$$

Hence if the distance between the point reference source and the
object is wider than the object area, the second term $\Gamma_3^{(2)}(\tau,0)$ in
Eq. (15) can be separated from the other terms on the τ axis and the
image of the object can be reconstructed by this term after squaring
operations as follows:

$$\left| \Gamma^{(2)}_3 (\tau,0) \right|^2 = \left| V_o (u_0 + v\tau) \right|^2 . \tag{16}$$

In the above derivations we have assumed ideal cases for both the reference source and the random phase mask. However, neither an ideal point reference source nor ideal δ correlated asymmetric random phase mask may be realized in real situations. Further details about the effects of these factors are given in Ref. 2.

Finally let us discuss a distinct feature of the present system which uses the 3rd order correlation function analysis. As is well known, the 3rd order correlation function of any Gaussian signal vanishes completely. Therefore it can be expected that the system, by using the 3rd order correlation function analysis, may not suffer from additive Gaussian noises which are independent of the signal, such as random background noises in the medium or noises added at the receiver. The reconstructed images which are obtained by the 2nd order correlation analysis suffer from ghost images resulting from the correlation function of Gaussian noises.[2] Considering, for the sake of the central limit theorem, that noises become nealy Gaussian, we can expect that the imaging method using the 3rd order correlation function analysis is a very effective imaging system under fairly noisy circumstances.

III. EXPERIMENTAL CONSTRUCTION AND RESULTS

Experiments are carried out in Water by using ultrasonic waves at 3 MHz. As shown in Fig. 3, slits made of sound absorbers are used as objects and illuminated by coherent ultrasonic waves generated from a transducer. An asymmetric moving random phase mask, a part of an acrylic disk with random surfaces and 270 mm radius, is rotated just behind objects. The moving speed of the disk is almost uniform along the object plane and about 500 mm/sec. The waves scattered by the rotating disk are received by a point-like transducer at a distance of 400 mm from the disk. The received signal is heterodyned in electronic circuits and then processed in an on-lined mini-computer after A/D conversion as shown in Fig. 3. The power spectrum and 3rd and 2nd order correlation functions are calculated in the mini-computer by using FFT algorithm effectively. In calculating the power spectrum, the signal is first Fourier transformed, squared and then smoothed by means of a spectral window and ensemble averagings. A Hamming window is used as the spectral window. Ensemble averagings are taken ten times. The 2nd order correlation function is obtained by inverse Fourier transform of the power spectrum. In calculating the 3rd order correlation function, the cross spectrum of the signal and its square is first calculated. The signal and its square are Fourier transformed and the results are multiplied and then averaged by making use of a Hamming window

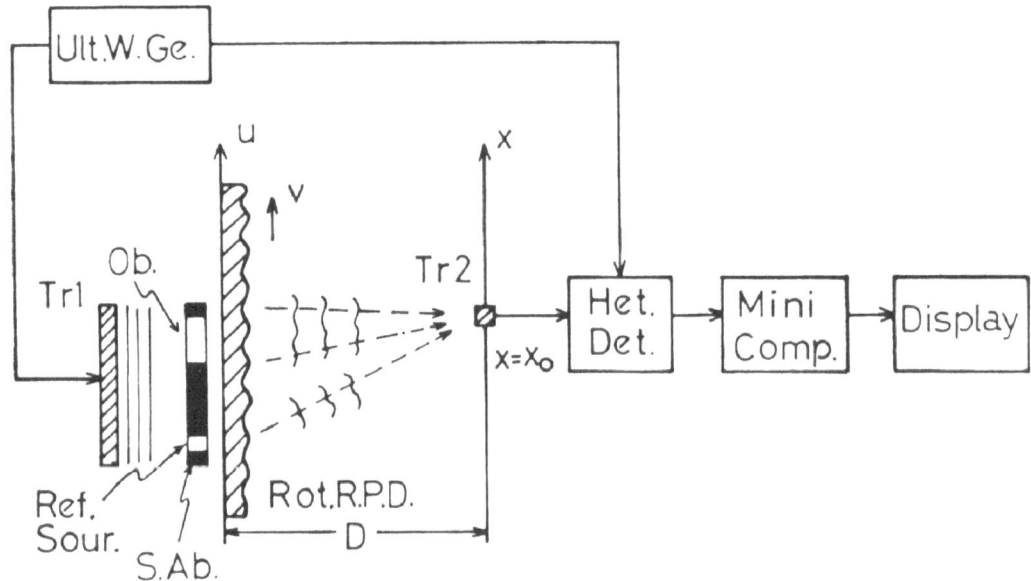

Fig. 3. Experimental construction. Ult. W. Ge., ultrasonic wave
         generator; Tr_1, object illuminating transducer; Ob., object;
         Ref. Sour., reference source; S. Ab., sound absorber;
         Rot. R. P. D., rotating random phase disk made of acryl;
         v, velocity of the random disk along the object plane;
         Tr_2, point-like transducer used as receiver.

and ten times ensemble averagings. Then the 3rd order correlation
function is obtained after the inverse Fourier transformation of
the cross spectrum.

Firstly, images are reconstructed by means of the power spectrum
analysis. The results are displayed in Fig. 4. The solid lines in
Fig. 4-(a),(b) are reconstructed images and the dotted lines in these
figures are the intensity distributions just in front of the disk,
that is, the object to be imaged. The object in Fig. 4-(a) is a
slit-like object and chosen to satisfy the requirements given by
Eq. (10). The reconstructed image in this case agrees well with
the object. On the other hand, the reconstructed image in Fig. 4-(b)
does not agree well with the object because it has finer structures
than that in Fig. 4-(a) and does not satisfy the requirements of
Eq. (10). However, the reconstructed image in Fig. 4-(b) is in good
agreement with the shadow of the object as shown in Fig. 4-(c),
where the solid line indicates the reconstructed image of the object
and the dotted line indicates the shadow of the object.

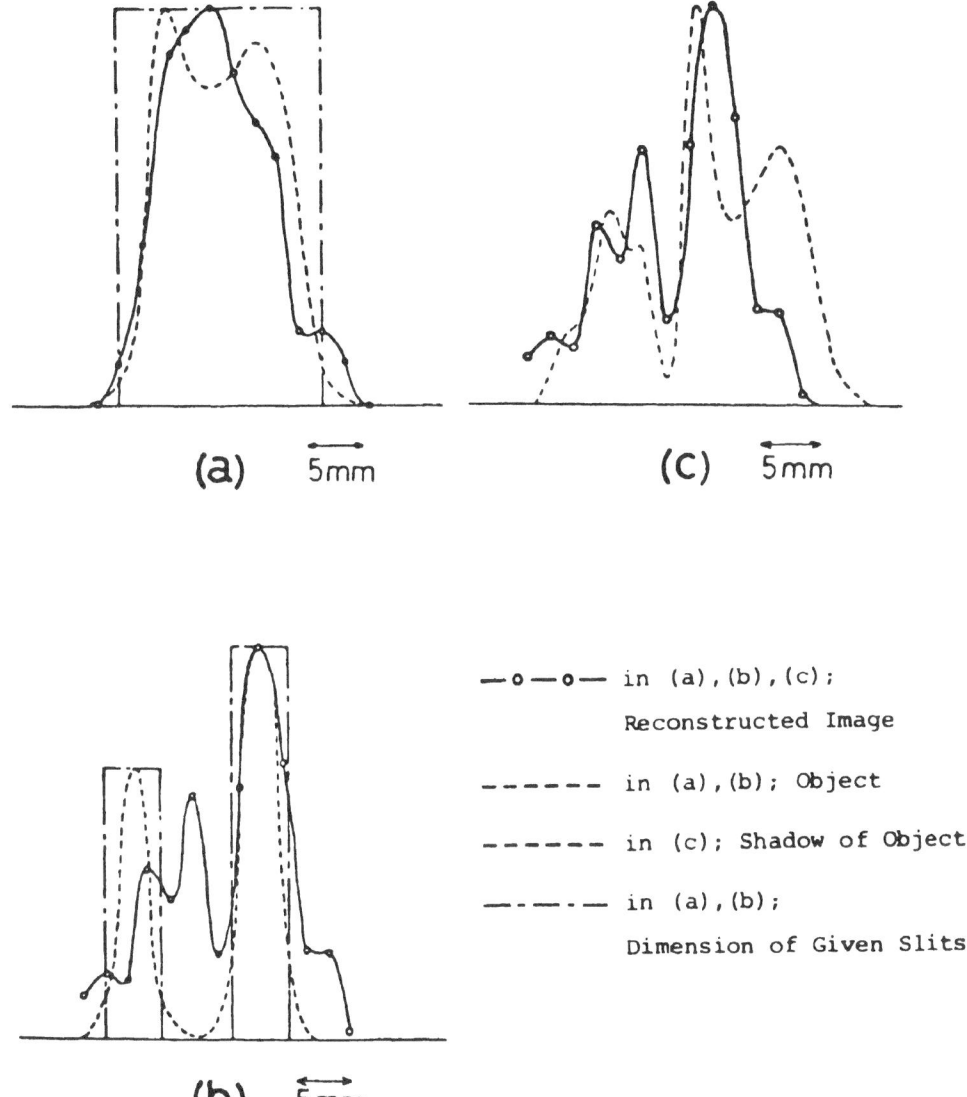

Fig. 4. Images reconstructed by power spectrum analysis.

Secondly, as a reference source, a 2 mm width aperture is placed near the object and illuminated by coherent waves generated by the transducer. The images of objects are reconstructed by means of the 3rd and the 2nd order correlation function analyses. The results are displayed in Fig. 5. The solid line with circles is the image reconstructed by means of the 3rd order correlation function analysis. The solid line with triangles is the image reconstructed by means of the 2nd order correlation function analysis. The broken line is the object to be reconstructed as the image. These results show that higher resolutions can be obtained by the methods of the 3rd and 2nd order correlation function analysis than that of the power spectrum analysis. And comparing the reconstructed images by means of the 3rd and the 2nd order cases, higher resolutions can be achieved in the 3rd order case than the 2nd order case. The second peak of the reconstructed images, which is reconstructed in the upper region on the τ axis, is less suppressed in the 3rd order case than the 2nd order case. Thus these experimental results support the theoretical discussions.

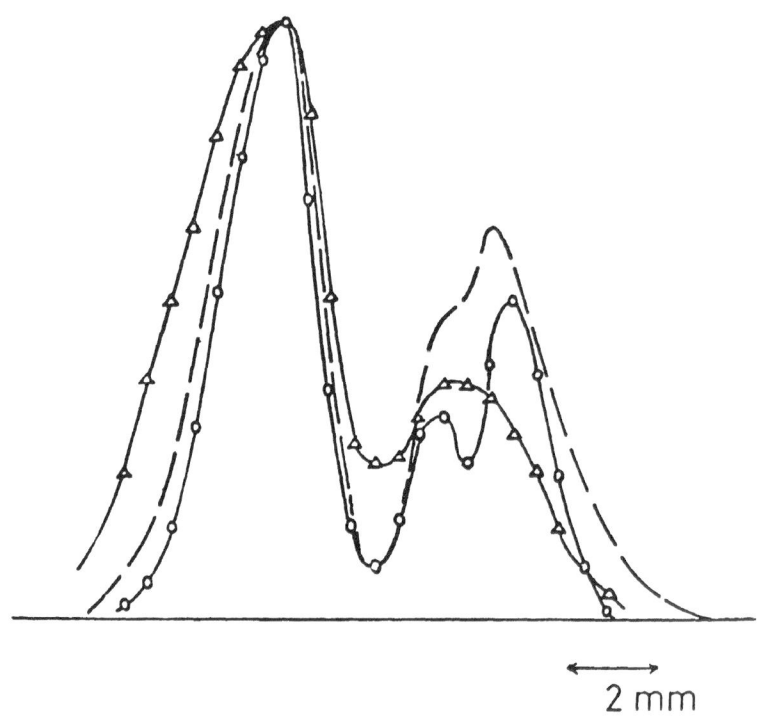

2 mm

Fig. 5.   Images reconstructed by correlation function analyses.
          —o—o— Image by 3rd order correlation function analysis
          —Δ—Δ— Image by 2nd order correlation function analysis
          — — — — — — Object

IV.   A METHOD BY USING A M-SEQUENCE RANDOM MASK [5]

In the imaging method by using the 3rd and the 2nd order correlation function analyses, a point reference source near objects is required. In this section, we show a method by using a M-sequence random mask, which does not require a point reference source, and let us discuss its special features and advantages over the conventional methods.

First, let us consider one-dimensional case. A M-sequence mask consists of two basic elements, giving the relative phase shifts zero and π, and these two elements are arranged along a line on the mask according to a M-sequence. Then denoting the effects of the change of the complex amplitute on the object wave field caused by this mask by $M(u,t)$, the detected signal $V_s(t)$ is given as follows:

$$V_s(t) = \int_{-\infty}^{\infty} V_0(u)M(u,t)\exp[(jk/2D)(x_0-u)^2]du \ , \tag{17}$$

where a constant factor is omitted. Then the correlation function $\Gamma(\tau)$ of the detected signal $V_s(t)$ with electronically generated M-sequence $M(u_0,t)$ is given as follows:

$$\Gamma(\tau) = E\{V_s(t+\tau)M(u_0,t)\}$$
$$= \int_{-\infty}^{\infty} V_0(u)\exp[(jk/2D)(x_0-u)^2]E\{M(u,t+\tau)M(u_0,t)\}du \ , \tag{18}$$

where u_0 is a initial constant. Since a M-sequence has the δ correlated property,

$$E\{M(u,t+\tau)M(u_0,t)\} = \delta(u-v\tau-u_0) \ , \tag{19}$$

it is easily verified from Eqs. (18) and (19) that the image of the object can be reconstructed by squaring $\Gamma(\tau)$ as follows:

$$|\Gamma(\tau)|^2 = |V_0(u_0+v\tau)|^2 \ . \tag{20}$$

This imaging method has several advantages. Firstly, in the conventional method a reference source must be placed near an object apart from the object by the area of the object, that is, a redundant area between the object and the reference source is required, while in the present method by using a M-sequence mask such redundant area is not required and hence object area for the imaging can be extended comparing with the conventional method. Secondly, while the method by using the 3rd order correlation function analysis can remove the effects of only Gaussian noises independent on signal, in the present system, as easily verified, any noises independent on the M-sequence can be removed even in the case when the mean of the noises is not zero because the mean of the M-sequence gives the value of zero.

In the two-dimensional imaging case, a disk on which a M-sequence is partitioned into some groups and each group is arranged on a different concentric circle can be used. By using this M-sequence mask the same time analysis of the detected signal as in the one-dimensional case gives the image of a two-dimensional object separated along the each circle on a different region of the τ axis. Hence after combining these results suitably the image of the two-dimensional object can be reconstructed.

V. CONCLUSIONS

A new ultrasonic imaging system which uses neither scanning of a receiver nor beam forming by a phased array is presented. The fundamental idea is based on the transformation of the spatial distribution of the wave intensity, which is the object to be imaged, into information in the temporal domain by using a moving random phase mask and images of objects are reconstructed by analyses of the random time series, such as power spectral and 3rd and 2nd order correlation function analyses. Theoretical developments and resolution analyses as well as some experimental results clearly demonstrating the theoretical discussions are presented. Finally a method by using a M-sequence random mask is proposed and its special features and advantages over the conventional methods are discussed.

REFERENCES

1. T. Sato, S. Wadaka and J. Ishii, "A New Ultrasonic Imaging System by Using a Moving Random Phase Mask and a Stationary Point Receiver", submitted to J. Acoust. Soc. Am.

2. S. Wadaka and T. Sato, "An Ultrasonic Imaging System by Using a Moving Asymmetric Random Phase Mask and 3rd Order Correlation Function Analysis", submitted to J. Acoust. Soc. Am.

3. Acoustic Imaging, edited by G. Wade (Plenum, New York and London, 1976), Ch. 2, 3, 5 and 11.
4. ibid., Ch. 2, 3, 7, 8 and 11.

5. Our paper to be submitted to J. Acoust. Soc. Am.

6. L. E. Estes and R. Boucher, J. Opt. Soc. Am. 65, 760 (1975).

BISPECTRAL PASSIVE HOLOGRAPHIC IMAGING SYSTEM

Takuso Sato, Kimio Sasaki, Yoichi Nakamura and
Masaharu Nonaka

Faculty of Science and Engineering, Tokyo Institute of
Technology
Nagatsuda, Midori-Ku, Yokohama 227, Japan

ABSTRACT

A new ultrasonic and/or acoustical imaging system, which uses
non-Gaussian random signals and cross- and auto-bispectral analysis
is presented in this paper. It consists of a fixed point receiver
on a hologram plane and a scanning receiver on the same plane. The
non-Gaussian signals radiated from the object and detected by these
receivers are analyzed. By calculating their cross- and auto-
bispectra and taking the ratio between them the holographic
information, that is the distribution of the amplitudes and phases
of a certain wave length is derived from these results over the
plane. The following points will be presented; i) principle, ii)
theoretical developments, iii) computer simulations and the outline
of a prototype of a practical system for the diagnosis of machine
system at audio frequency region. Mechanical noises, such as the
noises from engines of a submarine, can be regarded as non-Gaussian
signals and ambiguous noises surrounding circumstances, such as the
ocean tide, may be regarded as Gaussian noises, so this method is
especially effective when it is used under fairly large additive
Gaussian noises. Because bispectrum of Gaussian noises vanishes
completely they do not disturb the hologram when it is obtained by
the bispectral analysis. Although the fundamental parts of this
method have been submitted in various journals (Refs. 1 and 2),
comprehensive discussions will be given in this article.

I. INTRODUCTION

The passive sonar system has been developed as an ideal means
for ranging or detecting objects without actively emitting any

signal from the detecting system. These kinds of systems are desired especially when the sound from the object is the only available signal for the ranging and the imaging of the object or in the cases where, for some reasons, the emission of a signal from the detecting system is limited.

The signal from the object is, in most cases, of random nature, so the statistical analysis of the signal must be applied to get informations which are required for the imaging or ranging of the object.

Conventional methods of the signal processing in passive sonar systems are based on the power spectral analysis of the signal which is detected at two or three or at most a limited number of points and the location and the range of the object are estimated.[3-8]

To improve the characteristics of these passive sonar systems the following two problems must be solved; i) how to suppress the disturbing noises, and ii) how to get detailed structure of the object.

As for the first problem, the disturbing noises resulting from many independent components may be considered as almost Gaussian noises in as far as the generalized central limit theorem is concerned. Hence, if the Gaussian random signals are suppressed, most part of the noise problem may be solved.

In this paper, we adopt the bispectral analysis of signals[9] for this purpose. The signals from the objects, such as engines or driving systems, are also random, but main components of these signals may be considered to be periodical. Since these objects have rotating or vibrating parts of a fixed period, as the result the signals have fairly large bispectrum. As is well known the bispectrum of any Gaussian signal vanishes completely, so if the desired information is obtained from the bispectral analysis of the detected signal, a noise free passive sonar system may be constructed.

Following in this line, first, we show that the hologram signal over a scanning plane can be obtained from the cross- and auto-bispectral analyses of the detected signals at a fixed receiver and a scanning receiver which are put on this plane.

Then by using this hologram signal, the image of the object, that is the details of the structure of the object[8] is reconstructed through the data manipulation according to the conventional holographic imaging algorism. Thus, the solution to the second problem is given by this holographical imaging.

Moreover, methods of the improvement of the signal-to-noise ratio and more precise estimation of hologram signal from data of limited length are proposed by using various combinations of the three frequency conponents of the bispectrum.

The principle, theoretical developments, and some results of computer simulations which show the effectiveness of the bispectral passive holographic imaging system are presented.

And the outline of a prototype of a system designed for the diagnosis of machine system at audio frequency region is also discussed.

II. PRINCIPLE [1,2]

The schematic diagram of bispectral passive holographic imaging system is shown in Fig. 1. In this figure, without loss of generality, it is considered that M kinds of signals $\{S_i(\rho,t)\}$ are emitted from the objects and can be expressed by

$$S_i(\rho,t)=A_i(\rho)s_i(t) \qquad (i=1,\cdots,M), \tag{1}$$

where $s_i(t)$ and $A_i(\rho)$ are the i-th emitted random signal and the distribution of emission, respectively. And also let $\{s_i(t)\}$ be 3rd-order stationary processes with spectral representation,[10]

$$s_i(t)=\int_{-\infty}^{\infty} \exp(j\omega t)dZ_i(\omega) \qquad (i=1,\cdots,M) \tag{2}$$

Then, the joint 3rd-order or cross-bispectral density function $b_{s:ijk}(\omega_1,\omega_2)$ of $\{s_i(t)\}$ are defined by [9]

$$C_3[dZ_i(\omega_1),dZ_j(\omega_2),dZ_k(\omega_3)=b_{s:ijk}(\omega_1,\omega_2)\delta(\omega_1+\omega_2+\omega_3)d\omega_1 d\omega_2 d\omega_3$$

$$(i,j,k=1,\cdots,M), \tag{3}$$

where C_3 is a 3rd-order cumulant operation and δ is a Dirac's delta function. The surrounding noise $n_s(\rho,t)$ and the additive noises, $n_1(\mathbf{r}_1,t)$ and $n_2(\mathbf{r}_2,t)$, at the two receivers are assumed to be Gaussian and independent with respect to $\{s_i(t)\}$.

Under the above conditions, the received signals at the two receivers can be expressed as follows

$$u_q(\mathbf{r}_q,t)=\int_{\Pi_1} [\Sigma_{i=1}^M S_i\{\rho,t-\frac{d(\rho,\mathbf{r}_q)}{v}\}+n_s\{\rho,t-\frac{d(\rho,\mathbf{r}_q)}{v}\}]d\rho+n_q(\mathbf{r}_q,t),$$

$$=\Sigma_{i=1}^M\int_{-\infty}^{\infty}[\int_{\Pi_1}\frac{G(\omega)A_i(\rho)}{j\lambda d(\rho,\mathbf{r}_q)}\exp\{-jkd(\rho,\mathbf{r}_q)\}d\rho]\exp(j\omega t)dZ_i(\omega)$$

$$+N_q(\mathbf{r}_q,t) \qquad (q=1,2), \tag{4}$$

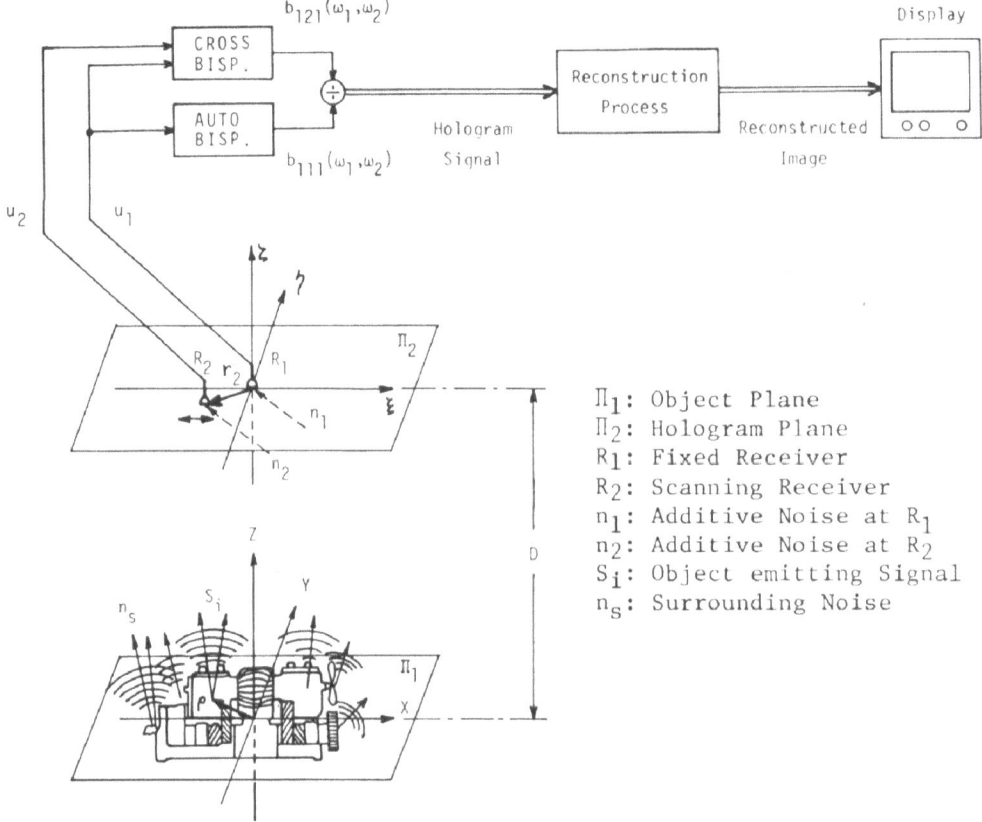

Π_1: Object Plane
Π_2: Hologram Plane
R_1: Fixed Receiver
R_2: Scanning Receiver
n_1: Additive Noise at R_1
n_2: Additive Noise at R_2
S_i: Object emitting Signal
n_s: Surrounding Noise

Fig. 1. Schematic Block Diagram of Bispectral
Passive Holographic Imaging System

where λ and k denote the wave length and the wave number of a wave
with angular frequency ω, v is the wave velocity, $d(\rho, r_q)$ represents
the distance between ρ on Π_1 and r_2 on Π_2, $G(\omega)$ is the amplitude
transmitting factor of the medium, and $N_q(r_q, t)$ is the overall
corrupting Gaussian noise at the receivers and independent with
respect to the object signals.

From the definition of bispectrum, the cross-bispectral density
functions of the received signals are expressed by (See Ref. 2 for
details of the derivation.)

$$b_{u:121}(\omega_1,\omega_2)=\Sigma_{1,m,n=1}^{M}\, b_{s:lmn}(\omega_1,\omega_2)\,\frac{G(\omega_1)G(\omega_2)G(-\omega_1-\omega_2)}{-j^3\lambda_1\lambda_2(\lambda_1^{-1}+\lambda_2^{-1})^{-1}}$$

$$\times\int_{\Pi_1}\frac{A_1(\rho_1)}{d_1}\exp(-jk_1d_1)d\rho_1\cdot\int_{\Pi_1}\frac{A_m(\rho_2)}{d_2'}\exp(-jk_2d_2')d\rho_2$$

$$\times\int_{\Pi_1}\frac{A_n(\rho_3)}{d_3}\exp\{j(k_1+k_2)d_3\}d\rho_3, \qquad (5)$$

where $d_i=d(\rho_i,r_1)$ $(i=1,3)$ and $d_2'=d(\rho_2,r_2)$. Replacing d_2' by $d_2=d(\rho_2,r_1)$ in Eq. (5), the auto-bispectral density function $b_{u:111}(\omega_1,\omega_2)$ of the received signal $u_1(r_1,t)$ at the fixed receiver is obtained directly.

If object emitting signals have different shape of bispectra, we can find a suitable combination of $\hat{\omega}_1$, $\hat{\omega}_2$, $\hat{\omega}_3$ on bispectral plane $\omega_1+\omega_2+\omega_3=0$, at which only the value of $b_{s:mmm}(\hat{\omega}_1,\hat{\omega}_2)$ dominates. At this point only the m-th object which emits $A_m(\rho)s_m(t)$ can be discriminated from the others. Then, the hologram signal for the m-th object can be derived as follows

$$H_m(r_2,\hat{k}_2)=b_{u:121}(\hat{\omega}_1,\hat{\omega}_2)/b_{u:111}(\hat{\omega}_1,\hat{\omega}_2),$$

$$=const.\times\int_{\Pi_1}\frac{A_m(\rho_2)}{j\hat{\lambda}_2d_2'}\exp(-j\hat{k}_2d_2')d\rho_2. \qquad (6)$$

Thus, the hologram signal is obtained. The conventional method of the image reconstruction in acoustical holography [11] can be used directly to obtain the final image. That is, the amplitude distribution of the object is reconstructed by carrying out the following operation

$$|I_m(\rho)|=|\int_{\Pi_2}\frac{H_m(r_2,\hat{k}_2)}{j\hat{\lambda}_2d(r_2,\rho)}\exp\{j\hat{k}_2d(r_2,\rho)\}dr_2|, \qquad (6)$$

where $|I_m(\rho)|$ is the reconstructed image at an image point ρ.

In this way it is shown that the bispectral passive holographic imaging system has the ability to reconstruct images, completely free from any Gaussian noise.

The effectiveness of this system and the outline of the concrete system designed for the diagnosis of a machine system at audio frequency region will be shown in the following sections.

III. RESULTS OF COMPUTER SIMULATIONS [2]

In this section some typical results obtained by computer simulations are presented to approve the principle given in the previous section. Further details are given in Ref. 2.

A. Fundamental Simulation

The simple model used for the simulation is shown in Fig. 2, where a point object p and a point noise n are used as the object. Only a surrounding noise n_s is taken into account.

Fig. 3 shows the characteristics of the object signal[12] and the corrupting noise. At a point (380 Hz, 380 Hz) on bispectral plane the value of bispectra of the signal is about 50 times that of the noise, so the noise may be considered to be Gaussian at this combination of frequencies with respect up to the 3rd-order spectrum. As for the details of bispectral estimation see Ref. 13.

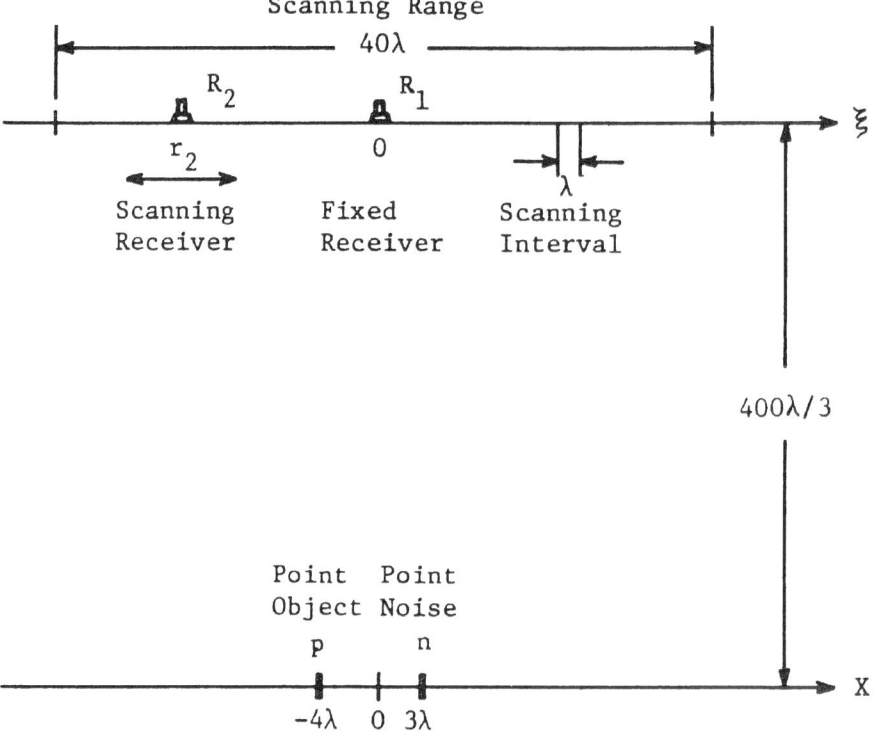

Fig. 2. Schematic Model Used for Simulation.
λ: wave length of a wave with frequency 380 Hz.

Fig. 4 shows the results of image reconstruction by using a wave of 380 Hz. From this result it is seen that the bispectral passive holographic imaging system discriminates the object from the surrounding noise clearly even in the case of small signal-to-noise ratio. While the imaging which uses the power spectral analysis fails to distinguish the object from the corrupting noise. These fundamental results agree with the discussions given in the previous section.

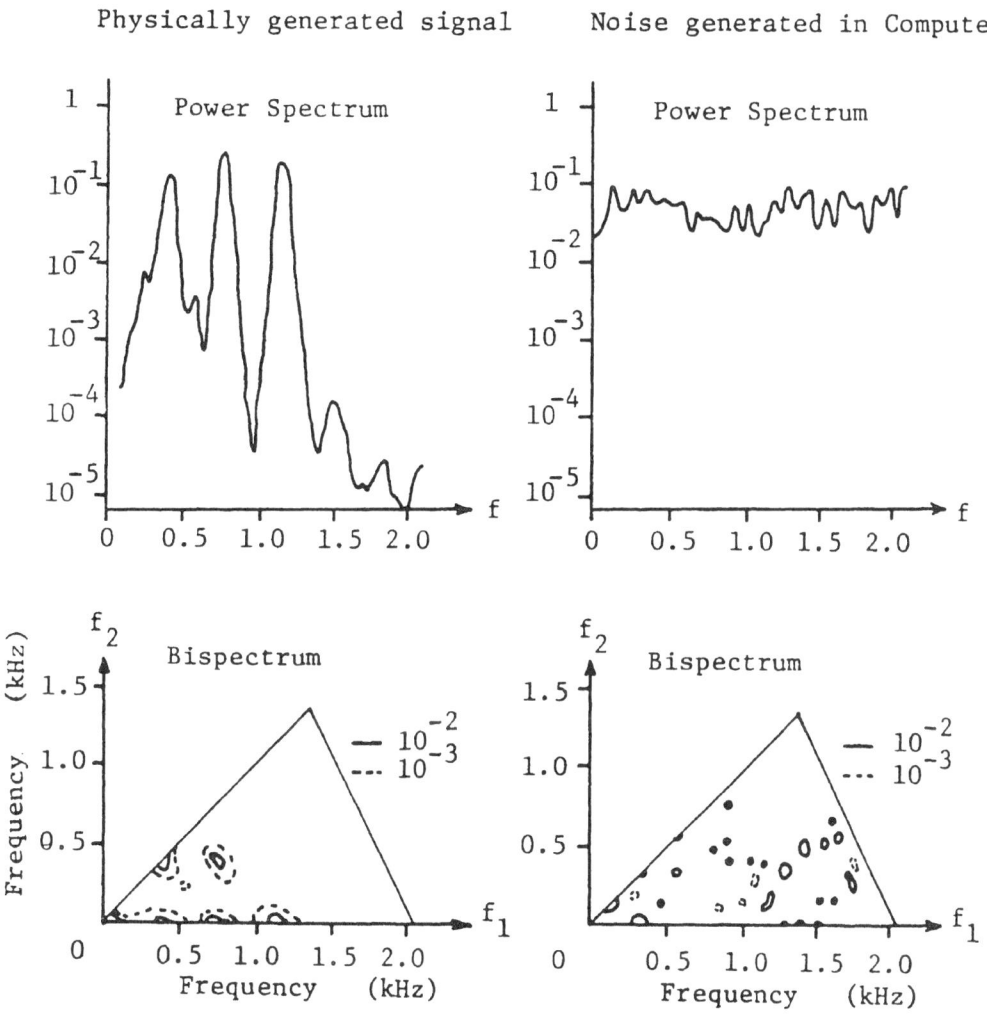

Fig. 3. Characteristics of Used Signal and Noise.
Bispectral values are represented with the
difference of every decimal unit.

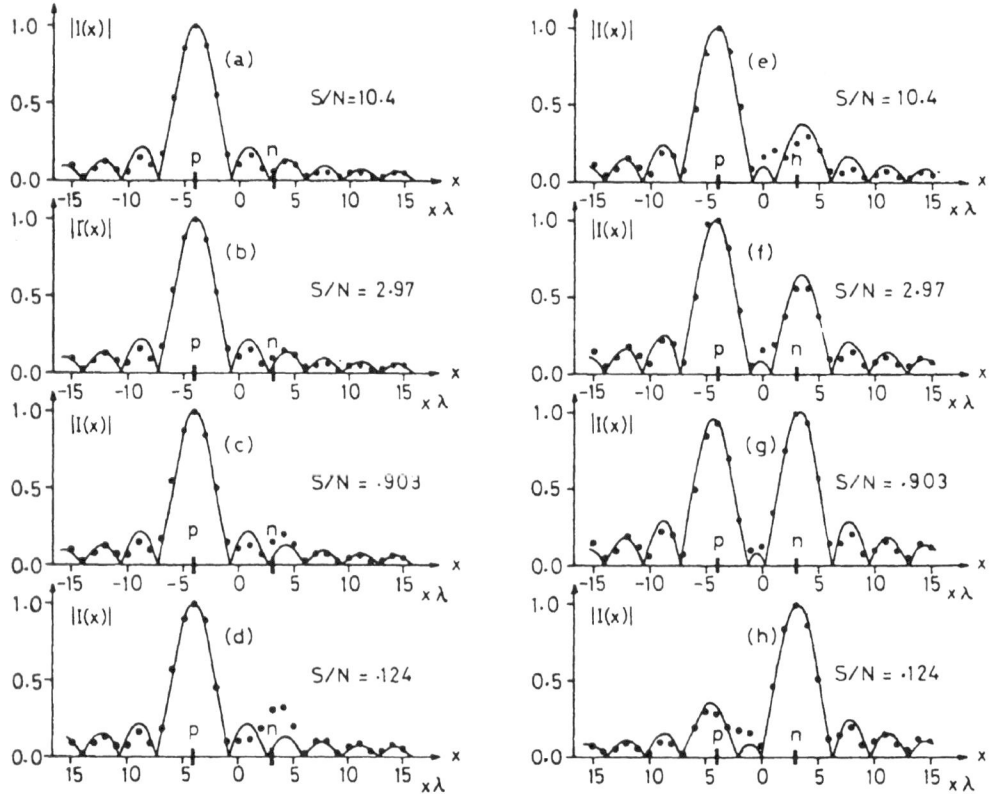

Fig. 4. Results of Image Reconstruction of a Point Object p and a
Point Noise n. (a),$\cdots$,(d), results through bispectral
analysis; (e),$\cdots$,(h), results through power spectral
analysis, where S/N and λ denote the signal-to-noise ratio
and the wave length used for imaging, respectively.

B. Application of Proper Combinations of
Triple Frequency Components

To get high resolution the use of a wave with high frequency
is required, in general. However, noises may have a fairly large
power in high frequency region. Hence, in these cases the passive
imaging system is forced to operate under a condition of a low
signal-to-noise ratio. If the noise is Gaussian, the difficulty is
overcome by using bispectral or higher order spectral analysis as
discussed previously. In some cases, however, noises may not be
necessary to be Gaussian, and this problem must be solved by some
means.

The fundamental idea to overcome this difficulty is the
effective use of broad band nature of the random signal. That is,
for fixed f_3 many combinations of three frequencies which satisfy

$f_1^{(i)}+f_2^{(i)}=f_3$, are used, and the holographic information for a wave
with f_3 is obtained as the results of the utilization of rather
large powers at $f_1^{(i)}$ and $f_2^{(i)}$. Thus, even when the bispectrum at
(f_1,f_2) for certain combinations of f_1, f_2 and f_3, is small compared
to that of the noise, the information of the object can be recovered

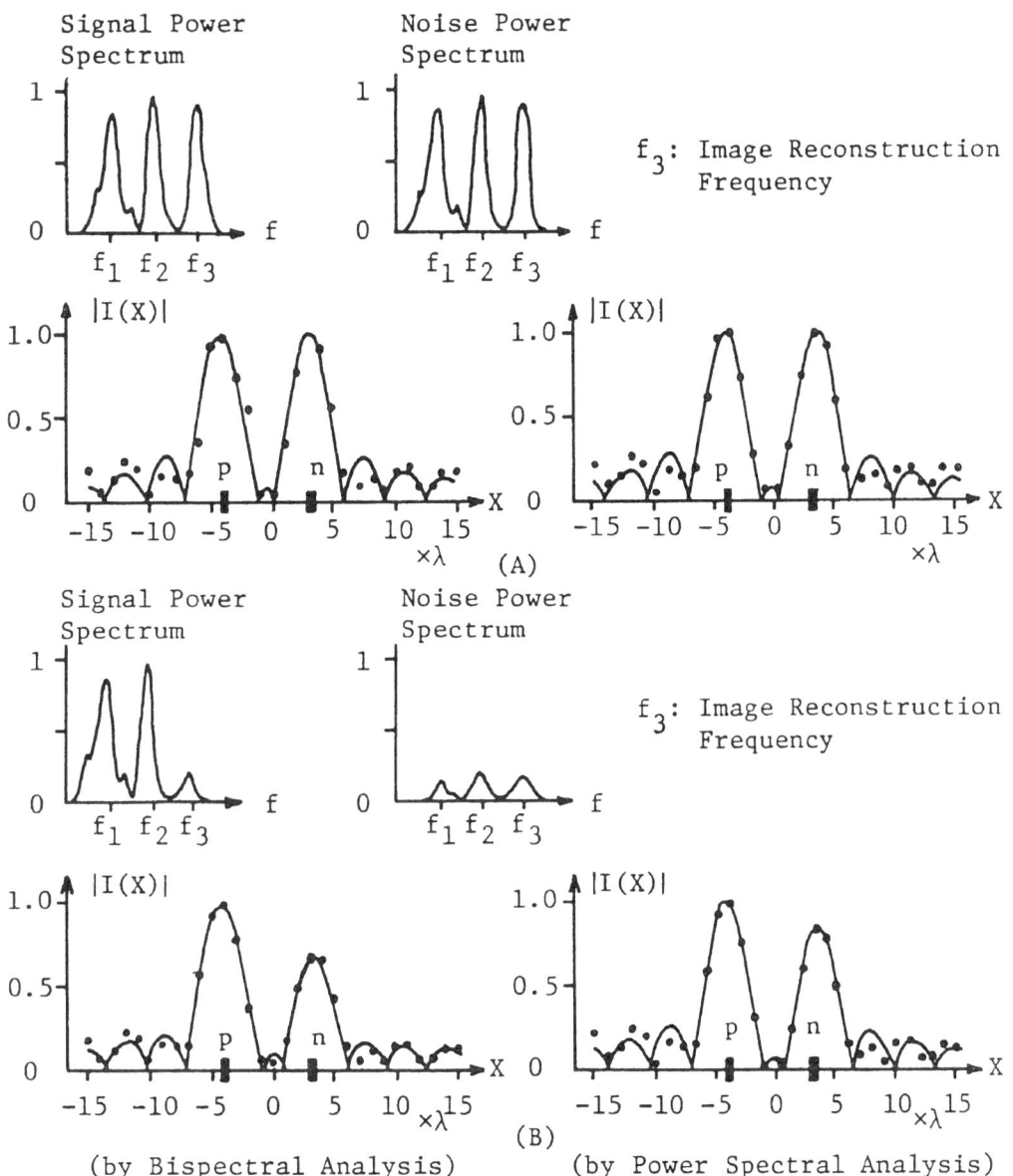

Fig. 5. Results of Image Reconstruction of a Point Object p and
a Point Noise n. (A) and (B) are the results in the case of
small signal-to-noise ratio and in the case of proper
combination of high power frequency components, respectively.

from other proper combinations of $f_1^{(i)}$, $f_2^{(i)}$, and f_3.

The results of a simulation is shown in Fig. 5. In this case the physically generated noise shown in Fig. 3 is employed as the noise and the object signal for the model in Fig. 2. Image reconstruction is carried out by using a wave of 1140 Hz. Fig. 5(B) shows the improvement of the signal-to-noise ratio of the object signal when the bispectral analysis is adopted.

IV. EXPERIMENTAL RESULTS

The bispectral passive holographic imaging system can be applied directly to the diagnosis of noisy mechanical systems. That is, the distribution of the intensity of the noise can be obtained as the image.

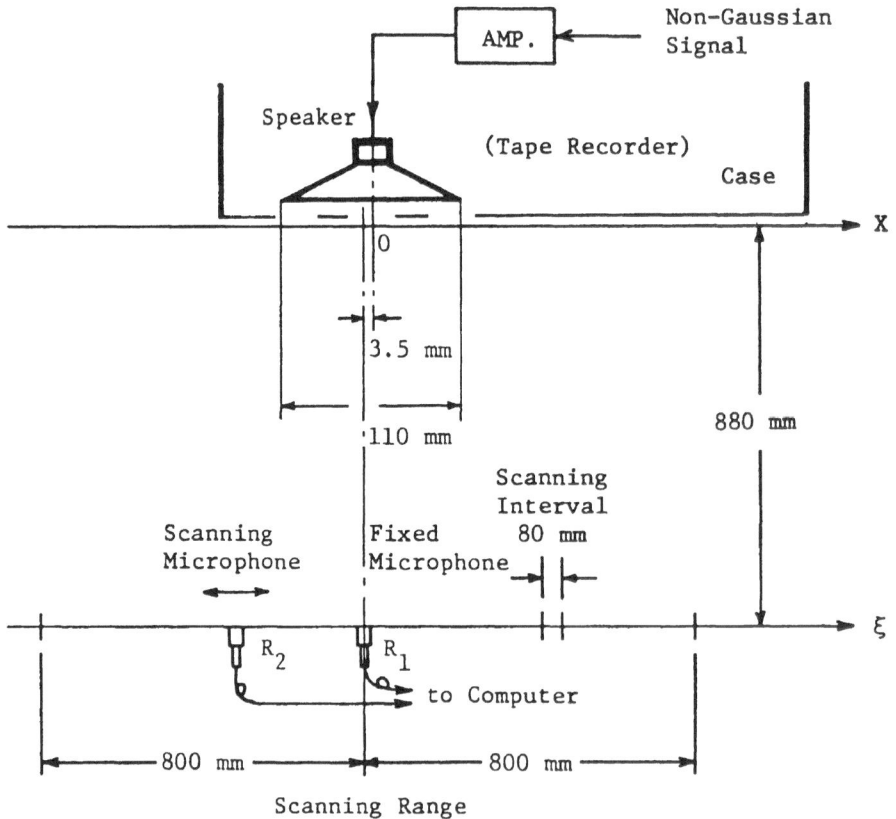

Fig. 6. Concrete Construction for Imaginf of
Noisy Mechanical System.

As a simple application, the sound from a tape recorder is examined. Fig. 6 shows the details of the concrete construction of the system. This system is under construction and only one dimensional image can be reconstructed.

An experimental result is shown in Fig. 7. In this case the non-Gaussian signal is generated through frequency multiplication from the signal used for the simulation. The sound field, about 15 mm away from the case of the recorder, is reconstructed by using a wave of 4394 Hz. Fairly good coincidence between the reconstructed images and the actual field can be observed. These kinds of imaging systems seem promising as a new type of the passive diagnostic system of noisy machines at audio frequency region.

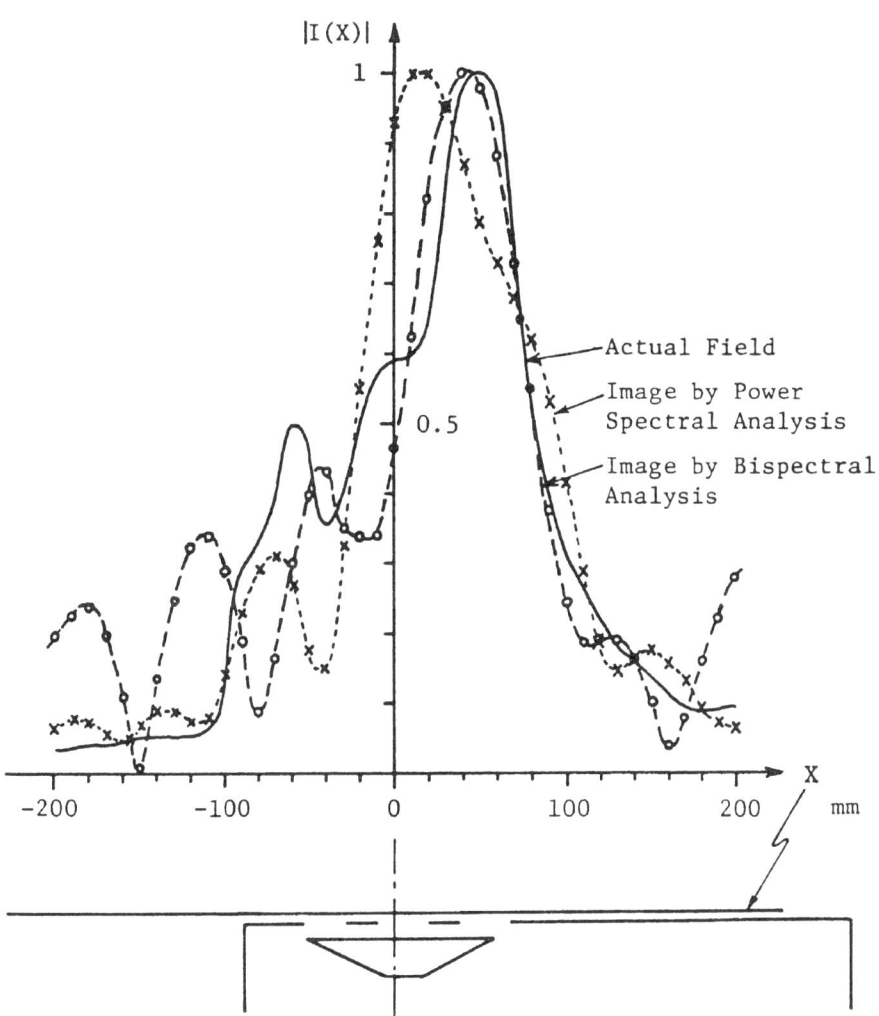

Fig. 7. Experimental Result of Image Reconstruction.

V. CONCLUSIONS

A new type of passive imaging system, which gives the images through the bispectral analysis of the detected random signals, is discussed. Some typical results of computer simulations and the experiment show the effectiveness of the system and its wide applicability to actual fields.

Random signals such as noises emitted from complicated mechanical systems, blowers, gear systems, and seismic waves, may be considered as non-Gaussian signals because of their quasi-periodic characters. In these cases the bispectral passive holographic imaging system may become a powerful means for imaging these object because the system is completely free from additive Gaussian noises as is shown.

The extension to the two dimensional imaging system and its applications for more practical cases are in the process of study.

REFERENCES

1. T. Sato and K. Sasaki, "Bispectral Holography", J. Acoust. Soc. Am. (accepted for publication).

2. K. Sasaki, T. Sato and Y. Nakamura, "Holographic Passive Sonar", IEEE Trans. Sonics and Ultrasonics (accepted for publication).

3. V. H. MacDonald and P. M. Shultheis, J. Acoust. Soc. Am. 46, Part 1, 37 (1968).

4. P. Heimdal and F. Bryn, in Signal Processing, J. W. R. Griffiths and P. L. Stocklin Eds. (Academic Press, New York, 1973) P. 261.

5. W. S. Liggett, Jr, ibid., P. 327.

6. W. R. Hahn and S. A. Tretter, IEEE Trans. Inf. Theory IT-19, 608 (1973).

7. W. R. Hahn, J. Acoust. Soc. Am. 58, 201 (1975).

8. S. Ueha, M. Fujinami, K. Umezawa and J. Tsujiuchi, Appl. Opt. 14, 1478 (1975).

9. D. R. Brillinger and M. Rosenblatt, in Advanced Seminar on Spectral Analysis of Time Series, B. Harris Ed. (wiley, New York, 1967) P. 153.

10. A. Blanc-Lapiere and R. Fortet, Theory of Random Functions (Gordon and Breach, New York, 1965) Vol. II, Chap. 8.

11. B. P. Hildebrand and R. B. Brenden, An Introduction to Acoustical Holography (Plenum, New York, 1972) Chap. 4.

12. K. Kiriyama and T. Sato, Bull. of Tokyo Institute of Technology 112, 9 (1972).

13. K. Sasaki, T. Sato and Y. Yamashita, J. Sound Vib. 40(1), 139 (1975).

COMPUTER SIMULATION OF LINEAR ACOUSTIC DIFFRACTION

John P. Powers

Department of Electrical Engineering
Naval Postgraduate School
Monterey, California 93940

ABSTRACT

Computer-aided acoustical imaging systems and computer simulations of other acoustic imaging techniques frequently require simulation of linear acoustic diffraction of large complex-valued data arrays. Computation efficiency requires the use of fast Fourier transform techniques. This paper compares two Fourier transform formulations of the propagation problem: the Fresnel integral and the spatial frequency domain approach. The following features are compared: restrictions on maximum and minimum propagation distances, sample sizes and number of samples required, adaptability to image processing techniques, and computational requirements.

INTRODUCTION

The use of computers in computer-aided acoustic imaging has become increasingly popular in recent years. The use of the computer in obtaining images by such techniques as backward wave propagation[1] offers such advantages as the elimination of the reconstruction wavelength scaling problem[2] obtained with optical reconstruction techniques, reference-free holography that uses the linear detection properties of piezoclectric transducers, and the possibility of incorporating image enhancement to improve the image obtained. Additionally computer simulation of the holographic process has been a useful tool in studying such novel techniques as phase-only holograms[3] and kinoforms[4]. In most computer-aided imaging techniques it is necessary to mathematically simulate the scalar wave diffraction process. This paper compares two techniques for this simulation, describing their features and some of their advantages and disadvantages.

In representing the acoustic diffraction formulations we will implicitly assume that the propagation medium is linear and homogeneous. We also will require that the resulting diffraction integrals must be amenable to computer solution in a reasonable amount of time which at the present infers that the integrals must be in the form of Fourier transforms so that the speed and efficiency of the Fast Fourier Transform[5] (FFT) can be brought to bear on the problem. The general problem then is: given a complex scalar wave $\underline{U}_i(x_i, y_i, 0)$ at some input plane, find an expression for the wave $\underline{U}_o(x_o, y_o; z)$ at some parallel output plane a distance z away, subject to the wave equations. Two forms of the solution incorporate the Fourier transform and will be considered after a short review of the features of the analog Fourier transform and the discrete Fourier transform (DFT).

The two dimensional analog Fourier transform is defined by the relationships

$$\underline{U}(x,y) = \int\int_{-\infty}^{\infty} \underline{A}(u,v) e^{j 2\pi (ux+vy)} du\,dv = \mathcal{F}^{-1}\{\underline{A}(u,v)\} \tag{1}$$

$$\underline{A}(u,v) = \int\int_{-\infty}^{\infty} \underline{U}(x,y) e^{-j 2\pi (ux+vy)} dx\,dy = \mathcal{F}\{\underline{U}(x,y)\} \tag{2}$$

where $\underline{U}(x,y)$ is the complex function in the space domain; $\underline{A}(u,v)$ is the complex transform in the spatial frequency domain; u,v are spatial frequencies (dimensions of cycles/meter); and $\mathcal{F}, \mathcal{F}^{-1}$ are symbolic operators for the transform and inverse transform operations respectively.

The discrete version of the Fourier transform (assuming an equal number of samples and sample spacing in both dimensions) is given by

$$\underline{f}(ma,na) = \frac{1}{N^2} \sum_{k=0}^{N-1} \sum_{\ell=0}^{N-1} \underline{A}(k\Omega, \ell\Omega) e^{j 2\pi \left[(ma \cdot k\Omega)+(na \cdot \ell\Omega)\right]}$$

$$= F^{-1}\{\underline{A}(k\Omega, \ell\Omega)\} \tag{3}$$

$$\underline{A}(k\Omega, \ell\Omega) = \sum_{m=0}^{N-1} \sum_{n=0}^{N-1} f(ma,na) e^{-j 2\pi \left[(ma \cdot k\Omega)+(na \cdot \ell\Omega)\right]}$$

$$= F\{f(ma,na)\} \tag{4}$$

where $0 \leq m, n \leq N-1$;

$0 \leq k, \ell \leq N-1$;

$\underline{f}$ (ma,na) is a complex valued sequence of samples in the
    space domain;

$A(k\Omega,\ell\Omega)$ is a complex valued sequence of samples in the
    spatial frequency domain;

N is the total number of samples in one dimension in the
    space or frequency domain (N x N total sample values);

a is the sample spacing in the space domain;

Ω is the sample spacing in the spatial frequency domain
    (and is equal to 1/Na);

F,F^{-1} are symbolic operators for the discrete Fourier
    transform and the inverse transform operation respec-
    tively.

The fast Fourier transform (FFT) is an efficient algorithm that uses
symmetry properties to compute this discrete transform. The efficiency of this algorithm for large N(>8) makes it the only practical
method of processing large two-dimensional complex valued data arrays, as is required in computer-aided acoustic imaging. Several
properties of the DFT are mentioned here as they have important re-
percussions later.

1. As mentioned above, if the sample size in the space domain
    is a, then the sample size in the spatial frequency domain
    is 1/Na where N is total number of samples in one dimension.

2. There are N x N samples in the space domain covering a
    region Na x Na; there are also N x N samples in the fre-
    quency domain covering a region 1/a x 1/a.

3. The DFT assumes that the input sequence is periodic in both
    the x and y dimensions. Hence the input is considered an
    infinite two-dimensional periodic array (with period Na).
    Similarly the inverse DFT requires that the sequence in the
    spatial frequency domain also be periodic in both dimen-
    sions (with a period of 1/a).

4. The scaling factor of $1/N^2$ in the inverse transform should
    be noted to ensure computational accuracy.

5. The wave fields in diffraction patterns are usually centered
    on the propagation axis (as in Fig 1a) to take full advan-
    tage of symmetry. The usual DFT algorithm however usually
    works on a wave-field that lies in the first quadrant and
    produces the spectrum also in the first quadrant with the
    (0,0) frequency component at the origin (as in Fig. 1b). In
    order to apply the usual DFT to the centered wave without
    getting a linear phase shift in the transform domain that
    accompanies simple translation, a data shuffle [10] can be
    used. Based on the assumed periodicity of the input wave

DFT

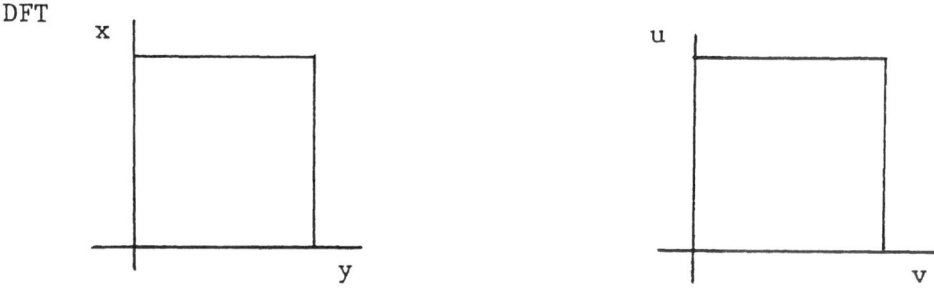

Input and output waves

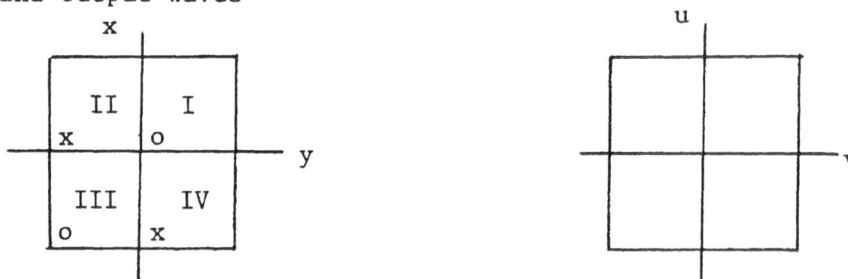

Figure 1   a) Geometric arrangement of input data and transform
required by most discrete Fourier transform algorithms

b) Geometrical arrangement of input data and transform
desired for diffraction problems. (Quadrant numbers
and data location marking refer to data shuffle described in text.)

and the resulting symmetry, this shuffle applied both before and after taking any transform or inverse transform
allows one to work with waves that are centered on the
axis and FFT routines that work only in the first quadrant
The shuffle routine exchanges the data in quadrants II and
IV (see Fig 1b) and the data in quadrants I and III. (The
alignment of the shuffle is indicated by the fact that the
data of position x in quadrant II of Fig 1b is exchanged
with the data of position x in quadrant IV. Similarly the
data of the positions marked by the o's of quadrants I and
III are also exchanged. Since this data shuffle is a mere
exchange of data locations, the additional computation time
is minimal except for the largest of arrays.

THE FRESNEL INTEGRAL

The first form of the solution of the propagation problem that
incorporates the Fourier transform representation is the Fresnel
integral[6]. Using the notation of the discrete Fourier transform
this expression is:

$$\underline{U}_o(k\Delta x_o, \ell\Delta y_o) = \frac{e^{j\frac{2\pi z}{\lambda}}}{j\lambda z} \ e^{j2\pi\left[(k\Delta x_o)^2 + (\ell\Delta y_o)^2\right]}$$

$$\cdot F\left\{\underline{U}_i(ma, na) e^{j\frac{\pi}{\lambda z^2}\left[(ma)^2 + (na)^2\right]}\right\} \tag{5}$$

$$k\Omega = k\Delta x_o$$
$$\ell\Omega = \ell\Delta \chi_o$$

where $\underline{U}_i(ma, na)$ is the sampled input function (sample spacing
= $\overline{a}$);

z is the propagation distance;
Δx_o, Δy_o are sample spacings of the output wave and are each
equal to $\lambda z/Na$;
$\underline{U}_o(k\Delta x_o, \ell\Delta y_o)$ is the sampled value of the output wave at the
plane a distance z from the plane of the input wave.

This expression for $\underline{U}(k\Delta x_o, \ell\Delta y_o)$ is valid only for propagation distances such that

$$z^3 >> \pi N^4 (\Delta x_o - a)^4 |16\lambda \tag{6}$$

The important properties of this formulation of the diffraction integrals are noted below.

1. There is a limitation on the minimum propagation distance
    due to the inequality of Eq. 6. There is no limitation on
    the maximum propagation distance; in fact the Fresnel inte-
    gral becomes the much easier to solve Fraunhofer diffraction
    integral[6] at very large propagation distances.

2. The sample spacing of the output becomes larger with in-
    creasing propagation distance since $\Delta x_0 = \lambda z / Na$. This is
    helpful for diverging waves since the sample spacing spreads
    at the same rate as the diverging wave ensuring complete
    coverage of the diverging beam with a minimum number of
    sample points.

3. The diffraction operation requires N^2 complex multiplica-
    tions, one data shuffle, one FFT operation, another data
    shuffle, and another N^2 complex multiplications.

4. The required sample spacing is determined by frequency
    aliasing considerations[7]. Since the exact derivation of
    the number of samples depends on the object, we choose a
    representative one-dimensional object, a slit of width 2a,
    and find an estimate of the sample number as a rough guide-
    line. The object is a slit of width 2a in a region 2W wide
    (see Fig 2a); the transform of this object is 2a since 2au
    (see Fig 2b). The spacing in the frequency domain will be
    $\Delta u = 1/2W$ and, if there are N samples, the maximum fre-
    quency will be

$$\text{Umax} = \frac{N}{2} \Delta u = \frac{N}{4W} \qquad (7)$$

The energy contained in the frequencies above Umax will be aliased
back into the frequencies below Umax. The fraction, ε, of the total
energy that is aliased is

$$\varepsilon = \frac{1}{a} \int_{\text{Umax}}^{\infty} |A(u)|^2 du \simeq \frac{2W}{\pi^2 Na} \qquad (8)$$

and hence the sample width is

$$\Delta x = \frac{2W}{N} = \pi^2 a\varepsilon \qquad (9)$$

and the number of samples is

$$N = \frac{2W}{\pi^2 a\varepsilon} \qquad (10)$$

Typically $\varepsilon \sim$ 5% or 10% is used. Hence the amount of aliasing tol-

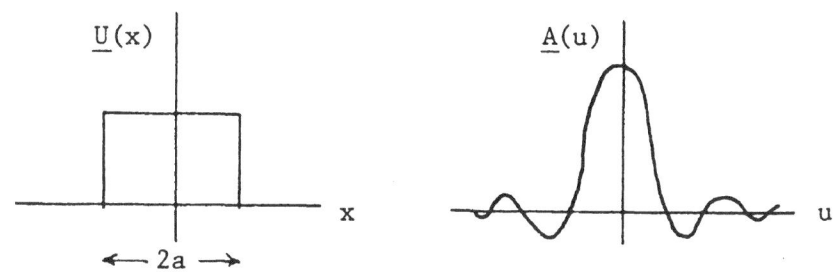

Figure 2   a) Hypothetical one dimensional object (a slit)
           b) Transform of object

erated determines the sample spacing and the number of samples. As
indicated previously this result is based on a particular input object (i.e. a slit). Objects with smaller scale discontinuities or
perturbations might require finer sample spacing while objects with
more gradual discontinuities can be analyzed with coarser sampling.
However the results of Eqs. 9 and 10 give approximate values for
sample spacing and total number of samples.

SPATIAL FREQUENCY DOMAIN APPROACH

The spatial frequency domain solution[6] to the diffraction problem relates the transform of the output wave to the transform of the
input wave by a simple complex multiplication:

$$\underline{A}_o(k\Omega,\ell\Omega;z) = \underline{A}_i(k\Omega,\ell\Omega;0)\,e^{j\frac{2\pi z}{\lambda}\sqrt{1-(\lambda k\Omega)^2-(\lambda k\Omega)^2}} \tag{11}$$

Realizing that the square root expression is imaginary for $(k\Omega)^2 +
(\ell\Omega)^2 \geq 1/\lambda^2$ and that the contribution from these terms (the "evanes-
cent waves") will be negligible if the propagation distance is more
than several wavelengths we can simplify this expression to

$$\underline{A}_o{}^{\cdot}(k\Omega,\ell\Omega;z) = \begin{cases} \underline{A}_i(k\Omega,\ell\Omega;0)\,e^{j\frac{2\pi z}{\lambda}\sqrt{1-(\lambda k\Omega)^2-(\lambda \ell\Omega)^2}} \\ \qquad\qquad\qquad \text{when } (k\Omega)^2+(\ell\Omega)^2 \leq \frac{1}{\lambda^2} \quad (12) \\ \\ 0 \qquad\qquad\qquad \text{when } (k\Omega)^2+(\ell\Omega)^2 \geq \frac{1}{\lambda^2} \end{cases}$$

where

$$\underline{A}_i(k\Omega, \ell\Omega; o) = F\{\underline{U}_i(ma,na)\};$$

$$\underline{A}_o(k\Omega, \ell\Omega; z) = F\{U_o(ma,na)\};$$

a is the spatial sample spacing;

Ω is the frequency domain sample spacing (equals $1/Na$).

The properties of this frequency domain appraoch are as follows:

1. The sample spacing in the output plane is the same as that of the input wave x ($\Delta x_o = \Delta x_i = a$). For a diverging wave we would need many more samples in the output plane to adequately describe the wave because of its larger size, so we must either have numerous zero valued samples at the input plane to adequately cover the output wave or we must restrict our coverage of the output wave to only a small portion of its breadth. Fortunately a remedy has been found to this dilemma by Sziklas and Siegman.[7] Reference 7 presents a wave transformation that converts a diverging wave diffraction problem into a collimated wave problem. The collimated wave problem is adequately handled in the frequency domain approach by the coordinate systems having equal spacings in the input and output planes. The solution to the collimated beam problem may then be used to easily find the solution to the diverging beam problem. The net effect is to obtain an effective output plane sample spacing that expands with propagation distance so that a conservative number of sample points can adequately describe both the input wave and the output wave.

2. Because the DFT assumes that the input wave samples are repeated in a periodic two-dimensional array, at some propagation distance L, the waves from the other "objects" will overlap the wave from the original object, thereby limiting the maximum propagation distance for which the frequency domain approach can be used. Figure 3 and the following analysis uses the one-dimensional slit as an example of this effect and an estimate of the maximum propagation distance. Using an approximation to the Fresnel integral for this specific object, Sziklas and Siegman[7] show that allowing ϵ_1% of the total wave energy in the overlapping fields at a propagation distance L requires a guard band or region of zero valued samples (as in Fig. 3) of value

$$G \geq 1 + \frac{L\lambda}{2\pi^2 a\epsilon_1} \tag{13}$$

This equation can also be used to find the maximum propagation distance L, given an object with a certain guard band value G and an allowed amount of energy overlap (e.g. 5%). Again this result is based on the slit object and would have to be increased for objects

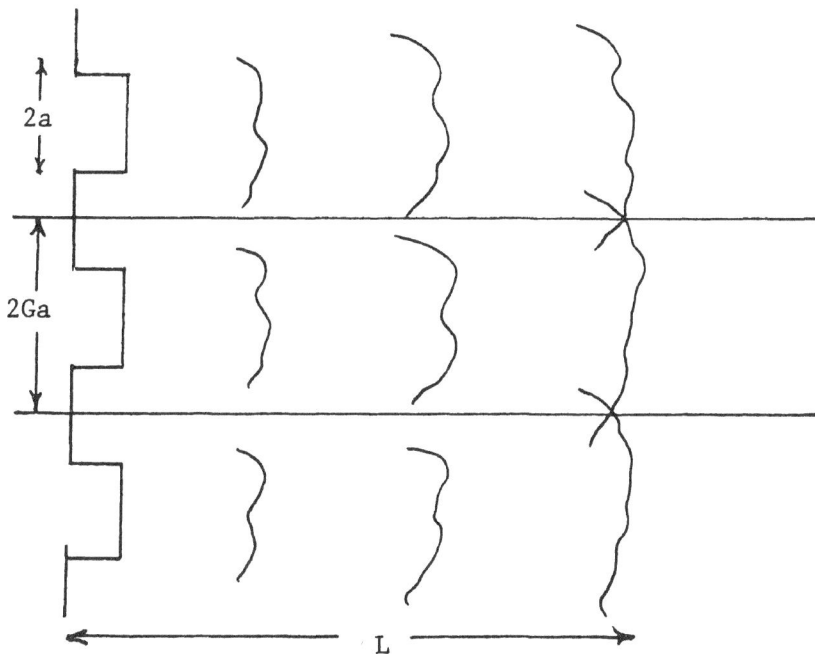

Figure 3 Hypothetical one dimensional slit object and its
          nearest periodic neighbors. At distance L the
          diffracted wavefronts are significantly over-
          lapped.

with smaller features that would have to be resolved or decreased
for objects with more tapered features than a slit.

      3. It is noted from Eq. 12 that the output frequency spec-
trum is band-limited i.e. all frequency samples lying outside of a
circle in the frequency domain with a radius of $1/\lambda$ are equal to
zero (for propagation distances longer than several wavelengths).
Hence only those samples of the input wave spectrum that lie within
this same circle must be taken. This leads to a determination of
the optimum spatial sample spacing and the fact that there is no
frequency aliasing in the spatial frequency approach to the diffraction problem. Rather than restricting our frequency samples to
those lying within the circle of radius $1/\lambda$ it is geometrically
simpler to consider those lying in the rectangle $|u|<1/\lambda$ and $|v|<1/\lambda$
(as in Fig 4). This leads to an oversampling by 12% of the absolute
minimum number of samples.

Choosing this sampling limit gives a maximum frequency in the

$$N = \frac{2Ga}{\Delta x} = \frac{4\,Ga}{\lambda} \tag{18}$$

It should be noted that many FFT routines require that the number of samples be an integral power of 2. Hence the final value of N may be that power of two above or below the calculated value of Eq. 18. If larger the field will be oversampled (with a resulting loss of computing efficiency and longer running times) or undersampled (with shorter running time but a loss of some resolution).

5. Another property of working in the frequency domain to handle the diffraction problem is that this approach is easily amenable to frequency domain image processing techniques[8]such as Weiner filtering, edge enhancement, deconvolution of the point spread function of the receiver, etc.). Here the spectrum of the diffracted wave can be manipulated by multiplication with a filter function:

$$\underline{A}_0'(u,v) = \underline{A}i(u,v)\ \underline{H}\ prop(u,v)\ \underline{H}\ filter(u,v) \tag{19}$$

where

   $\underline{A}_0'(u,v)$ is the spectrum of the processed output wave;

   $\underline{H}\ prop(u,v) = A_0(u,v)\,\big|\,Ai(u,v)$ is the "transfer function" for linear scalar diffraction and is found by dividing Eq. 12 by $Ai(u,v)$;

   $\underline{H}\ filter(u,v)$ is the filter function to perform the desired operation.[9]

Since one of the primary advantages of computer-aided imaging is the flexibility and capability to enhance the image and extract information, the ease of incorporating this class of frequency domain operations is a major advantage of the frequency domain approach.

SUMMARY

Table I summarizes the relative advantages and disadvantages of the two diffraction approaches. With an awareness of these strengths and weaknesses the researcher attempting computer-aided imaging will be able to choose the technique most suitable for his application.

ACKNOWLEDGEMENTS

This research was supported by the Foundation Research Program of the Naval Postgraduate School and the U.S. - France Scientific Exchange program administered by the National Science Foundation and the Centre National de la Recherche Scientifique. The author would also like to acknowledge the hospitality of Professor Pierre Alais and his colleagues of the Laboratoire de Mecanique Physique of the University of Paris VI during his work there.

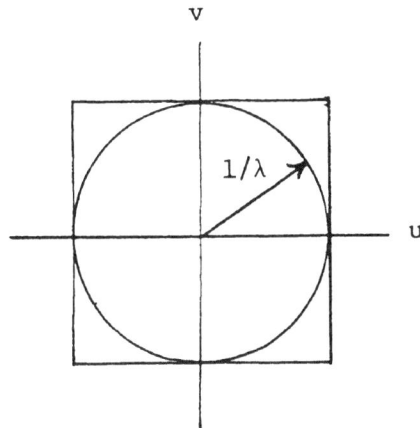

Figure 4   Frequency domain representation of bandlimited
           propagation showing circular limit (radius
           equals $1/\lambda$). Circumscribed rectangle shows
           geometrically simpler bandlimiting (with 12%
           oversampling).

U direction of

$$u_{max} = \frac{N}{2} \Delta U = \frac{1}{\lambda} \tag{14}$$

where Δu is the spatial frequency sample spacing. Hence,

$$\Delta u = \frac{2}{N\lambda} \tag{15}$$

and is also given by

$$\Delta u = \frac{1}{N \Delta x} \tag{16}$$

where Δx is the spatial sample spacing. Therefore

$$\Delta x = \frac{\lambda}{2} \tag{17}$$

is the optimum sample spacing of the object. It is noted that this
sample spacing is that required to give resolution of $\lambda/2$, the "diffraction limited" resolution.

     4. Knowing the required quard band size and the optimum
sample spacing from properties 2 and 3 above it is now possible to
compute the total number of samples by dividing the total object
width (including the guard band) by the optimum spacing:

TABLE I Summary of comparison between diffraction techniques

FRESNEL INTEGRAL	FREQUENCY DOMAIN APPROACH
1. Expanding sample spacing for diverging wave.	1. Expanding sample spacing can be made to occur by wave transformation.
2. Limited minimum diffraction distance.	2. Unlimited minimum diffraction distance.
3. Unlimited maximum diffraction distance.	3. Requires large guard band (and more samples) for longer diffraction distances.
4. Does not predict diffraction limited diffraction.	4. Correctly predicts diffraction limited resolution.
5. Some frequency aliasing.	5. No frequency aliasing due to bandlimited nature.
5. Image processing is separate operation.	6. Frequency domain filtering techniques are easily incorporated.
7. Requires two N^2 complex multiplications, two data shuffles, and one FFT operation.	7. Requires N^2 multiplications, four data shuffles, and two FFT operations.

REFERENCES

1. M.M. Sondhi, "Reconstruction of objects from their sound diffraction patterns", J. of the Acoustical Society of America, 46 (5):1158-1164, 1969.

2. F.L. Thurstone and A.M. Sherwood, "Three dimensional visualization using acoustical fields", Acoustical Holography, Vol 3, A.F. Metherell, Ed., Plenum Press, New York, pp. 317-331, 1971.

3. J. Powers, J. Landry and G. Wade, "Computed reconstructions from phase-only and amplitude-only holograms", Acoustical Holography, Vol 2, A.F. Metherell and L. Larmore, Ed., Plenum Press, New York, pp. 185-202, 1970.

4. L. B. Lesem, P.M. Hirsch and J.A. Jordan, Jr., "The kinoform: a new wavefront reconstruction device", IBM Journal of Research and Development, 13:150, 1969.

5. E. O. Brigham, The Fast Fourier Transform, Prentice Hall, Englewood Cliffs, New Jersey, 1974.

6. J. W. Goodman, Introduction to Fourier Optics, Chapts 3 and 4, McGray-Hill, New York, 1968.

7. E. A. Sziklas and A. E. Siegman, "Mode calculations in unstable resonators with flowing saturable gain. 2:Fast Fourier transform method", Applied Optics, 14(18):1874-1889, 1975.

8. M. Takagi, et al, "Image enhancement of acoustic images", Acoustical Holography, Vol 5, P.S. Green, Ed. Plenum Press, New York, pp. 541-550, 1974.

9. H. Andrews, Computer Techniques in Image Processing, Academic Press, New York, 1970.

10. D. E. Mueller, "A computerized acoustic imaging technique incorporating automatic object recognition", Engineer's degree thesis (unpublished), Naval Postgraduate School, Monterey, California, 1973.

ON-LINE INTERACTIVE COMPUTER PROCESSING OF ACOUSTIC IMAGES

C.H. Lee

Department of Electrical and Computer Science
Syracuse University
Syracuse, New York 13210

G. Heidbreder, G. Wade, A. Coello-Vera

Department of Electrical Engineering and Computer Science
University of California
Santa Barbara, California 93106

ABSTRACT

An interactive on-line computer image processing system designed for use in acoustic image restoration and enhancement has been developed. Signals from a Bragg-diffraction ultrasonic imager have been accessed on-line by a minicomputer and digitized. Signal and noise can be measured quantitatively, and digital restoration and enhancement techniques can be employed to minimize the effects of the different kinds of noise which influence image quality.

Systematic noise has been removed by employing two different approaches: 1) band rejection filtering and 2) subtractive background compensation. Results of their relative performance will be presented in this paper. Much of this noise is time-invariant and stems from the background insonification of the sound cell. Because of its time-invariant nature, time averaging is of little or no value, as has been shown in experiment.

Speckle noise, which is present in all coherent imaging systems, has been effectively reduced by neighborhood averaging. A similar type of interference phenomenon, also common to all coherent imaging systems, is known as "ringing" and has been studied using this system. Ringing, due to sharp cutoffs in the spatial frequency domain

of the images, has been significantly reduced by using windowing
techniques.

INTRODUCTION

An interactive, digital image-processing system of relatively
low cost has been designed and implemented. This system operates
on acoustic images that are accessed on-line. The system is based
on a System Engineering Laboratory (SEL) 810B minicomputer for which
interactive, graphic software was used. The software is of the
Culler-Fried type and was developed by Retz(1). With this system,
we have operated on images obtained from our Bragg-diffraction
imager. A diagram describing the entire system is shown in Fig. 1.

The processed images can be displayed in several different ways.
A Tektronix 611 Storage Oscilloscope or a TV monitor are available.
Either is controlled by means of a keyboard console. The TV camera
is connected to the computer through scan converter to accomodate
the different data rates of the TV system and the computer. A third
type of display is one which shows half-tone images on a plasma display panel which is connected to a Modem interface of the minicomputer system. More details about the hardware and software of the computer processing portion of the system can be found in references
(2), (3), (4) and (5).

The acoustic images were obtained from a Bragg-diffraction
imaging system. This kind of imaging system has been described
extensively in the literature, (6), (7) and (8), and we shall not
delve into it here in any detail. The Bragg-diffraction system
converts an acoustic absorption pattern of an object into a corresponding light pattern through Bragg diffraction. A real image is
then formed, after optical processing, from the first-order diffracted beam. This image is focused onto the photosensitive surface of
a TV camera. The TV camera lens is replaced by the lens system of
the Bragg-diffraction system. The light intensity so recorded is
consequently digitized into a 256 x 256 array, with 8 bits of gray
scale, and fed to the computer. The signal is severely affected by
several different kinds of noise. The relative effect and the different corrective techniques which can be employed are described in
the following sections.

BACKGROUND SYSTEMATIC NOISE

The first kind of noise considered here is the background
systematic noise. By that is meant the stationary (i.e., time-
invariant) light intensity spatial variation at the image plane of
the Bragg-diffraction system when no object is present and the
acoustic insonification throughout the imaging region of the acoustic
cell is uniform(9). This kind of noise is due to system imperfections and is very difficult to eliminate by lens adjustments, etc.

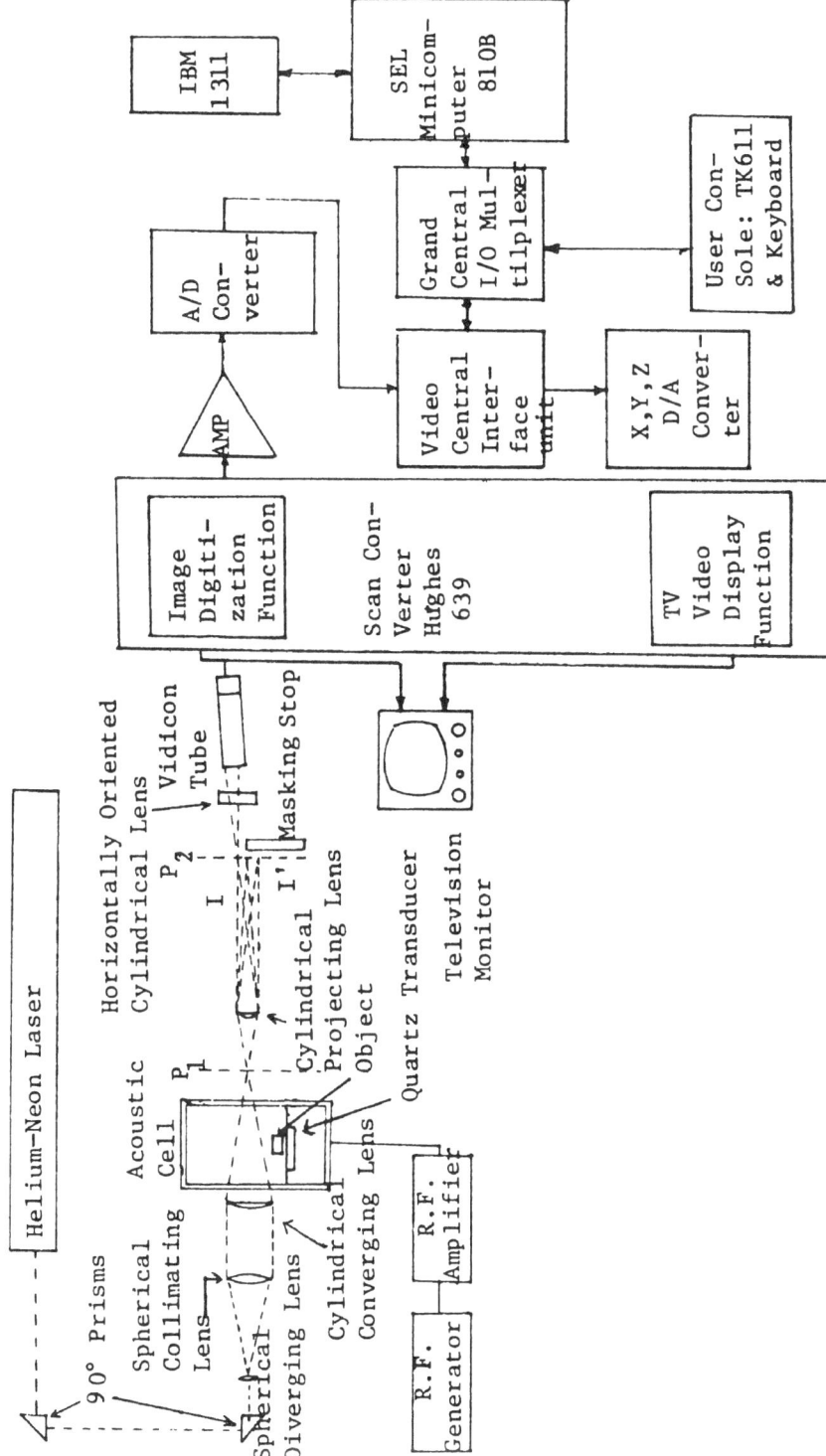

Fig. 1 Bragg-diffraction imaging system and the associated digital processing system.

Digital processing techniques, on the other hand, which have no
obvious optical equivalents, make it possible to remove this kind
of noise. Fig. 2 shows a typical picture of this noise without any
kind of correction. The logarithm of the magnitude of its Fourier
transform is shown in Figures 3 and 4. An ideal white background
without systematic noise should have an impulse at the center of
its spatial spectrum ($f_x = 0$, $f_y = 0$).

In order to evaluate the results of the different techniques for
noise removal, a test object, shown in Fig. 5, was used. An acoustic
image (without computer processing) of such an object is shown in
Fig. 6 and its Fourier spectrum is shown in Figures 7 and 8. A comparison of the object spectrum with the noise spectrum shown in
Fig. 3 reveals the signal energy distribution. A two-dimensional
filter was designed to reject the strong noise revealed in Fig. 3.
Such a filter is shown in Fig. 9 In order to reduce the side-lobes
in the impulse response, the edges of the filter between the pass-
bands and the reject-bands were graded. The result of filtering the
image of Fig. 6 is shown in Fig. 10. Much of the systematic noise
does not appear in this filtered image. Also, of course, some signal
energy distributed at the same spatial frequencies as the systematic
noise is removed. But, because the object in the image has a smoother
spatial frequency energy distribution than the noise, narrow-band
spatial frequency noise rejection eliminates little signal energy.

The time-invariance of the systematic noise suggests a different
removal technique. In order to carry out this technique, two light-

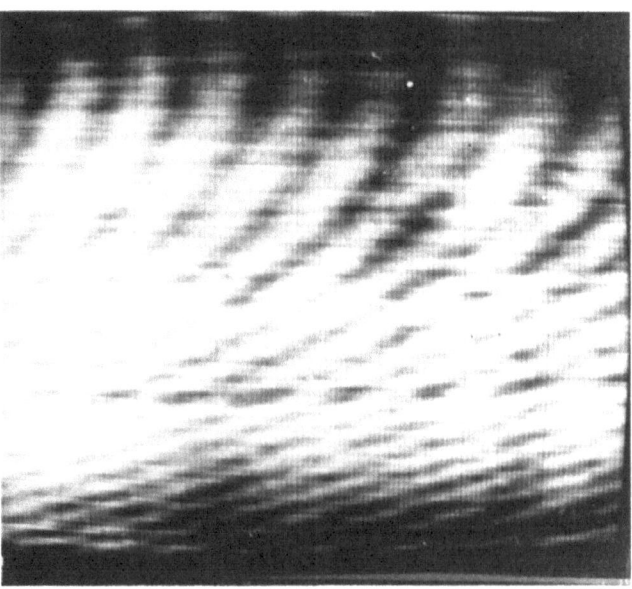

Fig. 2 White background image of the acoustic Bragg-diffraction
        imaging system.

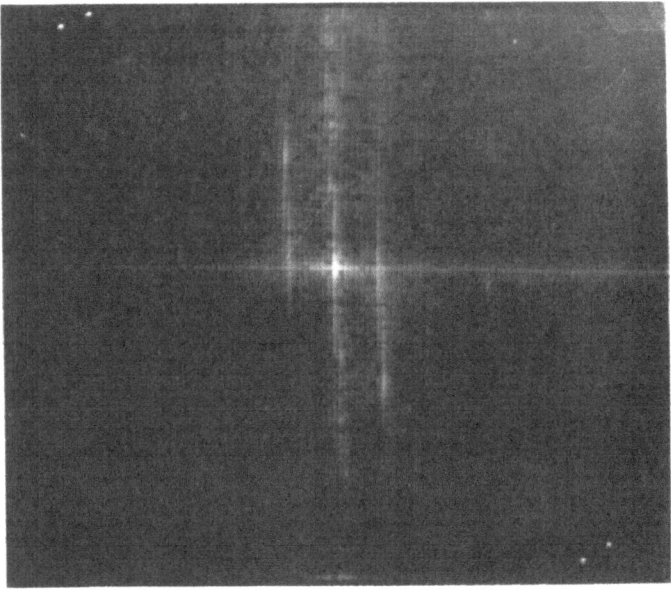

Fig. 3   Fourier transform of the white background of Fig. 2
         (logarithm of the magnitude).

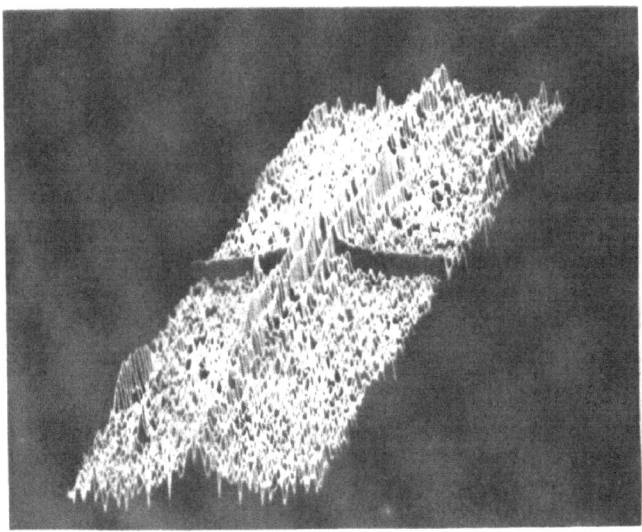

Fig. 4   Fourier transform of the white background of Fig. 2 in
         perspective display (logarithm of the magnitude).

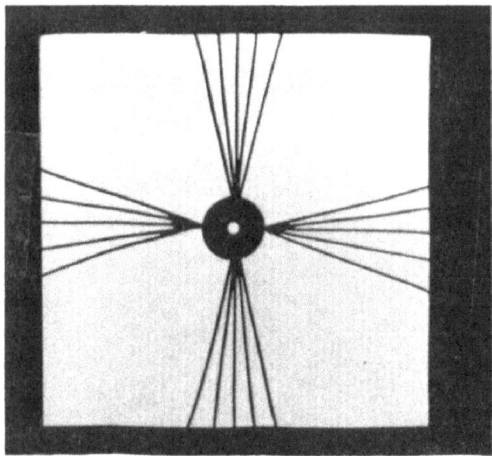

Fig. 5   Photograph of a test pattern which is used as an object in
         the acoustic Bragg imaging system.

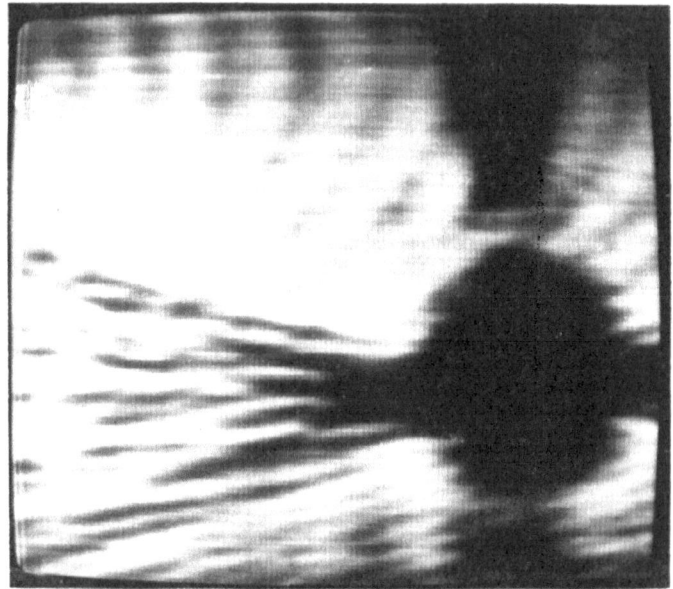

Fig. 6   Acoustic image of a test pattern from the Bragg-diffraction
         system.

Fig. 7   Fourier transform of the test pattern of Fig. 5 (logarithm of the magnitude).

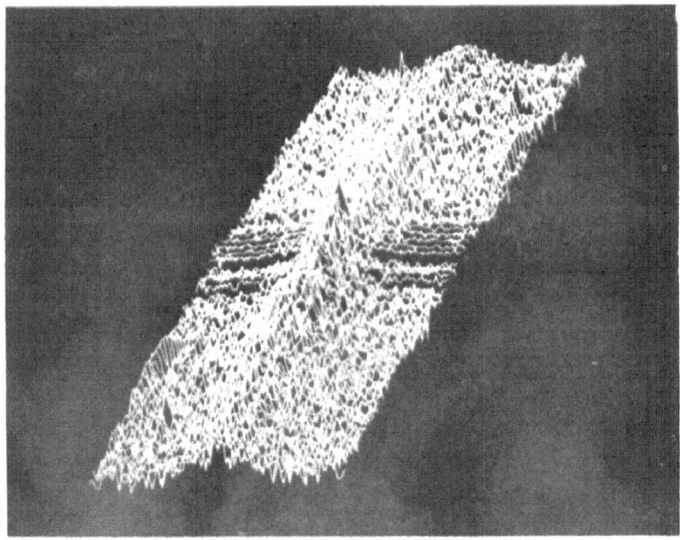

Fig. 8   Fourier transform of the test pattern of Fig. 5 shown in perspective display (logarithm of the magnitude).

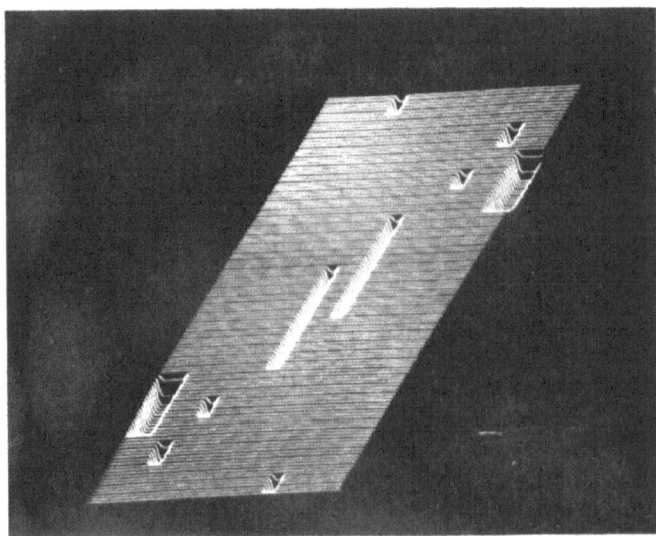

Fig. 9   Transfer function of a two-dimensional band-rejection filter
for systematic noise removal.

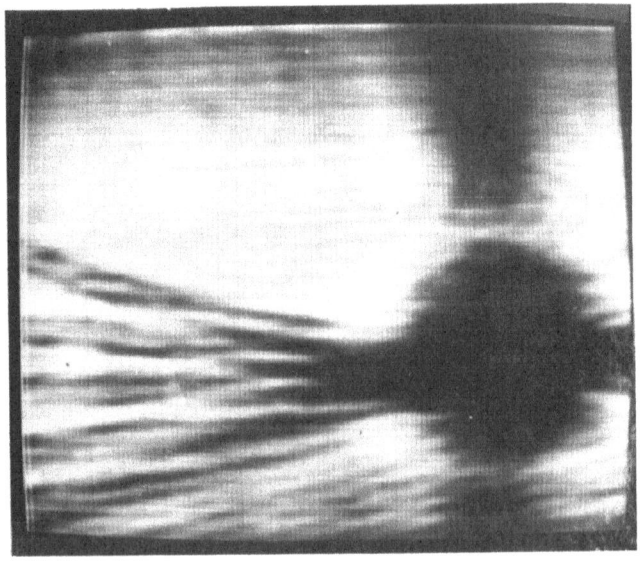

Fig. 10   Band-rejection filtered image of the test pattern in Fig. 6.

intensity patterns at the image plane must be obtained, in addition
to the pattern corresponding to the image to be enhanced. First,
the sound in the acoustic cell is turned off. The light intensity
at the image plane then corresponds to the dark background. This
pattern is digitized and stored in the computer as the signal $I_B(\ell,m)$.
Second, the sound is turned on with the object absent from the
acoustic cell. The resulting light-intensity pattern is digitized
and stored in the computer as the signal $I_W(\ell,m)$. Finally, the
object to be imaged is placed in the acoustic cell and its image is
digitized as signal $I(\ell,m)$. Then, the following calculation is
performed to obtain the background-compensated image signal $\hat{I}(\ell,m)$:

$$\hat{I}(\ell,m) = \begin{cases} 0 & \text{if } I < I_B \\[2mm] \dfrac{I(\ell,m) - I_B(\ell,m)}{I_W(\ell,m) - I_B(\ell,m)} \times 255 & \text{if } I_B \leq I \leq I_W \\[2mm] 255 & \text{if } I > I_W \end{cases}$$

The resulting image is shown in Fig. 11. A considerable reduction
in systematic noise is evident.

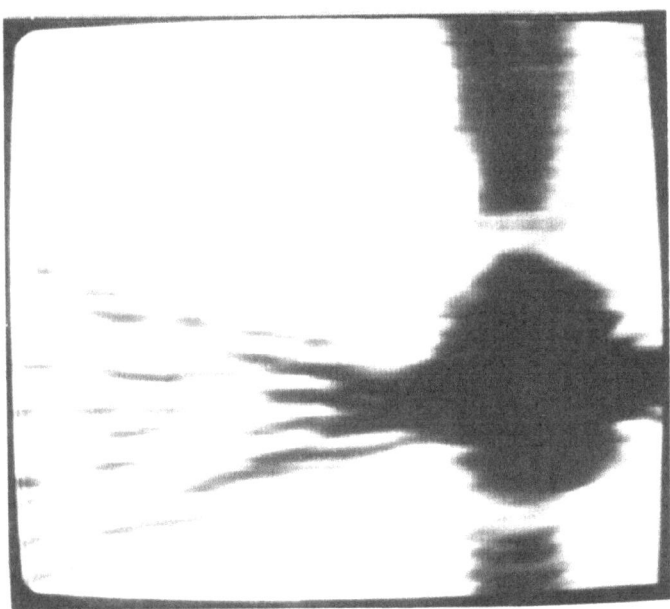

Fig. 11 Test-pattern image of Fig. 6 after background compensation.

A comparison of Figures 10 and 11 shows that the subtractive background-compensated image has somewhat better vertical-edge resolution. This is because in the filtered image, we have filtered out some of the signal information along with the noise. The situation could be improved upon if we were to use a more complex filter design. The problem with doing this, however, is that the processing time would increase. Even with the simple filtering operation described in this paper, it takes approximately 40 minutes of computing time to accomplish the work. On the other hand, the subtractive background compensation takes only about 1 minute. This fact indicates that in applications where processing time is to be minimized, the latter technique is more desirable than the former.

SPECKLE NOISE

Another kind of noise present in our system is speckle. This noise is due to interference of coherent light (or sound) scattered from inhomogeneities in the light (or sound) path. Gabor(10) has pointed out that the speckle can be smoothed out by averaging over a larger area than that of the noise textural features. This suggestion is indeed useful; speckle noise has been greatly reduced in our images by neighborhood averaging. When an image is stored in the scan converter memory, a 1024 x 1024 array is addressable. Neighborhood averaging may be accomplished by averaging subgroups of the array to obtain the elements of the digitized array. Fig. 12 shows the acoustic image of a metal grid with 2-mm spacing from center to center. The speckle can be seen to greatly impair the quality of the image. The result of 8 x 8 neighborhood averaging is shown in Fig. 13. The laser speckle has been reduced, but at the cost of a loss in resolution. Another approach would be to perform the neighborhood averaging in the form of a spatial convolution as shown in Figures 14 and 15. Fig. 14 shows a digitized acoustic image of the metal grid, and Fig. 15 shows the result of convolving that image with the impulse response given by the following array:

$$
\begin{array}{ccccc}
0 & 0 & 0 & 0 & 0 \\
0 & 1 & 1 & 1 & 0 \\
0 & 1 & 1 & 1 & 0 \\
0 & 1 & 1 & 1 & 0 \\
0 & 0 & 0 & 0 & 0
\end{array}
$$

Because speckle varies slowly with time, a background subtractive compensation technique, as used for background systematic noise, could also be employed to remove most of the speckle.

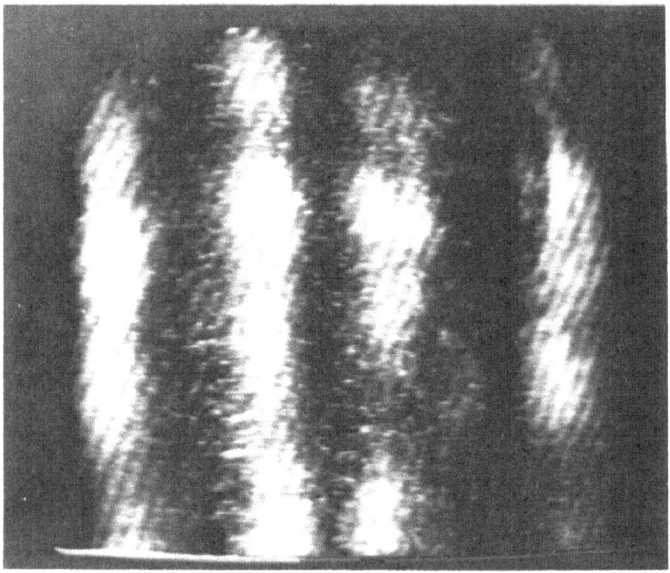

Fig. 12   Image of a metal grid with center spacing of 2 mm obtained
          from the Bragg-diffraction imaging system with an acoustic
          lens.

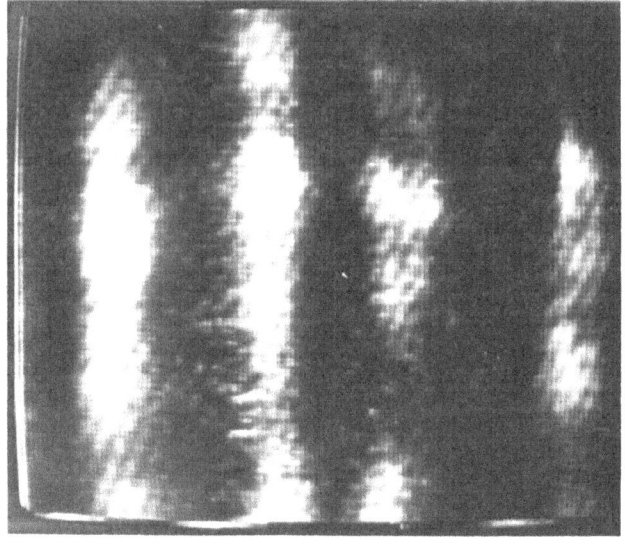

Fig. 13   Image of Fig. 12 after 8 x 8 neighborhood averaging.

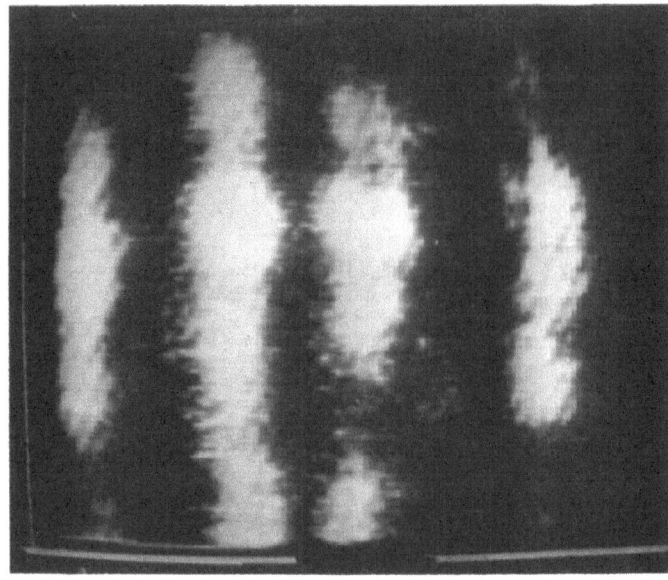

Fig. 14   A digitized image obtained from the photograph in Fig. 12.

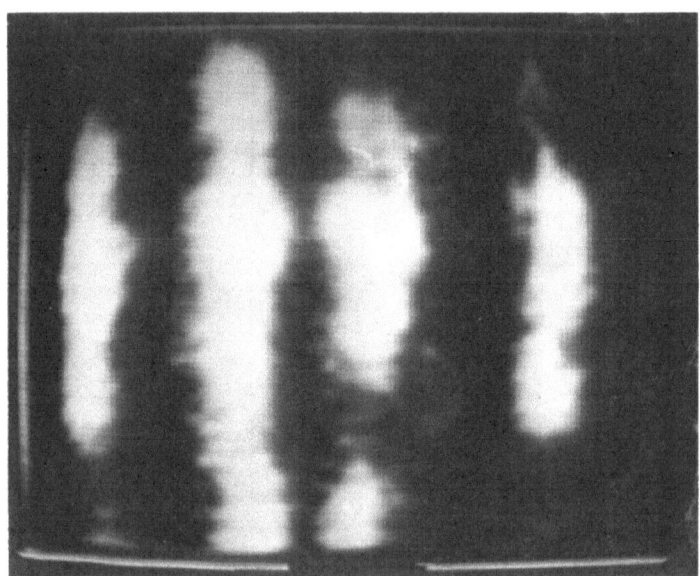

Fig. 15   Result of convolution of the image of Fig. 14 and the
          impulse response described in the text.

RINGING

In the Bragg-diffraction imaging system, sharp edges in the object generate oscillatory responses in the images. The amplitude of this oscillation decreases with increased distance from the edge. This effect is called "ringing" and is common in coherent imaging systems(11). It is associated with the sharp cut-off in spatial frequency of the transfer functions of coherent imaging systems.

Figure 16 shows an acoustic image of an aluminum block with two holes drilled inside. Ringing is particularly evident along the bottom edge of the block.

The nature of ringing suggests a method for its removal. Because it is associated with sharp cut-offs in the spatial frequency domain, a remedy would be a weighting of the transfer function to provide for a gradual cut-off in frequency. In digital image processing, this can be accomplished by using appropriate weighting functions. In our experiments, circularly symmetric hanning and Hamming windows were used. These functions are:

$$W_h(f_x, f_y) = \frac{1}{2}\left[1 + \cos 2\pi\left(\frac{1}{B}\sqrt{f_x^2 + f_y^2}\right)\right] \quad \text{hanning}$$

$$W_H(f_x, f_y) = 0.54 + 0.46 \cos 2\pi\left(\frac{1}{B}\sqrt{f_x^2 + f_y^2}\right) \quad \text{Hamming}$$

for $\sqrt{f_x^2 + f_y^2} \leq B$

and: $W_h(f_x, f_y) = W_H(f_x, f_y) = 0$

for $\sqrt{f_x^2 + f_y^2} > B$

In these expressions, f_x and f_y are the spatial frequencies, given in cycles per frame; and B is the cut-off frequency of the transfer function of the imaging system including the digital processor.

Figure 17 shows a hanning window with B = 96 cycles per frame, and Figure 18 shows the resulting acoustic image after windowing. In this image the ringing has been greatly reduced. Figure 19 shows the Hamming window, also for B = 96, and Figure 20 shows the processed acoustic image. The ringing is again removed. Comparison with the corresponding result for the hanning window shows no visible, significant difference.

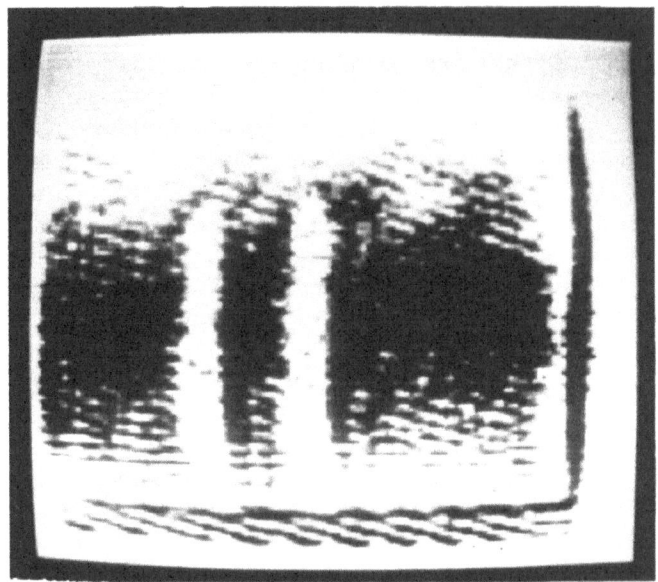

Fig. 16   Acoustic Bragg-diffraction image of aluminum block with
          holes.

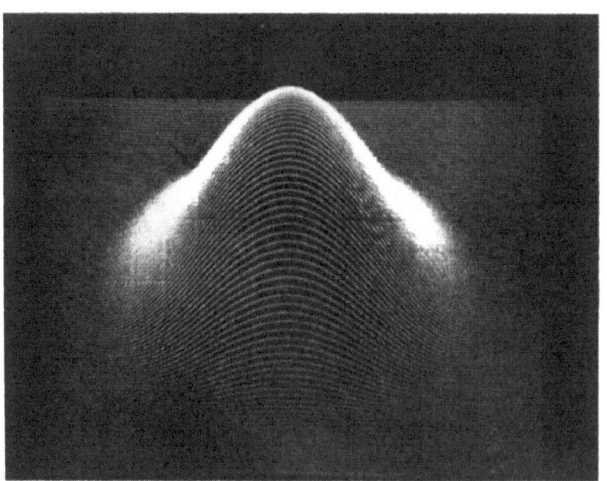

Fig. 17   Hanning window of size 96.

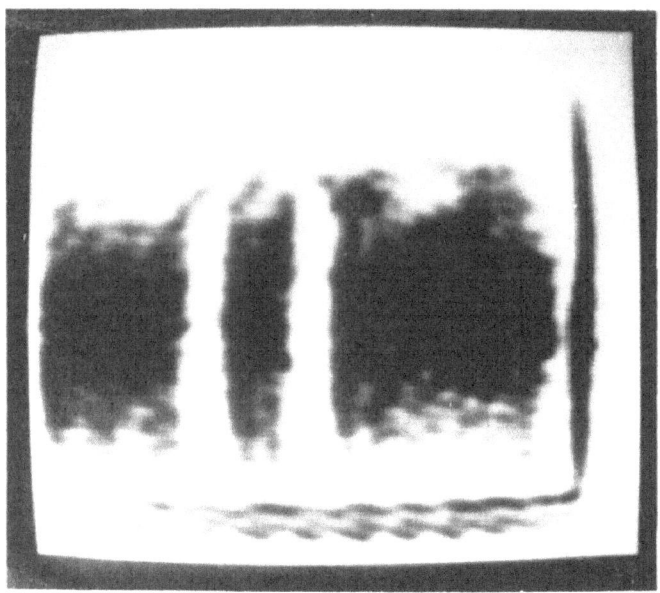

Fig. 18 Aluminum block after 96-point hanning windowing in spatial
frequency.

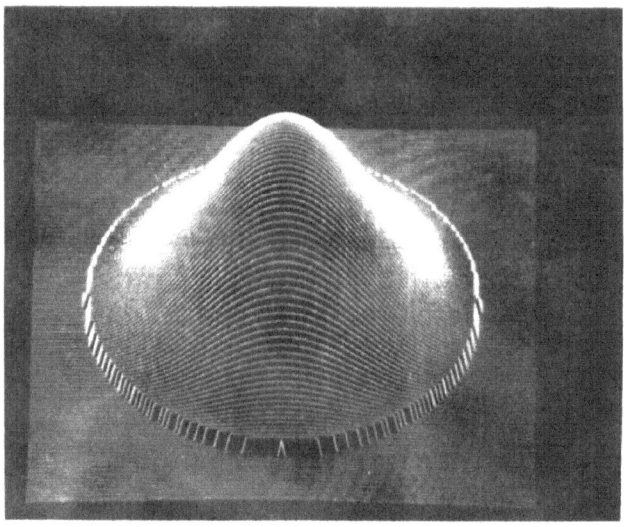

Fig. 19 Hamming window of size 96.

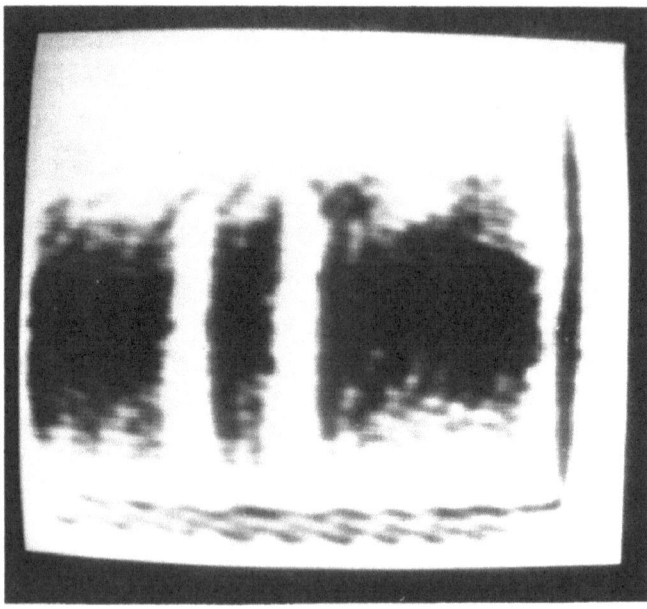

Fig. 20 Aluminum block after 96-point Hamming windowing in spatial
         frequency.

CONCLUSION

In the work reported in this paper, novel attempts were made to
minimize the effects of systematic noise, speckle, and ringing in
acoustic images. The relative magnitudes of the noise components
and other properties of the Bragg system were measured, and the
effectiveness of some digital processing techniques has been demonstrated. The work accomplished reveals some otherwise unavailable
quantitative information about the acoustic Bragg-diffraction imaging
system and also demonstrates well the convenience and capability of
an interactive image-processing system.

REFERENCES

1. David Lester Retz, "An Interactive System for Signal Analysis:
    Design, Implementation and Applications," Ph.D. Thesis, Univer-
    sity of California at Santa Barbara, published as Report CSL-25
    by the Computer Systems Laboratory, University of California,
    Santa Barbara, CA 93106, 1972.

2. Chin-Hwa Lee, "An Interactive On-Line Digital Image Processing
    System for Acoustic Image Enhancement," Ph.D. Thesis, University
    of California, Santa Barbara, CA 93106, 1975.

3. C.H. Lee, G.R. Heidbreder, and A.C. Berggreen, "Digital Image
    Enhancement Technique Applied to Acoustic Images," Ultrasonic
    Symposium Proceedings, IEEE Cat. #73, 1973, pp. 807-850.

4. C.H. Lee, M. Takagi, N.B. Tse, G.R. Heidbreder, and G. Wade,
    "Computer Enhancement of Acoustic Images," Acoustical Holography,
    Vol. 5, P. Green (ed.), 1974, pp. 541-551, Plenum Press, New York.

5. C.H. Lee, A.C. Berggreen, J. Pickens, and R. Bryan, "Linking
    Resources to Support an Interactive System for Acoustic Image
    Processing," presented at the 4th Annual Symposium of the EIA
    Committee on Automatic Imagery and Pattern Recognition, January
    1974, published as report CSL-33 by the Computer Systems Labora-
    tory, University of California, Santa Barbara, CA 93106.

6. H. Keyani, J. Landry, and G. Wade, "Bragg-Diffraction Imaging:
    A Potential Technique for Medical Diagnosis and Material Inspec-
    tion, Part II," Acoustical Holography, Vol. 5, P. Green (ed.),
    1974, pp. 25-41, Plenum Press, New York.

7. G. Wade, A. Coello-Vera, L. Schlussler, and S.C. Pei, "Acoustic
    Lenses and Low-Velocity Fluids for Improving Bragg-Diffraction
    Images," Acoustical Holography, Vol. 6, N. Booth (ed.), 1975,
    pp. 345-363, Plenum Press, New York.

8. G. Wade, "Bragg-Diffraction Imaging," Acoustic Imaging, G. Wade
    (ed.), 1976, pp. 189-226, Plenum Press, New York.

9. R. Smith and G. Wade, "Noise Characteristics of Bragg Imaging,"
    Acoustical Holography, Vol. III, A.F. Metherell (ed.), 1971,
    pp. 93-125, Plenum Press, New York.

10. D. Gabor, "Holography, 1948-1971," Proc. of the IEEE, Vol. 60,
    No. 6, June 1972, pp. 666-668.

11. J. Goodman, Introduction to Fourier Optics, 1968, pp. 131-133,
    McGraw-Hill, New York.

`

IMAGE PROCESSING FOR ABERRATION REMOVAL

J.C. STAMM and R. PRIEMER

Information Engineering Department

University of Illinois, Chicago Circle

ABSTRACT

Results in estimation theory are applied to the problem of removing aberrations caused by imaging systems. Recursive models for the object image brightness function and the image sensor performance are developed.

For the processing of NxN pixel images, known techniques require the formulation of an N^2 order image degradation model to fully incorporate the point spread function of the aberration. It is shown that a recursive formulation of the aberration results in an image degradation model of order N as opposed to N^2.

An algorithm for estimating the object image brightness given the sensor output is presented. Also, accounting for effects such as sensor noise and motion are demonstrated. Interpolation is readily achieved with little additional computational effort.

Examples (which demonstrate the effectiveness of the proposed techniques) are included.

I.   INTRODUCTION

Research and development of high quality acoustic imaging systems has been the subject of recent literature [1]. Although optical image reconstruction techniques [2] are widely used, they have the disadvantage of inflexibility when degradation occurs in the imaging system. Current technology has made available trans-

225

ducer arrays incorporating integrated circuit and large scale
integration readout schemes [3], [4], [5]. Such an array, inter-
faced to a digital computer, allows for flexible acquisition and
processing of acoustical image data [6], [7]. When holographic
techniques are employed, digital reconstruction from the sampled
diffraction pattern may be approached by a variety of methods [8],
[9], [10], [11].

In many applications, degradation occurs during the image
recording process: additive noise due to sampling and quantization; non-linearities in the recording system; spherical abbera-
tion; relative motion between sensor and object; interelement
crosstalk in the image sensor array; etc. For some types of degradation, special methods have been developed to improve image
quality. When imaging in a turbulent medium, Sato et al. [12] have
shown that incoherent imaging yields improved results. For focused
beam arrays, Collins et al. [13] have done an extensive analysis of
errors due to position and motion inaccuracies. Weak signal
enhancement in the presence of a spatially localized noise source
has been examined by Steinberg et al. [14] with excellent results.

A more general treatment of the problem of degradation removal
is the Fourier technique of apodization (Kramer et al. [15]). However, the work of Norton and Beer [16] indicates that great care
must be taken not to destroy the properties of the signal.

The present work approaches the problem of degradation removal
in the spatial domain, that is, post detection/reconstruction processing to improve image quality. The degradation process is
characterized by its point spread function (PSF). Previous work
by Takagi et al. [17] has shown successful improvement in image
quality for acoustically generated images for a simple PSF. In
general, for an image of dimension (NxN) pixels, the PSF has dimension $(N^2 \times N^2)$. The PSF from some types of degradation is analytic
[18], [19], although suitable algorithms for PSF identification
exist [20]. With the exception of a few special cases, processing
the PSF by known techniques requires an N^2 order image degradation
model, thus resulting in a large computational burden. Recent work
by Andrews and Patterson [21] has the goal of reducing the computational requirements for processing the PSF.

In the spatial domain, statistical properties are often sufficient to represent a class of images [22]. In section II, a
model is developed for transducer array sensors, and a recursive
model for the image brightness function is developed based on the
autocorrelation function of a class of images. It is shown that
partitioning the PSF into (NxN) submatrices allows for the development of an order N image degradation model. The effects of the
choice of scanning direction on the nature of the model are discussed.

Then, the image degradation and image brightness models are implemented with an order N recursive algorithm for estimating the object
image brightness given the post detection/reconstruction output.
The computational requirements for estimation and interpolation are
discussed. Section III presents examples demonstrating the practi-
cability of the proposed method. Conclusions and further discussion
are given in section IV.

II.   RESULTS

In this section, recursive models are developed for transducer
arrays and for post detection/reconstruction image brightness.
Next, a recursive formulation of the PSF is proposed. Choice of
scanning direction is shown to have significant effect on the predictive nature of the formulation. The models are then implemented
in an order N linear scheme for estimating the original image
brightness given the degraded image. It is further shown that
interpolation (smoothing) can be accomplished with little additional
computation. An analysis of the computational requirement is
included.

II.1 Transducer Array Model

Acoustic sensing arrays have been developed in a variety of
designs, materials, and readout schemes (see for example [3], [4],
[5]). However, test results have shown that they all exhibit
element-to-element coupling distortion through mechanical and
electro-mechanical interaction. For acoustic sensors, measurements
are possible only at discrete spatial points and represent the field
incident on the local area of acoustic-sensitive material. The
effects of coupling distortion on measurements can be illustrated
by a lattice structure joining these points (see figure 1).

For acoustic transducer arrays, a nearest neighbors model is
proposed. The output $\ell_{i,j}$ at the (i,j) spatial coordinate is

$$\ell_{i,j} = \sum_S d_{k,\ell} \, b_{i-k,j+\ell} + w_{i,j} \tag{1}$$

where $b_{i,j}$ is the acoustic field at the (i,j) element, S represents
the spatial sensitivity extent of the image sensor, the $d_{k,\ell}$
$(k,\ell \neq 0,0)$ represent coupling distortion of neighboring elements
on the (i,j) element, $d_{0,0}$ accounts for the contribution of the
local incident field to the output, and $w_{i,j}$ is zero mean white
noise.

The magnitude of coupling distortion effects depends on the
type and grid size of the sensor, the readout scheme, and the

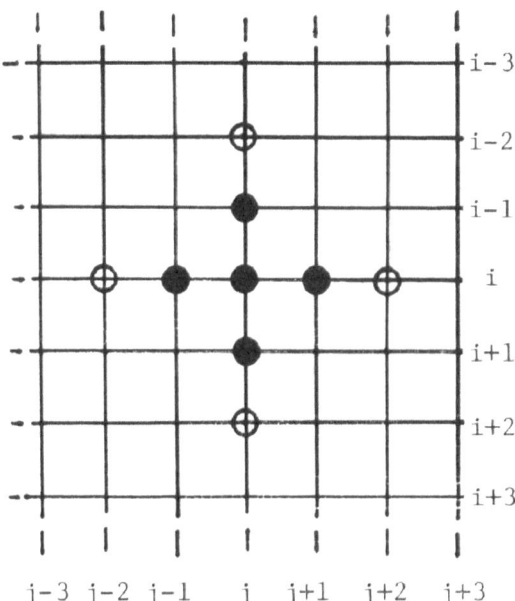

Fig. 1. Lattice structure of
acoustic transducer array.

relative field strength incident on the sensor. Nigam [3] has pointed out the need to reduce crosstalk by signal processing techniques. Experimental work [3], [4], [5] has shown that only first neighbors contribute significantly to element-to-element distortion. Consequently, a first order model for acoustic measurement will be used, i.e.,

$$\ell_{i,j} = d_{0,0} \, b_{i,j} + d_{0,1} \, b_{i,j+1} + d_{1,0} \, b_{i+1,j} + d_{0,-1} \, b_{i,j-1}$$
$$+ d_{-1,0} \, b_{i-1,j} + w_{i,j} \tag{2}$$

For focused acoustic imaging and shadowgrams, equation (2) accounts for blurring and loss of contrast in the displayed image. For holographic systems, equation (2) shows a smearing of the diffraction pattern in two dimensions. If the approximation of the diffraction pattern to a z-transform may be made, equation (2) can be written as

$$L(z_1,z_2) = d_{0,0} \, B(z_1,z_2) + d_{0,1} \, B(z_1,z_2+\Delta z_2) +$$
$$d_{1,0} \, B(z_1+\Delta z_1,z_2) + d_{0,-1} \, B(z_1,z_2-\Delta z_2)$$
$$+ d_{-1,0} \, B(z_1-\Delta z_1,z_2) + W(z_1,z_2) \tag{3}$$

where z_1 and z_2 are the discrete spatial frequencies, and Δz_1 and Δz_2 are the sample spacing frequencies, in the i and j directions respectively. (A comprehensive discussion of sample spacing frequencies and holographic reconstruction techniques may be found in Keating et al. [11].) Assuming $d_{1,0} = d_{-1,0}$ and $d_{0,1} = d_{0,-1}$, taking the inverse z-transform of equation (3), and using Euler's identity yields

$$\ell_{i,j} = [d_{0,0} + 2d_{1,0} \, \cos(i\Delta z_1) + 2d_{0,1} \, \cos(j\Delta z_2)] \, b_{i,j} +$$
$$w_{i,j} \tag{4}$$

From equation (4), it can be seen that coupling distortion of the acoustic transducer array produces a spatially-varying distortion in the reconstructed image. The magnitude of the distortion depends on the resolution (inter-element spacing) and the corresponding amount of crosstalk. The terms $2d_{1,0} \cos(i\Delta z_1)$ and $2d_{0,1}$ $\cos(j\Delta z_2)$ demonstrate the distortion trade-off between resolution and crosstalk. As the resolution is increased (decreasing Δz), the coupling distortion ($d_{1,0}$ and $d_{0,1}$) will increase, and vice-versa. With knowledge of the relationship between inter-element spacing (Δz) and crosstalk ($d_{1,0}$ and $d_{0,1}$), values could be selected to minimize the distortion effects in equation (4).

II.2 Image Brightness Model

Consider an image of dimension (NxN) pixels (discrete picture elements). For simplicity, assume that the proportionality constant between the incident acoustic field at the transducer array (or the reconstructed image for holographic systems) and the displayed gray level is one. Then, $b_{n,m}$ represents the brightness of the image at the spatial coordinate (n,m) with n,m = 1, 2, ..., N. A spatially stationary nearest neighbors model for the random variable $b_{i,j}$ is

$$b_{i,j} = \sum_I \alpha_{k,\ell} \, b_{i+k,j+\ell} + \beta \, u_{i,j}, \quad (k,\ell \neq 0,0) \tag{5}$$

where $u_{i,j}$ is a zero mean uncorrelated random variable and I represents the extent of the image. With this two-dimensional model for the image brightness, the state at any point is correlated to the state of every other point. The coefficients $\alpha_{k,\ell}$ are determined by the simultaneous solution of the set of linear regression equations

$$E[b_{i,j} \, b_{n,m}] = E[\sum_I \alpha_{k,\ell} \, b_{i+k,j+\ell} \, b_{n,m}], \quad (k,\ell \neq 0,0) \tag{6a}$$

where E denotes expectation, n,m = 1, 2, ..., N, and $E[\beta \, u_{i,j} \, b_{n,m}] = 0$. The coefficient β is found by squaring (5) and taking expectations

$$\beta = \left| \frac{1}{E(u_{i,j}^2)} \, E[(b_{i,j} - \sum_I \alpha_{k,\ell} \, b_{i+k,j+\ell})^2] \right|^{\frac{1}{2}}, \quad (k,\ell \neq 0,0) \tag{6b}$$

where it is arbitrarily imposed that $E[u_{i,j}^2] = E[b_{i,j}^2]$. Noting that $E[b_{n,m} \, b_{n+i,m+j}]$ is the autocorrelation function for the image, it will be shown that equations (5), (6a), and (6b) are a basis for a procedure for constructing a Gauss-Markov image brightness model for approximating the statistical nature of a given class of images.

Note this approach does not require separability or special forms for the image autocorrelation function. Further note that for repetitive imaging situations (e.g., medical applications or production testing) model construction need only be done once, since the resulting model represents the statistical properties for a given class of images.

A first neighbor form for (5) has been successfully employed by Jain and Angel [23] and Stamm and Priemer [24]. In equation (5) choose I such that

$$b_{i,j} = \alpha_{0,1} \, b_{i,j+1} + \alpha_{1,0} \, b_{i+1,j} + \alpha_{0,-1} \, b_{i,j-1} +$$
$$\alpha_{-1,0} \, b_{i-1,j} + \beta \, u_{i,j} \tag{7}$$

Now define N vectors of order N as $x_i = (b_{i,j})$, $j = 1, 2, \ldots, N$ for $i = 1, 2, \ldots, N$. Then (7) becomes

$$x_{i+1} = Q_0 x_i + Q_1 x_{i-1} + A u_i \tag{8}$$

where $A = -\beta/\alpha_{1,0}$, $Q_1 = -\alpha_{-1,0}/\alpha_{1,0}$ and Q_0 is the tridiagonal matrix

$$Q_0 = \frac{-1}{\alpha_{1,0}} \begin{bmatrix} -1 & & \alpha_{0,1} & & & 0 \\ & \cdot & & \cdot & & \\ \alpha_{0,-1} & \cdot & \cdot & & \cdot & \\ & \cdot & & \cdot & & \cdot \\ & & \cdot & & \cdot & \alpha_{0,1} \\ 0 & & & \alpha_{0,-1} & \cdot & -1 \end{bmatrix}$$

The model in (8) is a spatially one-dimensional form, and the state x_i is the brightness level for a row (or column) of the image model. Note that models incorporating other specific spatial correlations can readily be constructed.

II.3 Degradation Modeling

In many applications, degradation occurs during the acoustic field recording and image detection/reconstruction processes. In addition to the crosstalk mentioned in (II.1), degradations arise from: sampling; quantization; spherical aberration; relative motion between sensor and object; etc. The degradation process is characterized by its PSF. For an image of dimension (NxN) pixels, the PSF has dimension ($N^2 \times N^2$) and is traditionally modelled as:

$$Y = HX + V \tag{9}$$

where Y and X are (N^2) dimensional vectors representing a lexicographically ordered degraded image and original image respectively, H is the ($N^2 \times N^2$) PSF, and V (an (N^2) dimensional vector) represents additive white noise.

Considerable literature in both optical and acoustical imaging has been devoted to obtaining estimates of the original image X, given the degraded image Y, the PSF H, and the second-order statistics of the noise [18], [19], [21], [22], [23]. The problem remains a difficult and computationally complex one that is often dependent on the form or sparsity of the PSF. Fourier transform approaches often fail because of noise amplification for certain spectral regions.

The present approach is to partition the PSF into N^2 (NxN) submatrices

$$H = \begin{bmatrix} h_{11} & h_{12} & \cdot & \cdot & \cdot & h_{1N} \\ h_{21} & h_{22} & \cdot & \cdot & \cdot & h_{2N} \\ \cdot & \cdot & \cdot & & & \cdot \\ \cdot & \cdot & & \cdot & & \cdot \\ \cdot & \cdot & & & \cdot & \cdot \\ h_{N1} & h_{N2} & \cdot & \cdot & \cdot & h_{NN} \end{bmatrix} \qquad (10)$$

Next, separate equation (9) into N equations of order N, to yield

$$y_i = \sum_{j=1}^{N} h_{ij} x_j + v_i, \quad i = 1, 2, \ldots, N, \qquad (11)$$

where y_i and x_i represent the ith row (or column) of the degraded and original images respectively. Further separate the terms to

$$y_i = \sum_{j=1}^{i-1} h_{ij} x_j + h_{ii} x_i + \sum_{j=i+1}^{N} h_{ij} x_j + v_i,$$
$$i = 1, 2, \ldots, N, \qquad (12)$$

where it is assumed that the limits of the sum cannot be greater than N or less than 1. While processing the ith row (or column): the first term on the right side of (12) represents the relation of previously scanned rows (or columns) of the original image to the present row (or column) of the degraded image; the second term, the presently scanned row to the present output; and the third term, the relation of future rows to the present row.

As will be shown in the sequel, recursive schemes estimate the present input (i^{th} row of the original image) based on the present output (i^{th} row of the degraded image) and estimates of the previous inputs (previous rows of the original image). Therefore, if it were possible to eliminate the predictive terms in equation (12), the formulation would be ideal for recursive estimation schemes. The approach to this is as follows: First, examine the form of the PSF under varying conditions of scanning (left to right, right to left, top to bottom, bottom to top, rows or columns) to determine which scanning direction yields the fewest predictive terms. This is equivalent to finding the best 'lower triangular' form for H, i.e.,

$$
H_{\substack{\text{optimal} \\ \text{recursive} \\ \text{formulation}}} = \begin{bmatrix} h_{11} & 0 & \cdots & & 0 \\ h_{21} & h_{22} & & & \vdots \\ \vdots & \vdots & \vdots & & \vdots \\ \vdots & \vdots & \vdots & & 0 \\ \vdots & \vdots & \vdots & & \\ h_{N1} & h_{N2} & \cdots & & h_{NN} \end{bmatrix} \tag{13}
$$

Secondly, when predictive terms are unavoidable, they may be eliminated in a statistical sense with appropriate matrix products from the state equation (8), i.e.,

$$
x_{i+1} = Q_0 \, x_i + Q_1 \, x_{i-1} + A \, u_i \tag{14}
$$

$$
\therefore \quad h_{i,i+1} \, x_{i+1} = (h_{i,i+1} \, Q_0) x_i + (h_{i,i+1} \, Q_1) x_{i-1}
$$
$$
+ (h_{i,i+1} \, A) u_i \tag{15}
$$

$$
x_{i+2} = Q_0 \, x_{i+1} + Q_1 \, x_i + A \, u_{i+1} \tag{16}
$$

$$
\therefore \quad x_{i+2} = (Q_0^2 + Q_1) x_i + (Q_0 \, Q_1) x_{i-1} + (Q_0 \, A) u_i + A \, u_{i+1} \tag{17}
$$

The above can be continued for as many terms as are needed. The resulting terms are additive in equation (12), e.g., one predictive term would yield

$$
y_i = \sum_{j=1}^{i-2} h_{ij} \, x_j + (h_{i,i-1} + h_{i,i+1} \, Q_1) x_{i-1} +
$$
$$
(h_{ii} + h_{i,i+1} \, Q_0) x_i + v_i + (h_{i,i+1} \, A) u_i \tag{18}
$$

An important observation is that the noise terms increase with each additional predictive term. Thus, this formulation is limited by the effective signal to noise ratio after the predictive terms are eliminated.

II.4   Recursive Estimation Scheme

The image estimation problem now involves a post detection/reconstruction acoustic sensor array model (equation (2) or (4)), a statistical model for the object image brightness (equation (8)), and an Nth order recursive model for the PSF of the degradation due to the imaging system (equation (12)). The objective is to estimate the object image brightness given the post detection/reconstruction output.

Prior to implementation of the recursive estimation scheme, the acoustic array model will be incorporated into the degradation model. For focused acoustic imaging and shadowgrams, equation (2) is modified as in equation (8) by defining N vectors of order N; $\ell_i = (\ell_{i,j})$, $j = 1, 2, \ldots, N$, for $i = 1, 2, \ldots, N$, resulting in

$$\ell_i = d_{1,0}\, x_{i+1} + P\, x_i + d_{-1,0}\, x_{i-1} + W_i \tag{19}$$

where P is the tridiagonal matrix

$$P = \begin{bmatrix} d_{0,0} & & d_{0,1} & & 0 \\ & \ddots & & \ddots & \\ d_{-1,0} & & \ddots & & d_{0,1} \\ & \ddots & & \ddots & \\ 0 & & d_{-1,0} & & d_{0,0} \end{bmatrix}$$

Then, the components of $h_{i,i+1}$, $h_{i,j}$, and $h_{i-1,i}$ are determined to account for the effects of the coupling distortion $d_{1,0}$, P, and $d_{-1,0}$ respectively, for $i = 1, 2, \ldots, N$. For holographic systems, care must be taken since a change of coordinates is necessary to accomodate this modeling procedure. However, accounting for the coefficient of $b_{i,j}$ in equation (4) for the PSF model requires determining the components along the major diagonal of H.

After eliminating all predictive terms in the PSF formulation, the composite models for image brightness and degradation are

$$x_{i+1} = Q_0\, x_i + Q_1\, x_{i-1} + A\, u_i \tag{20}$$

$$y_i = \sum_{j=1}^{i} g_{i,j}\, x_j + C_i\, n_i \tag{21}$$

where the components of $h_{i,j}$ are modified to account for crosstalk distortion and predictive terms, C_i represents the composite (NxN) noise coefficient matrix, and n_i represents the composite N dimensional additive noise vector. The estimation problem is to estimate the x_i given the actual y_i with (21) as the degradation model. Equations (20) and (21) contain "delay" terms. If M is the maximum number of "delay" terms, a transformation to a representation of order NM is required to implement Kalman filtering techniques. However, the computational burden associated with a filter of order NM may be prohibitive. An alternative is to use the results of [25] to implement a filter of order N directly for (20) and (21). With the resulting filter, interpolation (smoothing) can be readily achieved.

The algorithm for obtaining the minimum mean square error estimate (MMSE) $\hat{x}_{j/i}$, $j = 1, 2, \ldots, i$, given $y_1, \ldots, y_i$ for the system (20) and (21) is

$$\tilde{y}_{i/i-1} = y_i - \sum_{j=1}^{i} g_{i,j} \hat{x}_{j/i-1} \tag{22}$$

$$K_{j/i} = \sum_{m=1}^{i} V_{j,m/i-1} g'_{i,m}$$

$$[\sum_{m=1}^{i} \sum_{k=1}^{i} g_{i,m} V_{m,k}/_{i-1} g'_{i,k} + C_i V_{n_i} C'_i]^{-1} \tag{23}$$

$$\hat{x}_{j/i} = \hat{x}_{j/i-1} + K_{j/i} \tilde{y}_{i/i-1} \tag{24}$$

$$V_{j,m/i} = V_{j,m/i-1} - \sum_{k=1}^{i} K_{j/i} g_{i,k} V_{k,m/i-1} \tag{25}$$

$$V_{j,i+1/i} = V_{j,i/i} Q'_0 + V_{j,i-1/i} Q_1 \tag{26}$$

$$V_{i+1,i+1/i} = Q_0 V_{i,i/i} + Q_1 V_{i,i-1/i} + A V_u A' \tag{27}$$

$$\hat{x}_{i+1/i} = Q_0 \hat{x}_{i/i} + Q_1 \hat{x}_{i-1/i} \tag{28}$$

for j and $m = 1, 2, \ldots, i$, where covariances such as $V_{j,m/i}$ are defined by

$$V_{j,m/i} = E [(x_j - \hat{x}_{j/i}) (x_m - \hat{x}_{m/i})']$$

Without any output, the initial condition for the estimator is $\hat{x}_{0/0} = E [b_{i,j}]$ and covariances such as $V_{j,m/0}$ are obtained using the object image autocorrelation function.

Examining equations (23) and (24) reveals that obtaining $\hat{x}_{j/i}$ for $j < i$ (interpolation) only slightly increases the computational burden since the matrix inversion in (23) and most covariances must be computed for filtering. Although the present work does not deal with reducing the computational requirements of this algorithm, considerable reductions in computation time can be obtained by taking advantage of matrix sparsity and symmetry in multiplication and inversion. The processing of y_i to obtain an estimate of x_j, for the system model (20) and (21) requires

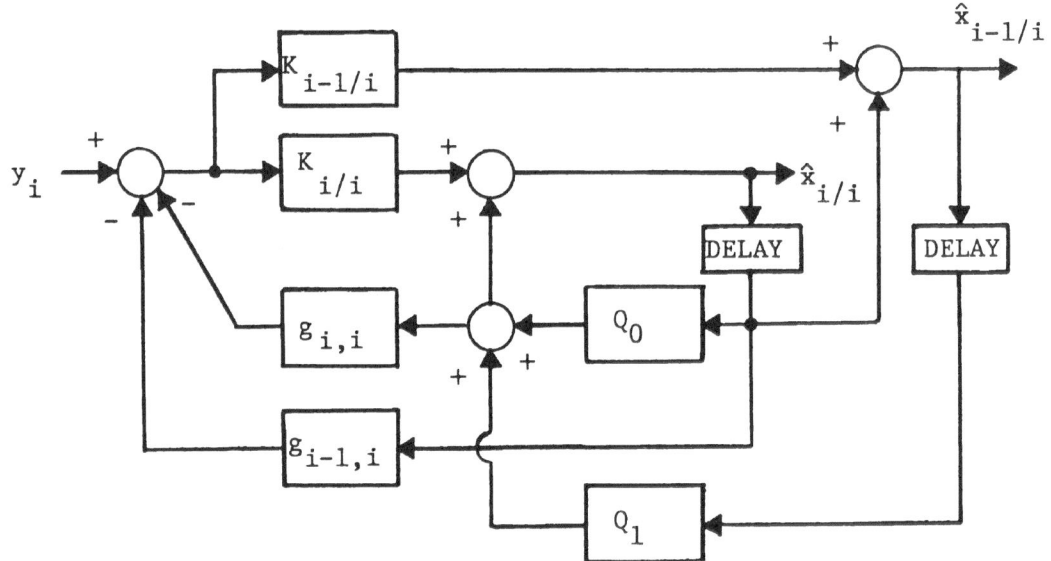

Fig. 2. Algorithm for obtaining $\hat{x}_{i/i}$ and $\hat{x}_{i-1/i}$ on-line.

$$((i - j)^3 + 2(i - j)^2 + 2.5(i - j) + 7.5) \ N^3 +$$
$$((i - j)^2 + 3.5(i - j) + 8.5) \ N^2 + N$$

multiplications per iteration. As will be shown in the next section, excellent results for processing several common types of degradation can be obtained for small values of (i - j), i.e. with filtering or one step interpolation. This results in an overall computational effort on the order of N^4 multiplications for processing an entire image, whereas processing as in equation (9) requires on the order of N^6 multiplications. Thus, the recursive formulation for degradation removal proposed herein has a considerable computational advantage over previous techniques.

Another feature of the proposed method is its potential use in "on-line" aberration removal applications. Since all parameters in the model are assumed known or can be computed a priori, filtering and interpolation gains ($K_{j/i}$) for each iteration may be calculated (equations (23), (25), (26), and (27)) before measurements are initiated. Thus, "on-line" processing requires implementing only equations (22), (24), and (28), and has a computational requirement of

$$(2(i - j) + 3) \ N^2 + N$$

multiplications per iteration to process a row (or column) of output. Figure (2) shows a block diagram of the proposed recursive scheme for "on-line" processing to obtain one-step interpolated estimates of the rows of the object image given the rows (columns) of the degraded post detection/reconstruction output.

III.   EXAMPLES

As in [22], suppose the image to be considered belongs to the class of images the autocorrelation function of which is

$$E \ [b_{n,m} \ b_{n+i,m+j}] = R_{i,j} = \sigma^2 \ \delta_1^{|i|} \ \delta_2^{|j|} \tag{29}$$

With this symmetric form for $R_{i,j}$, which is spatially stationary and separable, $\alpha_{0,-1} = \alpha_{0,1}$ and $\alpha_{-1,0} = \alpha_{1,0}$ in (7), and the matrix to be inverted in (6a) is symmetric. The parameter σ^2 is the average signal power.

The image used for these examples is shown in Figure (3). To facilitate image display, a line printer was used, and consequently, the spacing and width of pixels is much smaller in the horizontal direction than in the vertical direction. Intensity is simulated by character overprints for 16 gray levels. For this two brightness

level image, the light (not white) level was arbitrarily assigned
the value 5, and the dark (not black) level the value 1. For $N = 32$,
the use of (29) yielded $\sigma^2 = 7.12$, $\delta_1 = 0.919$, $\delta_2 = 0.914$. In (7),
$\alpha_{0,1} = 0.253$, $\alpha_{1,0} = 0.268$, and $\beta = 0.213$. The parameters are used in the
image model (equation (8)).

For the acoustic transducer array and model (2), suppose the
coefficients $d_{k,\ell}$ k,ℓ = ±1 (the element-to-element coupling) to
account for crosstalk distortion are 0.05 and $d_{0,0} = 1.0$. Therefore, coupling distortion (four nearest neighbors) contributes an
additional 20% to the array output. To simulate the sensor with (2),
a standard uniform distribution white noise generator was used.

In example 1, a focused acoustic imaging system is simulated
where both coupling distortion and white transducer array noise
($V_w = 0.27$) are present. The coupling distortion tends to wash out
bright texture and creates false contrast in the detected image.
An exposure is shown in Figure (4). The image to distortion power
ratio $I/D = 13.38$.

The algorithm (equations (22)–(28)) was used to obtain $\hat{x}_{i-1/i}$
(one-step interpolation) with Figure (4) (the post detection array
output) as the input y_i and (20) and (21) as the system model. The
I/D for the algorithm output increased to 30.78 yielding the image
in Figure (5). Thus, the effects of coupling distortion with additive noise can be successfully removed by the proposed recursive
image processing algorithm.

As a second example, relative motion between the transducer
array and the object was simulated for focused systems. It is
assumed that the motion is uniform during an exposure, so that the
PSF is spatially invariant. When the scanning direction is chosen
perpendicular to the motion, a diagonal form for H results. The
components of the $h_{i,i}$ are $d_{0,0}$ divided by the number of pixels of
blurring. Figure (6) shows the simulated output over 5 pixels of
motion during an exposure, and the nearest neighbors model for the
transducer array output is

$$\ell_{i,j} = 0.20\ b_{i,j} + 0.25\ b_{i,j+1} + 0.05\ b_{i+1,j} +$$

$$0.05\ b_{i,j-1} + 0.05\ b_{i+1,j} + 0.20\ b_{i,j+2} +$$

$$0.20\ b_{i,j+3} + 0.20\ b_{i,j+4} + w_{i,j}$$

where $V_w = 0.27$ and the coefficients have been modified to account
for coupling distortion. Then, the model (21) was constructed and
implemented as in example 1. The results of interpolation are shown
in Figure (7) where the I/D has improved from 3.49 (Fig. (6)) to
9.23 (Fig. (7)).

As a final example, a holographic acoustic imaging system was simulated by z-transforming the original image, synthesizing coupling on the z-transform, and taking the inverse z-transform of the result. For coupling coeff. of 0.05, the reconstructed image is shown in Figure (8). Note the spatial variation introduced into the bright texture, although the noise ($V_w = 0.27$) is unnoticeable. The PSF for this example is diagonal and the components of H are spatially varying as in equation (4). Smoothing results are shown in Figure (9) Some restoration of texture can be seen, however the I/D ratio improved from 2.89 up to 6.36, which is an improvement not as significant as was achieved in examples 1 and 2.

IV.   CONCLUSIONS

A nearest neighbors model has been proposed to account for element-to-element coupling distortion in acoustic transducer arrays. For focused acoustic imaging systems, crosstalk results in blurring and loss of contrast in the detected image. For holographic acoustic imaging systems, crosstalk introduces spatially varying distortion in the reconstructed image.

Recursive modeling of the point spread function for aberration in imaging systems was proposed. The resulting system model allows for processing of N measurements simultaneously. Choice of scanning direction in measurement taking was discussed with respect to the form of the point spread function. Proper choice was shown to be crucial in limiting both the effective signal-to-distortion ratio and the computational complexity in recursive estimation. Modification to include the coupling distortion model into the point spread function model was discussed.

An order N recursive estimation scheme for the proposed system model was presented. The computational requirements of the estimation scheme were given. "On-line" recursive image degradation removal was shown to be computationally feasible when filtering and interpolation gains are computed "off-line" a priori.

Examples were presented showing successful processing of crosstalk distortion and motion distortion in the presence of noise for focused acoustic imaging systems. Results of removing the spatially varying degradations in holographic acoustic imaging systems were less satisfactory. Further work in developing recursive processing models is currently underway, and results will be reported at a later date.

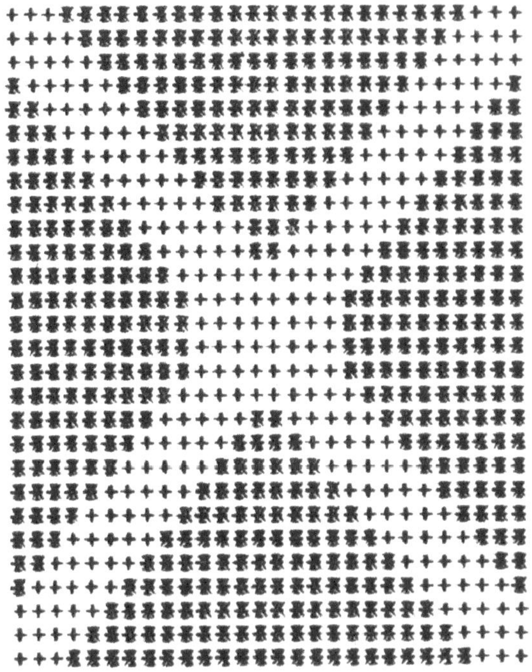

Fig. 3. Original image.

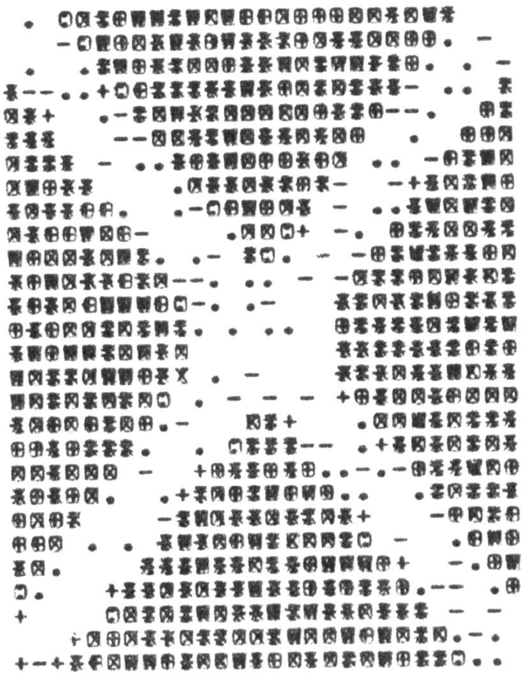

Fig. 4. Coupling and noise distortion.

Fig. 5. Interpolated image.

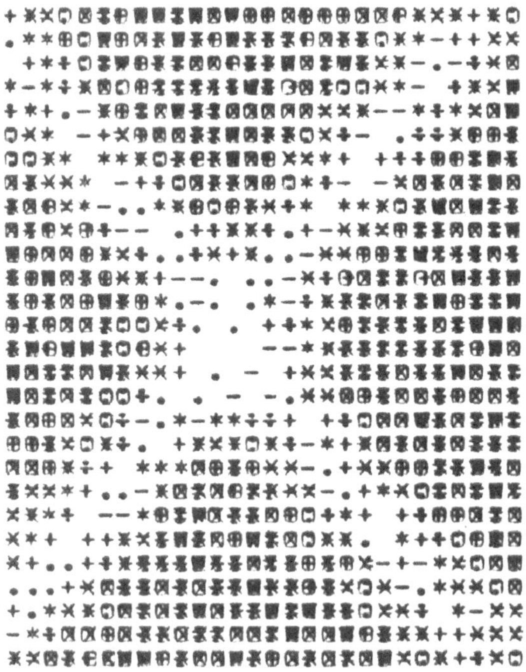

Fig. 6. Motion, coupling and
noise degradation.

Fig. 7. Interpolated image.

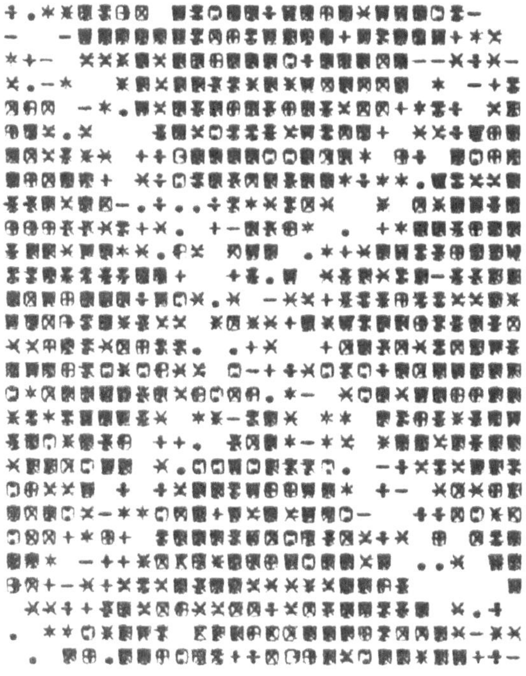

Fig. 8. Diffraction coupling
distortion.

Fig. 9. Interpolated image.

Acknowledgement

Computing services used in this research were provided by the Computer Center of the University of Illinois at Chicago Circle. Their assistance is gratefully acknowledged.

REFERENCES

[1] Kessler, L.W., "Review of Progress and Applications in Acoustic Microscopy," J. Acoust. Soc. Am., vol. 55, pp. 909-918, May 1974.

[2] Aoki, Y., "Acoustical Holograms and Optical Reconstruction," Acoustic Holography, vol. 1, Plenum Press, pp. 223-247, 1969.

[3] Nigam, A.K., "Condenser-Microphone Arrays in Acoustical Holography - A Review," J. Acoust. Soc. Am., vol. 55, pp. 978-985, May 1974.

[4] Knollman, G.C., Weaver, J.L., and Hartog, J.J., "Linear Receiving Array for Acoustic Imaging and Holography," Acoustical Holography, vol. 5, Plenum Press, pp. 647-658, 1974.

[5] Takagi, N., Kawashima, T., Ogura, T., and Yamada, T., "Solid-State Acoustic Image Sensor," Acoustical Holography, vol. 4, Plenum Press, pp. 215-236, 1972.

[6] Johnson, S.A., Greenleaf, J.F., Duck, F.A., Chu, A., Samayoa, W.R., and Gilbert, G.K., "Digital Computer Simulation Study of a Real-Time Collection, Post-Processing Synthetic Focusing Ultrasound Cardiac Camera," Acoustical Holography, vol. 6, Plenum Press, pp. 193-211, 1975.

[7] Goldstein, A., Ophir, J., and Templeton, A.W., "Research in Ultrasound Image Generation: A Computerized Ultrasound Processing, Acquisition, and Display (CUPAD) System," Acoustical Holography, vol. 6, Plenum Press, pp. 57-70, 1975.

[8] Goodman, J.W., "Digital Image Formation From Detected Holographic Data," Acoustical Holography, vol. 1, Plenum Press, pp. 173-185, 1969.

[9] Huang, T.S., "Digital Holography," Proc. IEEE, vol. 59, pp. 1335-1346, September 1971.

[10] Powers, J.P., "A Computerized Acoustic Imaging Technique Incorporating Object Recognition," Acoustical Holography, vol. 5, Plenum Press, pp. 527-539, 1974.

[11] Keating, P.N., Koppelman, R.F., and Mueller, R.K., "Maximization of Resolution in Three Dimensions," Acoustical Holography, vol. 6, Plenum Press, pp. 525-539, 1975.

[12] Sato, T. and Wadaka, S., "Incoherent Ultrasonic Imaging System," J. Acoust. Soc. Am., vol. 58, pp. 1013-1017, November 1975.

[13] Collins, H.D. and Hildebrand, B.P., "The Effects of Scanning Position and Motion Errors on Hologram Resolution," _Acoustical Holography_, vol. 4, Plenum Press, pp. 467-501, 1972.

[14] Steinberg, R.F., Keating, P.N., and Koppelman, R.F., "Experimental Implementation of Advanced Processing in Acoustic Holography," _Acoustical Holography_, vol. 6, Plenum Press, pp. 539-556, 1975.

[15] Kramer, D.B. and Anand, D.K., "Apodization in Numerical Holography," J. Acoust. Soc. Am., vol. 56, pp. 1545-1550, November 1974.

[16] Norton, R.H. and Beer, R., "New Apodizing Functions for Fourier Spectrometry," J. Opt. Soc. Am., vol. 66, pp. 259-264, March 1976.

[17] Takagi, M., Tse, N.B., Heidlreder, G.R., Lee, C.H., and Wade, G., "Computer Enhancement of Acoustic Images," _Acoustical Holography_, vol. 5, Plenum Press, pp. 541-550, 1974.

[18] Sondhi, M.M., "Image Restoration: The Removal of Spatially Invariant Degradations," Proc. IEEE, vol. 60, pp. 842-853, July 1972.

[19] Sawchuck, A.A., "Space-Variant Image Motion Degradation and Restoration," Proc. IEEE, vol. 60, pp. 854-861, July 1972.

[20] Ekstrom, M.P., "A Numerical Algorithm for Identifying Spread Functions of Shift-Invariant Imaging Systems," IEEE Trans. Computers, vol. C-22, pp. 322-328, April 1973.

[21] Andrews, H.C. and Patterson, C.L., "Singular Value Decompositions and Digital Image Processing," IEEE Trans. Acoust. Speech & Signal Proc., vol. ASSP-24, pp. 26-53, February 1976.

[22] Nahi, N.E., "Role of Recursive Estimation in Statistical Image Enhancement," Proc. IEEE, vol. 60, pp. 872-877, July 1972.

[23] Jain, A.K. and Angel, E., "Image Restoration, Modelling, and Reduction of Dimensionality," IEEE Trans. Computers, pp. 470-476, May 1974.

[24] Stamm, J.C. and Priemer, R., "Recursive Image Enhancement with a Nearest Neighbors Image Sensor Model," Proc. 13th Allerton Conf. Ckts. Syst. Theory. pp. 467-470, 1975.

[25] Biswas, K.K. and Mahalanabis, A.K., "Optimal Smoothing for Continuous-Time Systems with Mulitple Time Delays," IEEE Trans. Auto. Contrl., pp. 572-574, August 1972.

COMPUTORIZED RECONSTRUCTION OF ULTRASONIC VELOCITY FIELDS FOR MAPPING OF RESIDUAL STRESS*

B. P. Hildebrand and D. E. Hufferd

Battelle Pacific Northwest Laboratories

Richland, Washington 99352

It is well known that the velocity of sound propagation through a solid is altered when a stress is applied. The velocity change is small, and dependent upon the type of wave being propagated as well as the magnitude of the stress. Sensitivity is greatest to shear wave sound with the polarization vector parallel to the direction of stress. In this case, velocity changes as great as 0.6% have been measured. This paper describes preliminary work aimed at evaluating computerized reconstruction of velocity fields from velocity profiles to map residual stress concentrations in thick metal sections. We are currently using the Grid Technique of Sweeney and Vest for reconstructing data taken from liquid models. This method limits the number of rays and views that can be used; hence we intend to use the ART technique at a later date. We shall describe the experiment, electronic measuring system, the reconstruction method, and show reconstructions of 2% and 0.2% velocity anomalies in a liquid model and preliminary data in a steel model.

* This research was sponsored by the Electric Power Research Institute, 3412 Hillview Ave., Palo Alto, CA under contract #RP504-1.

INTRODUCTION

One of the outstanding problems of non-destructive testing is
the location and measurement of areas of residual stress. In the
manufacture of large structures, such as pressure vessels, a number
of very large welds are required. The heat affected zone surrounding the welds will contain residual stresses due to uneven cooling
rates. In practice, these stresses are relieved by heating the
entire vessel to some appropriate temperature and then carefully
controlling the cooling rate. This procedure is quite successful
in relieving any residual stress induced by the welding process.

There is, however, no satisfactory test for actually measuring
the success of the stress relief procedure. Thus, it sometimes
happens that high residual stress regions exist in the pressure
vessel. If these regions occur in critical areas, such as nozzle-
to-vessel joins, cracks may develop, with subsequent plant shut-
downs and expensive repair. If it were possible to detect and map
residual stress, local stress relief could be applied to avoid
future problems.

Standard NDT tests consist of radiography and ultrasonic pulse-
echo. Neither of these techniques can reveal the presence of residual stress; the former because it shows only density variations,
and the latter because these regions are not sharply defined and
hence do not reflect much sound. It is known, however, that the
velocity of sound is affected by stress, although this effect is
small. The purpose of the research reported in this paper is to
use computerized tomographic procedures to map residual stress areas
by reconstructing ultrasonic velocity changes. Since the reconstruction algorithms are identical to those developed for x-ray
scanners, the major problems to be solved are measurement sensitivity, geometry and speed of data acquisition.

ACOUSTICS

Acoustoelasticity is analogous to photoelasticity in that the
wave velocities are stress dependent, but since sound waves can be
longitudinal or transverse, acoustoelasticity is more complex. The
expressions for stress dependence of ultrasonic velocities in a
material which is initially isotropic are given in the coordinate
system shown in Figure 1.[1]

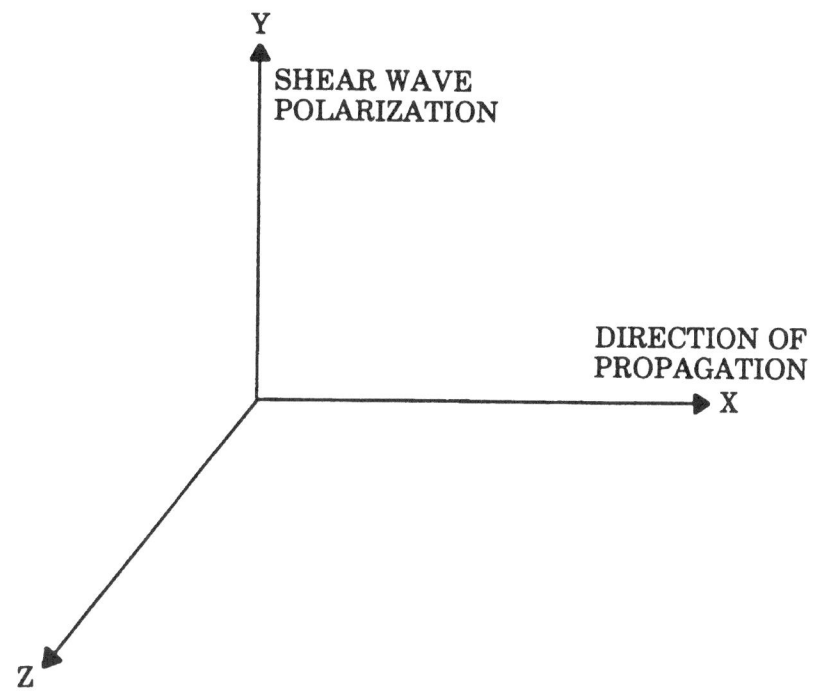

FIGURE 1.   Coordinate System for Stress Equations

$$\rho_0 \, V_{\ell p}^2 = \lambda + 2\mu - \frac{P}{3K_0} \left(7\lambda + 10\mu + 6\ell + 4m \right) \tag{1}$$

$$\rho_0 \, V_{sp}^2 = \mu - \frac{P}{3K_0} \left(3\lambda + 6\mu + 3m - \frac{n}{2} \right) \tag{2}$$

$$\rho_0 \, V_{\ell x}^2 = \lambda + 2\mu + \frac{T}{3K_0} \left[\frac{\lambda+\mu}{\mu} \left(4\lambda + 10\mu + 4m \right) + \lambda + 2\ell \right] \tag{3}$$

$$\rho_0 \, V_{\ell y}^2 = \lambda + 2\mu - \frac{T}{3K_0} \left[\frac{2\lambda}{\mu} \left(\lambda + 2\mu + m \right) - 2\ell \right] \tag{4}$$

$$\rho_0 \, V_{sx}^2 = \mu + \frac{T}{3K_0} \left(4\lambda + 4\mu + m + \frac{\lambda n}{4\mu} \right) \tag{5}$$

$$\rho_0 \, V_{sy}^2 = \mu + \frac{T}{3K_0} \left(\lambda + 2\mu + m + \frac{\lambda n}{4\mu} \right) \tag{6}$$

$$\rho_0 \, V_{sz}^2 = \mu - \frac{T}{3K_0} \left(2\lambda - m + \frac{n}{2} + \frac{\lambda n}{2\mu} \right) \tag{7}$$

Where ρ_0 = density of unstrained material
$\quad\quad$ V = ultrasonic velocity
$\quad\quad \lambda,\mu$ = second order Lame' elastic constants
$\quad \ell,m,n$ = third order Murnaghan elastic constants
$\quad\quad$ P = hydrostatic pressure
$\quad\quad$ T = uniaxial tension
$\quad\quad K_0 = 1/3 \left(3\lambda + 2\mu \right)$ =   bulk modulus

1st subscript ℓ refers to longitudinal wave
1st subscript s refers to shear wave
2nd subscript p refers to hydrostatic pressure
2nd subscript x, y or z refers to uniaxial tension, in the x,
y or z direction respectively

For compression, the sign of T is reversed. By setting P and T equal to zero we obtain the well known expressions

$$\rho_0 V_\ell^2 = \lambda + 2\mu \tag{8}$$

$$\rho_0 V_s^2 = \mu \tag{9}$$

Since the stress-strain relationship is non-linear, the elastic constants change with a resulting change in velocity. Consequently, if the change in velocity can be measured and mapped it should be possible to label this map in terms of stress through Equations 1-7.

As seen from Equations 1-7, stress changes all the velocities. These changes are very small since the non-linearity in the stress-strain relationship is small. Crecraft,[2] for example, determined experimentally that the change in longitudinal velocity in nickle-steel was only 0.012% for 2.5 MHz and 6000 psi compressive stress. Since the yield limit occurs at about 120,000 psi a maximum of 0.24% velocity change can be expected. Aluminum shows 0.025%/1000 psi, copper 0.008%/1000 psi in contrast to nickel-steel, 0.002%/1000 psi.

In general, shear wave velocity changes are dependent upon the polarization. If the particle motion is parallel to the direction of stress, the effect is about three times that on the longitudinal wave. If the particle motion is perpendicular to the direction of stress, the velocity change becomes very small.

We express each velocity in the form

$$V = V_0 \left(1 + \frac{\Delta V}{V_0} \right)$$

where V_0 = unstressed velocity and ΔV is the change due to stress.

Since $\frac{\Delta V}{V_0}$ is very small

$$V^2 = V_0^2 \left(1 + \frac{\Delta V}{V_0} \right)^2 \cong V_0^2 \left(1 + \frac{2\Delta V}{V_0} \right) \tag{10}$$

If we substitute this expression into Equations 1-7, we find that the relationship between stress and velocity change is linear. For example, Equation 1 becomes

$$\frac{\Delta V_{\ell p}}{V_{0\ell p}} = \frac{1}{2\rho_0 V_{0\ell p}^2} 2\left[\lambda + 2\mu - \frac{P}{3K_0}\left(7\lambda + 10\mu + 6\ell + 4m \right) \right] - \frac{1}{2} \tag{11}$$

For very high stress, where the stress-strain relationship becomes quite non-linear, we will no longer have a linear variation between stress and velocity change. In this case the exact expression for V^2 must be used.

The shear velocities can be used in a manner analogous to optical birefringence. By manipulating Equations 6 and 7 we can obtain the expression

$$V_{sy}^2 - V_{sz}^2 = \frac{T}{4\mu\rho_o}\left(4\mu + n\right) \tag{12}$$

Factoring the left-hand side of this equation we have

$$V_{sy} - V_{sz} = \frac{T\,(4\mu + n)}{4\mu\rho_o(V_{sy} + V_{sz})} \tag{13}$$

Hence, even though the elastic constants may not change much, birefringence can be used to measure stress.

From the above discussion it is evident that there are two distinct ways to compute residual stress from velocity changes. The simplest method is to measure the velocity change and use Equation 11 or its equivalent to calculate stress. This can be done for any mode of wave propagation. The alternative method is to use two perpendicularly polarized shear waves propagating in the same direction, measure their respective velocities, and then use Equation 13 to calculate stress.

A number of studies have shown that it is possible to measure velocities with sufficient accuracy to detect and calculate residual stress.[3,4,5] Typically, for steel, the stress acoustic constant is in the neighborhood of 0.5 nsec/inch/1000 psi. This constant means that a 1000 psi increment in stress produces a 0.5 nsec change in travel time through a 1 inch thick specimen. Since it is relatively easy to measure time-of-flight changes to this accuracy the sensitivity of this method should be about 1000 psi.

RECONSTRUCTION

At the present time, a number of researchers have made velocity measurements for the computation of stress. These have been spot measurements, or profiles. No one has yet attempted to make a number of profiles at different angles in order to reconstruct a cross section of a velocity anomaly, and hence a cross section of residual stress.

The technique of cross section mapping from a multiplicity of profiles has undergone a remarkable renaissance in recent years.

The EMI x-ray scanner was the first commercial device to exploit a mathematical technique first studied by Radon in 1917. This device works essentially as shown in Figure 2.

A highly collimated beam of x-rays is scanned in synchrony with a detector in a linear fashion. The detector records a profile of x-ray intensity. The scanner is then rotated by a fixed amount and the scan repeated. Usually 360 profiles are taken at 1° intervals. A computor is then asked to reconstruct a cross-section of the object traversed by the x-ray beam. It proceeds to view the problem as shown in Figure 3. Object space is divided up into cells. Knowing the angle θ , the computor can calculate the path length, L_{ij}, of each ray, i, through each cell, j. Each cell is assigned a value of x-ray attentuation, α_j. Thus, the attenuation along each ray is a linear sum of attenuation coefficient times cell path length. That is

$$a_i = \sum_j \alpha_j L_{ij} \tag{14}$$

where i denotes i^{th} ray
     j denotes j^{th} cell
     α_j = attenuation coefficient in j^{th} cell
     L_{ij} = path length of ith ray through j^{th} cell
     a_i = attenuation along i^{th} ray path

If enough profiles are taken so that there are as many knowns $\{a_i\}$ as unknowns $\{a_j\}$, the set of simultaneous equations can be solved for the a_j.

There are many algorithms that can be used to solve the set of equations. The most direct method is matrix inversion. It is also

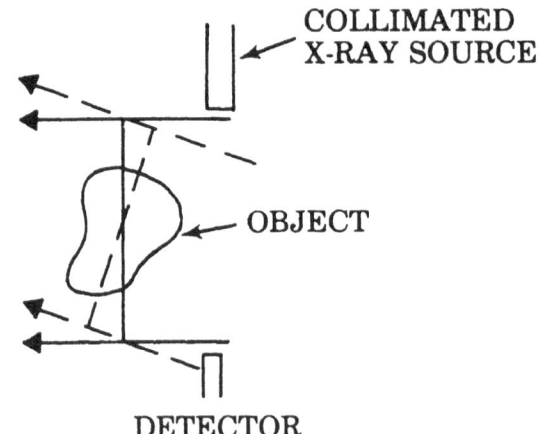

FIGURE 2. Principle of the CAT Scanner

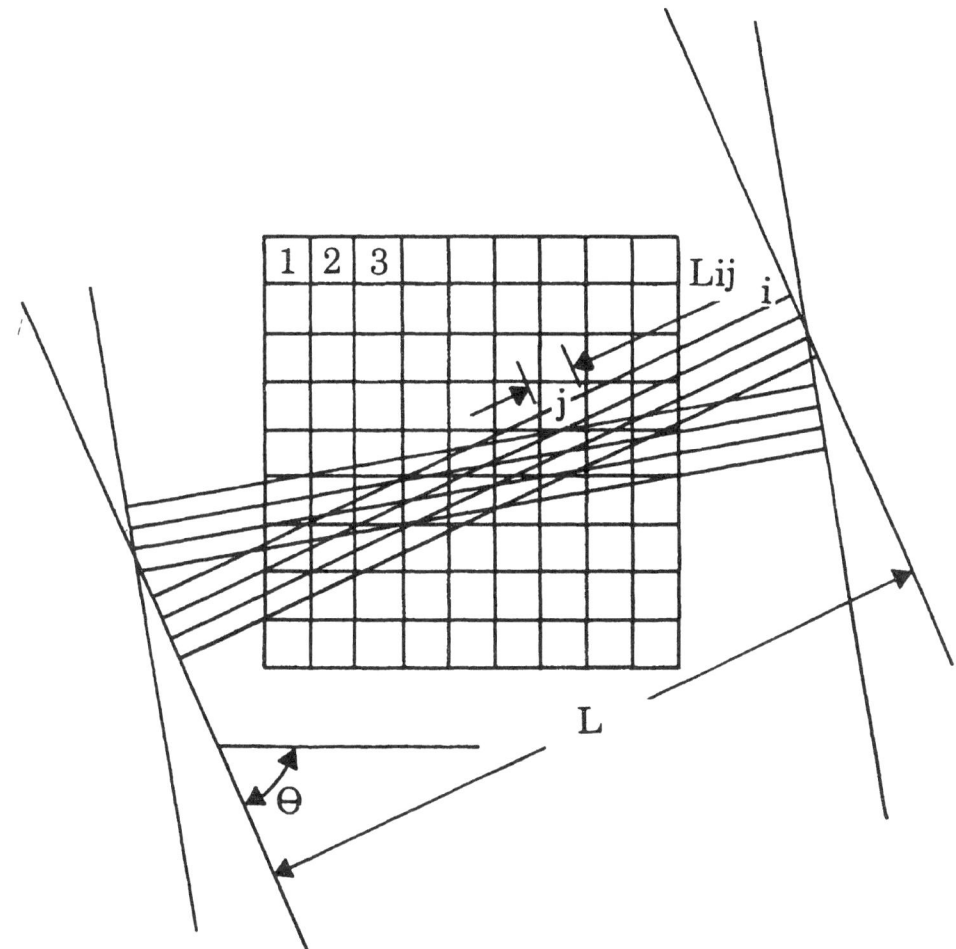

FIGURE 3.   Reconstruction Procedure

the most time consuming since no attempt is made to simplify the
matrix.   This method is the one we use since at present we are
reconstructing only a 9 x 9 image.   Other algorthims are ART
(arithmetic reconstruction technique) which is an iterative method;
convolution, which is based on Radon's original work; Fourier
Transform, which is the spatial frequency version of convolution.
All of these basic methods have variations available.   A good source
of information on the various algorithms and their advantages and
disadvantages can be found in a recent symposium proceedings.[6]

The only difference between the x-ray technique and the ultrasonic velocity technique is in the quantity to be measured and
reconstructed.   The pertinent set of equations to be solved are

$$\{T_i\} = \sum_j \frac{L_{ij}}{V_j} + \frac{1}{V_\omega}\left(L - \sum_j L_{ij}\right) \tag{15}$$

where T_i = total time of flight along i^{th} ray
$\quad\quad\quad V_j$ = velocity in j^{th} cell
$\quad\quad\quad V_\omega$ = velocity in water
$\quad\quad\quad L$ = total geometric path length

MEASUREMENT SYSTEM

The system we are using for the measurement of propagation time is shown in Figure 4.

The pulse generator, a Velonex 570, generates a 0.5 µsec 600 volt pulse with a repetition rate of 2KHz. This pulse starts the time interval counter and energizes a Nortec V-Z-16-2.25-0, 1 inch wide-band transducer. A Nortec R-Z-0.025-10-0, 0.025 inch wide-band receiver senses the pulse, a sense amplifier amplifies and thresholds it and sends it to the time interval counter. The counter, a Hewlett Packard 5345A, measures the time between the start pulse and the received pulse by counting the internal clock pulses generated in the interval. The instrument can be set to average over 10 nsec to 20,000 sec. Its worst case accuracy is ±0.7 nsec and worst case resolution is ±11 psec. With a 1 sec averaging time we consistantly see a repeatability better than ±0.3 nsec.

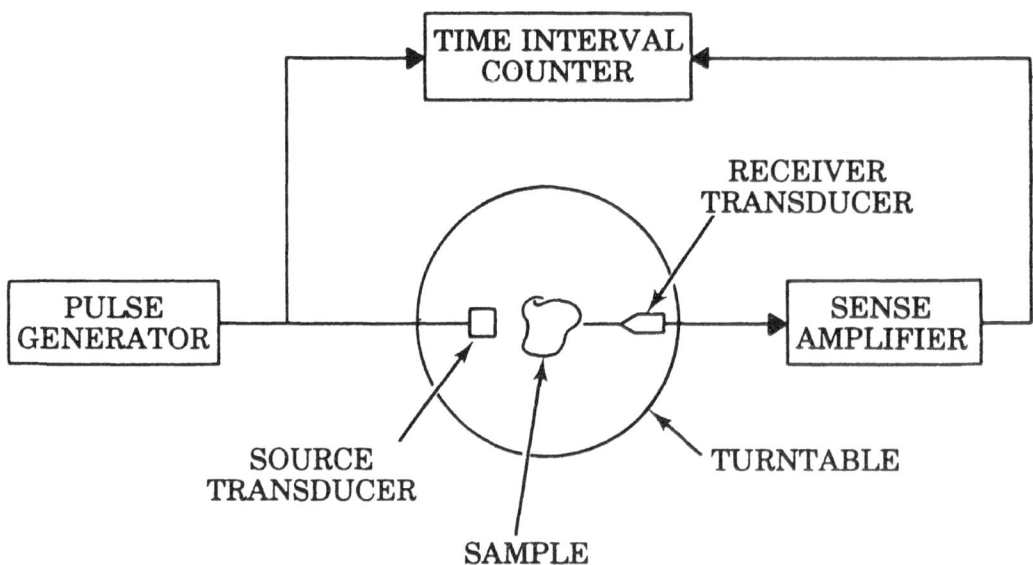

FIGURE 4. Experimental Arrangement for Velocity Profiling

APPLICATION

In the introduction we enunciated the basic reasons for attempting this research. We must now be more specific since the application has a profound influence on the experimental design. For the most part, we are interested in mapping stress anomalies in thick metal sections (4-10 inches thick). This in itself, restricts the total field of view available to 180°. Practical difficulties brought on by the large extent of these sections further restricts the field of view to about 90°. Although, in some cases it may be possible to have access to both sides of the section, it is usually inconvenient and sometimes impossible. Consequently, we must devise a method for gathering the required data when only one surface is accessible. Figures 5 and 6 serve to illustrate the limitations placed on us by the geometry.

Figure 5 illustrates the case when both sides are accessible. The only way in which different angular profiles can be taken is to launch the waves at a different angle by tilting the source. Theoretically it is possible to launch waves over a ±90° field. However, note that the receiver would need to be moved further away with increasing angle. Hence, the practical restriction of ±45°.

Figure 6 illustrates the dilemma produced by the requirement for single surface inspection, and the solution. The back surface

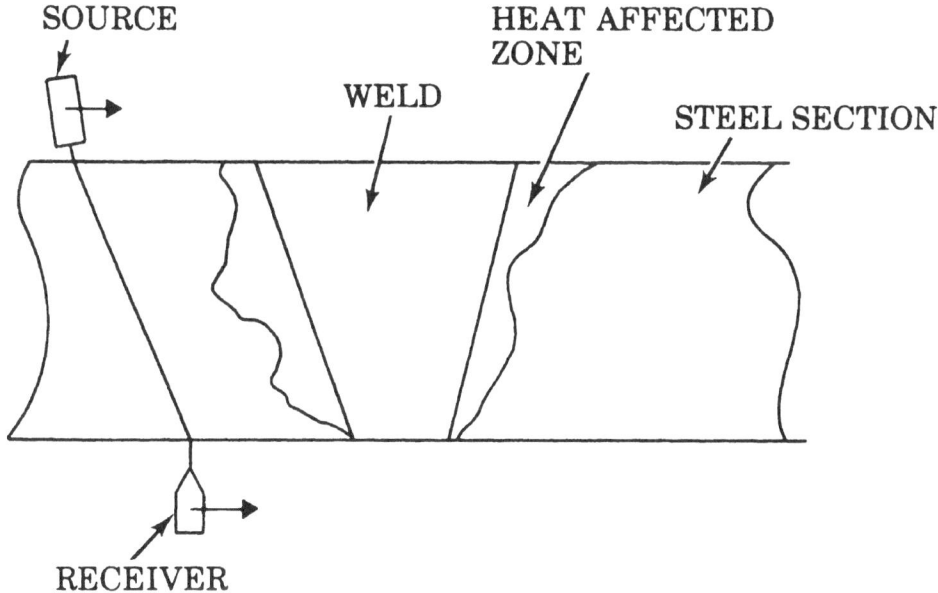

FIGURE 5. Scanning of Thick Sections in Transmission

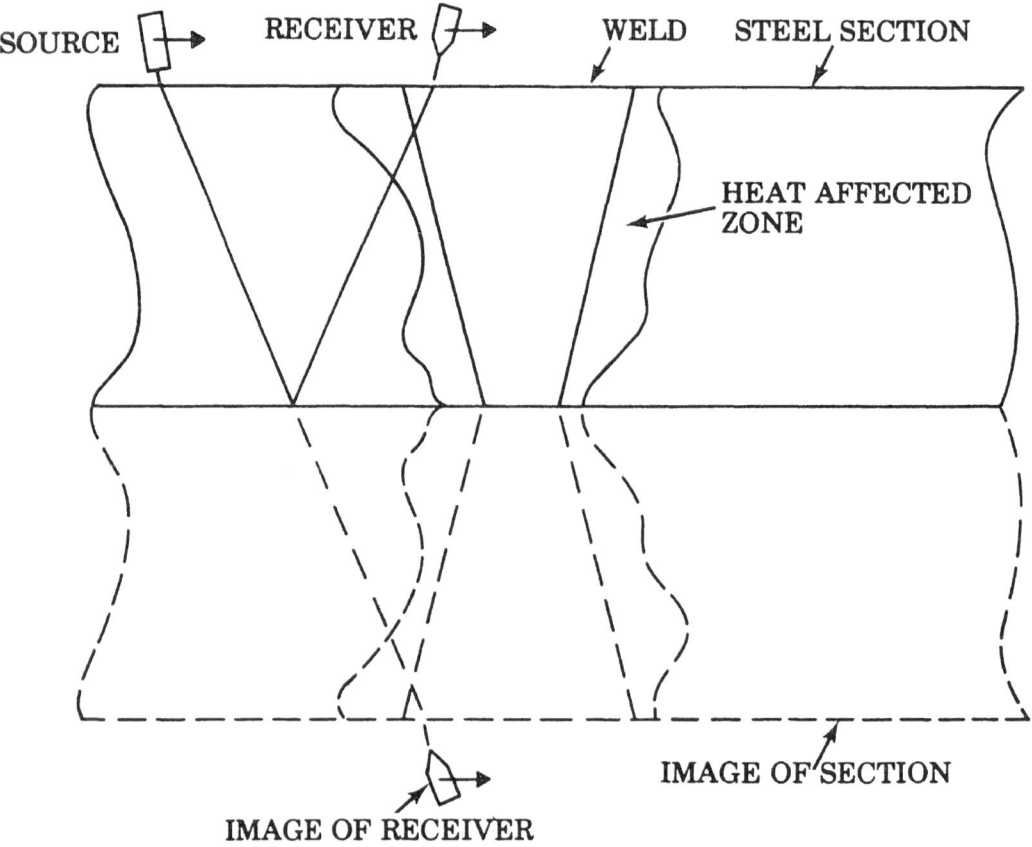

FIGURE 6. Scanning of Thick Sections by Reflection

of the section is simply used as a reflector and the receiver
placed to receive the reflected signal. Note, however, that the
reconstruction will now include a mirror image as well as the object
itself.

Note that in both cases, the total geometric path length
changes as a function of angle of view. This must, of course, be
taken into account in the reconstruction algorithm.

EXPERIMENTAL RESULTS

Experiment 1 Sensitivity

This experiment was designed to investigate the sensitivity of the
measurement-reconstruction system. Two fingers of a rubber glove

were filled with a mixture providing a velocity increase of 2%
and 0.2%. The physical arrangement was as shown in Figures 2 and
4; 21 rays over a span of 2.65 inches were measured in each profile and 19 profiles at 5° intervals were made, for a total field
of view of 90°. The first profile was taken parallel to the plane
of the fingers and the last perpendicular to it. Figure 7 shows
the 0° and 90° profiles and Figure 8 shows the reconstruction for
the 2% case.

The experiment was repeated with a 0.2% velocity increase in
the glove yielding the reconstruction shown in Figure 9. These
reconstructions are computed in terms of velocity change. With
the application of Equation 11 it would be easy to convert this to
stress.

Experiment 2   Reflection Method

This experiment was a simulation of the procedure sketched in
Figure 6. We built a goniometer arrangement as shown in Figure 10.
In this way the geometrical path length was kept constant. The
mirror was a 1/4 inch thick microflat glass plate, which we found
to be superior to anything else we tried. The object was one finger
of the rubber glove filled with a solution yielding a 2% velocity
increase.

The resulting reconstruction is shown in Figure 11. This figure clearly shows the mirror image generated by this procedure.

Experiment 3   Metal Model

After considerable thought and experiment we developed a model
that provided a realistic test in steel as well as a well controlled
stress field of known shape. As shown in Figure 12, a slab of type
1018 mild steel 2 x 1/2 x 3 inches was used as the base metal. A
1/4 inch hole was drilled as shown. A pin of diameter 0.2516 inches,
was made from the same steel stock. The pin was cooled in liquid
nitrogen (-190°C), the block heated to 1000°C, and the pin inserted
in the hole. After the model stabilized to room temperature the
sides were ground to remove scale and to assure constant thickness.
The model was then mounted on the turntable and scanned. In this
case, due to refraction, the angle through which the model was
rotated, was non-linear. That is, the angular increments were calculated to achieve 5° increments in the steel. Fifty-one measurements were made over a 0.9 inch span centered on the pin.

Rough computations show that the pin is subjected to about
100,000 psi compressive stress. The surrounding metal must balance

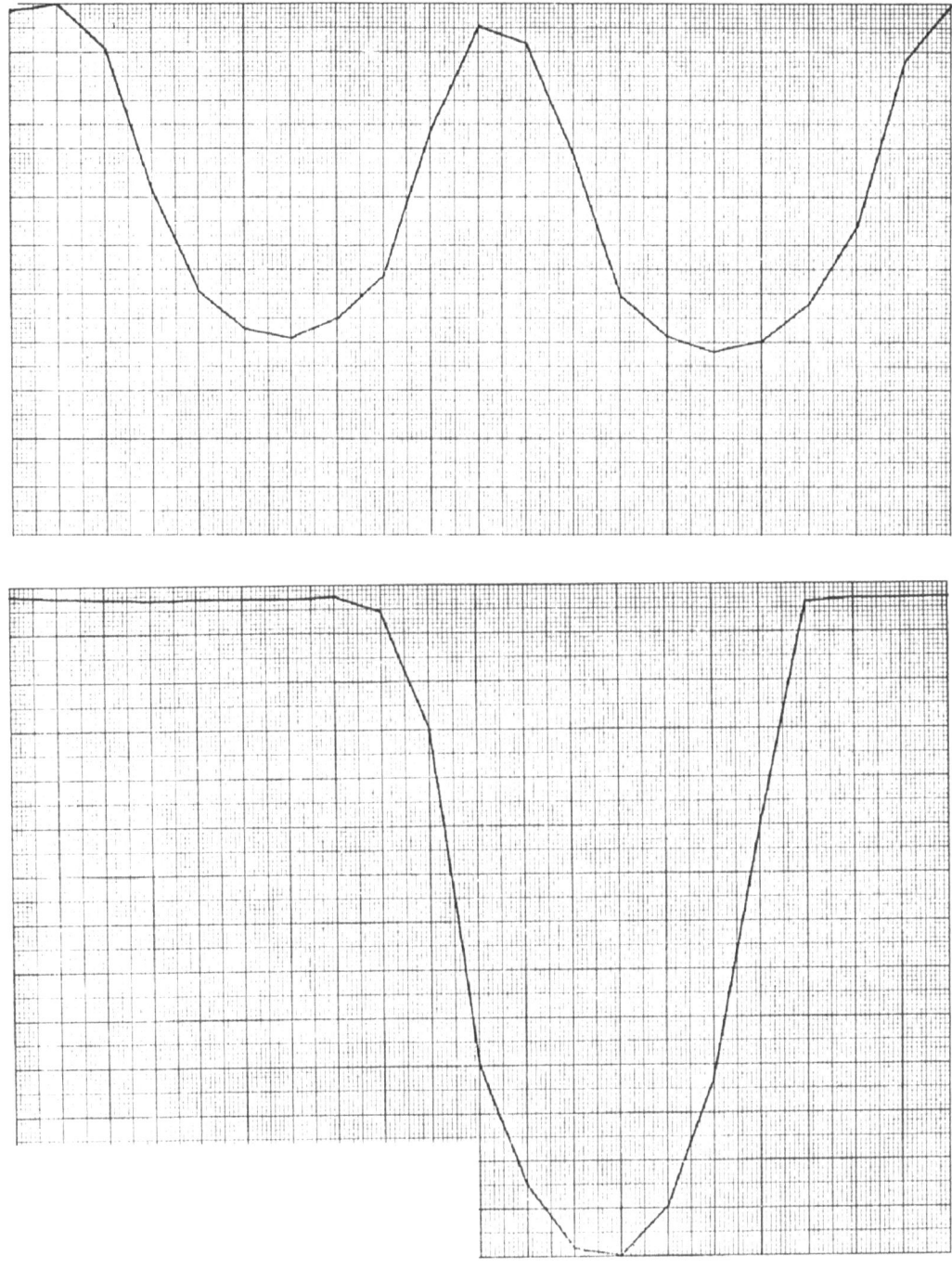

FIGURE 7. 0° and 90° Profiles of the 2% Liquid Model

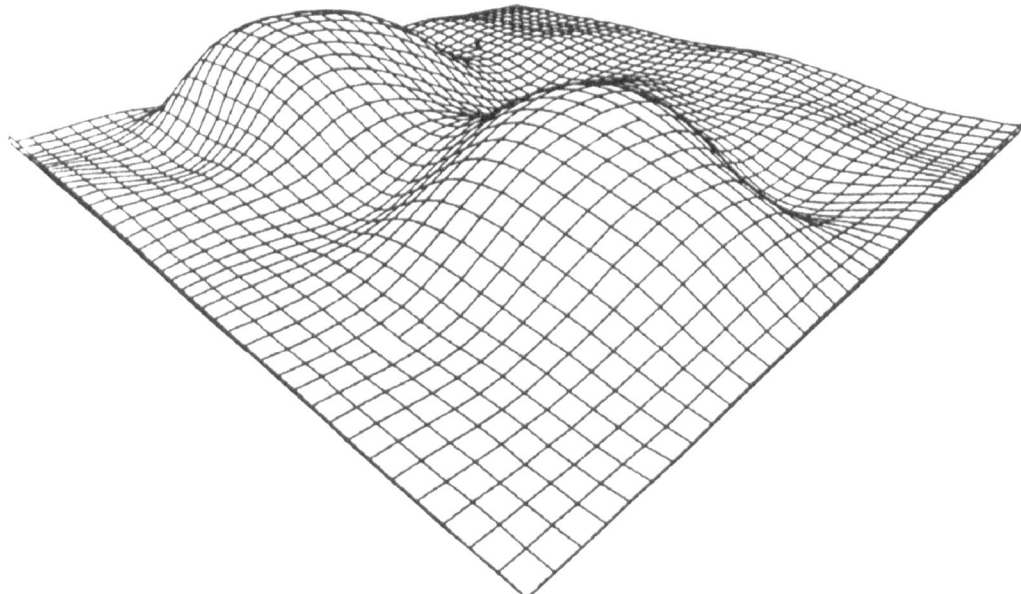

FIGURE 8.   Reconstruction of the 2% Liquid Model

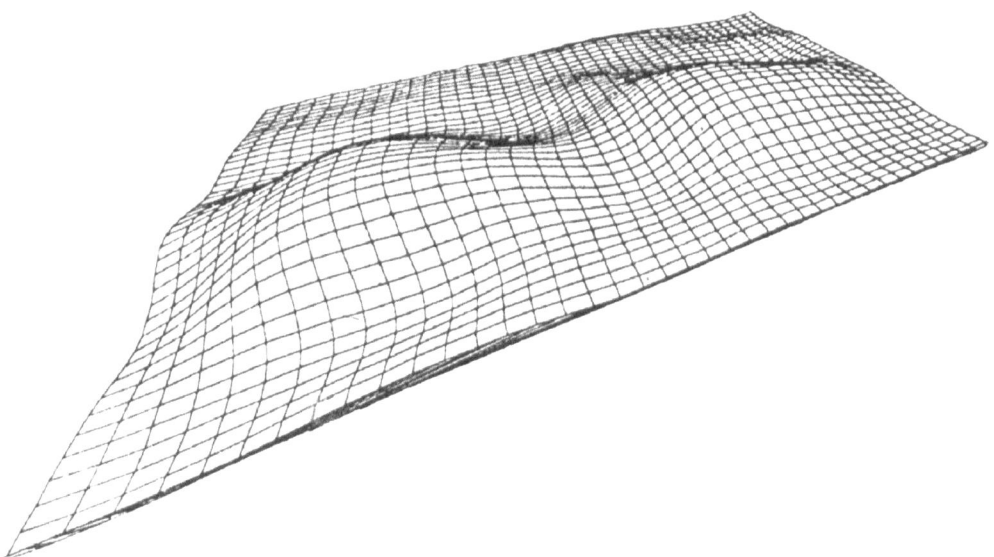

FIGURE 9.   Reconstruction of the 0.2% Liquid Model

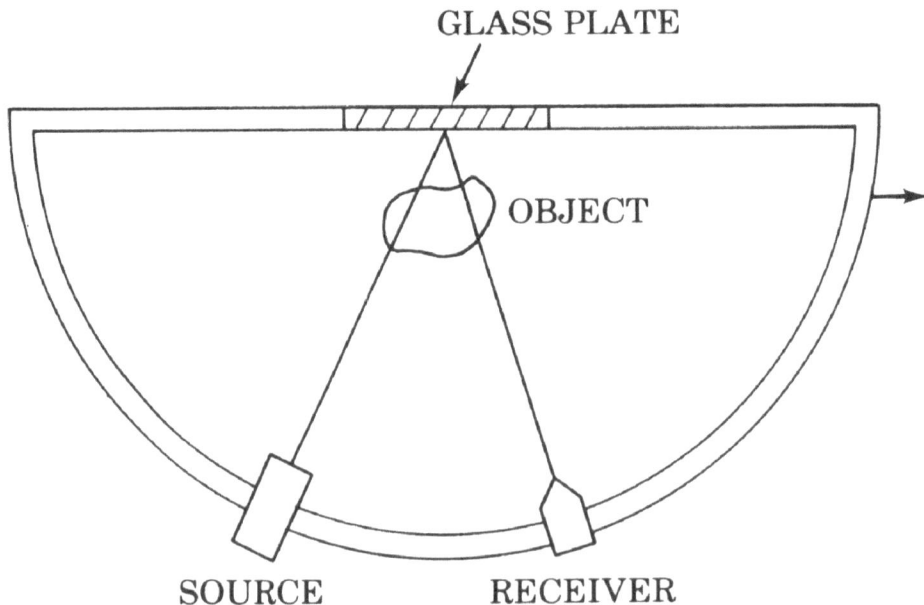

FIGURE 10.   Experimental Arrangement for Reflection Profiling

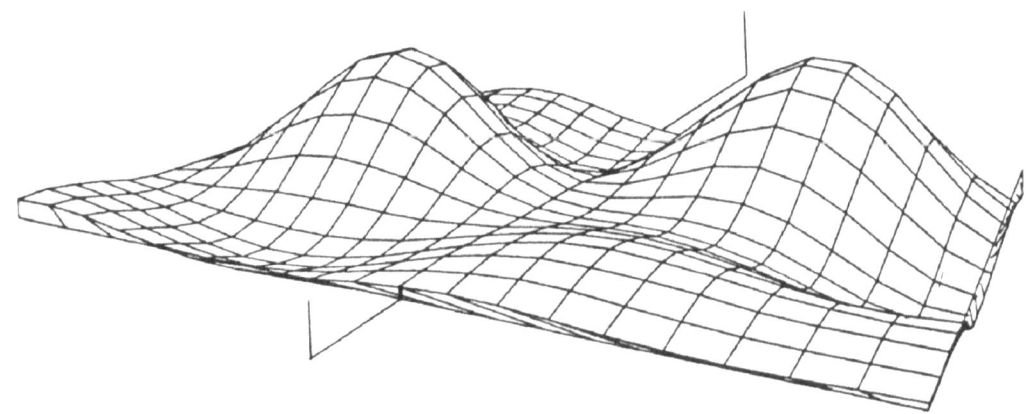

FIGURE 11.   Reconstruction of the 2% Liquid Model
             Profiled by Reflection

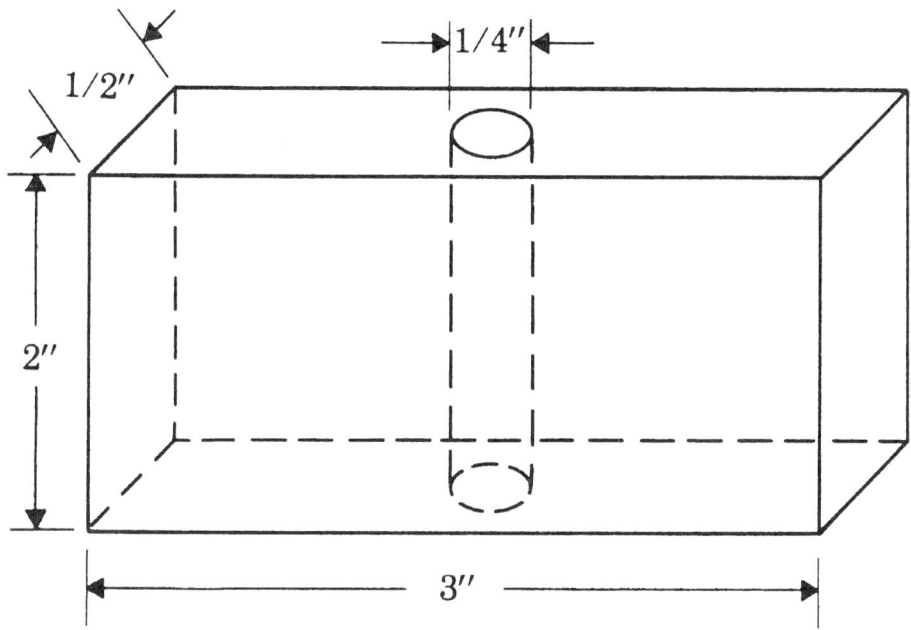

FIGURE 12. Drawing of the Steel Model

this with a distributed tensile stress. Figure 13 shows the 0°
profile. This is a plot of time-of-flight referred to the unstressed
part of the model. That is, an increase from baseline denotes velocity decrease and hence compressive stress. The total scan length
is 0.9 inch so that the pin itself encompasses rays 19-33. Within
this span the velocity has decreased while outside the pin velocity
has increased indicating tensile stress.

The reconstructed velocity section is shown in Figures 14 and 15.
These are two views of the velocity surface from top and bottom.
Since the program we are using is a direct matrix inversion we were
unable to use all the data we accumulated. Hence, we used only every
other ray measurement. Because of this, the reconstruction is not as
accurate as it could have been.

CONCLUSION

We have demonstrated that velocity fields can be imaged with
computerized axial tomographic techniques with sufficient resolution
to map residual stress concentration. We experimentally mapped
velocity anomalies of 0.2% and feel that 0.05% is technically feasible. These velocities translate to a sensitivity of 1000 psi in
a 1 inch thick region.

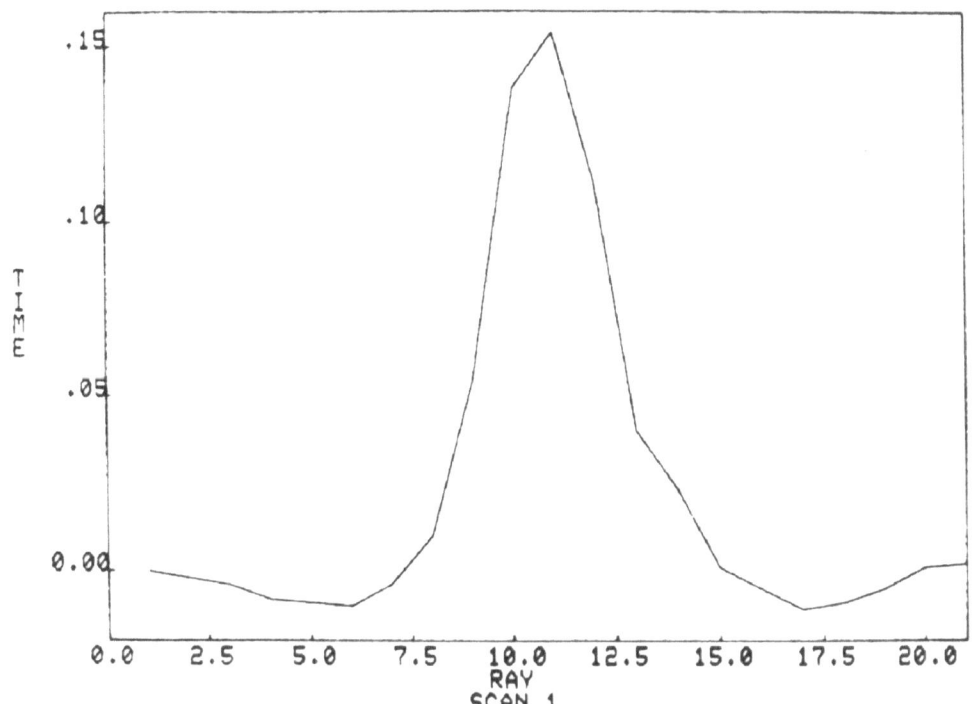

FIGURE 13.   0° Profile of the Steel Model

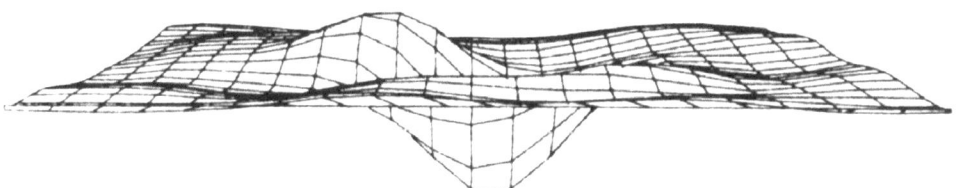

FIGURE 14.   Reconstruction of the Steel Model (from top)

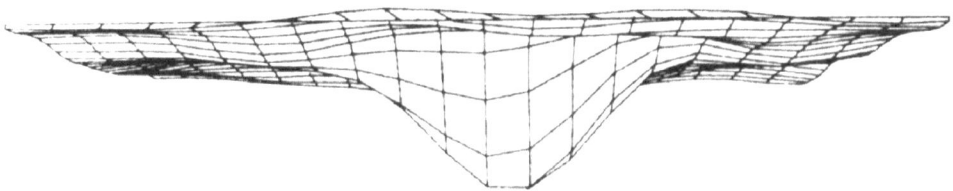

FIGURE 15.   Reconstruction of the Steel Model (from bottom)

We devised a protocol for obtaining the necessary time-of-flight profiles with only one accessible surface by using the back surface as a reflector.

A steel model made by shrink-fitting an oversized pin into a hole was designed, built and tested. The particular design we used turned out to provide gigantic stresses. Rough computations, based upon a linear stress-strain law, show that 100,000 psi compression stress is present at the surface of the pin. Our velocity measurements indicate either a much higher stress than that, or a fundamental change in velocity between the pin and the block due to the difference in treatment. A subsequent experiment with a sample of steel from the same billet showed the latter conjecture to be correct. A velocity change of 3.8% was measured after the sample was given the heat treatment. However, we feel that within the region surrounding the pin, a true stress effect is evident.

FUTURE PLANS

Our future plans call for further work on the steel model to make it more useful for calibration. For example, we will shrink fit the pin without heating the block. This will assure that the steel is not altered in any fundamental way. This will also assure that the stresses remain within the proportional limit. Computor calculations of stress fields will be plotted for comparison to the reconstruction from velocity data.

The measurement system will be improved so that the data can be taken automatically. Eventual plans call for an array so that the data can be taken in real time much as is done in the latest CAT scanners.

The ART algorithm will be implemented to speed up the reconstruction and provide a higher resolution image.

Finally, a prototype instrument for field testing on nuclear reactors will be designed and tested.

ACKNOWLEDGMENTS

We wish to acknowledge the help of a number of people; L. L. Kopf for the experimental work, C. M. Vest of the University of Michigan for his help in setting up the reconstruction program, S. A. Johnson and J. F. Greenleaf for their generous advice and provision of the ART program, and P. L. Carson of the University of Colorado for provision of the convolution algorithm. Finally we wish to thank K. Stahlkopf of EPRI for his encouragement and enthusiasm for this project and Professor K. Davis of Reed College for his continued interest and assistance during his tenure as a NORCUS fellow.

REFERENCES

1. B. J. Ratcliffe, "A Review of the Techniques Using Ultrasonic Waves for the Measurement of Stress Within Materials," British Journal of N.D.T., September 1969.

2. D. I. Crecraft, "Ultrasonic Measurement of Stresses," Ultrasonics Vol. 6, April 1968.

3. N. H. Hsu, "Acoustical Birefringence and the Use of Ultrasonic Waves for Experimental Stress Analysis," Experimental Mechanics, May 1974.

4. R. T. Smith, "Stress-Induced Anisotropy in Solids - The Acousto-Elastic Effect," Ultrasonics, July-September 1963.

5. P. J. Noronha and J. J. Wert, "An Ultrasonic Technique for the Measurement of Residual Stress", J. of Testing and Evaluation, Vol. 3, March 1975.

6. Technical Digest of the Topical Meeting on Image Processing for 2-D and 3-D Reconstruction from Projections, August 1975, sponsored by Opt. Soc. Am.

REFRACTIVE INDEX BY RECONSTRUCTION: USE TO IMPROVE COMPOUND B-SCAN RESOLUTION

J. F. Greenleaf, S. A. Johnson, W. F. Samayoa, and C. R. Hansen

Biophysical Sciences Unit, Department of Physiology and Biophysics, Mayo Foundation Rochester, Minnesota 55901

ABSTRACT

Reconstructions of two-dimensional distributions of acoustic speed were utilized to correct digitized compound B-scan images for aberrations caused by inhomogeneous refractive index. Profiles of propagation delays of acoustic pulses obtained during compound transmission scanning at 60 angles of view separated by 6° were used to reconstruct the distribution of acoustic speeds within a 64 x 64 element grid in the scan plane within which 8 to 16 digitized B-scans were obtained for views separated by 22.5° or less. Each B-scan contained 1000 to 1300 pulses, digitized at 10 or 20 megasamples/s. Values of calculated acoustic speed within those elements of the reconstruction grid which were intersected by the locus of each pulse trajectory were utilized to map the temporal sequence of echoes within each echo signal into the correct spatial sequence of echoes within the B-scan image. Straight line approximations to the loci of the acoustic beams were used. The set of corrected B-scans were summed to obtain high resolution compound B-scans of 128 x 128 or 256 x 256 picture elements. This method seems particularly suited to breast imaging although enhancement of abdominal scans may be possible as well.

INTRODUCTION

Several ultrasonic imaging techniques utilize methods of obtaining echoes from multiple viewpoints (compound scanning) in order to minimize various aberrations caused by the intrinsic properties of ultrasonic imaging. Examples of such techniques are compound B-scanning (1), synthetic aperture imaging (2), and synthetic focusing (3), all of which obtain images utilizing acoustic echoes obtained from many angles of view around the subject of interest.

Since echoes from each point reflector are received from a number of angles, determination of the precise position in space of each reflector requires knowledge of the acoustic speed intervening between the transmitter, the point reflector, and the receiver. In the general case, the distribution of acoustic speed is unknown and therefore the spatial position of the point reflectors responsible for each echo in the return signal is not known accurately. The resulting image aberrations are strongest when the phase of the signal is required in the imaging process, as in the cases of synthetic aperture and synthetic focus imaging. In these cases, the phase of the signal can be shifted many radians by inhomogeneous distributions of acoustic speed, and therefore summation of these echoes from many angles can cause serious errors in the resulting image.

This laboratory has recently developed techniques of obtaining quantitative distributions of refraction index within tissues utilizing computer assisted ultrasonic tomography (4). In principle, by utilizing these two-dimensional quantitative distributions of acoustic refraction index, one can correct the echoes obtained during a compound B-scan or synthetic focus or synethetic aperture imaging and obtain higher resolution and aberration-free images (5). Since the arrival times of pulses propagated through materials of interest, such as tissues, can be measured to within 5 to 10 nanoseconds, echo arrival times in compound B-scans, and echoes from other types of data acquisition, can be corrected to within a quarter wavelength for signals containing frequencies up to 5 MHz or more.

The purpose of this paper is to describe results of correcting compound B-scan aberrations utilizing

reconstructions of the distribution of acoustic speed
obtained by computer assisted tomography (4).

METHODS OF PROCEDURE

Compound B-scan data were obtained by divergent
fan beam scanning utilizing a 1 cm diameter 5 MHz
transducer with an F 10 lens made of polystyrene. The
B-scan data were obtained by digitizing the received
pulses at a rate of one sample per 100 nanoseconds at a
resolution of 8 bits per sample. Up to 1300 samples
were digitized from the signal representing each return
echo. The echoes for each B-scan view were obtained
with a divergent mechanically scanned fan beam containing 150 digitized echo paths equispaced over an
angular range of 22.5 degrees. Up to 16 such B-scan
views were obtained, equispaced over 360°. Data required for reconstructing the distribution of index of
refraction within the region of study were obtained
with the same fan beam scanner and were analyzed in a
manner similar to that described previously (4).

Reconstructions of acoustic velocity within a
two-dimensional region divided into 64 x 64 elements
were utilized to correct the arrival time of the pulses
within the compound B-scans.

The arrival time in a B-scan of a pulse echo from
a reflector at r, which has traveled through media of
acoustic speed C(r) is:

$$t(r) \approx 2 \int_o^r (\frac{1}{C(1)}) \, dl \tag{1}$$

where the 2 takes into account the double path length,
and the path of integration is approximated by a
straight line from the transmitter (at r = 0) to the
reflector at r.

Given the distribution of acoustic speed C and
the path along which to integrate, t(r) is the time
within the echo signal at which to look for an echo
arriving from a reflector at r.

By performing the integral indicated by Equation
1 along the line in space represented by each B-scan
echo path, the true position in the image for each
return echo in the B-scan can be calculated.

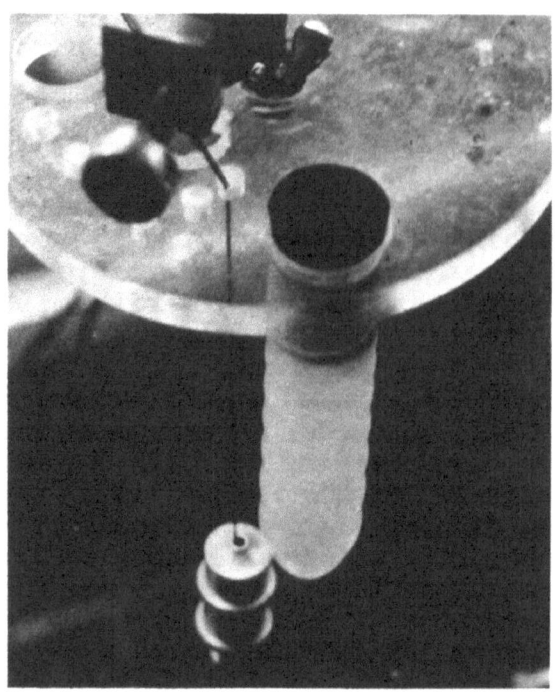

Figure 1 Photograph of the test object ("phantom")
utilized to obtain compound B-scans to be corrected by
reconstructions of the distribution of acoustic speed
in a transverse plane through the phantom. The phantom
consists of a finger cot filled with saline whose
acoustic velocity is 5% higher than that of the surrounding water (not shown) and a string plumb line at
the geometric center of the compound B-scan region.

EXPERIMENTAL PROCEDURE

    A test object (phantom) was utilized for the
experiments to be described. The phantom consisted of
a rubber finger cot filled with distilled water containing salt such that the speed of sound within the
saline-filled finger cot was approximately 5% higher
than that of the surrounding water. In addition to the
finger cot a string held taught by a plumb bob was
centered precisely in the scan region (Figure 1).

    Reconstructions of the distribution of velocity
within this phantom were calculated from 60 fan beam
scans, 6° apart, each containing 150 echo paths equispaced over 22.5°. The resulting reconstruction, shown

Figure 2 Reconstruction of the distribution of the
reciprocal of the acoustic speed within the phantom
shown in Figure 1. Circular region represents the
finger cot filled with saline of acoustic speed approximately 5% higher than the surrounding tap water (white
background area). The high velocity of the string
suspended at the center of the image can also be seen.
Reconstruction data were obtained from 60 views separated by 6°, each view consisting of 150 time of flight
samples obtained in a fan beam over 22.5°. Image was
calculated from data utilizing a fan beam convolution
algorithm.

in Figure 2, was obtained by a divergent convolution
technique described elsewhere (6). Compound B-scans
were obtained of this phantom by scanning divergent
B-scans at 16 points of view equally spaced around
360°. Each B-scan contained 150 digitized pulses.

A computer generated compound B-scan utilizing the
raw data obtained from 16 equispaced views around the
phantom is shown in Figure 4. Note that the echoes
from the string in the center of the image are blurred
toward the finger cot. This effect is caused by the
higher acoustic speed within the saline finger cot
which alters the position in time of the received
echoes. Those echoes received from the string after
having traversed only water will of course arrive at
the correct time in the signal. Echoes from the string

which have traversed the saline region will arrive up to
1.5 μseconds early causing their positions in the
B-scan image to be displaced toward the finger cot.

Figure 4 illustrates the corrected B-scan, in
which the spatial positions of the various echoes have
been corrected utilizing Equation 1 and the acoustic
speed of the path over which they have traveled. The
acoustic speed C(r) was measured from the reconstruction
of acoustic speed shown in Figure 2. The central spot
representing the compound echoes from the string in the
center of the image is now in sharp focus. Also, note
that the border of the finger cot is much sharper than
in Figure 3. Figures 3 and 4 have been constructed
utilizing the detected (rectified) signal rather
than the raw signal as digitized by the A-D converter.
For extremely high resolution, compound B-scans should
be obtained by adding the return signals after deconvolution of the echoes with the transient response of
the transducer (7).

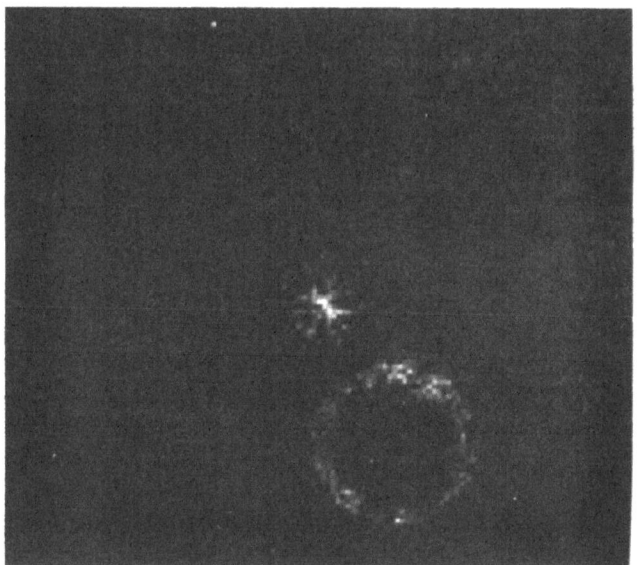

Figure 3 Computer generated compound B-scan of phantom
shown in Figure 1. Sixteen views equispaced around
360°, each view consisting of 150 pulses equispaced
over a fan beam covering 22.5° were utilized for this
compound B-scan image. Echo signals were digitized at
a rate of 10 megasamples/s.

Figure 4   Compound B-scan shown in Figure 3 has been
corrected for arrival times utilizing the reconstructed
distribution of acoustic speeds illustrated in Figure 2.
Note increased resolution of echoes from the string in
the center of the image and sharper focus of the echoes
around the perimeter of the finger cot. Blurring shown
in Figure 3 has been greatly decreased.

     An example of the degree to which B-scans are
blurred in real tissue is indicated in Figure 5, in
which perfect B-scans were simulated in the computer
such that a single echo was received in each B-scan
view from a point reflector at the exact center of the
image. These echoes were then back-projected as a
compound B-scan utilizing the distribution of acoustic
speeds obtained by reconstructing the two-dimensional
distribution of acoustic speeds within an excised
fresh breast. The reconstruction of the acoustic speeds
within the excised breast are shown in Figure 6. Note
that echoes received at almost any angle from a reflector at the center of the breast will have altered
arrival times due to differences in acoustic speed
encountered by the pulse traveling through about the
same amount of tissue - half of the breast. Arrival
times of echoes from point reflectors near the surface
of the breast, however, will be displaced a greater
amount since echoes will traverse a small amount of
tissue for transmitter positions on the side of the

Figure 5 Computerized compound B-scan of theoretically
perfect pulse echo as if obtained from a perfect reflector at the center of the breast whose velocity reconstruction is shown in Figure 6. Perfect reflector
is blurred by up to 1.5 microseconds or about 1 to 2 mm
due to inhomogeneous distribution of acoustic speed
within the breast. Blur from the reflector at edge of
breast would be expected to be greater since inhomogeneity of acoustic speeds encountered by echoes from
the various directions would be greater. Arrival time
of echo within each B-scan was calculated by integrating
elemental delays along intersection of the pulse trajectory with grid of elements of acoustic speed reconstruction shown in Figure 6.

breast near to the reflector, but the arrival time
may be greatly altered for echoes received from the
reflector through the diameter of the entire breast.
Therefore, the calculated degree of blurring caused by
the point reflector at the center of the breast, as
shown in Figure 6, is probably a conservative estimate.
Arrival times are altered as much as 1.5 microseconds
within this image and thus the radius of the blurring
is on the order of 1 to 2 mm. The blurring for point
reflectors at the edge of the breast could be expected
to be twice this amount and most of the blur will be in
the radial direction.

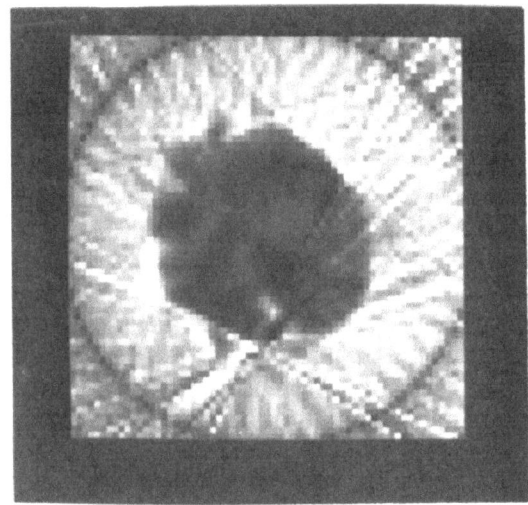

Figure 6 Reconstruction of distribution of acoustic
speed within an excised breast obtained by subcutaneous
mastectomy postmortem. Reconstruction was calculated
from 60 views of the breast separated by 6°. Each
view contained 150 samples equispaced over an angle of
45°.

DISCUSSION

     Recent developments of techniques for obtaining
distributions of refraction index within transverse
planes through tissues allow the accurate correction of
compound B-scans for inhomogeneous distributions of
refraction index. Since arrival times of pulses transmitted through tissue can be measured to within 5 to 20
nanoseconds, and since it is the total error of arrival
time which is utilized to correct B-scan data, it is
possible to correct B-scan data to within a fraction of
a wavelength even in signals containing frequencies
as high as 5 MHz. Whether or not such corrections can
be applied to synthetic aperture and synthetic focusing
imaging techniques is yet to be seen. However, before
true synthetic aperture imaging can be accomplished in
tissue, the phase of the various received echoes
utilized to construct a high resolution image must in
some manner be corrected for inhomogeneous distributions of refraction index.

     Certainly compound B-scans of the breast can be
obtained simultaneously with the time of flight profiles required for computer assisted tomography. Such

B-scans could be corrected utilizing the method described in this report. A device which scans the
breast and obtains reconstructions of two-dimensional
distributions of refraction index and attenuation coefficients (4) along with high resolution compound
B-scans of multiple transverse two-dimensional planes
covering the anatomical extent of the intact breast
could be built in which all three modes of data could
be acquired at the same time during the same scan.

The availability of several independent quantitative acoustic properties of tissue should greatly aid
the development of mass population screening techniques
utilizing automated tissue characterization methods.

ACKNOWLEDGEMENTS

These investigations were supported in part by
Research Grants HV 42904, HL 00060, HL 00170,
RR 00007, and HL 04664 from the National Institutes of
Health, U.S. Public Health Service.

REFERENCES

1. Kossoff, G., W. J. Garrett, and G. Ranavanovich:
    Grey scale ecography in obstetrics and gynecology.
    Australasian Radiology 28(1) March, 1974, pp 63-111.

2. Keating, P. N., R. F. Koppelmann, T. Sawatari, and
    R. F. Steinberg: Holographic aperture synthesis
    via a transmitter array. Acoustical Holography, 6
    (Ed: Newell Booth), Plenum Press, New York, 1975,
    pp 485-506.

3. Johnson, S. A., J. F. Greenleaf, F. A. Duck,
    A. Chu, W. F. Samayoa, and B. K. Gilbert:
    Digital computer simulation study of a real-time
    collection, postprocessing synthetic focusing
    ultrasound cardiac camera. Acoustical Holography,
    6 (Ed: Newell Booth), Plenum Press, New York,
    1975, pp 193-211.

4. Greenleaf, J. F., S. A. Johnson, W. F. Samayoa, and
    F. A. Duck: Algebraic reconstruction of spatial
    distributions of acoustic velocities in tissue
    from their time of flight profiles. Acoustical
    Holography, 6 (Ed: Newell Booth), Plenum Press,
    New York, 1975, pp 71-90.

5. Johnson, S. A., J. F. Greenleaf, W. F. Samayoa,
    F. A. Duck, and J. D. Sjostrand: Reconstruction of
    three-dimensional velocity fields and other para-
    meters by acoustic ray tracing. 1975 Ultrasonic
    Symposium Proceedings. IEEE Cat. #75, CHP 994-4SU.

6. Lakshminarayanan, A. V.: Reconstruction from
    divergent ray data. Technical Report #92, January,
    1975, Department of Computer Science, State Uni-
    versity of N. Y. at Buffalo.

7. Kak, A. C., L. R. Beaumont, and J. Wolfley: Signal
    processing in the determination of acoustic impe-
    dance profiles. TREE 75-7, December, 1974, Depart-
    ment of Electrical Engineering, Purdue University,
    West Lafayette, Indiana.

APPLICATION OF ACOUSTICAL HOLOGRAPHY TO NOISE DATA ANALYSIS

W. S. Gan

Department of Physics, Nanyang University
Upper Jurong Road, Singapore 22
Republic of Singapore

ABSTRACT

A measure basic to nearly all noise data analysis works is filtering. Filtering is the extraction of useful information from data. Holography is information and so it can be used to extract certain information contained in a signal. Here we are interested to extract the repetitive or periodic signals from noise. The technique used is acoustical correlation filtering. We first record the periodic signal with a reference signal and this acoustical hologram is the complex filter function. The experimental method of recording the hologram follows that of Graham's thesis. The filtered signal is the convolution of the composite noise spectrum to be analyzed and the reconstruction from the holographic filter function. The next step is to obtain the reconstruction of the filter function in digital form. Digitization is done from the photographic form of the hologram. The digitized form serves as an input for the computer. The filtered signal then can be given as the output of the computer.

INTRODUCTION

In a previous paper[1] I pointed out the use of acoustical holography in noise source identification. It extends optical correlation filtering to the

acoustical case. Here we show the possibility of
applying acoustical holography to noise data analysis.
Again we use the technique of acoustical correlation
filtering. In optical works, we can filter out the
word we want from a page in a book. This is analogous to filter out the signal we are interested from
a complex noise spectrum. The application of these
papers is to noise control works. The advantage of
the holographic method over conventional method is
time saving, when a large number of noise signals
have to be analyzed and correlated.

THEORETICAL   BACKGROUND

      In optical correlation filtering,[2] the filtered
image is given by
                  f ⊛ h, which is the convolution
of the two functions.
                  f = Input = text being filtered,
h = reconstruction of holographic filter of word to be
filtered. Extending this to the acoustical case,
f will become the complex noise spectrum to be analyzed and h the reconstruction of the holographic filter of the periodic signal to be filtered. We shall
illustrate this by the following examples:
(1)An input square pulse: The shape is given in Fig.1.

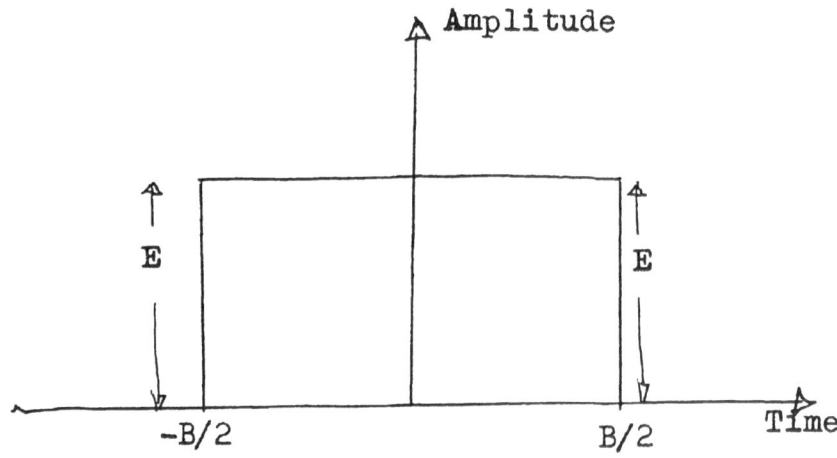

Fig. 1. A Square Pulse.

The Fourier Transform of the signal to be analyzed is

$$= F(\omega) = \int_{-B/2}^{B/2} E\, e^{-j\omega t}\, dt$$

$$= \frac{2E}{\omega}\, \sin\left(\omega \frac{B}{2}\right) \tag{1}$$

for t = -B/2 to B/2 and
$F(\omega) = 0$ for the rest of time.
Now we want to filter out the periodic signal $E\cos\omega t$
from the square pulse. We add on the sinuisoidal
wave $\cos\omega t$ for recording the acoustical hologram.
Then the holographic filter function

$$= h(t) = (E + 1)\cos\omega t \tag{2}$$

Then the Fourier Transform of h(t)

$$= H(\omega) = \int_{-\infty}^{\infty} h(t)\, e^{-j\omega t}\, dt$$

$$= \int_{-B/2}^{B/2} (E + 1)\cos\omega t\, e^{-j\omega t}\, dt$$

$$= (E + 1)\left(\frac{B}{2} + \frac{\sin\omega B}{2\omega}\right) \tag{3}$$

Hence $F(\omega) \times H(\omega) = (E + 1)\dfrac{2E}{\omega}\sin\left(\omega\dfrac{B}{2}\right)\left(\dfrac{B}{2} + \dfrac{\sin\omega B}{2\omega}\right)$

From the Convolution Theorem, which states that
Fourier transform of the Convolution of two functions
is equal to the product of the transform of the two
functions; we have

$$f(t) \circledast h(t) = \frac{1}{2\pi}\int_{-\infty}^{\infty} e^{j\omega t}\left[\,F(\omega)\times H(\omega)\right]\, d\omega$$

$$= \frac{1}{2\pi}\int_{\omega_1}^{\omega_2} e^{j\omega t}\,(E + 1)\frac{2E}{\omega}\,\sin\left(\omega\frac{B}{2}\right)$$

$$\left(\frac{B}{2} + \sin\omega\frac{B}{2\omega}\right)\, d\omega \tag{4}$$

which can be shown to be the function $E\cos\omega t$.

(2)A damped free vibration: The shape is given in
    Fig.2.

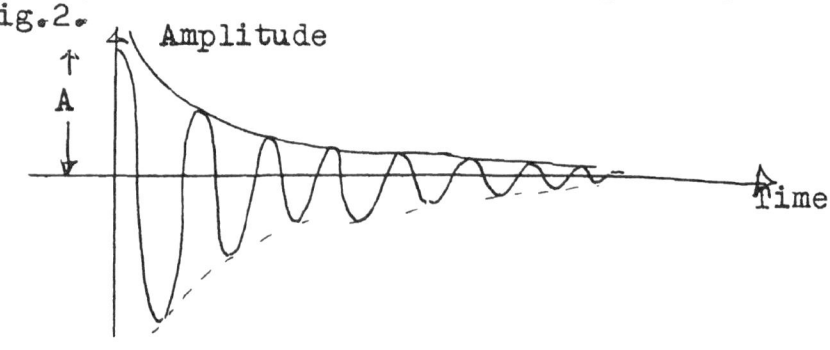

Fig.2. A Damped Free Vibration.

        The input signal to be analyzed is given by
$f(t) =$
        $A \cos(\omega t + \phi) e^{-\alpha t}$.

The Fourier Transform of this function is

$= F(\omega) = \int_0^{t_1} A\cos(\omega t + \phi) e^{-\alpha t} e^{-j\omega t} \, dt$

$\qquad = \text{Real Part of } \int_0^{t_1} A e^{j(\omega t + \phi - \omega t) - \alpha t} \, dt$

$\qquad = A \dfrac{\cos\phi}{\alpha} (1 - e^{-\alpha t_1})$ \hfill (5)

(t_1 = time taken for the input signal to die out to
a very small value)
Now we want to filter out the periodic signal
$A\cos(\omega t + \phi)$ from the damped vibration. We add on
the sinuisoidal wave $\cos \omega t$ for recording the acoustical hologram. Then the holographic filter function

$= h(t) = A\cos(\omega t + \phi) + \cos \omega t$ \hfill (6)

The Fourier Transform of $h(t)$

$= \int_0^{t_1} \left[A\cos(\omega t + \phi) + \cos \omega t \right] e^{-j\omega t} \, dt$

$= \dfrac{A}{2} \dfrac{\sin(2\omega t_1 + \phi)}{2\omega} + \dfrac{A}{2} \cos\phi \, t_1 + \dfrac{t_1}{2} + \dfrac{\sin(2\omega t_1)}{4\omega}$ \hfill (7)

keeping only the real parts. $- \dfrac{A}{2} \dfrac{\sin\phi}{2\omega}$

Hence $F(\omega) \times H(\omega) = A \dfrac{\cos\phi}{-\alpha}(e^{-\alpha t_1} - 1) \times$

$$\left[\frac{A}{2}\frac{\sin(2\omega t_1 + \phi)}{2\omega} - \frac{A}{2}\frac{\sin\phi}{2\omega} + \frac{A}{2}\cos\phi\, t_1 + \frac{t_1}{2}\right.$$

$$\left. + \frac{\sin 2\omega t_1}{4\omega}\right] \tag{8}$$

Similarly as the first example,

$$f(t)\circledast h(t) = \frac{1}{2\pi}\int_{\omega_1}^{\omega_2} \cos\omega t\; A\frac{\cos\phi}{-\alpha}(e^{-\alpha t_1} - 1)\times$$

$$\left[\frac{A}{2}\frac{\sin(2\omega t_1 + \phi)}{2\omega} - \frac{A}{2}\frac{\sin\phi}{2\omega} + \frac{A}{2}\cos\phi\, t_1 + \frac{t_1}{2}\right.$$

$$\left. + \frac{\sin 2\omega t_1}{4\omega}\right]\, d\omega \tag{9}$$

which can be shown to give the periodic function
$A\cos(\omega t + \phi)$.

EXPERIMENTAL SET-UP

The experimental method for recording the acoustical hologram follows the way of Graham[3]. Here the hologram is recorded electronically and not like the liquid surface method. We need an acoustic field scanner to carry microphone and the light over a plane in the sound field of the acoustic source. The arrangement is shown in the following diagram:

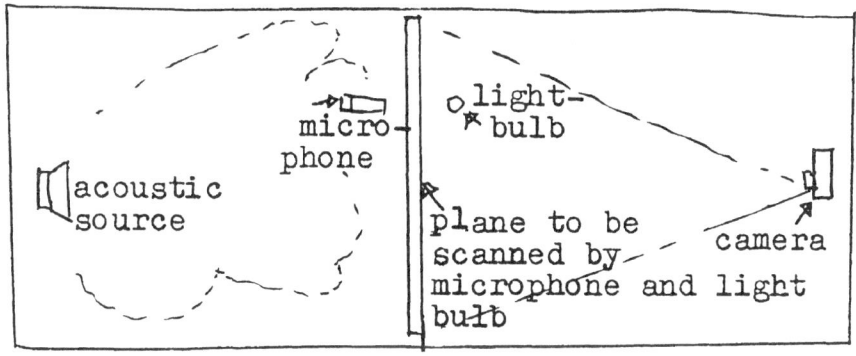

Fig. 3. Experimental Set-Up for Recording
the Acoustical Hologram.

DIGITIZATION OF THE PROCESS

This is done in the following way: The reconstructed form of the holographic filter function
is generated by the two-dimensional convolution of
the hologram with a complex valued propagation function. The process treats the hologram as a wave
field in a plane and computes the form that field
assumes after propagating some specified distance
normal to the hologram plane. The double integral
of the convolution is then programmed and fed into
the computer and this is the input to the computer.
The filtered signal is then given by the output of
the computer. So now we have the reconstructed form
of the holographic filter function as

$h(t) \circledast g(t)$ where $g(t)$ is the reference wave.
For our first example of the square pulse, this will
be

$$\frac{1}{2\pi} \int_{\omega_1}^{\omega_2} e^{j\omega t} \frac{2E}{\omega} \sin\left(\frac{\omega B}{2}\right) \times \left(\frac{B}{2} + \frac{\sin\omega B}{2\,\omega}\right) d\omega$$

which can be programmed.
For our second example of the damped free vibration,
the required integral for the reconstructed filter
function is

$$\frac{1}{2\pi} \int_{\omega_1}^{\omega_2} e^{j\omega t} \frac{A}{2} \frac{\sin(2\omega t_1 + \phi)}{2\,\omega} - \frac{A}{2} \frac{\sin\phi}{2\,\omega}$$

$$+ \frac{A}{2}\cos\phi\, t_1 + \frac{t_1}{2} + \frac{\sin 2\omega t_1}{4\omega} \times \left(\frac{B}{2} + \frac{\sin\omega B}{2\omega}\right) d\omega$$

which can be programmed as the input to the computer.

APPLICATIONS

The holographic method has the advantage of time
saving over the conventional methods, when a large
number of noise signals have to be analyzed. Another
advantage is that the results can be given as three-
dimensional plots instead of two-dimensions as given
by ordinary technique. Our method has potential
applications in analyzing seismic data from oil exploration works. The experimental system can be designed to perform the filtering simultaneously for
dozens, hundreds and in principle thousands of chan-

nels, and then to provide a very simple means for correlating the signals represented in the different channels , e.g. 120-channel analysis has been carried out in some oil-industry for analyzing the seismic data in oil exploration. Each of the channels may represent the signal from the seismograph, in this case. In order to extend our method to this case, we must have a holographic filter function for each channel. All these filter functions can be stored on the hologram. This holographic filtering of all signals simultaneously can save a great deal on the computing time. This will be extension of this paper.

REFERENCES

1. W. S. Gan, "Application of Holography to Noise Source Identification," Proceedings of the 1974 International Conference on Noise Control Engineering, p. 113, Institute of Noise Control Engineering, U. S. A. (1974).

2. G. W. Stroke, An Introduction to Coherent Optics and Holography, Academic Press, Second Edition, New York, (1969), p.83-85.

3. T. Graham, Ph.D. Thesis (1969), "Long Wavelength Acoustic Holography".

4. R. L. Cohen and M. S. Lang, to be published.

5. L. J. Cutrona, E. N. Leith, C. J. Palermo, and L. J. Porcello, ITE Trans. Inform. Theory, $\underline{6}$, 386(1960).

SCAN CONVERTER FOR ULTRASONIC MEDICAL IMAGING

J. Deschamps
Thomson- CSF, Electron Tube Division

R. J. Kleehammer
Dumont Electronics Corporation

1 - INTRODUCTION

There is a growing requirement in scientific, medical, civil
and military fields for a means of storing the information
contained in an image with up to 2000 lines resolution. The
solution adopted must be economical, compact and reliable.

The subject of this paper is the TH 7505, a scan converter
that has been specially designed to satisfy this need. It
permits signals from various sensors, particularly type-B
echography images from ultrasonic probes, to be displayed
as TV images.

It is designed around a TH 8811 high-resolution, double-ended
storage tube whose characteristics, fully exploited by the
associated circuitry, give the TH 7505 some important capabilities :

- continuous image readout, even during writing, because of the
  use of a storage tube having two electron guns. There is no
  crosstalk between the writing and reading sections.
- faithful halftone reproduction for all writing speeds and
  modes.
- the possibility of accurately erasing selected parts of the
  recorded information.
- image storage time greater than 30 minutes ; possibility of
  long-term retention (information storage for a prolonged period
  of time when the electron beams are cut off).

- high resolution, and zoom capability for detailed examination
of selected parts of the image.

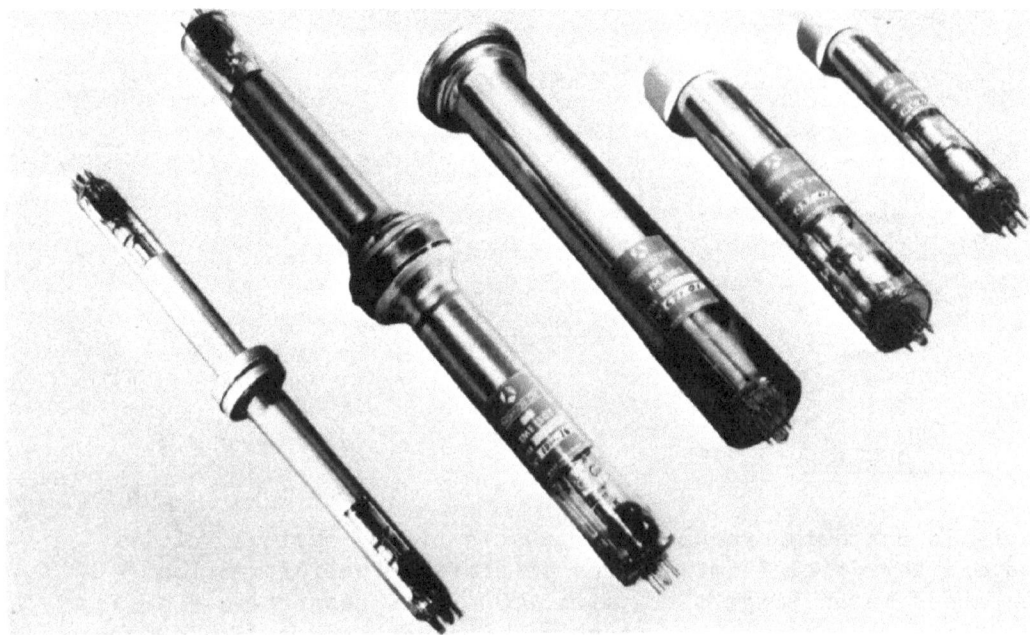

Fig. 1.   The TME Storage-tube Family

2 - THE TH 8811 DOUBLE-ENDED STORAGE TUBE

A storage tube is an electron tube having a target that can store
data, and one or several electron guns that are used to write and
read this data. In single-ended tubes (one gun) reading and writing
is sequentiel ; with two guns (see Figure 2) these two functions
can be performed simultaneously.

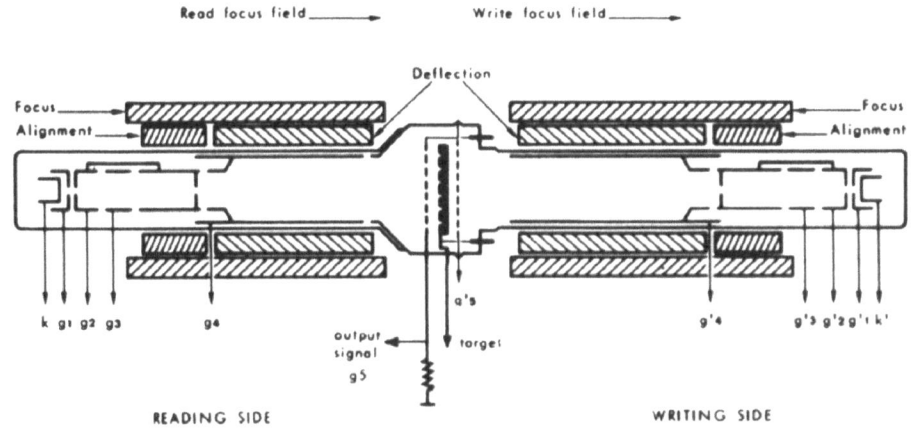

Fig. 2.   Structure of the TH 8811 high-resolution double-ended storage
tube.

The TH 8811 is a high-resolution double-ended storage tube. Its structure is shown schematically in Figure 2. The main components are two electron guns (reading and writing guns) located on opposite sides of the storage target assembly. Both guns use electromagnetic focusing and deflection. Each one consists of : a cathode, k ; a control grid, g1 ; an accelerating grid, g2 : an erasing grid, g3 ; a focusing grid, g4 and a field mesh, g5 The field meshes are very fine, and are mounted near and parallel to the target.

The target is of the membrane type, consisting of a continuous, thin dielectric film deposited on a conducting mesh. Its structure permits the arrival of reading beam electrons to be controlled by means of negative charges deposited on the surface of the dielec-tric (a technique known as coplanar control). The principle of operation of the reading side is analogous to that of single-ended tubes (silicon-target storage tubes) and permits continuous readout for several minutes without degradation of the information; the readout is non-destructive.

The important, and special property of this tube's target is that charge deposition can be performed on either of its surfaces ; this is because of the strong capacitive coupling effect, due to the thinness of the dielectric film. This means that the three operations (writing, reading, erasing) can be carried out by either of the two guns. This versatility permits various combi-nations of operating modes to be used, depending on the time available for each operation in the given application. The tube's structure eliminates crosstalk between the reading and writing signals : the target forms a screen between the two parts of the tube, preventing the writing-beam electrons from penetrating into the reading section. This property, together with the fact that the output signal is taken from the field mesh, results in complete isolation between the reading and writing sections, permitting the two electron guns to be used simultaneously.

3 - DESCRIPTION OF THE TH 7505

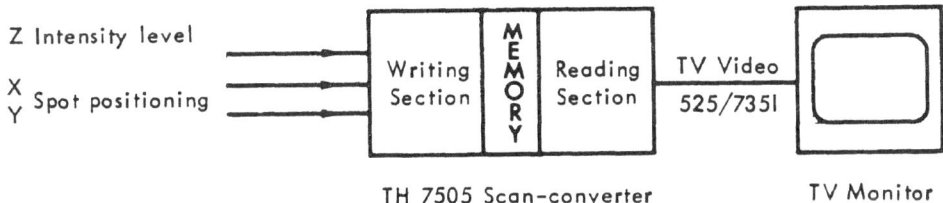

Fig. 3. Block diagram of TH 7505 scan-conversion unit operation.

TH 7505 is a scan-conversion unit that can be represented schematically as shown in Figure 3.

The sensor, in this case an ultrasonic probe, supplies an intensity modulation signal (Z) and X and Y spot-positioning signals to the writing section. The electron beam, controlled by these three signals, deposits a charge distribution on the dielectric of the storage target.

The reading section consists of circuits that cause the reading beam to scan the storage target following a TV standard. This beam is modulated by the charge distribution on the target, and gives a TV video signal that can be applied directly to a standard monitor.

The conversion of the echo signal from the probe into a TV video signal has thus been made possible by a target that can store information.

We shall now examine some points of interest in the operation of the TH 7505.

3.1 Continuous Readout

Because the reading beam continuously scans the storage target, the recorded information can be displayed without interruption on the TV monitor. The characteristics of the storage target are such that readout is possible for several minutes without any perceptible image degradation.

TV readout is performed at 50 or 60 Hz (50 or 60 fields per second). It can even be carried out during information refreshing (writing by the other gun). In this way, even rapid movements can be reconstituted, as will be explained later.

By varying the amplitude of the reading sweeps, the dimensions of the part of the stored image displayed on the full screen of the TV monitor can be changed. This useful technique ("zoom") permits interesting parts of the image to be studied in detail, a linear magnification of up to X4 being available.

3.2 Fidelity of the Signal Level

An important characteristic of the TH 7505 is that the signal which is stored and read out is always proportional to the amplitude of the input signal, being independent of the time for which this input signal is applied. Equipment using storage tubes frequently record signals whose amplitude are proportional to the (input signal X writing duration) product.

The fidelity of signal level in the TH 7505 is due to the use of "equilibrium writing", a technique which is well-known in the storage-tube field. It is obviously of great interest in applications where halftone rendition is important, or where the writing scans are erratic, as is the case when an ultrasonic probe is moved manually.

However, experiments have shown that a problem remained unresolved, due to variations in the energies of the electrons in the writing beam. The defect appeared at very low writing speeds, target equilibrium only being obtained assymptotically with time.

A technique known as "mode restitution" is used to correct this effect. As a detailed explanation of this operation falls outside the scope of this paper, we shall only indicate the basic principle. The writing beam performs an operation known as "selective erasure" (which will be described later), this being used to return the target's equilibrium potential to an accessible value within a finite time.

3.3 Selective Erasure

Selective erasure, effectively the inverse of a writing operation, is carried out with the writing gun. Writing is the deposition of negative charges on the target dielectric whereas selective erasure involves the deposition of positive charges. This inversion can be obtained by altering the biasing of the writing gun's electrodes so as to change the mechanism of secondary emission from the target.

To be a valid technique, selective erasure must fulfill the following requirements :

- switching from writing to selective erasure operation must not perturb the readout.

- switching between these two operating modes must be very rapid in both directions.

- the selective erasure speed must be variable, to suit the application.

- finally, and most important of all, the registration between the two modes must be perfect. In other words, for a given X, Y address, the writing beam must strike exactly the same part of the target in both modes.

These requirements have been satisfied in the TH 7505 by using
a tube that was developed for this application, the TH 8811,
and by careful design of the associated circuits.

3.4 Using the Selective Erasure Mode

An important use of selective erasure is in the "mode
restitution" technique mentioned previously. In addition to
this, it makes many other operating modes possible, such as
moving displays, or partial refreshing of the stored information.

As an example, we will explain how stored information can be
renewed rapidly while keeping a displayed image that an operator
can observe without fatigue (see Figure 4).

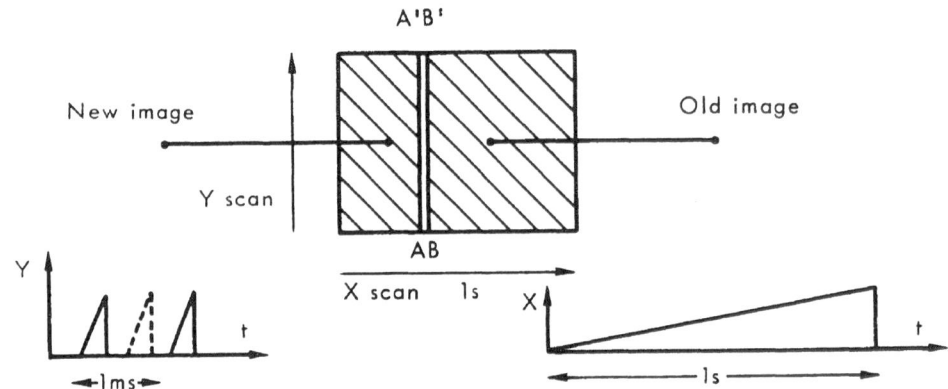

Fig. 4.   Example of the Use of Selective Erasure.

Consider an image formed by scanning in rectangular coordinates
(or equally in $\rho\theta$). The X sweep is a slow sawtooth, for example
one second (the time required to establish an image), whereas the
Y sweeps consist of a series of 300 μs sawteeth, recurring at
1 ms intervals (hence giving 1000 lines per image). In this case,
selective erasure can be effected by inserting a supplementary
scan between successive writing scans. During this supplementary
scan, the writing gun is biased for selective erasure operation.
In addition, the X sweep is offset slightly so that the erase
sweep BB' is slightly in advance of the writing sweep AA'. The
better the registration between the two modes, the smaller is the
offset required.
Using this technique, the old image can be replaced by a new one
in such a way that the observer has a continuous readout, the
only inconvenence being a fine line sweeping across the image.

CHARACTERISTICS AND PERFORMANCE OF TH 7505

In conclusion, here are the general characteristics and main performance figures for the TH 7505

General characteristics

Dimensions : standard 19" rack

Supply : 117 V/60 Hz or 220 V/50 Hz

Output signal : composite video

$$\left.\begin{array}{l} 525 \text{ TVL - 60 Hz} \\ 735 \text{ TVL - 60 Hz} \\ 625 \text{ TVL - 50 Hz} \\ 875 \text{ TVL - 50 Hz} \end{array}\right\} \quad 1 \text{ Vpp}/75\,\Omega$$

Input signals : - X,Y spot positioning
                 - Z video signal 1 V/75 Ω
                 - 1 TTL signal to control the selective erasure mode

Typical performance

- Limiting resolution : better than 2000 TV lines

- Halftones : 8 levels, comparable to broadcast TV.

- Storage time : better than 30 minutes.

- Writing speed : 1 target diameter in 300 μs

- Selective erasing speed : 1 target diameter in 60 μs

- Total erasure time : 400 ms

- Precision between writing and selective erasure modes : better than 3/1000 target diameter.

- Mode switching time : $\leq$ 10 μs

- Zoom : linear magnification of X 1 to X 4 along X and Y axes.

SPATIALLY AND TEMPORALLY VARYING INSONIFICATION FOR THE ELIMINATION OF SPURIOUS DETAIL IN ACOUSTIC TRANSMISSION IMAGING*

J. F. Havlice, P. S. Green, J. C. Taenzer, and
W. F. Mullen

Stanford Research Institute
333 Ravenswood Avenue
Menlo Park, California 94025

ABSTRACT

Acoustic transmission images obtained with spatially coherent insonification are often difficult to interpret since complex interference patterns are superimposed on pertinent object details.

In this paper we report results obtained with the SRI real-time ultrasonic imaging system, using a new insonification technique. Theoretical considerations for the design of a diffuse insonifier are presented along with experimental results demonstrating a significant reduction in the problems of nonuniform insonification and of frame-to-frame variations of the sound field. The diffuse insonification is essentially uniform in space and time, and free of near-field diffraction patterns. The effects of out-of-focus objects have, by and large, been eliminated. Images of biological objects exhibit less contrast with diffuse insonification but provide a more faithful representation of object structures. The implications of this result with regard to the dynamic range capabilities of the imaging system are discussed.

INTRODUCTION

In recent years there has been considerable progress in the development of real-time acoustic transmission imaging systems.[1,2] Part of the attractiveness of transmission imaging lies in its

---

*This work was supported by NIH Grant No. GM18780-5.

orthographic image presentation, which, in form, resembles that
obtained with x-ray fluoroscopy. The images are more familiar than
those obtained using B-scan techniques and therefore are more
readily interpretable. In addition, because the images are formed
by the attenuation properties of tissues they may contain new and
diagnostically significant information.

Much of the thrust of the research in transmission imaging has
been directed toward the development of sophisticated and novel
receiving systems. As a result, devices are available that operate
at a nominal frequency of 2 MHz and provide real-time images with
resolution of approximately 1 mm. One example of such an instrument
is the SRI ultrasonic camera, which is sufficiently sensitive to
image throughout the adult abdomen with incident acoustic intensities of less than 1 mW/cm^2. The image of the human elbow shown
in Figure 1 is typical of those obtained using the SRI ultrasonic
camera. Clearly identifiable are the humerus in the upper arm, and
the radius and ulna in the forearm. The interrelationship of these
and other structures in the arm is dramatically portrayed in the
real-time display, and emphasizes the utility of dynamic visualization. As encouraging and useful as these images are, there is one

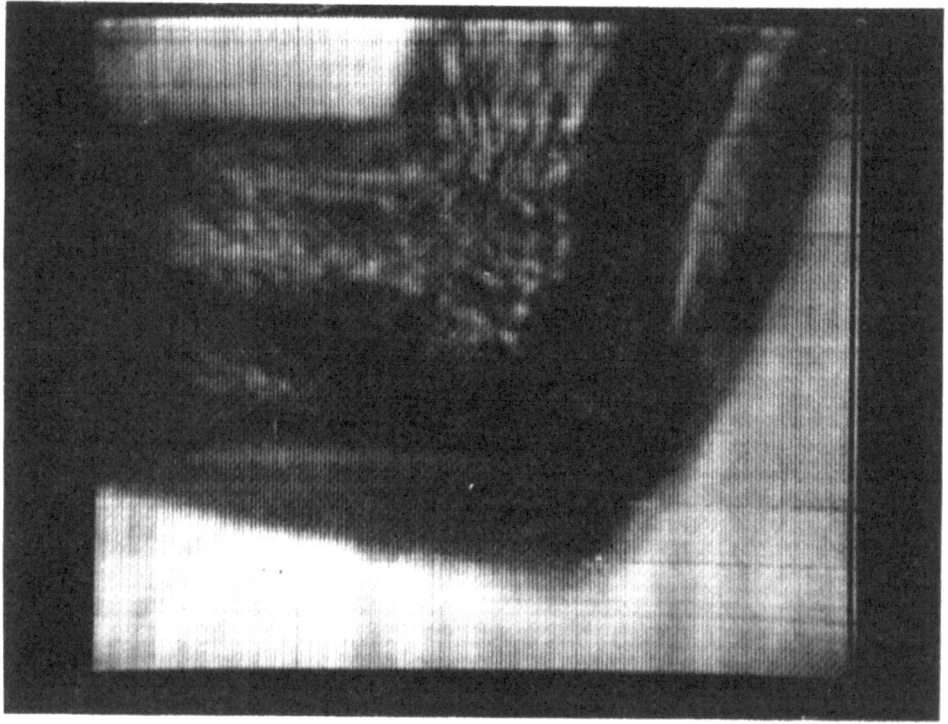

Figure 1.   Ultrasonic transmission image of the elbow obtained using
             coherent insonification. Note the high-contrast patterns
             in the soft tissue regions.

feature that inhibits direct interpretation. In the soft tissue
region, highly complex patterns appear. These patterns are tissue
related since, in the dynamic display, they move with the arm.
With a good deal of concentrated attention, it is possible to identify certain of the fine pattern structures as arising from blood
vessels, tendons, and muscle fascia. However, this interpretation
is often difficult and indirect. In addition, the complex patterns
tend to mask details that may have diagnostic importance.

These patterns are caused in part by near-field diffraction of
the sound emanating from the insonifying transducer and in part by
out-of-focus anatomical structures; they are present because the
insonification is both coherent and strongly directional. In the
sense that these patterns are not <u>directly</u> interpretable and tend
to confuse and mask the interpretation of in-focus structures, we
shall refer to them as spurious. In this paper we demonstrate the
advantages of using diffuse insonification to eliminate these
spurious image patterns. We will compare images obtained using
both "coherent" and "diffuse" insonification and comment on the
implication of these results to the design of the receiving system.

COHERENT vs DIFFUSE INSONIFICATION

Two aspects of the insonifying field are important in the discussion that follows: the spatial and temporal coherence and the
angular spectrum. The first determines the contrast of interference
patterns that appear in the image; the second, to be discussed in a
later section, affects the axial extent of the in-focus region and
the contrast of the image detail. A transducer excited by a con-
stant-amplitude, single-frequency electrical signal generates a
sound beam with the property that the relative phases of the Huygen
point sources across the transducer face are time invariant. Interference patterns from such totally coherent insonification may
be formed from the same wavefront or from wavefronts generated at
different times. Because the interference patterns that appear in
the image are nonvarying, they are independent of the temporal response characteristic of the detection system.

If the frequency of the applied electrical signal varies with
time, the phase relationship across the transducer face is still a
well-defined function. Provided that only one wavefront is used,
it is always possible to form interference patterns from such
spatially (but not temporally) coherent insonification. Depending
on the rate and total excursion of the frequency variation, it may
or may not be possible to form interference patterns from wavefronts
generated at different times. The temporal response of the detection
system now plays an important role in the characteristics of these
patterns in the image. In 1974 Korpel[3] and his co-workers

demonstrated the advantages of using frequency-modulated sound to eliminate Talbot images, spurious patterns that arise when coherent fields are used to image periodic structures. In general, bandwidth limitations permit the eradication of only fine-structured interference patterns by frequency modulation alone. Computer studies indicate that very large bandwidths (100% or greater) are needed for effective spurious signal suppression, whereas the SRI system has reasonable sensitivity over approximately 500 Hz (25% bandwidth at 2 MHz). In addition, because of the frequency dependence of sound attenuation in tissue, the body filters the frequency spectrum and further limits the effective bandwidth. In later paragraphs, where comparisons are made between coherent and diffuse insonification, the term coherent will refer to this spatially, but not temporally, coherent sound.

If the phases of the Huygen point sources of the insonification system are statistically independent (both spatially and temporally), the insonifying field is totally incoherent and no interference fringes are possible near the plane of the sources. At large distances from this plane, interference patterns may still be obtained since wave propagation tends to destroy total incoherence. In general, however, the interference effects are largely eliminated. The important feature of totally incoherent insonification is the consequent statistical independence of the impulse responses in the image plane. Because of this independence, the impulse responses add on the basis of intensity rather than amplitude. The two experimental systems described in later paragraphs use somewhat different methods to achieve incoherence. Whereas a totally incoherent technique adds the intensities of point response functions, both of these systems add the intensities of a sequence of entire images, each image being formed from a coherent addition of point response functions. Each contributing image displays the spurious detail that accompanies coherent insonification; but the summation image, as we shall demonstrate, is nearly free of this degradation. We will refer to the sound field generated by these systems as "diffuse."

EXPERIMENTAL SETUP AND RESULTS

There are two major considerations in the design of an effective diffuse sound source. First, the angular spectrum of the insonification must be large so that each point in the object plane is insonified from many directions. This reduces the shadowing effect of highly reflective or highly attenuating structures. Second, the insonification must change with time so that the interference patterns arising from spatially coherent sound are smeared out in the summation image.

In the first experimental system, a plane wave transducer was used to generate a sound field 15 cm by 15 cm in cross section. However, before reaching the object to be viewed, the sound passed through a diffuser consisting of two closely spaced styrene panels with a fluid in between. One side of each styrene sheet was flat; the other side consisted of a large number of small, random bumps. The two flat sides faced outward toward the water in the test tank; the two sides with bumps faced inward toward a liquid sealed between the two sheets. FC 75, fluorinated hydrocarbon, was chosen as the liquid between the styrene sheets in order to provide a significant velocity discontinuity. The sound emanating from the transducer entered the diffuser through one of the flat sides; upon emerging from the first styrene sheet into the FC-75, the sound was randomly refracted in many directions. A similar process occurred when the waves passed into the second panel. In this way, a plane wave from a standard transducer was converted into a beam with a large angular spectrum. The time variation of the insonification was obtained by linearly translating the diffuser. The images, which appeared on a cathode ray tube, were incoherently summed by time-exposing photographic film. Although the results shown in this paper were obtained with 10-second exposures (corresponding to 75 frames on the CRT), adequate results could be obtained with exposures as short as 1 second (about 7 to 8 frames).

To demonstrate the effectiveness of diffuse insonification, the experimental setup shown in Figure 2 was devised. We will be

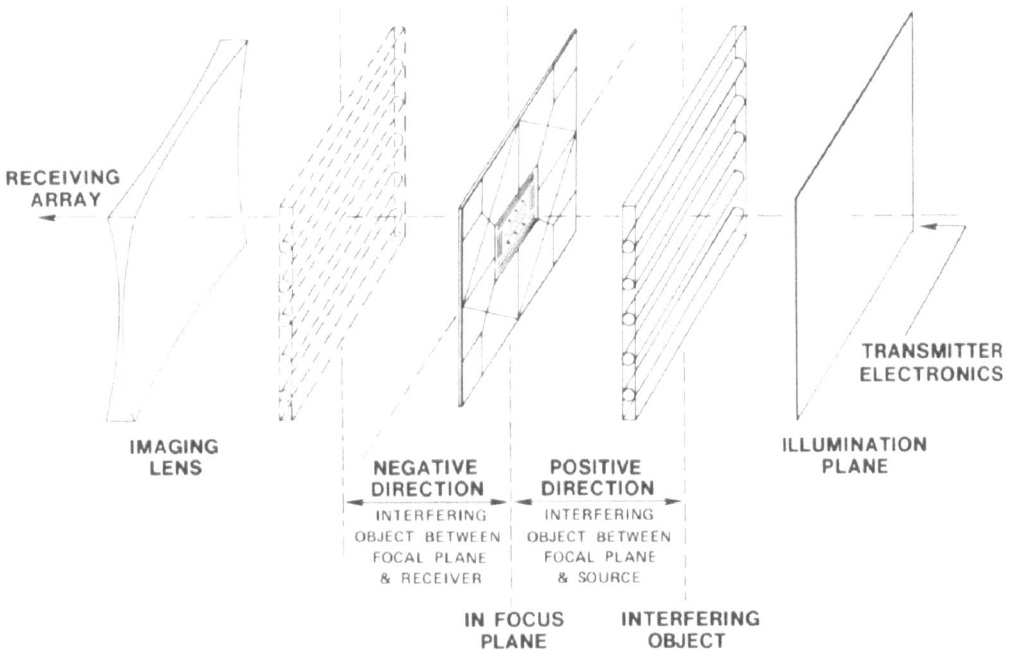

RECEIVING ARRAY

IMAGING LENS

NEGATIVE DIRECTION

POSITIVE DIRECTION

INTERFERING OBJECT BETWEEN FOCAL PLANE & RECEIVER

INTERFERING OBJECT BETWEEN FOCAL PLANE & SOURCE

IN FOCUS PLANE

INTERFERING OBJECT

ILLUMINATION PLANE

TRANSMITTER ELECTRONICS

Figure 2.   Experimental system for demonstrating the effects of diffuse insonification.

concentrating our attention on the region between the insonification
plane and the imaging lens.* An object-to-be-imaged, in this case
a simple resolution test pattern, was placed in the focal plane.
An interfering structure was placed a selected distance from the
focus either between the insonification plane and the focal plane
or between the focal plane and the imaging lens. In the first
experiments, the interfering structure was a series of horizontal,
parallel, polysytrene bars, 0.63 cm in diameter. With this interfering structure placed 4 cm away from the focal plane, between
the object and the insonifier, the images of Figure 3 were produced.
Note that in the case of coherent insonification, the presence of
the interfering structure is clearly evident as dark bands across
the image. The ability to resolve fine detail, as in the center
of the test pattern, can be seen to be dependent on the location
of the parts of the interfering structure. In the case of diffuse
insonification, the effect of the intefering structure on the
images has been greatly reduced. In Figure 4, the interfering
structure was placed 8 cm from the focal plane, in this case between the object and the imaging lens. Even at this relatively
large distance the quality of the image obtained using coherent
sound is reduced by the noticeable presence of the horizontal bars.
In the case of diffuse insonification, the effect of the interfering structure has again been eliminated.

As a next level of complexity, two interfering structures were
introduced, both 4 cm from the object (see Figure 5). The horizontal

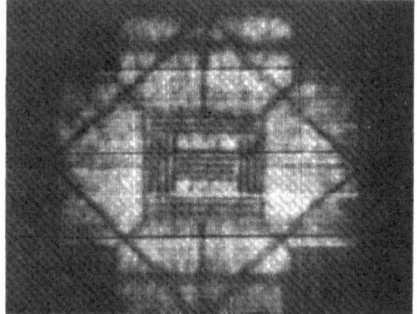

Figure 3. Comparison of ultrasonic transmission images of a resolu-
           tion test pattern with an interfering structure 4 cm from
           the focal plane, between object and insonifier. The
           image on the left was obtained with coherent insonifica-
           tion; the one on the right with diffuse insonification.

---

*The same principle would apply for imaging systems employing elec-
 tronic focusing, in which case the transducer array would be
 positioned in the imaging lens plane.

 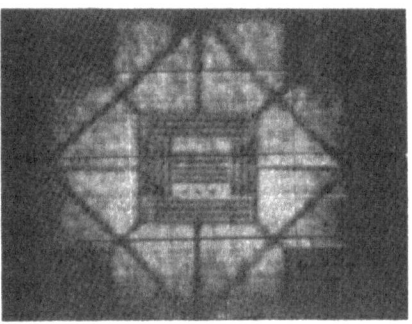

Figure 4. Comparison of ultrasonic transmission images of a resolution test pattern with an interfering structure 8 cm
from the focal plane, between object and imaging lens.

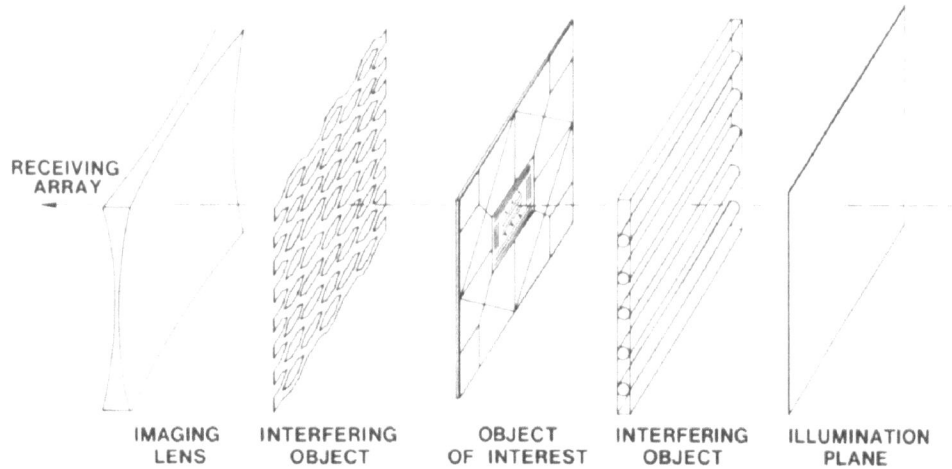

| RECEIVING ARRAY | IMAGING LENS | INTERFERING OBJECT | OBJECT OF INTEREST | INTERFERING OBJECT | ILLUMINATION PLANE |

Figure 5. Experimental system for demonstrating the advantages
of diffuse insonification with two interfering objects.

bars from the previous experiment were located between the transmitter and the object, and a coarse metal grating was added between
the object and the imaging lens. The image obtained with coherent
insonification is shown in Figure 6a. The presence of the object
has been almost completely masked by the diffraction patterns
contributed by the two interfering structures. This is not sur-
prising considering the combined complexity of the two gratings.
Yet, with diffuse insonification (Figure 6b) the interference has
been almost completely eliminated, and the object is imaged with
little degradation.

Images of anatomical structures also show a significant improvement when obtained with diffuse rather than coherent insonification. An example is shown in Figure 7 where, once again, the
image is that of the elbow. For comparison, Figure 1, the image
obtained with coherent insonification where the ultrasonic camera
is focused on the bone, has been reproduced on the left (Figure 7a).
In comparing the soft tissue images, note that the high-contrast,
spurious patterns arising from coherent interference due to out-
of-focus veins, muscles, and tendons have been eliminated by diffuse
insonification. The humerus, radius, and ulna are identifiable
in both photographs, but only with diffuse insonification can one
clearly distinguish, for example, the small dark bump that is
suggestive of the head of the radius. That information may be
present in the image obtained with coherent insonification, but
it is masked by the spurious patterns produced by out-of-focus
structures.

In comparing the photographs of Figure 7, it may appear at
first sight that information has been lost in the diffuse insonification case. However, many of the patterns that appear in the coherent image are not readily interpretable. Although the eliminated
patterns are related to the anatomy, they are due to out-of-the-
focal-plane anatomy that casts acoustic shadows. These patterns

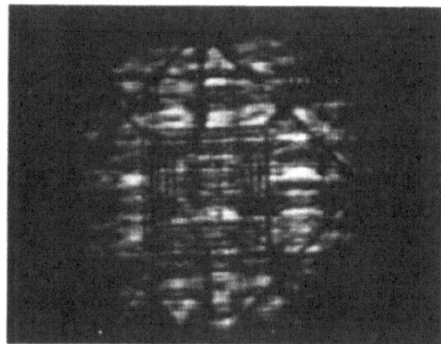

Figure 6.   Comparison of ultrasonic transmission images of a reso-
             lution test pattern with two interfering objects. Left:
             coherent insonification. Right: diffuse insonification.

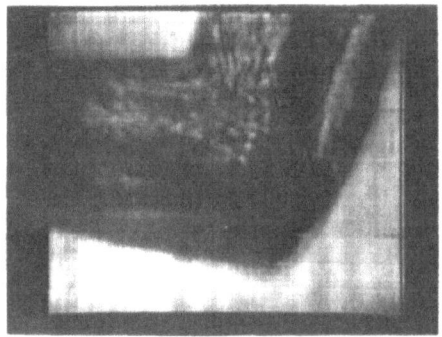 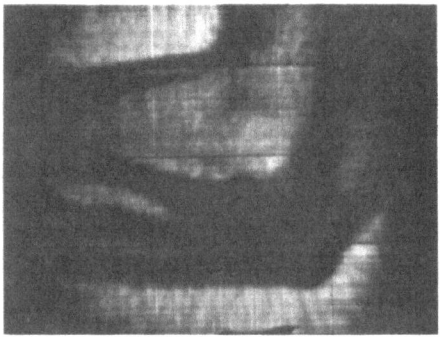

Figure 7. Comparison of ultrasonic transmission images of the el-
           bow. Note that in the image on the right, obtained with
           diffuse insonification, the soft tissue region is clear
           of the high-contrast, spurious patterns that appear in
           the image on the left, obtained with coherent insonifi-
           cation.

often mask the interpretation of the structures of interest. Those
anatomical structures, whose effects have been eliminated in the
diffuse case, can be imaged by moving them into the focal plane.
One of the characteristics of imaging with diffuse sound is the
narrow axial extent of the in-focus region. We will consider this
effect analytically in a later section, but experimental verification is shown in Figure 8. Both images were obtained with diffuse
insonification, except that on the left the bones were in focus
and on the right they had been moved 1 cm out of the focal plane.

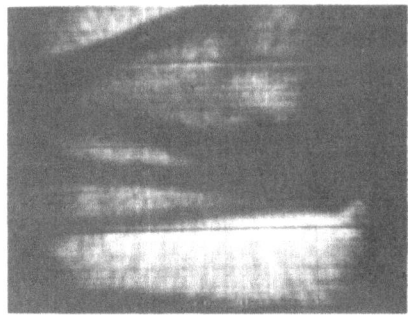 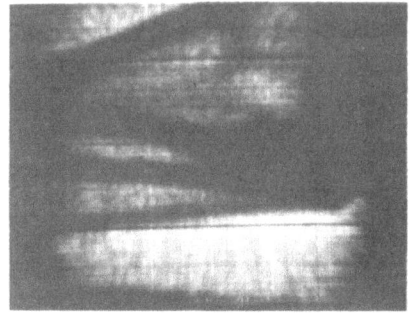

      Radius and Ulna   Radius and Ulna
      in Focal Plane   Approximately 1 cm
                                       From Focal Plane

Figure 8. Acoustic images of lower arm obtained using diffuse
           insonification.

Note that these relatively large structures contribute little interference when displayed only 1 cm from the focal plane. If the
bones were 2 cm from the focal plane, their presence would be
apparent only by a slight darkening of the image.

Another example of the advantages of using diffuse insonification is shown in the images of the human foot in Figure 9. The
elimination of spurious patterns in the soft tissue region allows
unobstructed visualization of the tibia, fibula, and, most dramatically, the Achilles tendon.

THEORETICAL CONSIDERATIONS

Diffuse insonification has two striking effects on acoustic
image formation. The images generally have much less contrast
than do the companion coherent images, and the apparent depth-of-
field is much smaller. In designing the SRI ultrasonic camera for
use with coherent insonification, logarithmic amplification was
used to compress 60 dB of acoustic information into approximately
20 dB for display on CRT monitors. This was necessary due to the
propagation of shadows resulting from highly reflective or attenuating structures both in and out of the focal plane. To visualize
detail lying "behind" such shadowing structures, large dynamic
ranges are required. Shadowing effects are largely eliminated in
diffuse insonification since sound waves now impinge from a large
number of directions. Objects that were previously shadowed are
now insonified from sound rays that propagate around the shadowing
structure. The soft tissue images obtained with diffuse insonification are representative of the attenuation properties of the

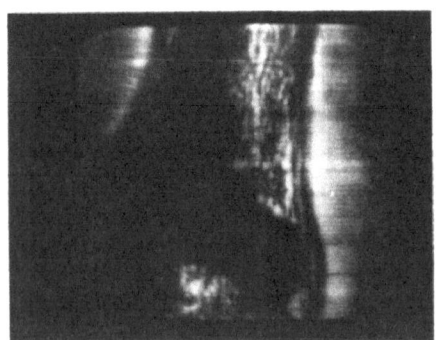

Coherent Insonification   Diffuse Insonification

Figure 9.   Comparison of acoustic images of the foot obtained using
            coherent and diffuse insonification. Note that the
            image obtained with diffuse insonification clearly
            displays the Achilles tendon.

various tissues. Since the differential attenuation between many anatomical organs is not very high, the contrast is generally low. Typically, the image information is contained within an 8–10 dB range. Therefore, instead of compressing image signals with a 60-dB range into 20 dB for display, the images presented here were produced by expanding the signals to display only the top 10-dB range. Signals below that level are displayed as black. When operated in this manner the system is quite sensitive to small signal changes; to ensure reliable interpretation of the images, the transducers and the subsequent signal processing circuits must be quite uniform from channel to channel. We were able to achieve a system uniformity of ±0.25 dB over all 192 receiving channels. Although the information is more subtle when diffuse sound is used, it more faithfully represents attenuation in the focal plane and is interpretable and reliable.

The narrow axial extent of the in-focus region arises from the time-varying character of the insonification and the intensity integration of the resulting time-varying image. Figure 10 illustrates a simple model of diffuse insonification. Consider a matrix of point acoustic sources that insonify some object in the focal plane. The sources are turned on sequentially and the image intensities (not amplitudes) are summed. It can be shown from geometrical considerations that the images of objects lying exactly

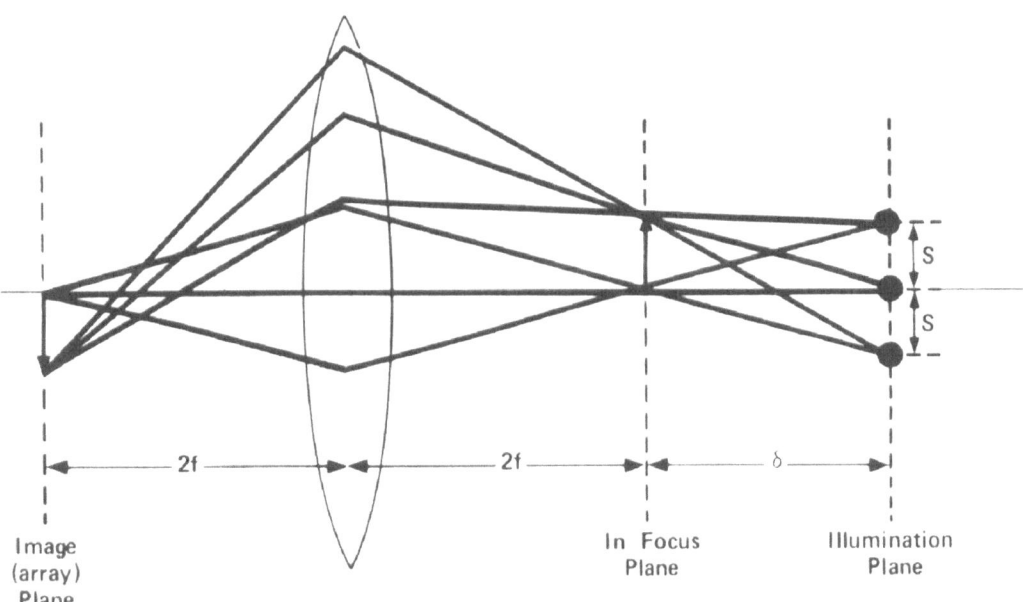

Figure 10. Ray paths and image location for objects in focal plane. Note that the images from each source are in correct registration.

in the focal plane do not move as the position of the acoustic
source is changed and may be summed without loss of in-focus detail.
Figure 11 illustrates the effect of summing images of objects that
do not lie in the focal plane. In this case it is apparent that
images resulting from different source positions are not in correct
registration. Hence, in the integrated image formed from many
point sources, the detail is smeared out, resulting in a generally
uniform background. The axial extent of the in-focus region is
determined by the number of sources, the angular extent of the
sources, relative size of the out-of-focus object, the distance of
the out-of-focus object from the focal plane, and the imaging lens.
As the angular spread of the insonifying field approaches the angle
subtended by the imaging lens aperture as seen from the focal plane,
the depth of the in-focus region approaches the focal depth of
the imaging lens.

An experimental illustration of this effect is shown in the
sequence of photographs of Figure 12a-d. The resolution test pattern was placed in the focal plane of the imaging system as before,
and the interfering metal structure used in Figure 7 was placed
4 cm from it, between the imaging lens and the focal plane. Figures
12a through 12d show images obtained, respectively, with one source,
four sources, 16 sources, and 25 sources placed in the matrix shown

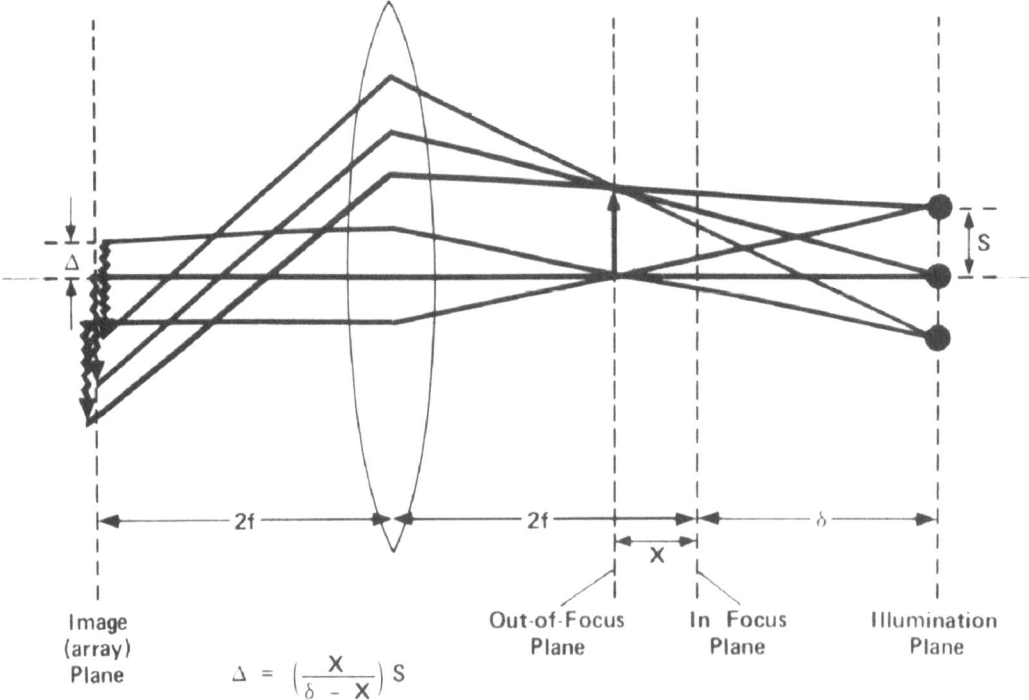

$$\Delta = \left(\frac{x}{\delta - x}\right) S$$

Figure 11. Ray paths and image location for objects not in focal
            plane. Note that images from the different sources are
            not in correct registration.

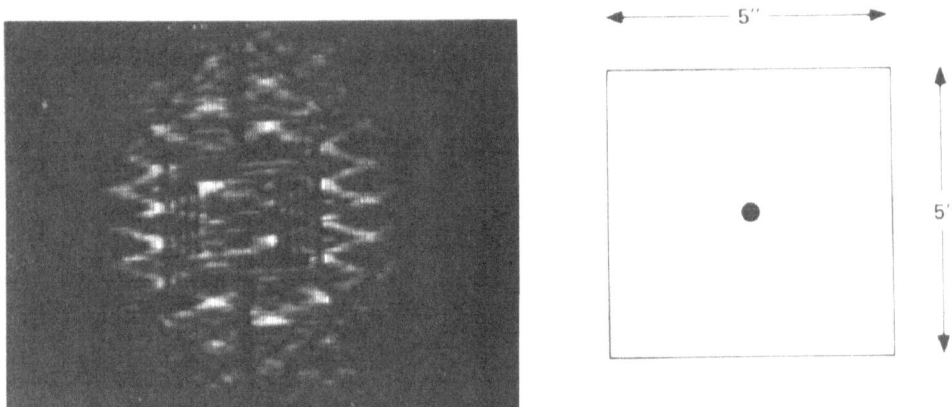

Figure 12a. Ultrasonic transmission image of a resolution test
pattern with a single interfering structure. One
source was used to insonify the object.

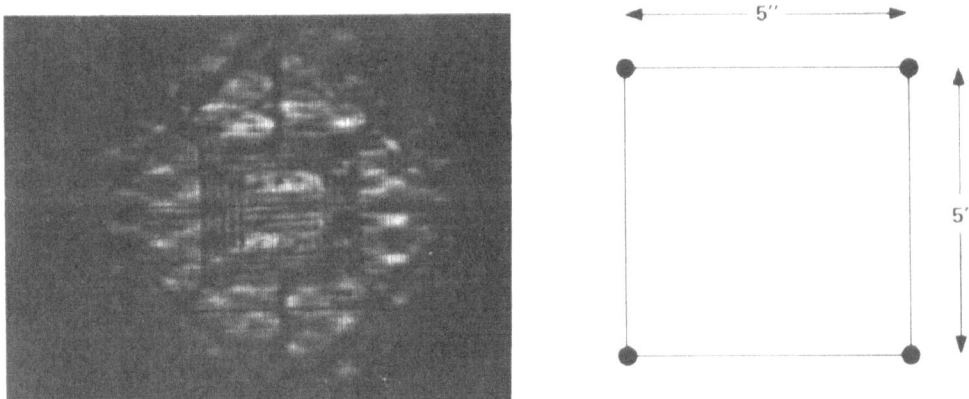

Figure 12b. Ultrasonic transmission image of a resolution test
pattern with a single interfering structure. Four
sources were used in the matrix shown on the right.

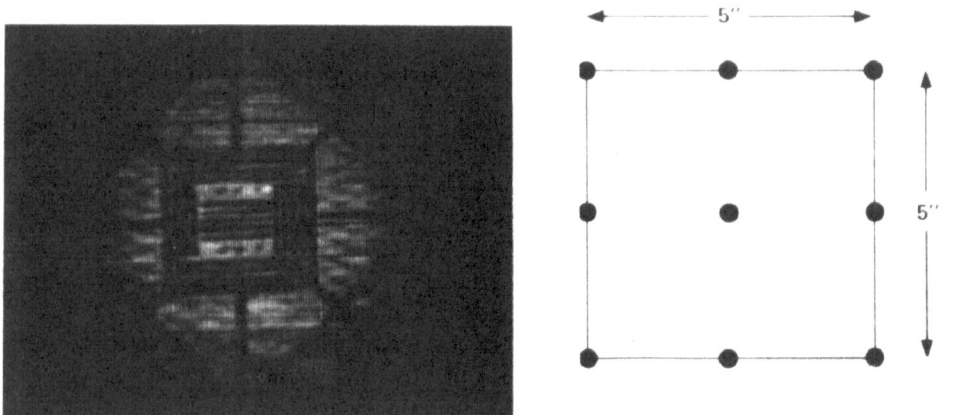

Figure 12c.   Ultrasonic transmission image of a resolution test
              pattern with a single interfering structure. Nine
              sources were used in the matrix shown on the right.
              Note that the effects of the interfering structure
              are diminished.

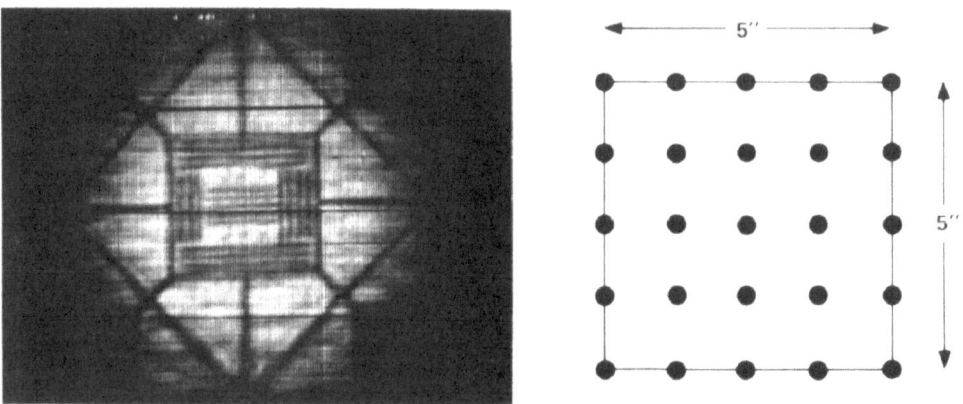

Figure 12d.   Ultrasonic transmission image of a resolution test
              pattern with a single interfering structure. Twenty-
              five sources were used in the matrix shown on the right.
              Note that the effects of the interfering structure are
              largely eliminated.

to the right of each photograph. Each source was turned on independently; the composite image was produced by multiple exposure
of the photographic film by the CRT display. It is clear that the
patterns resulting from the interfering structure, prominent in
Figure 12a and b, have been substantially diminished in Figure 12c
and nearly eliminated in Figure 12d.

SUMMARY

     In summary, a method for eliminating spurious patterns in
acoustic transmission imaging systems has been presented. The resulting images are a more faithful representation of true object
detail, and allow, for the first time, unambiguous interpretation
of the attenuation images of soft tissue. The dynamic range requirement for the receiving system is relaxed, but the requirement
for system uniformity is more stringent. Although the images in
this paper were not obtained in real time, a real-time imaging
system with diffuse insonification is under development.

REFERENCES

1. Green, P. S., Schaefer, L. F., Jones, E. D., and Suarez, J. R.:
    A New High Performance Ultrasonic Camera. In Acoustical Holog-
    raphy, Vol. 5, P. S. Green (Editor), pp. 493-503, Plenum Press,
    New York, 1974.

2. Mezrich, R. S., Etzold, K. F., and Vilkomerson, D. H.: System
    for Visualizing and Measuring Ultrasonic Wavefronts. In
    Acoustical Holography, Vol. 6, N. Booth (Editor), pp. 165-191,
    Plenum Press, New York, 1975.

3. Korpel, A., Whitman, R. L., and Ahmed, M.: Elimination of
    Spurious Detail in Acoustic Images. In Acoustical Holography,
    Vol. 5, P. S. Green (Editor), pp. 373-390, Plenum Press, New
    York, 1974.

RECONSTRUCTING THREE-DIMENSIONAL FLUID VELOCITY VECTOR FIELDS FROM ACOUSTIC TRANSMISSION MEASUREMENTS

S. A. Johnson, J. F. Greenleaf, C. R. Hansen,
W. F. Samayoa, M. Tanaka, A. Lent*, D. A.
Christensen** and R. L. Woolley***

Biophysical Sciences Unit, Department of
Physiology and Biophysics, Mayo Foundation,
Rochester, Minnesota

*Department of Computer Science, State University of New York/Buffalo, Amherst, New York

**University of Utah, Salt Lake City, Utah

***Brigham Young University, Provo, Utah

ABSTRACT

A theory with supporting experimental evidence is presented for reconstructing the three-dimensional fluid velocity vector field in a moving medium from a set of measurements of the acoustic propagation time between a multiplicity of transmitter and receiver locations on a stationary boundary surface. The inversion of the integrals relating the acoustic propagation path to the propagation time measurements is affected by linearization and discrete approximation of the integrals and application of an algebraic reconstruction technique (ART). The problem of the presence of certain invisible fluid flow functions is treated. Since this technique does not require the presence of scattering centers or the optical transparency of the medium, it may be applied in many cases (i.e., turbid, opaque, or chemically pure media) where Doppler or optical (e.g., laser holography) methods fail.

INTRODUCTION

Methods for the measurement and description of the flow of fluids (both liquids and gases) may be divided into two main classes: invasive techniques and non-invasive techniques. Invasive methods make use of devices such as probes, the introduction of markers, the measurement of pressure changes across restrictions, etc. The non-invasive methods make use of the external measurement of some flow-dependent property such as those relating to changes in optical properties, acoustic properties, electro-magnetic properties, etc. At the present time, several optical methods exist for gaining information about fluid flows. Laser Doppler, Schlieren optics, interferometric, holographic, etc., methods are known to the art. These methods require that the fluid be nearly transparent to the incident light. In addition to near perfect fluid transparency, Doppler methods require the presence of scattering centers or particles. Natural or artificial dust particles or fluctuations in density can serve this purpose. Low scattering center concentration provides longer ranges but sensitivity suffers. Higher scattering center concentration provides higher sensitivity but limits the range (depth) of measurement. The use of higher transmit power levels increases range and provides a stronger scattered signal but power is usually restricted by practical upper limits. Schlieren, interferometric, and holographic methods work well in general only for gas flows where density variations may be larger than for liquids.

Acoustic methods have also been proposed and applied which make use of either the Doppler scattered sound or transmitted sound along a single beam. The use of pulsed Doppler methods allows measurement of the component of fluid velocity (not the true fluid velocity) along the acoustic beam. The same trade off between sensitivity, which governs laser doppler measurements, also exist for acoustic Doppler techniques.

The average fluid velocity component along a transmitted beam can be determined from the measured time of propagation along the beam. Flow meters designed around this principle are well known.

All of these methods suffer from several common weaknesses: first, only the component of flow parallel to the beam is determined; second, only that region of

the flow traversed by the beam is sampled. Thus, none
of the previously mentioned methods in their simplest
form measure true three-dimensional flow.

This paper will demonstrate that three-dimensional
fluid flow can be determined by a new transmission
method which overcomes the range and sensitivity problems of Doppler methods and which overcomes the averaging problem (determining only the average parallel
component) of the single beam transmission delay method.

METHOD

It has been suggested that fluid flow within a
measurement region may be determined by transmitting
and receiving acoustic energy through the measurement
region along a plurality or rays such that each volume
element is traversed by a set of rays having components
in each direction for which flow components are to be
reconstructed (1). The propagation time of the acoustic
energy along the plurality of rays constitutes the only
measurements required by the method (1).

Each ray propagation time measurement is an integral of a function of acoustic speed and fluid velocity
along the ray and the set of such measurements constitute a simultaneous set of integral equations which
may be inverted (solved) to obtain the unknown fluid
velocity vector field (1). The equation which relates
the propagation time along each ray from a source a to
a receiver b is given by (see Figures 1 and 2)

$$t_{ab} = \int_a^b ds/\|\vec{\tau}_m C(\vec{r}) + \vec{V}(\vec{r})\| \tag{1}$$

where $\vec{\tau}_m$ is the unit tangent vector along the acoustic
ray as seen from the moving medium, $C(\vec{r})$ is the local
acoustic speed as seen in the moving media, and $\vec{V}(\vec{r})$ is
the fluid velocity vector as measured in the laboratory.

In general the actual ray path taken by acoustic
energy from the source a to the receiver b is not known
a priori even though the time of propagation may be
measured quite accurately. It will be shown that the
actual paths may be found either by appropriate simplifying assumptions or from proper mathematical consideration of the complete set of propagation times between
many sources and receivers on a boundary surface surrounding (but not necessarily enclosing) the flow.

This paper will proceed with the simplest case of fluid velocities much less (10% or less) than the speed of sound and then generalize to higher fluid velocities.

When the velocity of the fluid is everywhere much less than the speed of sound, the denominator in Equation 1 can be approximated by the expression $(C + \vec{\tau}_m \cdot \vec{V})$ where it is assumed that the ray tangent vector $\vec{\tau}_m$ as seen from a coordinate system embedded in the fluid is almost identical to the ray tangent vector $\vec{\tau}$ as seen in the laboratory coordinate system. In most fluids, when $|\vec{V}| \ll C$, the variations in $C(\vec{r})$ are also correspondingly small. With these assumptions, Equation 1 can be written as:

$$t_{ab} = \int_a^b (1/C)(1 - \vec{\tau} \cdot \vec{V}/C)\, ds \tag{2}$$

The assumption that $\vec{\tau}_m$ is nearly equal to $\vec{\tau}$ allows the further assumption that the acoustic rays are nearly straight lines. Thus the ray paths may be found in terms of the known source point a and receiver point b. The use of straight line rays provide a further benefit, namely, a simple method for separating the dependence of the time t_{ab} on both $C(\vec{r})$ and $V(\vec{r})$. This separation is obtained by forming the linear combinations $(t_{ab} + t_{ba})$ and $(t_{ab} - t_{ba})$ where now t_{ba} refers to the propagation time between b and a when b serves as the source and a as the receiver.

$$t_{ab} + t_{ba} = 2\int_a^b (1/C(\vec{r}))\, ds \tag{3}$$

$$t_{ab} - t_{ba} = -2\int_a^b (\vec{\tau} \cdot \vec{V}/C^2)\, ds \tag{4}$$

Thus, even if $C(\vec{r})$ is a function of position $\vec{r}$ it may be found by the inversion of the multiple set of equations (3) corresponding to a well chosen set of ray paths. This solution may be used to define C^2 in Equation 4 so that $\vec{V}$ may be found from the same or a similar set of ray measurements used to solve Equation 3.

It is important to choose a set of ray paths which provide a unique solution to Equations 3 and 4. One such good arrangement of transducers which provide paths with linear independent integrals is shown in Figure 1. Flow is assumed in a cylinder formed by translating a closed plane curve, nowhere convex inward, in a straight line (square and circular pipes are

members of this set) as shown in Figure 1. The source
and receiver transducers are placed on the intersection
of this cylinder and two parallel planes. These planes
are shown arranged such that the axis of the cylinder
is normal to both plane a and plane b. Ray paths are
used from any transducer a_i in plane a to any other
transducer a_j in plane a or to any transducer b_j in
plane b. As shown in Figure 1, both intraplane rays
and interplane rays are utilized in this arrangement.

Geometric Arrangement of Transducers

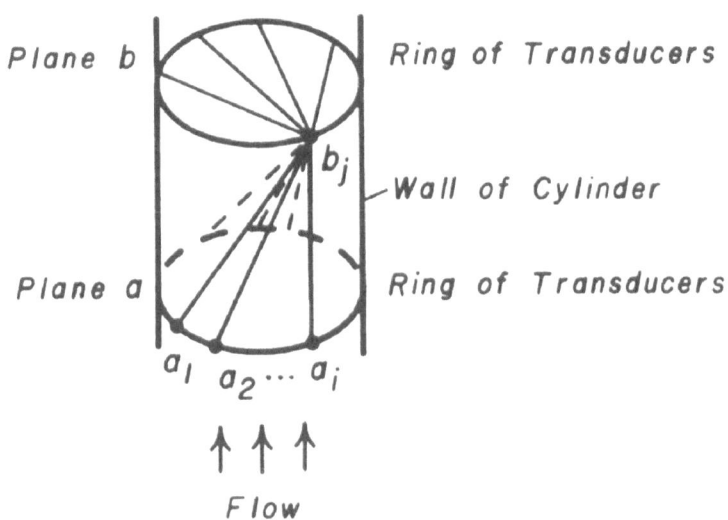

Figure 1 Illustrates a geometry in which three-
dimensional fluid flow can be reconstructed with a few
simplifying assumptions. Two of many planes a,b,c,...
containing transducers $a_1,a_2,...a_2$, $b_1,b_2,...b_2$, etc.
are shown.

The mathematical form of $\vec{\tau}\cdot\vec{V}$ can be derived by
reference to Figure 2 which shows a coordinate system
in the fluid flow field with the z axis in the direction of the axis of cylindrical symmetry. $\vec{\tau}$ is the unit
tangent vector to a acoustic ray at an arbitrary point,
i.e., $\vec{\tau} = d\vec{r}/ds$, while $\hat{i}$, $\hat{j}$, $\hat{k}$ are unit vectors along

the x, y, z axes which are fixed in direction. Let
$\vec{r} = \vec{R} + z\ \hat{k}$, then $\vec{R}_O = d\vec{R}/ds$ and then $\vec{R}_O$ and dR are the
projections of $\vec{\tau}$ and ds respectively onto the xy plane.
The magnitude of $d\vec{r}$ is ds, the magnitude of $d\vec{R}$ is dR.
Here α, β, γ are the direction angles and (R_O, θ) are
the polar coordinates of $\vec{R}_O$.

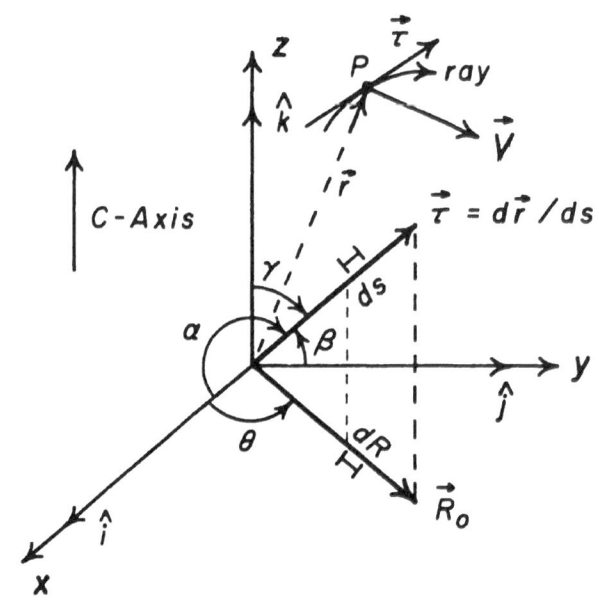

<u>Figure 2</u> Coordinate system in flow field with z-axis
in direction of cylindrical channel axis (C-axis).
$\vec{\tau} = d\vec{r}/ds$ is unit tangent vector to acoustic ray at an
arbitrary point P, but $\hat{i}$, $\hat{j}$, $\hat{k}$ unit vectors along
x,y,z axes are fixed in direction. Note $|d\vec{r}|$ = ds and
the projection of ds is dR, i.e., $|d\vec{R}|$ = dR. Note also
the fluid flow vector $\vec{V}(\vec{r})$ at point P.

With this notation, an expression for $\vec{V} \cdot \vec{\tau}$ may be
written in terms of components V_x, V_y, V_z of $\vec{V}$ and the
cylindrical coordinates of $\vec{\tau}$. Note that dR = (sin γ)ds
where $\vec{\tau} = \vec{R}_O + \hat{k}\cos\gamma$. With this substitution for
$\vec{\tau} \cdot \vec{V}$ Equation 4 becomes (note, in this paper the symbol
$\equiv$ will indicate a definition)

$$I_0(a_ib_j) \equiv t_{a_ib_j} - t_{b_ja_i} =$$

$$-2\int_a^b (1/C^2) \left[(V_x\cos\theta + V_y\sin\theta)\sin\gamma + V_z\cos\gamma \right] ds \qquad (5)$$

which may be written

$$I_2 \equiv [I_0(a_ib_j) - I_1(a_ib_j)] = -2\cos\gamma \int_{a_i}^{b_j} (1/C^2)V_z dR \qquad (6)$$

where

$$I_1(a_ib_j) \equiv -2\int_{a_i}^{b_j} (1/C^2)(V_x\cos\theta + V_y\sin\theta)dR \qquad (7)$$

The advantage of writing Equations 6 and 7 in place of Equation 5 is made more clear by noting that I_1 from Equation 7 is independent of γ. Thus, V_x and V_y may be found by taking a sufficiently large set of rays in either plane a or plane b (not from plane a to plane b) and solving the corresponding set of Equations 7 for V_x and V_y. In other words, $I_0(a_ia_j) = I_1(a_ia_j)$ when $\gamma = \pi/2$ and where a_j is the projection of b_j onto the a-plane.

The function V_z may be found without solving Equation 7 for V_x and V_y if the value of I_2 can be found for each a_i and b_j. If a_i and b_j are in different planes and if a_i and a_j are their projections along the cylindrical axis onto one plane, then $I_1(a_ib_j)$ is identical in value to $I_1(a_ia_j)$. With I_2 determined for each ray in a sufficiently large set of rays from I_0 and I_1, it is then possible for Equation 6 to be inverted. Thus, Equation 6 may be replaced by the following practical equation, relating V_z to actual time measurements.

$$I_2 = [(t_{a_ib_j} - t_{b_ja_i}) - (t_{a_ia_j} - t_{a_ja_i})] =$$

$$-2\cos\gamma \int_{a_i}^{a_j} (V_z/C^2)dR \qquad (8)$$

Thus, V_z is found from data consisting of measurements between transducer pairs both in the same plane and between distinct planes. The integral in Equation 7 or Equation 8 may be written in terms of an approximate discrete sum by subdividing planes a or b into finite elements or pixels. Then dR corresponds to the length of the ray in each pixel. Using this notation, Equations 7 and 8 become

$$I_1(a,i,j)_s = -2\Sigma_k (1/C^2) [V_x(k)\cos\theta_s + V_y(k)\sin\theta_s]L_{sk} \quad (9)$$

$$I_2(a,b,i,j)_s = -2\cos\gamma_s \Sigma_k (1/C^2)V_z(k)L_{sk} \quad (10)$$

where L_{sk} is the length of ray s in pixel k. Methods for solving Equations 9 and 10 are well known (2).

The case of nonconstant speed of sound C is also described by Equations 3, 6, and 7 or their extensions, such as Equation 9. Thus, the inclusion of $(1/C^2)$ inside the integral or summation is justified because the speed C can be determined by considering the set of equations in a plane as per Equation 3. Thus, both $\vec{V}$ and C can be reconstructed even in those cases where C is not a constant. The extension of Equations 3 and 4 to more complex geometries than covered by Equation 8 is possible and often straightforward but may result in a more complex data collection and reduction system.

In the example described by Equations 9 and 10 and illustrated in Figure 1, only two planes a and b of transducers were treated. The velocity vector $\vec{V}$ with x, y, and z components thus reconstructed is representative of the type of flow between planes a and b which is not a function of z. The case where flow changes rapidly with coordinate z can be treated by the use of more than two measuring planes, i.e., more planes than a and b, and by spacing those planes closer together.

In the preceding discussion, it was assumed that measurements were made of one intraplane data set (i.e., rays a_i to a_j for permutations of i and j) and one interplane data set (i.e., rays a_i to b_j for permutations of i and j). It is also possible to collect a second intraplane data set in the b plane (i.e., rays b_i to b_j for permutations of i and j). This extra data set may be used to average with the a plane to provide an improvement in signal to noise ratio. In this case, the fluid velocity may be considered to be reconstructed on an imaginary plane, halfway between a and b.

As has been mentioned, Equations 9 and 10 can be solved by a modification of the ART algorithm (1). The ART method is usually applied to reconstruct scalar quantities from their projections. In the operation of this technique, the reconstruction of vector quantities is required. This is accomplished by writing $(\vec{\tau} \cdot \vec{V})$ as $(V_x \cos \alpha + V_y \cos \beta + V_z \cos \gamma)$. At each point along the ray $\cos \alpha$, $\cos \beta$, and $\cos \gamma$ are known. The quantities V_x, V_y, and V_z are each sought in the N^2 pixels (picture elements) in a square array with N pixels on a side. One way to accomplish this task is to modify the ART algorithm to reconstruct a scalar U in a rectangular array of $3N^2$ pixels where the value of U is V_x in the first square sub array of N by N, where U is V_y in the middle square sub array of N by N, and where U is V_z in the last square sub array of N by N. In the case of cylindrical symmetry such as used in Equations 9 and 10, only two such square sub arrays are used to solve Equation 9 and only one such square sub array is required to solve Equation 10.

Fast Flows

The problem of reconstructing fluid flows where the velocity of the flow is <u>not</u> insignificantly small compared to the speed of sound requires a more accurate treatment of the ray paths than the straight line approximations of Equations 9 and 10 (1). Thus, the ray paths may bend significantly due to both the transformation from moving to laboratory coordinate systems and from acoustic speed variations due to density variations (the latter being more common in gases than liquids). If a complete set of acoustic ray paths could be determined, than Equation 1 could be inverted notwithstanding its nonlinear nature. Techniques are known for finding the ray paths when $\vec{V}(\vec{r})$ and $C(\vec{r})$ are known. One such technique is called ray tracing and has been previously described (1,3). Thus, it seems that either the ray paths or $\vec{V}$ and C can be found independently even in fast flows.

While many methods could be suggested for finding both the ray paths and $\vec{V}$ and C, for purposes of illustration only, an interactive technique will be described. First, linear ray paths are assumed on a sufficiently coarse grid (i.e., large pixels). Second, the speed C and velocity $\vec{V}$ are reconstructed. Third, more accurate ray paths are found by ray tracing. It

may be necessary to smooth the reconstructions obtained
from the previous step before ray tracing. Fourth,
Equation 1 is solved again for new values of C and $\vec{V}$
using the ray path from step three. Steps three and
four may be repeated many times as the grid size is
made smaller.

The case for supersonic flow requires special
treatment because there may be transducer pairs for
which it is impossible to find linking rays (sound
cannot travel upstream in supersonic flows). Although
it seems probable that a suitable set of invertable
equations may be obtained in many cases of this type
when proper consideration is given to transducer
placement, a proof of this conjecture is not presently
known to the authors.

Invisible Fluid Flow

Under some circumstances, Equations 3 and 4 cannot
be inverted. A simple example is the case of radial
flow in a plane from a point source in that plane. If
this flow has angular symmetry, than $t_{ab} - t_{ba} = 0$.
Any such flow in a plane which contains $\vec{\tau}$ (i.e., $\vec{\tau}$ has
no components perpendicular to this plane) is invisible.
However, the flow may be detected if $\vec{\tau}$ has components
perpendicular to the plane. See Figure 3 for the
geometry of this example.

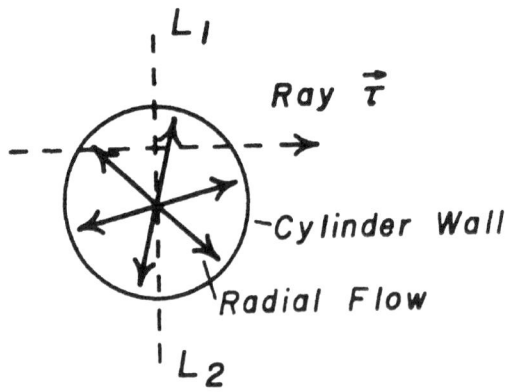

Figure 3 Invisible radial flow. See text for
explanation.

From Figure 3, it is seen that a decrease in speed
on the left of symmetry axis L_1 L_2 is equal to the increase in speed on the right of axis L_1 L_2. It should
be noted that radial flows are not invisible to the
second order terms in the expansion of Equation 2
since $(\vec{\tau}\cdot\vec{V}/C)^2$ does not change sign upon reflection
about L_1 L_2, although the contribution from the second
order term is usually negligible for $|\vec{V}| < C/5$.

Under some circumstances the integral in Equation
8 may not vanish, as it does in the case of angularly
symmetric radial flow, yet the system of Equations 7
and 8 are still not invertable because each integral in
a set corresponding to different rays is not linearly
independent. A simple example of this situation is the
case of two-dimensional flow in a channel with rays
taken between points on opposing walls of the channel.
The geometry of this case is given in Figure 4 below.

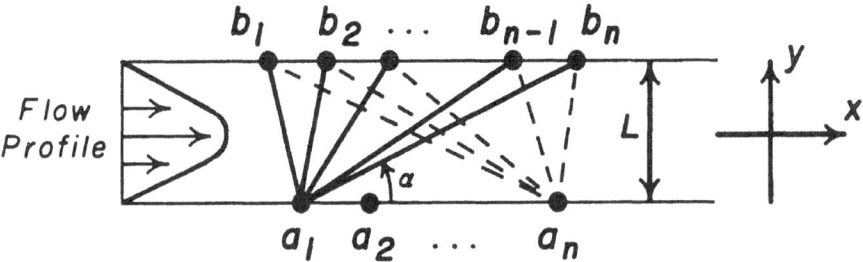

Figure 4 Two-dimensional flow in a channel cannot be
reconstructed by linear approximation with the geometry
shown. However, the average flow is determined by
forward and backward measurements along each ray.
See text for explanation.

In Figure 4 transducers $a_1 \ldots a_n$ are located on
the lower wall and transducers $b_1 \ldots b_n$ are located on
the upper wall. Let $\alpha(a_i, b_j) = \alpha_{i,j}$ be the angle

which the ray between a_i and b_j makes with the x-axis.
The x-axis is parallel to the parallel channel walls.
Let the shortest distance between the channel walls be
L. The integrals (Equation 4) can be evaluated by
changing the parameter of integration by the formula
$ds = (\sin \alpha_{i,j})^{-1} dy$ and noting that $\vec{V} \cdot \vec{\tau} = V_x \cos$
$\alpha_{i,j}$. Here we assume $V_y \equiv 0$ for steady state flow.

Thus equation 4 becomes

$$t_{a_ib_j} - t_{b_ja_i} = -2 \cot \alpha_{i,j} \int_{a_i}^{b_i} \frac{V_x}{C^2} \, dy \qquad (11)$$

If the fluid is incompressible then C^2 is constant.
The continuity equation and the assumption $V_y = 0$ im-
plies that $\frac{\partial}{\partial x} V_x = 0$. Thus the integral in Equation 11
is a constant independent of the path a_ib_j and all such
equations are linearly dependent. Note $t_{a_ib_j} - t_{b_ja_i} =$
$-2 < V_x > X_{a_ib_j}/C^2$ where $< V_x >$ is the average flow and
$X_{a_ib_j}$ is the x-axis distance between a_i and b_j.

Computer Simulation Studies

Digital computer simulation studies were conducted
in FORTRAN language on a CDC 3500 computer to test the
previously described theory. Computer simulation permits the testing of a wide range of flow configurations
and acoustic path geometries. In our case, both two-
dimensional parallel ray projection and fan beam projections were simulated. Such projection data were made
either by integrating Equation 2 by analytic or numerical means for a specific velocity function $\vec{V}(\vec{r})$. The
set of acoustic propagation times $\{t_{ab}\}$ thus obtained
was then used as measurement data to reconstruct the
velocity function $\vec{V}_R(r)$. The simularity of $\vec{V}_R(\vec{r})$ and
$\vec{V}(\vec{r})$ is a measure of the accuracy of the reconstruction
method. A modified algebraic reconstruction technique
(ART) with an underrelaxation parameter of 0.75 was
used (4).

The angular velocity of several vortex models with
angular symmetry were reconstructed from simulated
parallel projection data. The velocity of a natural
fluid vortex measured in a plane perpendicular to the
vortex axis may be modeled by

$$\vec{V}_{P1}(\vec{R}) = \frac{wA^2}{R} \quad (1 - e^{-R^2/A^2})\hat{\theta} \tag{12}$$

where $\hat{\theta}$ is a unit vector defined to be perpendicular to $\hat{R}$ and to $\hat{K}$. This function behaves like wR for very small R and like wA^2/R for very large R. The maximum value occurs near R = A.

In order to simplify the simulation and reconstruction algorithms it is desirable to study a function which vanishes outside a finite radius B and yet corresponds closely to the behavior of $\vec{V}_{P1}(\vec{R})$ defined in Equation 12. The function $\vec{V}_{P2}(R)$ defined by Equation 13 was chosen to meet these requirements

$$\vec{V}_{P2}(\vec{R}) = \begin{cases} (w - \frac{wA}{B})R\hat{\theta}, & R \le A \\[2mm] (\frac{wA^2}{R} - \frac{wA^2}{B})\ \hat{\theta}, & A < R \le B \\[2mm] 0, & B < R \end{cases} \tag{13}$$

The three-dimensional velocity $\vec{V}(\vec{R})$ for a point in a plane is given by $\vec{V}(\vec{R}) = \vec{V}_P(\vec{R}) + V_z(\vec{R})k$. In these simulations, the function $V_z(\vec{R})$ was defined to have a parabolic shape representing a downward flow given by

$$V_z(\vec{R}) = \begin{cases} -K[(R/B)^2 - 1], & R < B \\[2mm] 0, & R \ge B \end{cases} \tag{14}$$

The constants of geometry and flow were given values of A = 4.24 cm, B = 9.55 cm, w = 50 radians/sec, K = 91 cm/sec and simulated projection data were computed. A graph of $|\vec{V}_{P2}(R)|$ vs. R along a diameter and the corresponding generated fan beam projection data $(t_{ab} - t_{ba})$ are shown in Figure 5. Also shown in the same figure is the function $|V_R(\vec{R})|$ which is a reconstruction of $|V_{P2}(\vec{R})|$ from the set projection data $t_{ab} - t_{ba}$. The geometry of the flow and transducer locations (the geometry corresponds to that given in Figure 1) is also described in Figure 5.

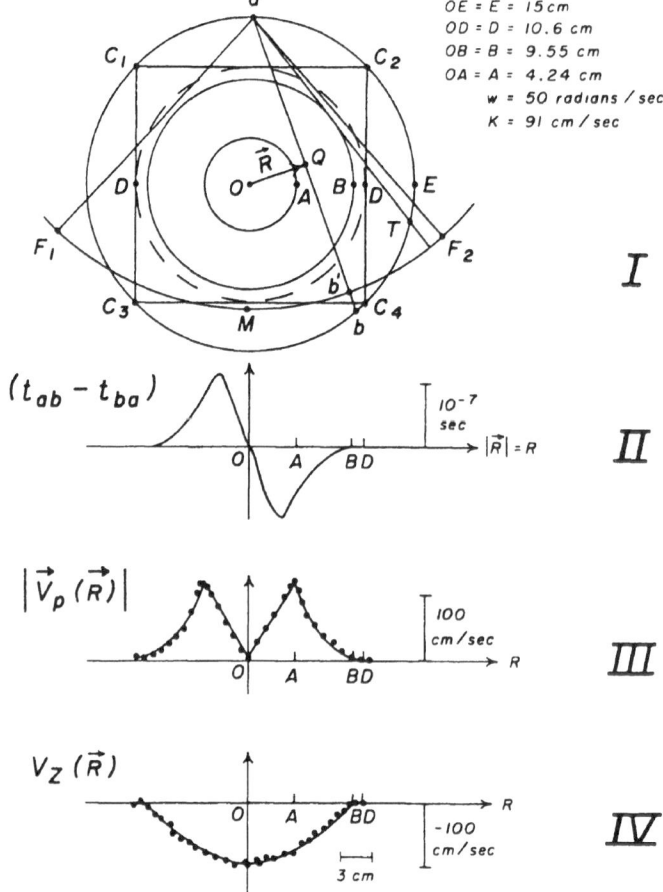

$$OE = E = 15 cm$$
$$OD = D = 10.6 \ cm$$
$$OB = B = 9.55 \ cm$$
$$OA = A = 4.24 \ cm$$
$$w = 50 \ radians \ / \ sec$$
$$K = 91 \ cm \ / \ sec$$

Figure 5 Experimental geometry for vortex reconstruction. Line I: shows the source and detector geometry transducers at a and b on circle of radius OE. Square $C_1 C_2 C_3 C_4$ is area reconstructed. Radius OB is 0.9 of OD. OA is 0.4 of OD. For simulation, transducers on radius Oa = Ob. For experimental data collection source at a and receiver at b' and receiver moves on arc of circle $F_2 b \ MF$, with a at center. Line II: shows the simulated time difference $(t_{ab} - t_{ba})$ ordinate calibration shown for values of A,B, w,k given in this figure. The abscissa R is the perpendicular distance from 0 to ray ab. Line III: shows assumed planar fluid speed function (solid line) and reconstructed values sampled along line ODE (solid dots). Line IV: shows assumed perpendicular (V_z) fluid component (solid line) and reconstructed values sampled along line ODE (solid dots).

Gray scale pictures of the three components V_x, V_y, V_z of the reconstructed function $\vec{V}_R(\vec{R})$ and of the planar magnitude $(V_x{}^2 + V_y{}^2)^{1/2}$ are shown in Figure 6. The picture of $(V_x{}^2 + V_y{}^2)^{1/2}$ shows the circular nature of the concentric flow lines. In these studies, 153 views each with 153 rays were used.

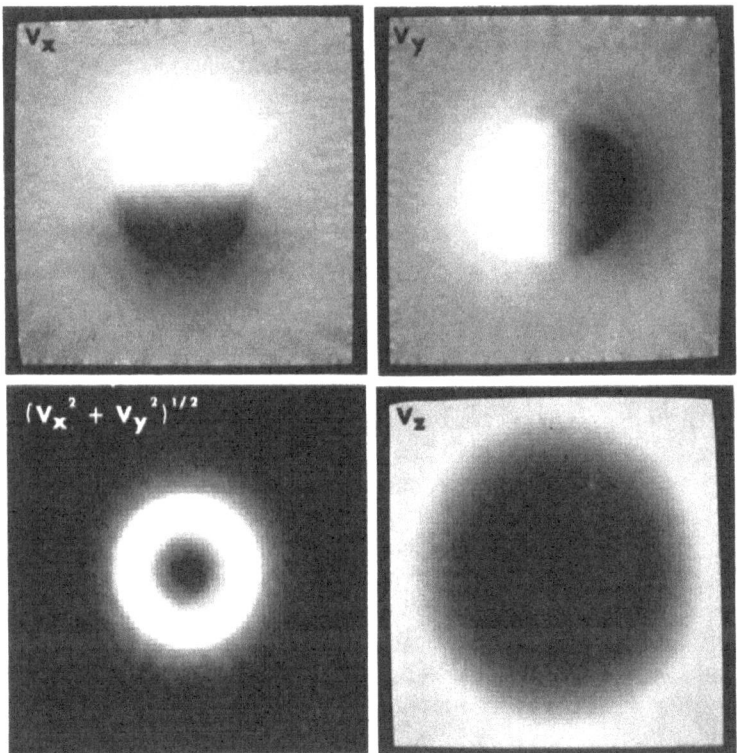

Figure 6 Reconstruction of vector components of simulated fluid vortex. Top left shows x component V_x. Top right shows y component V_y. Bottom right shows z component V_z. In these fluid velocity component reconstructions black is negative, gray is zero and white is positive in values. Bottom left is the magnitude of the xy planar component $(V_x{}^2 + V_y{}^2)^{1/2}$. Note circular equal speed contours. Images are 64 pixels per side.

Experimental Studies

The success of the computer simulation studies prompted an experimental verification of these results in the laboratory. Accordingly, an experiment was conducted with a flow and transducer geometry which matched the simulation studies as closely as possible. Although a circular transducer array like that shown in Figure 1 was not available, a fan beam geometry corresponding to the fan F_1aF_2MF, shown in Figure 5 was possible by using an ultrasound scanner designed for breast cancer detection studies (5). The source moved on a circle of radius 21 cm while the detector moved around the source on a circular arc of radius 26 cm on the opposite side of the center. All source intervals are equal.

A time of flight detector provided values of t_{ab} only and the difference $(t_{ab} - t_{ba})$ was computed from the relation $(t_{ab} - t_{ba}) = 2(t_{ab} - t_{oab})$ where t_{oab} is the time t_{ab} with zero flow velocity. The values of $(t_{ab} - t_{ba})$ were determined for 60 views with 150 rays per view and processed by the same algorithm used in the simulation studies. All ray intervals are equal.

The raw data for all rays in all views is shown in Figure 7. Also shown are the gray scale pictures for the components V_x and V_y and the magnitude (speed) $(V_x^2 + V_y^2)^{1/2}$. No measurements of the component V_z were made. It is seen that the data are contaminated with some noise and the reconstructions are not of the high quality as those of the simulation shown in the previous figure. This degradation is probably due to vibration during scanning about the source of the arm of length 26 cm, which holds the detector, and possibly to some instability in the time of flight detector. The images show the general features of a vortex inspite of the noise present.

Potential Applications

One application of this method is the design and construction of improved flow meters for liquids and gases. The ability to reconstruct the local speed of sound allows inference of local density and temperature changes. The combination of velocity and density reconstructions permits the reconstruction of local and net mass flow rates. The inclusion of percent mass per volume of trace materials or pollutants permits the

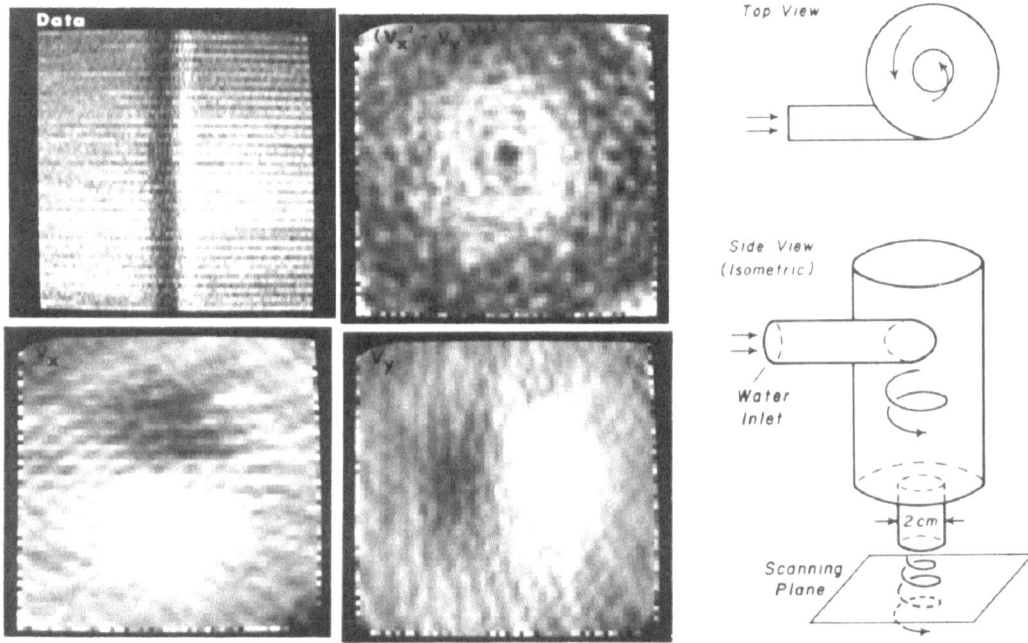

Figure 7 Experimental data and reconstruction of vector components of fluid vortex. Top left shows experimental data. Black is negative, gray is zero, white is positive. Top right shows reconstructed planar fluid speed $(V_x^2 + V_y^2)^{1/2}$, black is zero, white is positive. Bottom left shows x component of velocity V_x. Bottom right shows y component of velocity V_y. In V_x and V_y, black is negative, gray is zero, and white is positive. Reconstructed flow is maximum (73 cm/sec) at a radius of 0.62 cm. Reconstructions are 64 pixels per side. Geometry of vortex, scanning plane, and vortex generator are shown in top and side views in right margin (6).

calculation of total mass flow of such materials in pipes, flues, or smoke stacks. Also, the total energy flow may be calculated in regions around wind power extraction machines such as windmills, wind turbines, or vortex towers or vortex power devices. In some circumstances, the temperature of a gas flow may be reconstructed. That this is possible may be seen by observing that the speed of sound in a monoatomic gas is given by

$$C = \sqrt{(C_p/C_v)\,kt} \tag{15}$$

where C is the speed of sound C_p and C_v are the heat capacities at constant pressure and constant volume respectively, k is a constant proportional to the molecular weight of the gas, and t is the temperature of the gas in degrees Kelvin. Thus, a reconstruction of C when squared and rescaled provides a reconstruction of the temperature T (or the molecular weight, if T is known).

If the gas has no sources or sinks, then the continuity equation

$$\nabla \cdot (\rho \vec{V}) = 0 \tag{16}$$

where ρ is density and $\vec{V}$ is velocity, may be integrated throughout the region where C and $\vec{V}$ were previously reconstructed. This integration will require some assumptions on the value on some suitable boundary. The integral of Equation 16 is a reconstruction of the density function ρ. The pressure distribution may be obtained by using the gas law, $P = \rho\, RT/M$ (here, M is the gas molecular weight, R is the gas constant, P is the pressure, ρ = density, T = absolute temperature) and the reconstructions of ρ and T. Thus, a reconstruction of pressure is possible. The technique of measuring temperature may be extended to mixtures of gases using average molecular weights.

As has been seen, the data collected by the apparatus may be processed by algorithms of various levels of sophistication (e.g., straight lines or curved line reconstruction). One further level of improvement would involve the use of certain fluid dynamic equations such as the continuity equation, the momentum equation, and/or the energy equation (or their equivalents) as constraints in the solution of Equation 1. This may lead to greater accuracy in reconstruction or allow reconstruction with less data (1).

One further potential application is the reconstruction of the statistical moments of flow at each voxel (cubic volume element) in space.

Biomedical applications such as measuring flows of blood, respiratory gases, and other fluids both in and out of the body are possible when suitable measuring boundary conditions exist. For example, fluid carrying prosthetic devices could be tested by this noninvasive technique. However, where a suitable measurement collecting boundary does not exist (e.g., flows in the heart can only be viewed from a narrow view angle due to shielding of the lungs), Doppler methods will remain unchallenged.

SUMMARY

A mathematical theory and computational algorithms are presented for an acoustic method for reconstructing fluid velocity vector fields. Computational FORTRAN programs were written and tested with simulated and real data for the case where the fluid speed is much less than the speed of sound. These tests indicate the method has the capacity for reconstructing three-dimensional fluid flows and net flows.
It seems likely the method can be extended to reconstruct mass flow fields (variable density) and temperature fields. The method is capable of reconstructing flows with or without the presence of scattering centers (7).

ACKNOWLEDGEMENTS

The authors wish to thank Miss Pat Snider and associates for the secretarial and graphic assistance in preparing the manuscript. The photographic help of Mr. Leo Johnson is appreciated. The use of Mr. Willis VanNorman's gray scale and color computer graphic programs is appreciated. The first author thanks Professors Gary Flandro and Harvey Greenfield of the School of Engineering of the University of Utah and Professor Hans Weinberger of the Department of Mathematics, University of Minnesota for stimulating discussions. The support from Dr. Earl H. Wood, Chairman, Biophysical Sciences Unit, Mayo Clinic, and Dr. Erik Ritman, also at Mayo Clinic, is appreciated. This research was supported by Grants HL-00170, HL-00060, RR-00007, HL-04664, N01-HT-4-2904 from the National Institutes of Health, United States Public Health Service; N01-CB-64041 from the National Cancer Institute.

REFERENCES

1. Johnson, S. A., J. F. Greenleaf, W. F. Samayoa,
   F. A. Duck, and J. Sjostrand: Reconstruction of
   three-dimensional velocity fields and other parameters by acoustic ray tracing. 1975 Ultrasonics
   Symposium Proceedings, IEEE Cat. #75 CHO 994-4SU.

2. Johnson, S. A., J. F. Greenleaf, A. Chu, J.
   Sjostrand, B. K. Gilbert, and E. H. Wood:
   Reconstruction of material characteristics from
   highly refraction distorted projections by ray
   tracing. Image Processing for 2-D and 3-D
   Reconstruction from Projections: Theory and
   Practice in Medicine and the Physical Sciences.
   A Digest of Technical Papers, August 4-7, 1975,
   Stanford, California, pp TUB 2-1 - TUB 2-4.

3. White, R. W.: Acoustic ray tracing in moving
   inhomogeneous fluids. Journal, Acoustical Society
   of America, Vol. 53, No. 6, 1973.

4. Herman, G. T., A. V. Lakshminarayanan, and S. W.
   Rowland: The reconstruction of objects from
   shadowgraphs with high contrasts. Pattern
   Recognition, Vol. 7, pp 157-165, 1975.

5. Greenleaf, J. F., S. A. Johnson, W. F. Samayoa,
   and C. R. Hansen: Refractive index by reconstruction: use to improve compound B-scan resolution.
   (See this Proceedings, 1976).

6. Data shown in Figure 7, top left panel, is an image
   of the difference between the time of flight with
   flow and without flow (fast arrival = white, no
   change = gray, slow = black) vs scan position
   (left to right) vs angle of view (top to bottom).

7. Johnson, S. A.: Patent pending.

ULTRASONIC SYNTHETIC APERTURE IMAGING

M.L. Dick[1], D.E. Dick[1,2], F.D. McLeod[3], N.B. Kindig[1]

1. Dept. of Elec.Engr., Univ. of Colo., Boulder, Colo.
2. Physical Medicine & Rehabilitation, and Bioengineer-
   ing, Univ. Colo. Med. Ctr., Denver, Colo.
3. Dept.of Physiology,CRHL, Colo.St.U., Ft. Collins,Colo.

Currently available medical ultrasonic imaging systems are
limited in some applications by their resolution. In particular,
better resolution is needed to detect small lesions in the female
breast. The apertures required to improve resolution can be obtained using large transducers at the expense of depth of field.
Alternatively, synthetic aperture techniques offer the potential
for achieving the necessary resolution without sacrificing depth
of field. Advantages of synthetic aperture imaging are high focus,
independent of depth, coupled with the absence of near field aberrations. The goal of this study is to develop a synthetic aperture
ultrasonic imaging system specifically designed for breast scanning.

Improvement in ultrasonic imaging system resolution has been
investigated with a variety of approaches. Eggleton (1975) has
developed large aperture, single transducers with repeated scans
at different distances from the skin to compensate for the limited
depth of field. Green et al. (1974) have developed a long linear
array of transducers which are used in through-transmission mode
and have recently reported a technique to minimize artifacts resulting from out-of-focus information (Havlice et al., 1976).
Annular transducer arrays have also been used to improve resolution
by the incorporation of swept receiver focus using variable time
delays (Melton, 1971; Burckhardt et al., 1974; Vilkomerson, 1974;
Kossoff, in press). While some improvement in diagnostic accuracy
has been reported from these approaches, there is still a need for
improved resolution without degrading depth of field or introducing
artifacts from out-of-focus information, and including optimum
focus on both transmit and receive. Synthetic aperture imaging
offers the potential for providing these benefits.

Synthetic aperture imaging was developed originally for electromagnetic radar terrain mapping (Cutrona et al., 1961). It has been adapted to ultrasound by several groups (Flaherty et al., 1970; Burckhardt et al., 1974; Onoe, 1974; Tsujiuchi et al., 1974; Matzuk, 1976; Cribbs et al., 1976). Usually, these adaptations have followed the same lines as used in the radar systems with linearization of the distance function, use of photographic raw data storage, and optical reconstruction without apodizing. The use of electronic data storage and reconstruction techniques was mentioned by Flaherty et al., (1970) and there has been a recent implementation using analog scan converters (Cribbs et al., 1976). This work has demonstrated the capability of synthetic aperture imaging for improved resolution but has not led to a system suitable for extensive clinical testing.

The purpose of this paper is to develop the mathematical basis for synthetic aperture ultrasonic imaging in a form suitable for medical application. The design of such a system is also described. A minicomputer-based system has been developed for the purpose of studying and optimizing parameters. Preliminary results on estimation of resolution capability are presented along with some observations on the potential for future extensions of the synthetic aperture approach. The system under development in this project will be used initially for the detection of breast lesions.

MATHEMATICAL BASIS

The coordinate system used for this development assumes that a small transducer moves in a straight line parallel to the x-axis. The y-axis is vertical with the transducer located at $y = 0$ and pointing parallel to the y-axis. The transducer at any time is located at $(x_T, 0)$. Assume that there is a single point reflector located at (x_r, y_r) and that the transmitted signal is given by

$$s(t) = A_0 \sin(\omega_0 t) \cdot w(t)$$

$$w(t) = \begin{cases} 1, & 0 \leqslant t \leqslant T_0 \\ 0, & \text{otherwise} \end{cases} \tag{1}$$

where ω_0 is the center frequency of the ultrasound pulse
T_0 is the pulse duration, which is normally an integral number of half cycles.

The reflected signal is a function of the transducer position and is given by

$$r(x_T,t) = A_1 \sin[\omega_0(t-\tau_r)] \cdot w(t-\tau_r)$$

$$d_r(x_T) = \sqrt{(x_r-x_T)^2 + y_r^2}$$

$$\tau_r = \frac{2 \cdot d_r(x_T)}{c} \tag{2}$$

where c is the velocity of sound in tissue

A_1/A_0 is the attenuation due to absorption, scattering cross-section, and transducer inefficiency.

The received signal, $r(x_T,t)$, must be detected or demodulated by some scheme. Although previous implementations have used either AM detection or demodulation with a single reference sine wave, significant advantages result from the use of both quadrature components. This step is similar to the use of both quadrature components in Doppler ultrasound for the separation of forward from reverse flow and also to the separation of real and virtual images in conventional holography. This demodulation yields

$$u_c(x_T,t) = r(x_T,t) \cdot \cos(\omega_0 t)$$

$$= \frac{A_1}{2}\left\{\sin[\omega_0(2t-\tau_r)] - \sin(\omega_0\tau_r)\right\} \cdot w(t-\tau_r)$$

$$u_s(x_T,t) = r(x_T,t) \cdot \sin(\omega_0 t) \tag{3}$$

$$= \frac{A_1}{2}\left\{\cos[\omega_0(2t-\tau_r)] + \cos(\omega_0\tau_r)\right\} \cdot w(t-\tau_r)$$

The demodulation is followed by low-pass filtering to remove the sum frequency from both $u_c(x_T,t)$ and $u_s(x_T,t)$. This filtering yields

$$v_c'(x_T,t) = -\frac{A_1}{2}\sin(\omega_0\tau_r) \cdot w_1'(t-\tau_r)$$

$$v_s'(x_T,t) = \frac{A_1}{2}\cos(\omega_0\tau_r) \cdot w_1'(t-\tau_r) \tag{4}$$

where $w_1'(t-\tau_r)$ is a modification of the delayed original window, $w(t)$, due to the low-pass filter.

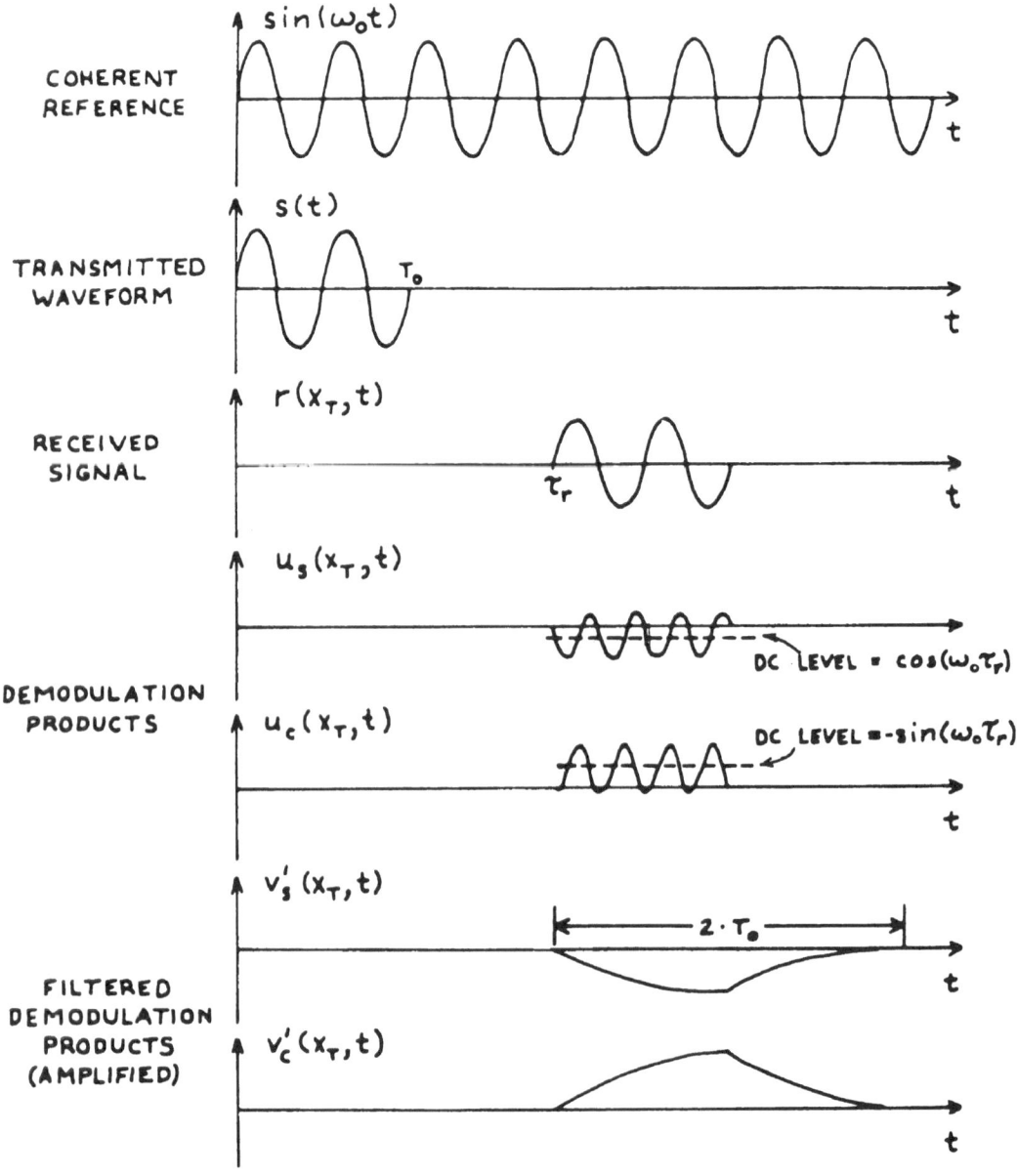

Figure 1. Typical waveforms resulting from the transmitter, receiver, and demodulation.

Figure 1 shows the waveforms expected for these various signals. Note that the low-pass filter tends to smear out the original box car function. This has the effect of integrating in depth. This integration will limit resolution but will provide compensating reduction of noise spikes.

If we assume that sound propagates at a constant velocity, c, then the time variation in Eq. 4 is equivalent to the distance to the spherical wavefront, ξ . Thus, using the substitution, $t = 2 \cdot \xi / c$ and $\tau_r = 2d/c$, we can rewrite Eq. 4 as

$$v_c(x_T, \xi) = -\frac{A_1}{2} \sin(4\pi d_r/\lambda) \cdot w_1(\xi - d_r)$$

$$v_s(x_T, \xi) = \frac{A_1}{2} \cos(4\pi d_r/\lambda) \cdot w_1(\xi - d_r) \tag{5}$$

where $\lambda = c/f_0$

$$w_1(\xi) = w_1'(2 \cdot \xi / c)$$

As given in Eq. 5, $v_c(x_T, \xi)$ and $v_s(x_T, \xi)$ represent the response to a point reflector a distance, d_r from the transducer. Figure 2 shows a plot of $\sin(4\pi d_r/\lambda)$ and $\cos(4\pi d_r/\lambda)$ as a function of $x_T - x_r$. Note that the spatial frequency of oscillation changes considerably due to the nonlinear distance function, $d_r(x)$.

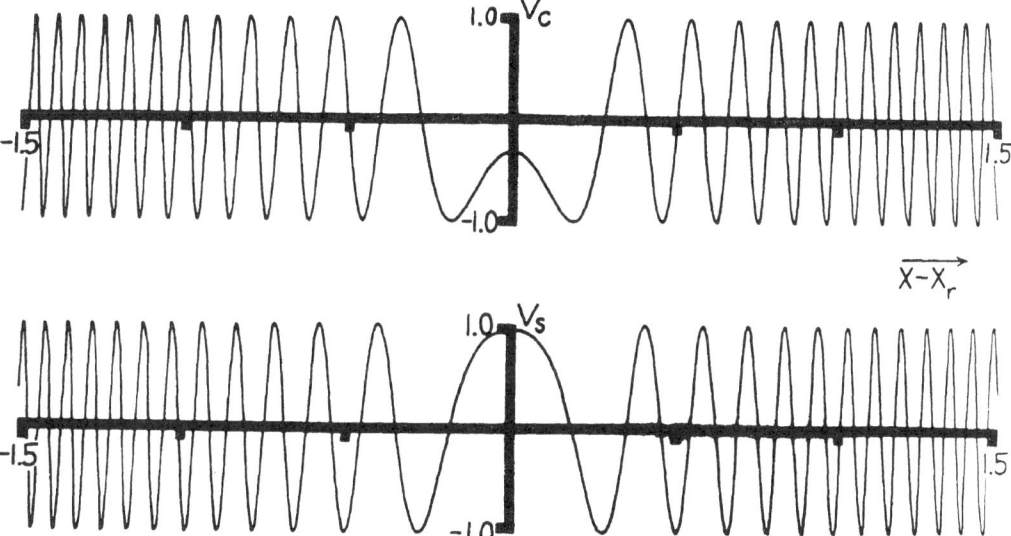

Figure 2. Plot of the two quadrature components, $\cos(4\pi d_r(x)/\lambda)$ and $\sin(4\pi d_r(x)/\lambda)$, which would result from scanning a point reflector located at a depth of 2.5 cm with an ultrasound center frequency of 2.25 MHz.

If the transducer is moved along the x-axis and the demodulated echoes are recorded, the result is a pair of two-dimensional arrays,

$v_c(x, \xi)$ and $v_s(x, \xi)$, with echoes lying on a hyperbolic arc given
by $\xi = d_r(x)$. These two arrays are equivalent to a spatial impulse
response; this concept will be used below to develop the reconstruction technique. In general, a collection of points or a
continuous distribution of a reflecting surface will result in very
complex patterns in the v_c and v_s arrays.

Once the arrays, $v_c(x, \xi)$ and $v_s(x, \xi)$, have been formed, the
final image is constructed. The reconstruction process can be explained as a matched filter using Eq. 5 as the impulse response.
This leads to an integral along an arc for each point in the reconstructed image. Thus, for each point (x_0, y_0) in the final image,
we must form the image, $g(x_0, y_0)$, according to the equation

$$g(x_0, y_0) = \frac{1}{X_I} \int_0^{\xi_{max}} \int_{x_0 - X_I}^{x_0 + X_I} [v_s \cdot \cos(4\pi d_0/\lambda) - v_c \cdot$$

$$\cdot \sin(4\pi d_0/\lambda)] \cdot \delta(\xi - d_0 - \Delta) \; dx \; d\xi \qquad (6)$$

where $d_0(x) = \sqrt{(x - x_0)^2 + y_0^2}$

  $\delta(\xi)$ is a delta function
  X_I is the symmetrical integration interval
  Δ is an offset to compensate for the delay in the
        response of the low-pass filter used to obtain the
        v_c and v_s arrays from u_c and u_s.

The double integral in Eq. 6 is required to describe the
hyperbolic arc that constitutes the integration path. Note that
the offset, Δ, in Eq. 6 should be chosen to coincide with the
maximum of $w_1(\xi - d_r)$. For most cases, this will be at $\Delta = T_0 \cdot c/2$.
The geometry of the reconstruction process is shown in Fig. 3.

To demonstrate that Eq. 6 is a matched filter formulation,
consider the reconstruction resulting if the input data are generated by scanning a point reflector. In this case, v_c and v_s are
given by Eq. 5 and substituting Eq. 5 into Eq. 6 yields

$$g(x_0, y_0) = \frac{1}{X_I} \int_0^{\xi_{max}} \int_{x_0 - X_I}^{x_0 + X_I} (A_1/2) \cdot [\cos(4\pi d_r/\lambda) \cdot \cos(4\pi d_0/\lambda)$$

$$+ \sin(4\pi d_r/\lambda) \cdot \sin(4\pi d_0/\lambda)] \cdot \delta(\xi - d_0 - \Delta) \cdot$$

$$\cdot w_1(\xi - d_r) \; dx \; d\xi \qquad (7)$$

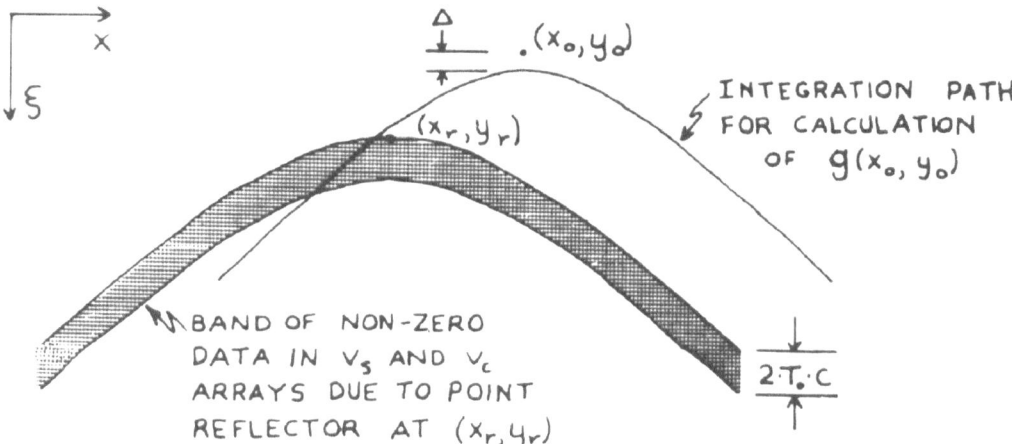

Figure 3. Geometry of the reconstruction process with a band of non-zero data resulting from a point reflector at (x_r, y_r) shown in relation to the integration path for estimating $g(x_0, y_0)$.

When $(x_0, y_0) = (x_r, y_r)$, we have $d_0 = d_r$ and the integration in ξ can be performed. This results in the evaluation of $w_1(\Delta)$ which should be close to unity. Assuming $w_1(\Delta) = 1$ and evaluating Eq. 7 at the point (x_r, y_r) yields

$$g(x_r, y_r) = \frac{A_1}{2X_I} \int_{x_0 - X_I}^{x_0 + X_I} [\cos^2(4\pi d_r / \lambda) + \sin^2(4\pi d_r / \lambda)] dx$$

$$= \frac{A_1}{2X_I} \cdot 2X_I = A_1$$

(8)

Thus, the integral in Eq. 6 correctly estimates the amplitude, A_1, of the received echo when it is evaluated at the location of the reflector. It should be noted that it is necessary to retain gray scale information in the raw data arrays, v_c and v_s, for this to be true. Some implementations of the synthetic aperture technique have used high contrast film to store these arrays; this destroys the gray scale information present in the original echo data. It is also important to include both quadrature components, v_c and v_s to distinguish amplitude from phase changes.

Synthetic aperture imaging offers the possibility of very good resolution in the x-y plane, but in the third dimension, z,

perpendicular to the x-y plane, there is a problem. The side looking radar technique used for terrain mapping is not appropriate
since medical imaging involves objects that are not constrained to
a surface. The effect of poor lateral resolution is to collapse
the image in the z-direction as is done with conventional X-ray
images and also with computerized axial tomographic equipment.
There are two possible solutions to prevent loss of resolution in
the z-direction:

1. Build a special oblong transducer or transducer array
   with good resolution in the z-direction but with wide
   acceptance angle in the x-y plane.

2. Use the synthetic aperture concept to obtain high
   resolution in both planes by performing a raster scan
   with a small transducer.

The first of these options will result in a bizarre near-field
antenna pattern. The second option is more appealing but the data
volume for a three-dimensional system is unmanageably large. Difficulties with these two options and the success of radiographic
CAT scanners suggest that the best approach is to allow poor
resolution in the z-direction

RESOLUTION ANALYSIS

Calculation of resolution capability requires evaluation of
Eq. 7 with $(x_o,y_o) \neq (x_r,y_r)$. This is very difficult due to the
nonlinear nature of both d_r and d_o. Effectively the integrand in
Eq. 7 reduces to $\cos[4\pi(d_o-d_r)/\lambda^o]$ which produces a slow oscillation over the interval of integration. This oscillation is gated
by the product, $\delta(\xi-d_o-\Delta) \cdot w_1(\xi-d_r)$, which is zero except where
the integration path overlaps the band of non-zero data generated
by the point reflector as was shown in Fig. 3. This factor makes
the integral small and the division by X_I reduces it still further.

Computer calculations of the integral in Eq. 7 could be done
to estimate the point spread function but insight into the limiting
factors would be lost. As an alternative, the following approximate analysis has been developed. Initially the assumptions may
seem quite unreasonable but it is easiest to follow the reasoning
by starting with crude approximations.

Initially, assume that the non-zero data resulting from scanning a point reflector lies along two straight lines as shown in
Fig. 4a. If this were true, the the integration arc for reconstruction will be the same shape. It is also shown in Fig. 4a. If we
neglect the response time of the low-pass filters and all phase

variation of the received data and modify Eq. 6 to reflect this assumption, then the point spread function is shown in Fig. 4b. The important aspects of this result include:

1. The width of the central lobe is determined by the depth resolution of the original ultrasound pulse, Δ.

2. The amplitude of the side lobe is determined by the ratio of the depth resolution, Δ, to the half aperture width, X_I.

3. The width of the side lobe is determined by the half aperture width, X_I.

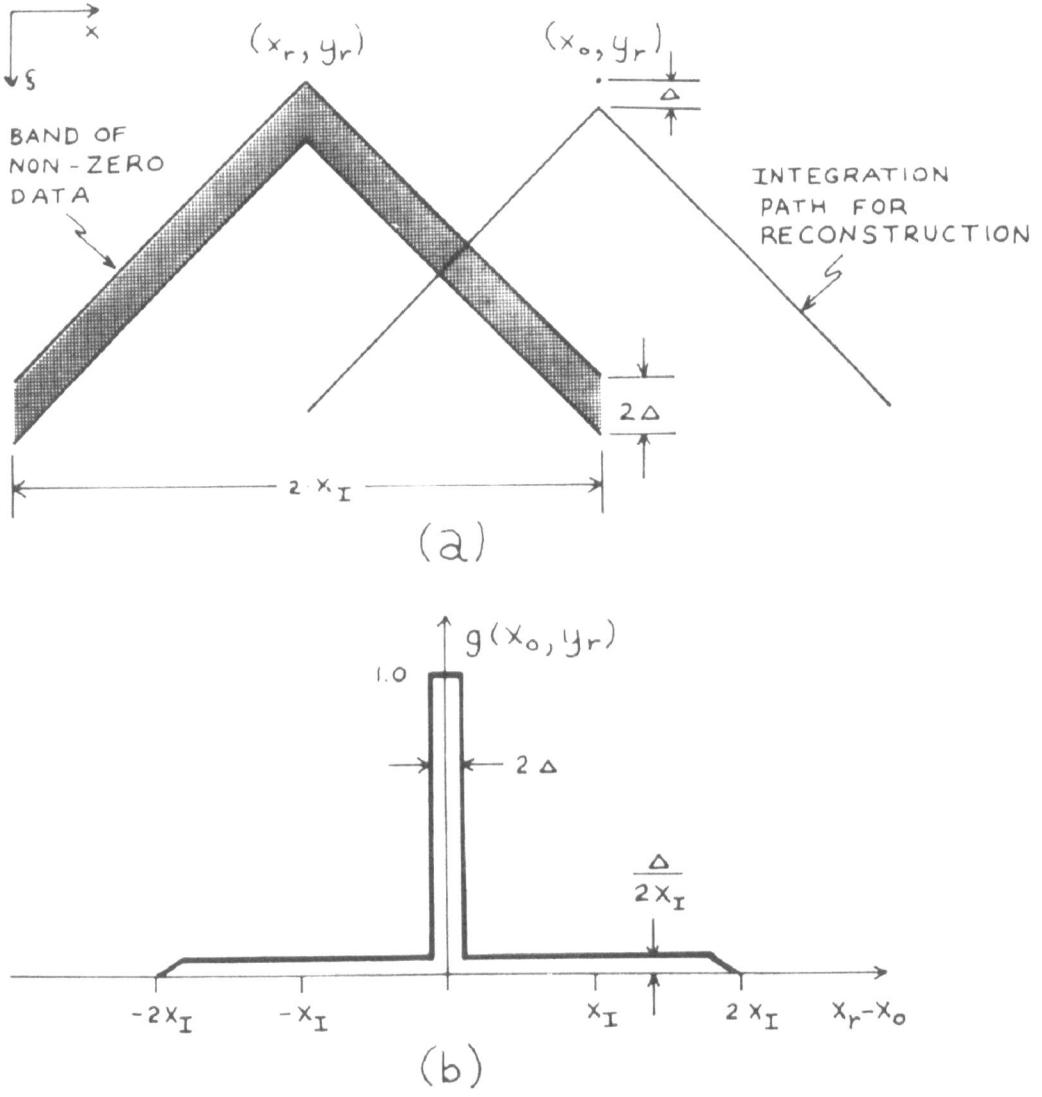

Figure 4. Simplest approximation to reconstruction geometry for a point reflector in (a), and resulting point spread function in (b).

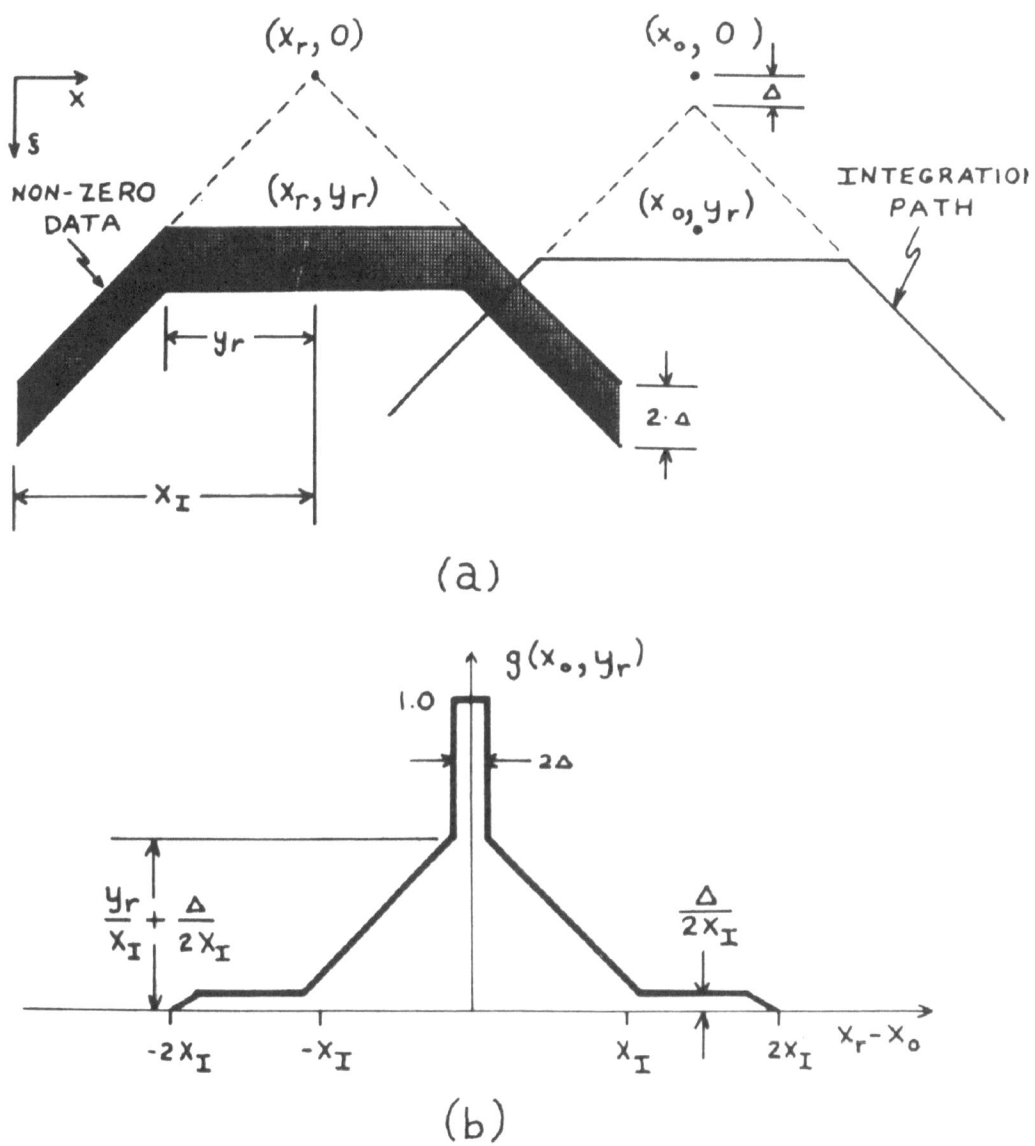

Figure 5. A second approximation to reconstruction geometry for a point reflector in (a), and resulting point spread function in (b).

Except when y_r = 0, the form of the arc of non-zero data will be significantly different than the assumed form of Fig. 4a. To obtain a more realistic estimate on resolution, assume that this shape is modified to have a flat top as shown in Fig. 5a. The length of this flat top depends on the range to the reflector, y_r, when the transducer is located at x_r. Note that the remainder

of the swath lies along the asymptotes of the hyperbolas that we expect in actuality. Making the same assumptions as before, the point
spread function of Fig. 5b can be derived. Here again, the same observations as for Fig. 4b will hold regarding width of the main lobe,
width of the side lobes, and amplitude of the side lobes. The major
difference is that the ratio of range, y_r, to half aperture, X_I, determines whether the sloping portion of the side lobe or the flat
portion of the side lobe will dominate. This implies that X_I must
increase linearly with the depth of reconstruction to maintain a
constant point spread function.

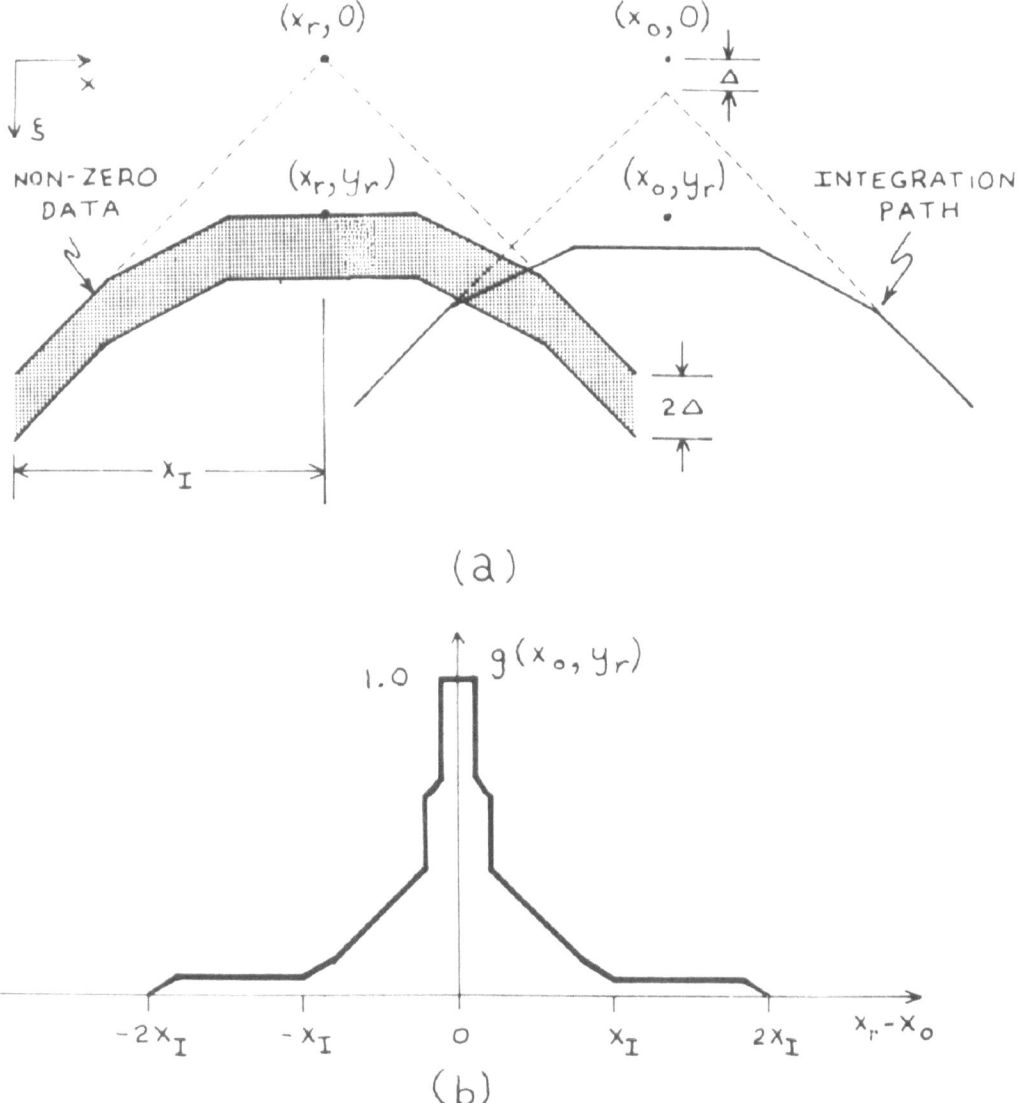

(a)

(b)

Figure 6. A more complex approximation to reconstruction geometry
for a point reflector in (a), and resulting point spread function(b).

One more refinement of the approximation to the hyperbolic
swath of non-zero data can be made and this is shown in Fig. 6a.
With the same neglect of phase as before, the point spread function
is shown in Fig. 6b. Again the observations made for Fig. 4b still
hold for this result although minor deviations can be seen.

Multiple straight line segments can be made to approximate the
actual hyperbola as closely as desired. The small changes between
Figs. 4b, 5b, and 6b suggest that the fundamental observations
following Fig. 4b should still apply as modified following Fig. 5b.
The main conclusion is that once the half aperture, X_T, exceeds the
range, then resolution is limited by the pulse duration, Δ. Since
usually $\Delta \approx \lambda$, this is consistent with ultimate resolution limits
encountered in optics and radar.

The effect of including phase variations in this analysis would
be to change Figs. 4b, 5b, and 6b into graphs of the envelope of the
point spread function. The actual point spread function would os-
cillate within this envelope with both negative and positive side
lobes.

HARDWARE DESIGN

The early synthetic aperture imaging systems have utilized
film, to store the raw data, combined with optical reconstruction.
This approach makes it difficult to store both quadrature components,
v_c and v_s. It is also difficult to optimize system performance by
experimenting with different apertures, windows, ultrasound frequencies, and scanning velocities. An alternative is to use electronic analog data storage on scan converters. This permits an all-
electronic system for maximum flexibility and ease of optimization.
A similar approach has been reported very recently (Cribbs et al.,
1976).

A block diagram of the resulting system is shown in Fig. 7.
Note that two separate scan converters are used to store the raw
data arrays and a third converter is used for formation of the final
image. It is possible to reduce the number of scan converters by
storing both raw data arrays on the same converter tube but then
scan converter resolution becomes a limiting factor. A microprocessor controls and synchronizes all aspects of the formation of
the image. The generation of the image is done in two distinct
phases as explained below.

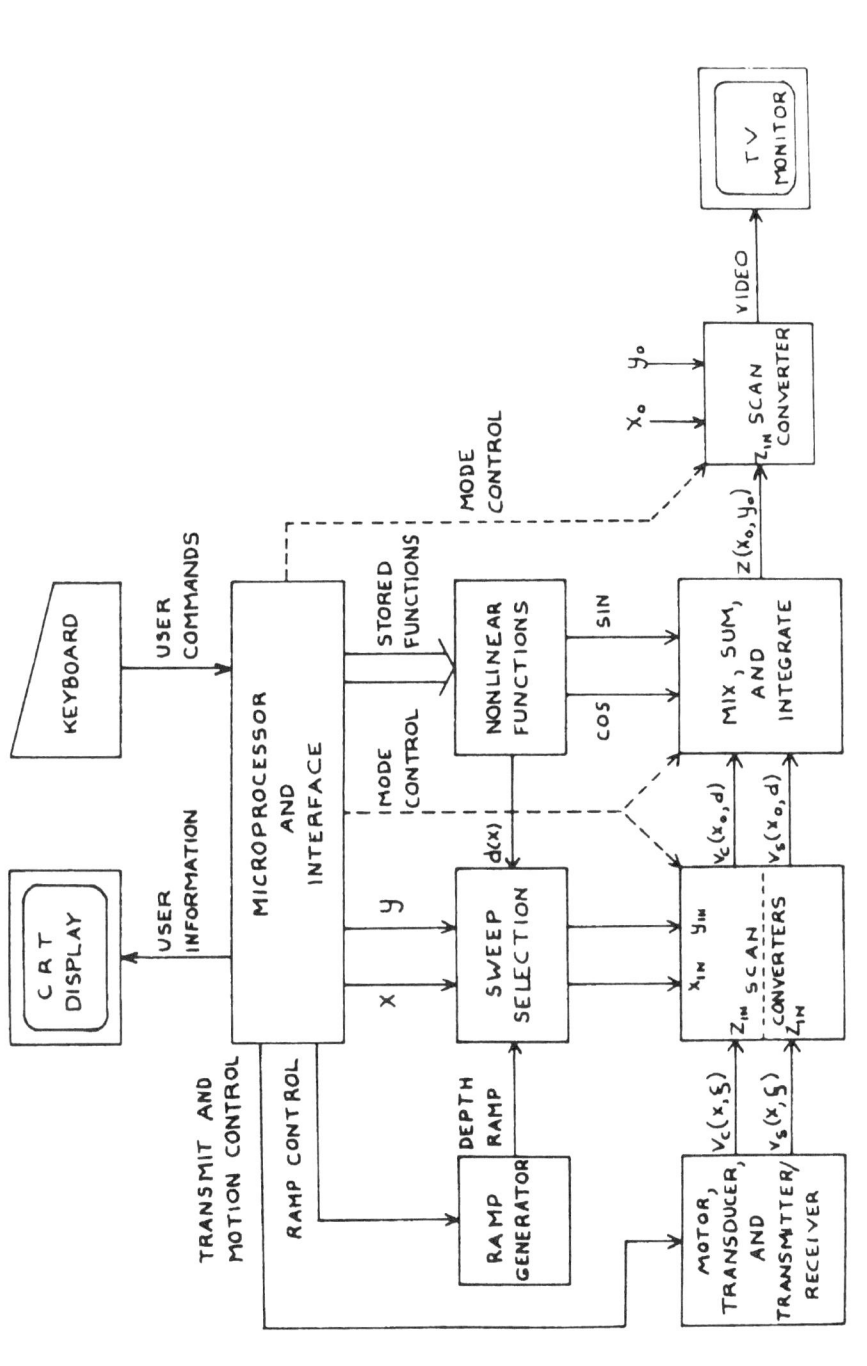

Figure 7. Block diagram of the hardware for a synthetic aperture system. Electronic analog data storage is provided by scan converters and a microprocessor provides overall synchronization and control.

Generation of Raw Data Arrays

Initially the transducer is positioned at $x = 0$ and all scan converters are erased. XIN for the two raw data storage converters corresponds to the position of the transducer and $YIN = \xi$ is a sweep corresponding to the propagation of sound. The received, demodulated echoes appear as $v_c(x, \xi)$ and $v_s(x, \xi)$ and are stored as vertical lines on the data storage converters. After each line of data is stored, the transducer moves a small increment, the XIN voltage increases to reflect this motion, and the YIN sweep repeats to store another line. This sequence repeats until the transducer has moved over the entire scan path.

Formation of Reconstructed Image

Once the v_c and v_s data arrays are complete, calculation of $g(x_0, y_0)$ for all x_0 and y_0 must be done. Again starting in the upper left corner by setting $x_0 = 0$ and $y_0 = 0$, the raw data scan converters are placed in the READ mode. XIN for these two converters is a sweep from $x_0 - X_I$ to $x_0 + X_I$; YIN must be $d(x)$. Simultaneously, $\cos(4\pi d(x)/\lambda)$ and $\sin(4\pi d(x)/\lambda)$ are generated and multiplied times the two outputs of the data scan converters to form the integrand in Eq. 6. The sum of the two products is integrated over the required interval and the result is stored on the third converter as a single point at (x_0, y_0). This process repeats for each value of (x_0, y_0) until the image is complete. The output of the third scan converter is fed directly into a television for viewing and photographing.

Hardware Considerations

Most of the hardware for this system is conventional and readily available. Since $d(x)$ will be different for each depth, y_0, it is recalculated for each line of the reconstructed image by the microprocessor and stored digitally in a RAM. These data are used repetitively for each value of x_0 by feeding a D/A converter directly to obtain $d(x)$ for YIN. The digital values of $d(x)$ are also used as an address for digital sin/cos resolver packages to generate the nonlinear sine and cosine functions used in Eq. 6. The resolver outputs are fed into multiplying D/A converters. Applying the scan converter outputs to the analog inputs of these converters yields the desired products for Eq. 6.

The system as described above will require considerably more time to form an image than is required with optical reconstruction.

This limitation is due to the sequential nature of the reconstruction on a point-by-point basis with electronic data storage as contrasted with simultaneous parallel reconstruction with optical processing. Assuming that 500x500 points are required in the reconstructed image and that each point requires 256μsec to form, the total time of formation will be 64 sec. This does not seem too large a penalty to pay for the flexibility provided by this approach.

SYSTEM DEVELOPMENT

The system as described above has been partially implemented using a minicomputer which had already been constructed for other purposes. The minicomputer includes a general purpose analog interface with A/D, D/A, and logic outputs, plus magnetic tape, CRT, and a keyboard. A version of BASIC also exists with ability to incorporate assembly language subroutines for controlling the analog interface.

Since the scan converter represents the critical element in this implementation, the development of the hardware was started using only a scan converter and the minicomputer. The minicomputer first generates simulated ultrasound data by calculating the net reflection, including phase, from a set of point reflectors. The D/A converters generate XIN, YIN, and v_c and v_s in the same way that they would be produced by a scanning ultrasound system. Since only a single scan converter was available, both v_c and v_s are stored on it, using the top of the tube for v_c and the bottom for v_s. The calculations involving sines, cosines, square roots and products are all done in BASIC.

Reconstruction follows a similar approach with the D/A converters supplying XIN and YIN along the hyperbolic arc associated with each (x_o, y_o). As the arc is traced out, the ouput of the scan converter is sampled with an A/D converter. The multiplication, summation and integration is done digitally using $\cos(4\pi d(x)/\lambda)$ and $\sin(4\pi d(x)/\lambda)$ as calculated in BASIC. With only one scan converter it is not possible to create an image for viewing but it is possible to store and plot individual lines of the reconstructed image. The result is the point spread function if a single point reflector is used to form the raw data arrays.

Figure 8 shows the two raw data arrays that were generated by this system for a point reflector located at a depth of 5 cm. The alternating light and dark pattern along the arcs corresponds to the chirp pattern plotted earlier in Fig. 2. Since the scan converter can only write by increasing the video, it is necessary to bias the video inputs so that zero corresponds to the middle gray level.

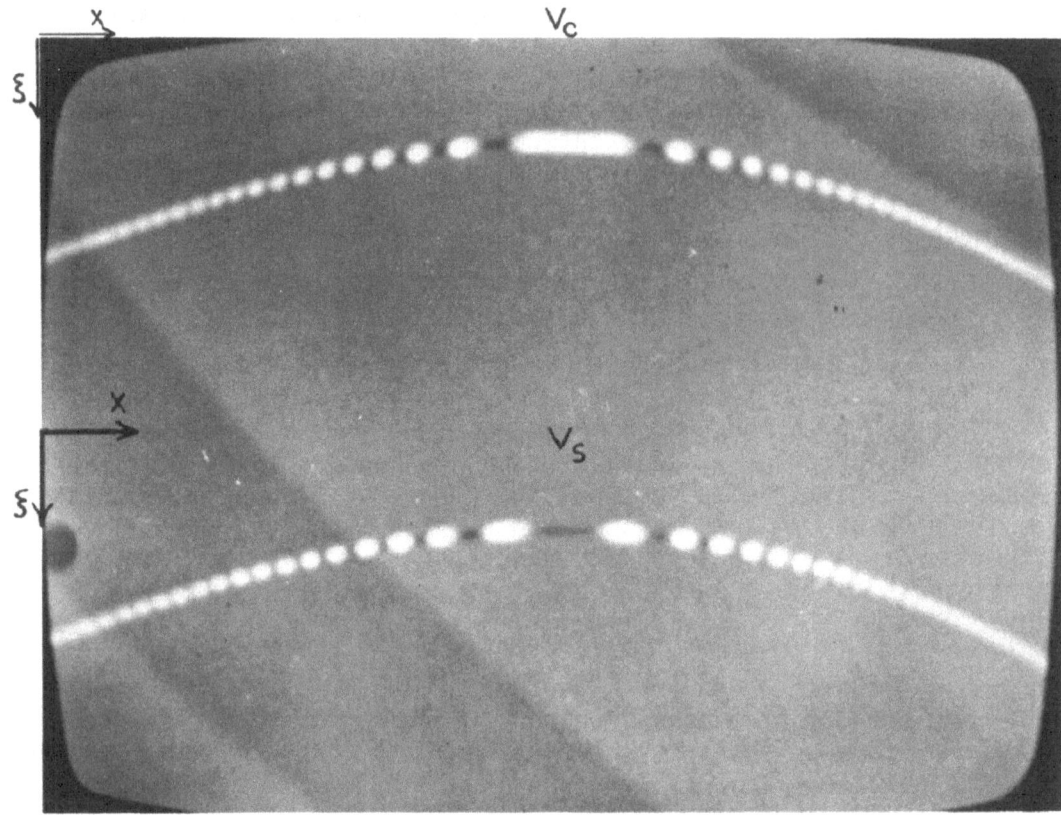

Figure 8. Raw data arrays v_c and v_s for scanning a point reflector. The v_c array is on top and the v_s array is the bottom half, X and ξ have different scales.

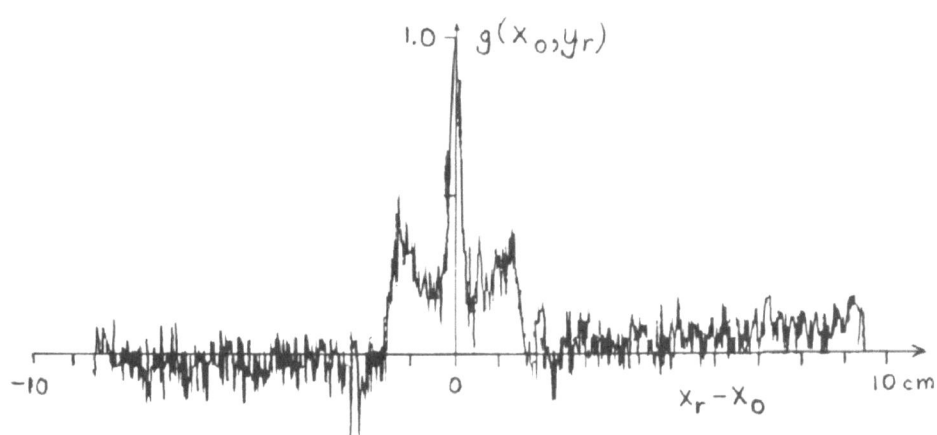

Figure 9. One line of a reconstructed image at the depth of the point reflector.

Figure 9 shows a point spread function that was generated by this system by reconstruction of raw data that represented a point reflector at 5.0 cm. The width of the swath, 2Δ, was 0.28 cm and X_I was 1.0 cm. Note that the point spread function of Fig. 9 has a main lobe width of 0.3 cm and the side lobes fall off to zero at ±2.0 cm. Both of these results check very closely with the predictions of the resolution analysis as presented earlier. The abrupt fall-off of the side lobes when the integration path no longer intersects the non-zero data can also be seen.

The noise superimposed on this point spread function is due to limitations on the settling time of the scan converter. Limitations in the speed of the general purpose minicomputer system analog interface require a point-by-point readout of the v_c and v_s arrays instead of a single continuous arc. This leads to misalignment between the stored data lines and the readout voltages and thus generates errors in estimating the point spread function.

FUTURE DEVELOPMENTS

The system described in this paper has been useful to check the predicted performance of the synthetic aperture imaging system. The next step is to complete the system and start both in vitro and clinical testing. A comprehensive clinical evaluation is being planned to evaluate the usefulness of this instrument for breast imaging.

In addition to the completion of this hardware, several subsequent developments are planned to extend the usefulness and range of applications. These plans are briefly described in the following paragraphs.

Analytical Optimization

The mathematics presented earlier can be extended to help determine optimum choice of parameters for the system such as the optimum spacing of data lines for the scanning system. The simplifying assumptions used in the resolution analysis are being refined to enable more quantitative estimates on resolution.

The mathematics necessary to study the artifacts produced in the reconstructed image are being developed. As in time series analysis, it may be possible to make significant improvements by introducing a window function in the reconstruction process. This is equivalent to apodizing in optics.

Computer Optimization

The minicomputer-based system will be used to study major problems such as the degradation that results from overlying distorting media. This is particularly a problem in the breast where wide variety of fat content may be encountered. Along similar lines, Jellins et al. (1973) have described significant problems resulting from a velocity differential between the water bath and breast tissue. This is a major problem for high resolution imaging systems and its effects can be studied with the computer-based system that has been developed.

Ultrasound Improvements

The initial resolution analysis indicates that lateral resolution is strongly affected by the depth resolution, particularly for large apertures. This makes it desirable to seek methods for decreasing this depth uncertainty. The most promising approach is the use of pulse compression as it was originally developed for radar applications. Preliminary work on the application of these principles to pulse echo ultrasound has already been reported (Jethwa et al., 1974; Furgason et al., 1975; Luque et al., 1976). Combination of the pulse compression approach with synthetic aperture imaging will require significant modification of the mathematical analysis presented here.

SUMMARY

An ultrasonic imaging system has been designed for medical application using the synthetic aperture approach. The mathematical basis for the resulting system includes significant changes from that developed for radar systems. An analysis of potential resolution has been performed and shows that lateral resolution is limited by the duration of the ultrasound pulse. A hardware system to implement the synthetic aperture approach has been designed using electronic, analog data storage. This design has been partially implemented using a general purpose minicomputer and a scan converter. The results from using this system to simulate scanning a point reflector verify the theoretical analysis of resolution. Completion of the system and future applications of ultrasonic synthetic aperture imaging for breast scanning is described. The potential improvement in resolution should provide significant medical benefits by the detection of smaller and hence earlier lesions.

REFERENCES

1. Burckhardt, C.B., Grandchamp, P.A., Hoffman, H., 1974, Methods for increasing the lateral resolution of B-scan, in Acoustical Holography Vol. 5, P. Green, ed., Plenum Press, p. 391-413.

2. Cribbs, R., Arnon, S.D., 1976, High resolution synthetic aperture ultrasound imaging, Proc. 1st World Fed. Ultrasound Med. Biol., San Francisco.

3. Cutrona, L.J., Vivian, W.E., Leith, E.N., Hall, G.O., 1961, A high-resolution radar combat-surveillance system, IRE Trans. Mil. Electron. MIL-6:119-133.

4. Eggleton, R.C., 1975, State-of-the-art of single transducer ultrasound imaging technology, presented at annual meeting of Amer. Assoc. of Physicists in Med., 37 pp., also submitted to Medical Physics.

5. Flaherty, J.J., Erickson, K.R., Lund, V.M., 1970, Synthetic Aperture Ultrasonic Imaging System, U.S. Patent #3,548,642.

6. Furgason, E.S., Newhouse, V.L., Bilgutay, N.M., Cooper, G.R., 1975, Application of random signal correlation techniques to ultrasonic flaw detection, Ultrasonics 13:11-17.

7. Green, P.S., Schaefer, L.F., Jones, E.D., Suarez, J.R., 1974, A new high-performance ultrasonic camera system, in Acoustical Holography Vol. 5, P.S. Green, ed., Plenum, pp. 493-504.

8. Havlice, J.R., Green, P.S., Taenzer, J.C., Mullen, W., 1976, Removal of spurious detail in acoustic images using spatially and temporally varying insonification, Proc. 1st WFUMB, San Francisco.

9. Jellins, R., Kossoff, G., 1973, Velocity compensation in water-coupled ecography, Ultrasonics, 11:223-226.

10. Jethwa, C.P., Olinger, M.D., Brenner, D.P., Fry, F.J., 1974, Ultrasonic transmission imaging and blood flow measurement using Gaussian noise source, Proc. 1974 IEEE Symp. Sonics Ultrasonics, 28-32.

11. Kossoff, G., Radovanovich, G., Garrett, W.J., Robinson, D.E., in press, Annual phased arrays in ultrasonic obstetrical examinations, Ultrasonic Diagnostics, Milan, Plenum Press.

12. Luque, J., Dick, D.E., McLeod, F.D., 1976, Transducer limita-
    tions on random signal ultrasound, Proc. 29th Ann. Conf. Eng.
    Biol., in press.

13. Matzuk, T., 1976, Low-cost real-time high resolution synthetic
    aperture imaging system, Proc. 1st WFUMB, San Francisco.

14. Melton, Jr., H.E., 1971, Electronic Focal Scanning for Improved
    Resolution in Ultrasonic Imaging, Ph.D. Thesis, Duke University.

15. Onoe, M., 1974, Computer processing of ultrasonic images, in
    Ultrasonic Imaging and Holography, Stroke, G.W., Kock, W.E.,
    Kikuchi, Y., Tsujiuchi, J., eds., Plenum, pp. 455-502.

16. Phillips, D.J., Smith, S.W., vonRamm, O.T., Thurstone, F.L.,
    1975, A phase compensation technique for B-mode echoencephalo-
    graphy, in Ultrasound in Medicine Vol. 1, D.N. White, ed.,
    Plenum, pp. 395-404.

17. Tsujiuchi, J., Ueha, S., Ueno, K., 1974, Holographic Synthetic
    Aperture Sonar System, in Ultrasonic Imaging and Holography,
    Stroke, G.W., Kock, W.E., Kikuchi, U., Tsujiuchi, J., eds.,
    Plenum, pp. 531-552.

18. Vilkomerson, D., 1974, Acoustic imaging with thin annular aper-
    tures, in Acoustical Holography Vol. 5, P.S. Green, ed., Plenum
    Press, pp. 283-316.

DOPPLER FLOW VISUALIZATION USING LARGE TIME-BANDWIDTH SIGNALS

M. Siegel, M. Olinger and B. Ho

Michigan State University, Department of Electrical

Engineering and Systems Science, East Lansing, Michigan

Our group is primarily interested in the imaging of blood flow, the construction of velocity profiles, and in determining volume blood flow in the cardiovascular system. In order to examine the factors limiting our ability to image velocity in blood vessels, it is necessary to interrelate system resolution constraints with the target model and its environment.

To help the reader develop an appreciation for the relative importance of these constraints, this paper will examine a traditional (though non-optimum) receiver for imaging velocities in a simplified blood-flow model. Additionally, we will point out some basic problems with this model and receiving system, and indicate possible directions to explore in order to correct these deficiencies.

To simplify the analysis two initial assumptions will be made.

1. The Doppler approximation is valid.

$$(1) \quad \frac{2 \cdot v_{max}}{c} \cdot T \cdot B < 1$$

v_{max} = maximum radial velocity

c = velocity of sound in the medium

T = signal duration

B = signal bandwidth

Note that for some signals this approximation may not be valid. When the approximation is valid the power spectrum of signals reflected from moving targets appear to undergo a simple frequency translation (the Doppler shift).

2.   Correlation detection will be used. (This may prove to be non-optimum because of the clutter (reverberation) present.)

If we assume a transmitted waveform of the form,

(2) $f(t) \cos w_c t = Re[\tilde{f}(t)e^{jw_c t}]$,

the received waveform for a simple target (assuming the Doppler approximation is valid) is of the form,

(3) $r(t) = Re[\tilde{f}(t-\tau)e^{j(w_c + w_d)t + \theta}]$
         $= Re[\tilde{f}(t-\tau)e^{jw_c t}]$

here w_c is the carrier frequency

w_d is the Doppler frequency $= \dfrac{2 \cdot v}{c} \cdot w_c$ (related to target velocity)

τ is the two way propagation time to target (related to target range)

the correlation receiver computes:

(4) $\left| \int \tilde{f}(t) \; \tilde{f}^*(t-\tau)e^{jw_d t} \; dt \right|^2$ Note: Here τ is the delay difference between the actual and assumed target delay and similarly for w_d:

         $= |\phi(\tau,w_d)|^2$

This is by definition $\theta(\tau,w_d)$, the ambiguity function (where $\phi(\tau,w_d)$ is the time frequency autocorrelation function). Figure 1 and 2 show a typical correlation receiver.

The ambiguity function, which is directly related to the output of our correlation receiver, provides a convenient way to examine the performance and resolution of our receiving system. If we examine the ambiguity function for various normalized signals, $(\int f(t) \cdot f^*(t)dt = 1)$, we can begin to approximate the range and velocity resolution of systems using these signals. Even though the construction of the entire ambiguity function is often difficult there are two important properties associated with it which are useful in estimating system resolutions. For a particular correlator output centered on a specific w_d,τ location, these properties may be stated as follows:

1.   Along the τ axis, $\phi(\tau,0)$ is the autocorrelation function of the complex envelope of the transmitted signal.

$\int \tilde{f}(t) \; \tilde{f}^*(t-\tau)dt$ = Fourier Transform of the power spectrum
                          (Weiner Khinchin)

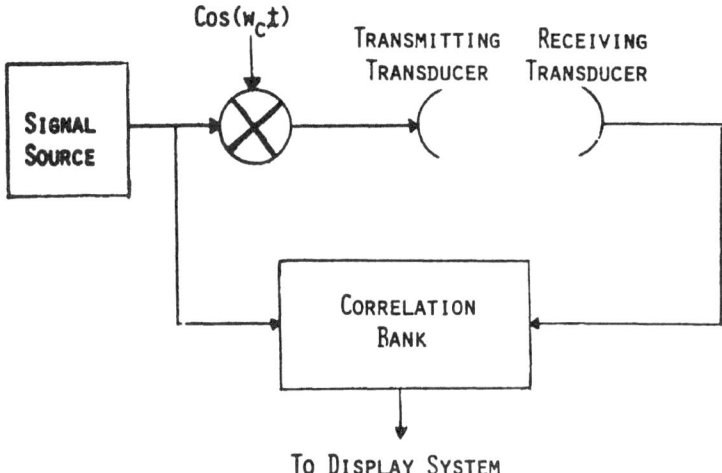

Fig. 1. Simplified block diagram of the velocity imaging system.

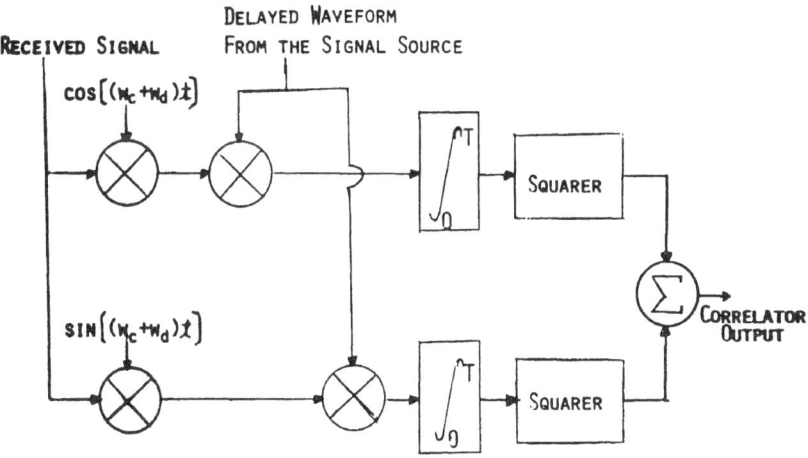

Fig. 2. A typical correlator in the correlation bank.

2. Along the frequency axis, $\phi(0,w)$ is the Fourier transform of

$|f(t)|^2$

From property 1 it can be seen that the ability to discriminate between targets at different ranges (i.e. at different τ's) when the velocity is known, varies inversely with the bandwidth of the power spectrum of the complex envelope, ($\Delta\tau \sim \frac{1}{B}$). From property 2 it can be seen that the ability to discriminate between targets at different velocities when the range is known (i.e. at different w's), varies inversely with the time duration of the complex envelope. Note that for amplitude modulated transmitted signals (no FM content), these resolutions may be applied simultaneously. Thus by examining the ambiguity surface for different transmitted waveforms, typical resolutions may be shown to be:

(5) Range resolution

$$\Delta R = \frac{\Delta\tau}{2} c \approx \frac{1}{B}\frac{c}{2}$$

Velocity resolution (assuming a maximum radial velocity of v_{max} and N possible velocity resolution cells):

$$\Delta f_d \approx \frac{1}{T}$$

(6) $\Delta f_d = \Delta(2\cdot\frac{v_{max}}{c}\cdot f_c) \approx \frac{1}{T}$ or $\Delta v = \frac{v_{max}}{N} \approx \frac{c}{2\cdot T\cdot f_c}$

Some examples of the ambiguity surface for transmitted waveforms used in typical systems are shown in figures (3), (4), (5) for a. a single pulse; b. a pulse train; c. a random signal.

In figure (3) illustrating the ambiguity surface for a single pulse we observe that the function has a single peak whose width along the τ axis is directly proportional to the pulse width and whose width along the frequency axis is inversely proportional to the pulse width. Thus we see that simultaneous improvement in range and velocity resolution is not possible using a single pulse.

The ambiguity surface for a pulse train, figure (4), shows that we can decrease the width of the major peak along the frequency axis (i.e. improve Doppler resolution) by increasing the time duration of the pulse train. The range resolution can be improved by decreasing the individual pulse widths (which increases the bandwidth of the signal). This transmitted waveform allows us to simultaneously improve range and velocity resolution at the cost of introducing ambiguous responses. Targets centered at these ambiguous peaks can cause the receiver to make an incorrect decision. Most of the present pulse echo Doppler bloodflow meter systems effectively employ this type of signal.

The ambiguity surface for the random signal shows that the width of the central peak along the τ axis (range resolution) is determined by the signal bandwidth. The width along the frequency axis (velocity resolution) is determined by the signal duration. Of particular importance is the fact that there are no significant ambiguous responses. For stochastic signals note that bandwidth and signal duration may be increased independently.

Now that an intuition for the time and bandwidth resolutions has been developed, some of the times involved in processing and imaging for various implementations of this correlation receiver will be examined. For convenience a random signal will be used and some typical parameters assumed:

Bandwidth B = 1 MHz
Number of velocity resolution cells N = 5
Signal duration T = 10^{-3} seconds

First, a receiving system which is conceptually simple to visualize but expensive to implement will be examined. Initially, consider an imaging system which would determine the velocity of blood flow across the diameter of a blood vessel; this will be extended to a system which would image velocity over the entire vessel crosssection.

One possible way to achieve 5 Doppler resolution cells at each range location is to implement a two dimensional array of correlators. Each element of the array would be similar to that shown in figure (2), each element being matched to a particular velocity and range. For a bandwidth of 1 MHz, a range resolution slightly less than 1 millimeter is obtained. For N = 5 and a vessel diameter of approximately one centimeter, 50 correlators are needed. When the output of any one of these correlators exceeds a threshold, the decision is made that a target is present at that specific range and velocity. Since we assume a signal duration time and therefore an integration time of 1 millisecond, a position-velocity update is possible approximately every millisecond (plus the two way propagation time of the transmitted signal). For typical target distances (assuming blood vessels close to the surface) we can shift the beam axis over a distance equivalent to one resolution cell approximately each millisecond. Note that the lateral resolution is determined by the beam width of the transducer - in our case approximately 1 millimeter.

Since we examine each beam axis position for 1 millisecond and there are 10 beam positions across the vessel, 10^{-2} seconds is required to scan across the vessel.

A simple display system can be utilized to present velocity

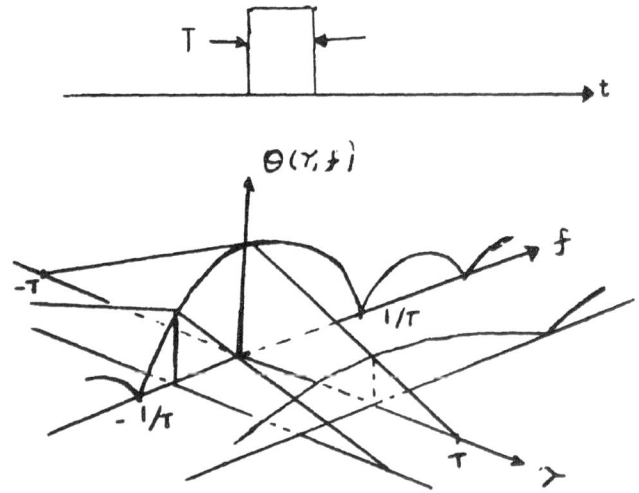

Fig. 3. Ambiguity function for a single pulse.

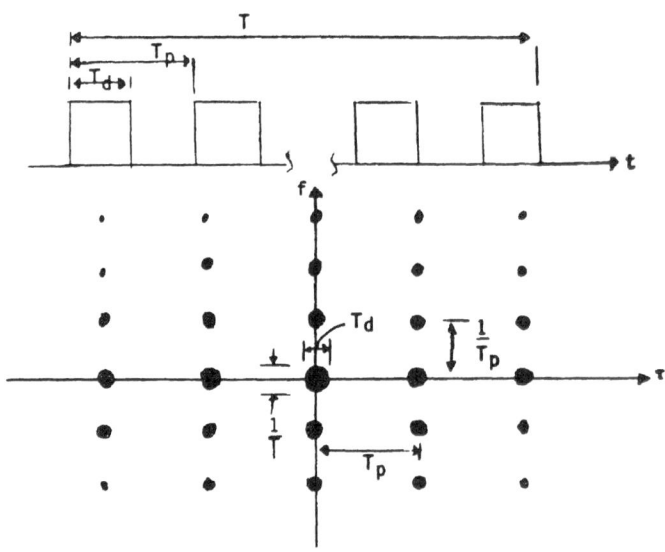

Fig. 4. Ambiguity function for a periodic pulse train.

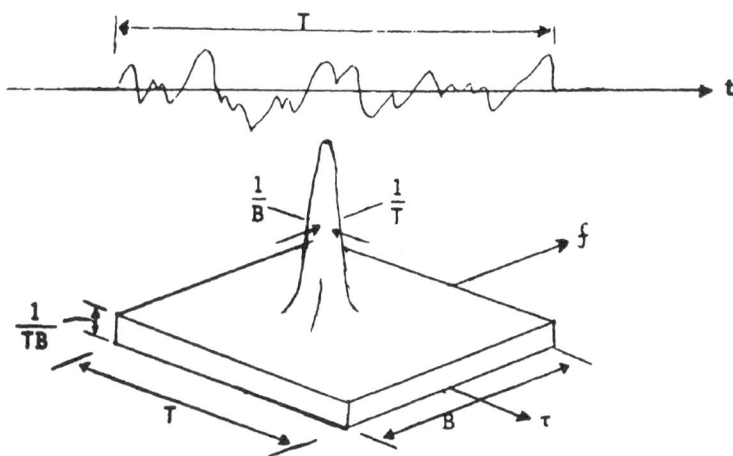

Fig. 5. Ambiguity function for a noise waveform.

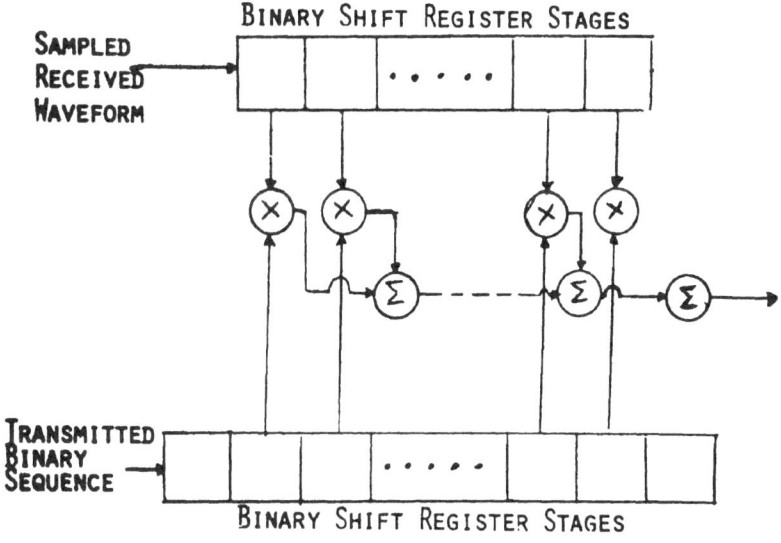

Fig. 6. Correlator implementation for binary sequences.

over the vessel crosssection, if at each range position, the five
Doppler outputs are examined and the greatest of the five used to
intensity modulate the screen proportional to the associated
velocity. If we assume normal heart rates, real time velocity
imaging can be achieved.

The previous system required 50 correlators and 10 delay
lines. An alternate technique, reducing the number of correlators,
is to use one variable delay line and step through the range in
1 mm. steps. This decreases the total number of correlators to
5 but increases the processing time by a factor of 10. In this
case, the 1 cm. diameter vessel with the same resolutions requires
0.1 second.

One of the limiting devices in the hardware implementation of
this "simplified" system is the design of a delay line capable of
delaying a 1 MHz bandwidth analog signal centered at 5 MHz for up
to 1/2 millisecond (the maximum expected two way propagation time).
The requirement for such a delay line with its tight constraints
is unique to transmitters using non-deterministic signals. For
deterministic signals other techniques, such as matched filter
processing may be employed. In the past, analog acoustic delay
lines were used in our experimental random signal system; however,
these "mechanically-varied" lines are too slow for use in velocity
imaging. Recently we have constructed a digital electronic delay
line consisting of a sampler and a series of shift registers for
use in our random signal imaging system. This delay line performs
polarity sampling (at an 1 MHz rate). Thus in place of multiplication in our correlators we can use "exclusive or" circuits
("polarity-coincidence correlation"). Low pass filters can be used
to perform the integration (with integration times ~1/Bandwidth).
Note that this method of processing is similar to transmitting a
random pulse sequence. The advantages of random pulse sequences
over the more conventional coherent sequences are the reductions
in range and velocity ambiguities.

Pseudorandom codes which are actually deterministic have been
developed (the Barker code, etc.), to approximate the random pulse
sequence - a major advantage of these codes is that the delay line
(which preserves the transmitted waveform) is not required. A
method occasionally used in modern radar systems to perform the
required correlation when binary sequences are used is to load the
binary samples of both the received signal and the original transmitted signal into shift registers, figure (6). By using these
registers in conjunction with "exclusive ors" and a "summer" the
cross correlation corresponding to specific delays and velocities
are computed. Unfortunately, for the imaging system example
discussed where a millisecond signal is transmitted, approximately
10^4 shift register stages and 10^5 exclusive or circuits would be
required.

   Thus far the velocity and range resolutions and hardware
implementation of this simplified system seem reasonable. However
a number of theoretical simplifications have been made and theoretical problems ignored. As an example of a situation where one
of these assumptions is not valid, let us try to achieve better
range resolution by increasing the bandwidth. As noted earlier,
the Doppler assumption is valid when equation (1) is satisfied.

   If, for example, v_{max} = 75 cm/sec, c = 1.5 x 10^5 cm/sec,
T = 1 msec and B = 2 MHz, then $(2 \cdot \frac{v}{c}) \cdot T \cdot B$ = 2.0 and the Doppler
approximation is not acceptable. If we considered v_{max} as the
maximum expected velocity of the target, then from equations (1),
(5) and (6) it can be seen that for the Doppler approximation to
be valid, the relationship between the number of desired velocity
resolution cells, N, and the desired range resolution, ΔR, which
must be satisfied is:

   (7) $N < \dfrac{2 \cdot \Delta R \cdot f_c}{c}$

If the Doppler approximation is not valid a different correlation
operation must be defined and implemented. The proper correlation
operation to be implemented is:

   (8) $\psi(\alpha, \tau) = \int \tilde{f}(t) \cdot \tilde{f}(\alpha t - \tau) \, e^{j(1-\alpha)t} dt$

      where $\alpha = (1 + \dfrac{2v}{c})$

Note that the Doppler approximation assumes that

   $\tilde{f}(\alpha t - \tau) e^{j w_d t} = \tilde{f}(t - \tau) e^{j w_d t}$

   The magnitude squared of $\psi(\alpha, \tau)$ is one form of the generalized
ambiguity function from which both resolution constraints and
ambiguous responses can be determined. The procedure required to
calculate these resolutions is generally more involved than when
the Doppler approximation is valid. Additionally, generating
$\tilde{f}(\alpha t - \tau) e^{j w_d t}$ is much more difficult than merely compensating for
a Doppler shift in the received signal.

   Regardless of whether the Doppler approximation is valid or
not another theoretical problem arises since we are assuming that
the received signal may be represented as:

   $\tilde{b} \; \tilde{f}(\alpha t - \tau) e^{j w_d t}$

where $\tilde{b}$ is a complex gaussian random variable. This form results
from the superposition of reflections from many point scatterers
(the red blood cells) and application of the central limit theorem.
However a number of assumptions are required in accepting this form.

One of the most controversial assumptions is that a large
fraction of the red blood cells illuminated by the ultrasonic beam
remain illuminated (i.e. do not pass out of the beam) for a time
corresponding to the signal duration. This has been referred to
in the literature as the transit time effect (5,8). Following the
work of VanTrees and Middleton, Brody and Meindl have developed a
statistical description of the flowing blood as a target. The
description derived is an equation for the autocorrelation function
of the received signal in a CW Ultrasonic blood flowmeter. The
extension of these results to arbitrary transmitted signals is not
difficult and the resulting correlation function may be used to
design an optimum receiver. The effect of finite transit time on
the so called Doppler spectrum has been treated but to our knowledge the effect of finite transit time on the resolution of an
optimum receiver for an ultrasonic velocity imaging system using
large time bandwidth signals has not been investigated. This is
a problem which we are presently studying. Additionally, we are
studying the possibility of a slightly different target model
which may be more suited to this specific application as well as
simplify the receiver design.

Another problem of concern, especially in Doppler imaging, is
that of target resolution in a dense target environment. In the
case of a dense environment, returns from targets other than that
for which the particular correlator was designed (i.e. clutter) may
contribute enough to the correlator output so as to completely mask
the presence or absence of the desired target (2). If one decides
to implement a correlation detector then system performance is
highly dependent on the choice of transmitted waveform. It should
be noted that for a dense environment correlation detection is not
optimum. Only when there is relatively little overlap of the
ambiguity surface with the target environment does the performance
of the correlation receiver approach the performance of the optimum
receiver.

At the present time a completely satisfactory characterization
of the target environment for medical imaging systems does not
exist. Using a crude model for the target environment which we
have already developed, our studies indicate that a pulsed random
signal is a reasonable choice of waveform for correlation detection
in Doppler imaging systems. Using pulsed random noise the advantages of low duty cycle, freedom from ambiguous responses, etc.,
can still be realized. We have been developing a statistical
characterization of the environment of targets in ultrasonic imaging systems and plan to use that model to investigate the feasibility of implementing an optimum receiver.

We have examined some of the theoretical problems and practical
system requirements necessary to implement velocity imaging in

ultrasonic bloodflow measurement. Although additional study is
needed in the areas of system performance, target and environment
modeling and receiver implementation, practical high-resolution
velocity imaging systems appear to be achievable.

REFERENCES

1.   C. W. Helstrom, Statistical Thy. of Signal Detection, London
     England, Pergamon Press.

2.   A. W. Rihaczek, Radar Signal Design for Target Resolution,
     Proceedings of the IEEE, February 1965.

3.   D. Middleton, A. Statistical Thy. of Reverberation and
     Similar First Order Scattered Fields, IEEE Trans. on Inf.
     Thy., IT-13.

4.   H. L. VanTrees, Optimum Signal Design and Processing for
     Reverberation Limited Environments, IEEE Trans. on Military
     Electronics, Mil-9.

5.   W. R. Brody, J. D. Meindl, Theoretical Analysis of the CW
     Doppler Ultrasonic Flowmeter, IEEE Trans. on Bio. Med. Engr.,
     BME-21.

6.   C. P. Jethwa, M. D. Olinger, Blood and Flow Measurement Using
     Random Signal Flowmeter, Proceedings of the 19th AIUM,
     October 1974.

7.   M. Siegel, M. Olinger, B. Ho, Velocity Profiles in the Presence
     of Obstructions Using a Random Noise Ultrasonic Blood Flow
     Meter, IEEE Sonics and Ultrasonics Symposium, 1975.

8.   V. L. Newhouse, L. W. Varner, P. J. Bendick, Transit Time
     Effects on Ultrasonic Doppler Blood Flow Measurement, 27th
     ACEMB, October 1974.

A TRANSMITTER FOR DIAGNOSTIC IMAGING--A PROGRESS REPORT

K. Wang, H. Shen and H. Chang

Dept. of Electrical Engineering

University of Houston, Houston, Texas 77004

G. Wade, K. Su, M. Lo, S. Elliott

Dept. of EECS

U.C.S.B., Santa Barbara, California 93106

ABSTRACT

An imaging approach utilizing a scanning-focused-beam and piezoelectric detection possesses the best inherent sensitivity with potential to produce an effective image with the least tissue exposure to ultrasound. A possible embodiment of a system based on this highly-desirable approach is being worked on. The key element is an opto-acoustic transducer (OAT) addressed by light carrying a focus-inducing pattern.

The OAT must be carefully designed and evaluated before construction. Two basic types of OAT structures have been investigated by means of equivalent circuit models and computer simulation, which takes into consideration the dependence of OAT performance on device geometry, amplitude and phase distortions, material properties and operating conditions. The simulations indicate that certain configurations with available materials have the proper characteristics to qualify as candidates for OAT construction. Designs for prototype OATs have been made. The construction and testing of the prototypes are initiated.

INTRODUCTION

To be practical for medical use, an acoustic imaging system must satisfy a list of requirements.[1] Of particular importance among these requirements are real-time capability and good sensitivity.[1-3] Real-time capability permits the object to move about and to be positioned within the system in such a way as to provide the best image. High sensitivity permits operation at power levels low enough to ensure patient safety.

Recently, many new real-time systems[4-7] have been devised and some have shown sufficient promise that substantial effort is being expended in their development. Although good images have been obtained of thin objects, there are still a number of important problems to be solved. In particular, the sensitivity of most systems needs to be improved before they can be considered practical for diagnostic use.

Existing and potential systems to produce real-time orthographic images can be divided into two categories: one utilizing laser beam read-out of the acoustic information, the other utilizing piezoelectric detection of the acoustic field.

An analytical procedure was established[3] to characterize the ideal performance of an imaging system in terms of its threshold acoustic contrast, defined as the ratio of the smallest difference in the acoustic transmittance of two adjacent resolution elements of the object when the corresponding two adjacent cells in the final image can just barely be distinguished from each other (as having different shades of intensity) to the larger transmittance of the two object elements. The analytical procedure was applied to systems (existing and hypothetical) in both categories that hold promise in real-time orthographic diagnostic imaging.[8]

Systems in the first category considered (laser beam read-out) include the static-ripple imaging system,[9] the dynamic-ripple imaging system,[10] and the Bragg-diffraction imaging system.[11] Systems in the second category (piezoelectric detection) considered include those using various scanning modes:[3] the positively-scanning-transmitter (PST) system, the positively-scanned-receiver (PSR) system, the negatively-scanning-transmitter (NST) system, and the negatively-scanned-receiver (NSR) system.

THE OPTIMUM SYSTEM CONFIGURATION

The threshold acoustic contrast has been derived for systems in both of these categories. It was found[8] that, with the compatible operating conditons assumed and the perfect components hypothesized, the inherent capabilities of the systems in the first

category can be considerably higher. Sample calculations[6] were
also carried out to compare the inherent sensitivity of a hypothetical system using the PST principle with that of an existing
system with impressive performance, the ultrasonic camera[12] of the
Stanford Research Institute (operating on a principle equivalent
to a PSR system with improved integration time).

The calculations show that the hypothetical system has
superior inherent capability because it permits more efficient use
of insonification for imaging. Examples demonstrate that when it
comes to imaging deep-lying targets, this hypothetical system may
be the only real-time system inherently capable of operating below
a low recommended maximum average insonification.

The above-mentioned highly-desirable system employs a
scanning-focused-beam. Its advantages stem from its method of
insonification. A scanning-focused-beam system operates on a
principle analogous to that of a scanning-electron-beam microscope.
The transmitter not only generates a beam that is focused down to
a small cross section but also steers or scans the focal spot in a
raster pattern at the target plane. The focused acoustic beam is
used to probe the object to obtain information on the relative
amount of absorption or scattering at each point of an orthogonal
section. The transmitted or scattered acoustic signal is picked
up by a piezoelectric receiver listening in the focused mode with
its focus coinciding with the transmitter focus and scanning in
synchronism with the transmitter.

ONE EMBODIMENT SCHEME OF A SCANNING-FOCUSED-BEAM SYSTEM

Several schemes of ultrasonic imaging using scanning-focused
beam insonification and focused receiver are being worked on.[6]
One possible scheme to produce a scanning-focused-beam requires
neither high-density transducer arrays nor acoustic lenses. The
key element is an opto-acoustic transducer (OAT) addressed by
light carrying a focus-inducing pattern. Receiving transducers
based on somewhat similar physical principles have been investigated by several workers[13,14] (laser spot readout).

An OAT is a sandwich structure that transforms an optical
input pattern into an acoustic output pattern. When a pattern of
light intensity is incident on the OAT, the OAT generates a
corresponding pattern in its acoustic pressure amplitude distribution. When light carrying a focus-inducing pattern (e.g. a zone-
plate pattern) is incident on the OAT the generated sound beam
converges to a focus induced by the focusing properties of the
addressing pattern. Thus, focusing can be accomplished by optical
projection of a focus-inducing pattern onto the OAT. Scanning of
the focused beam can be accomplished by setting the projected light

pattern into motion. Depending on the OAT design, either mechanical scan or all-electronic scan of the optical pattern can be used.

For all-electronic rapid scanning, an acousto-optic laser
scanner and a Fourier transform hologram can be used. First, a
Fourier transform hologram of the desired transparency (with the
focus-inducing pattern) is made with a point source reference.
The carefully fabricated transparency containing a focus-inducing
pattern (e.g., an acoustic zone-plate pattern) with a pinhole
aperture in the center is placed in the object plane of an optical
Fourier-transforming lens. With laser light illumination, a
hologram is recorded at the Fourier-transform plane. Next, an
array of pinholes is made in an opaque mask. Each pinhole should
be close in optical properties to the one used for recording the
hologram. The periodic spacing of the pinholes should be scaled
to match the laser scanner output.

During reconstruction, a laser beam goes through an X-Y
acousto-optic scanner. The optical setup for Fourier-transform
holography is such that the scanning laser beam perpendicularly
addresses the pinholes in a raster sequence. The laser beam
passing through one of the pinholes constitutes the reconstruction
beam, and projects the focus-inducing pattern onto the OAT, which
in turn generates a focused sound beam. As the laser beam moves
from one pinhole to the next, the reconstructed pattern at the OAT
is translated spatially by an amount determined by the pinhole
spacing. Thus the acoustic focus at the object plane is translated
from one resolution element toward the next in synchronism with
the laser scan.

DESIGN OF THE KEY ELEMENT

The opto-acoustic transducer (OAT) is the key element in the
proposed scanning-focused-beam system. When illuminated by light,
the OAT produces a near-field acoustic pressure amplitude pattern
in accordance with the incident optical intensity pattern. A
simple example of a focus-inducing pattern with desirable characteristics is a Gabor zone-plate (GZP). Since a zone-plate and its
negative have the same focusing properties, either positive or
negative correspondence between the OAT acoustic output pattern
and the optical input pattern is acceptable. This suggests two
types of OAT structures for consideration.

In one type of OAT structure, two layers of materials, one
photoconductive and the other piezoelectric, are sandwiched between
two thin electrodes. One of the electrodes is transparent. The
transparent electrode is adjacent to the photoconductive layer,
and faces the incident light. A spatially uniform ac voltage is
applied across the electrodes. Let the parameters be such that

under dark conditions most of the potential drop appears across the layer that is both piezoelectric and photoconductive; the OAT sends out strong radiation. Light deactivates the structure by reducing the voltage across the photoconductive-piezoelectric layer. This reduction weakens the sound output in the illuminated regions of the OAT. When deactivated, the generated acoustic near-field pattern has a negative correspondence to the incident optical intensity pattern. This kind of structure will be referred to as the negative-type OAT (N-OAT).

An ideal OAT would generate a near-field pressure pattern that is either a positive or a negative replica of the incident light pattern. With incident light carrying a focus-inducing pattern, an ideal OAT would invariably produce a beam that converges to a good focus. An actual OAT, however, will not faithfully transform the focus-inducing pattern in the input light into a replicated pattern in the output sound, whether it be P-OAT or N-OAT. Amplitude and phase distortions are introduced. Whether an OAT is useful in our system depends on its ability to produce a focused beam when light carrying the focus-inducing pattern is incident.

The optoacoustic transducer (OAT), the key element in our proposed system, has to be carefully designed and evaluated before construction for experimental testing. A great deal of effort has been spent on computer simulation of various OAT configurations to take into consideration the dependence of an OAT's focusing capability on its device geometry, on the amplitude and phase distortions that arise in using the OAT to transform a light intensity variation into an acoustic complex amplitude variation at a given frequency, and on the material properties and operating conditions. It was found from our preliminary considerations and simulations[15,16] that certain available materials with proper characteristics and dimensions can qualify as candidates for OAT construction. Preliminary design of several OATs were obtained.[16] Construction (a major phase in our work due to the sophisticated solid state procedures required) and testing of scaled down OATs are initiated.

The focusing ability of an OAT depends not only on the configuration and the piezoelectric and photoconductive properties of the materials used, but also on the thicknesses of the layers and the operating parameters. Equivalent circuits for both type (P-OAT and N-OAT) of structures and the Fresnel diffraction formula for acoustic propagation have been used in the CW calculation. The sandwich is assumed air-backed and radiating into water as the propagating medium.

The focus-inducing pattern we plan to use first is a zone-plate pattern because of its simplicity. Between the two types of zone-plate patterns, Gabor zone-plate (with smooth continuous radial transmittance variations) is preferred to Fresnel zone-plate (with binary abrupt radial transmittance variations) because of the problems that might arise with multiple foci.

The first prototype system we plan to build is to operate near 3MHz with f/2 focusing capability and a focal length of 10 cm. This calls for a pattern approximately 12 cm. in diameter at the OAT. When assuming mechanical or holographic 1:1 projection, an acoustic zone-plate pattern 12 cm. in diameter must be recorded optically. Under the assumption that their effects can be isolated, the effects of defects in recording the acoustic Gabor zone-plate pattern onto an optical medium on focusing have been investigated.[15] For example, fringe shift and film nonlinearity are inevitable during the fabrication of the optical zone-plate transparency. The effect of deviation from ideal zone-plate recording due to these and other practical limitations on focusing was calculated. From the theoretical results, upper limits of the tolerances for the fabrication of the zone-plate transparency were set and incorporated into the design of our optical setup for recording zone-plates. The intensity transmittance function of the optical transparency $\tau(r)$ is to follow

$$\tau(r) = (1/2 + 1/2 \cos \alpha r^2) \text{circ}(r/\ell)$$

where r is the cylindrical radial coordinate, α and ℓ constants.

An optical system[15] was set up to record the desired acoustic zone-plate pattern onto K649F plates. Controlled radial exposure was obtained via an area-modulated precision mask and a rotational system. Careful alignment and calibrations were exercised to reduce the fringe shift to within the allowed tolerance. An inverse mapping method was employed to partially compensate for film-nonlinearity lest the recording should degenerate toward resembling a binary pattern. The intensity transmittance pattern was measured with a radial scanning system and compared with the ideal pattern. Good transparencies have been obtained. This recording system can be used at a later stage in the project to record a "trained" apodization pattern for best system focusing. Depending on the scheme chosen for optical projection and scanning, the transparency recorded in this manner is used as the slide for projection onto the OAT or as the object for making Fourier transform holograms.

Among II-VI compounds, based on photoconductive response, our preliminary choice for the photoconductive material is basically CdS. As has been suggested to us by Fraser and Land, we contemplate using CdS with small concentration of ZnS to obtain the desired speed and photoconductive characteristics. For the dielectric material, calcium flouride (CaF_2) is arbitrarily chosen.

Its transmittance is greater than 80% in visible light range and its relative dielectric constant is 6.8. The materials for the main piezoelectric in the P-OAT structure considered include quartz, $LiNbO_3$, $BaTiO_3$, PZT 4, and PZT 5.

Based on our choices, the equivalent circuit for P-OAT is more complicated than that for N-OAT because both layers are piezoelectric. In both types of structures, the OAT activation or deactivation is done via photoconductive switching. The photoconductive resistance R_p is a variable resistance. It varies spatially across the OAT according to the transmittance function pattern $\tau(r)$ of the focus-inducing transparency as

$$R_p = \frac{K}{\dfrac{1}{s-1} + \tau(r)}$$

In the above equation, the parameter K is a function of properties of the photoconductive layer: its thickness, quantum efficiency, carrier lifetimes and mobilities. It also depends on the intensity I_o of the light carrying the pattern $\tau(r)$ and the elemental area sectioned for calculation. The quantity S is the switching ratio of the photoconductive layer, defined as the ratio of maximum conductivity under illumination (by I_o) to the conductivity under dark conditions.

With choice of materials for an OAT configuration and with the main piezoelectric layer (the layer adjacent to the opaque electrode in both P-OAT and N-OAT sandwiches) operating at fundamental resonance, values of most of the components in the equivalent circuit can be calculated. To completely specify the equivalent circuit, the thickness t' of the other layer (the layer adjacent to the transparent electrode) and detailed photoconductive properties of the switching layer must be specified. To facilitate the calculation and to reduce the number of variables, we specify K and S in the preceding equation after the thickness t' is chosen. We then calculate the acoustic distribution at the focal plane when such an OAT is illuminated by a light intensity pattern $I_o \tau(r)$.

For such an OAT with fixed choice of basic materials and thicknesses, calculation is performed for a wide range of values for K and S. The ability of the OAT to produce focused sound is then evaluated. The focusing evaluation is based on the normalized peak intensity (peak acoustic intensity in the main lobe normalized by the square of the voltage across the OAT) and the lobe ratios (the ratios of the peak intensity in the first and subsequent side lobes to the peak intensity in the main lobe). It was found as expected that for OAT configurations showing satisfactory focusing, there exists for each of these configurations an optimum K value

(K_{opt}) for best focusing. When K is away from K_{opt} by more than one order of magnitude, severe degradation in focusing usually results with loss in peak intensity. For a fixed K, the switching ratio S does not impose sensitive influence on focusing in each case, as long as it is greater than 100. When the switching ratio becomes low, focusing disappears due to increase in lobe ratios and lack of contrast.

To recapitulate, for a particular configuration the OAT must be operated near K_{opt} with a sufficient switching ratio. However, the photoconductive parameters K and S are interdependent. They are related to each other through material properties. When they are both specified, the photoconductive material is almost completely characterized. Whether such characteristics is practical and achievable with the specified photoconductive layer thickness under the operating conditions has to be examined before the configuration can be accepted as an OAT design for the system.

For real-time operation, the photoconductive response of the OAT must be sufficiently fast. A frame time of 1/30 sec., a field of 200 x 200 spots, and the operating frequency of 3MHz would require the photoconductive relaxation time be no longer than 1 μsec. This speed requirement adds to the requirement on photoconductive material characteristics. Although most CdS films in great use have slower photoconductive response, proper film deposition and addition of Zn are expected to shorten the response time into our desired range.

With the initial choice of thickness t' of P-OAT's, focusing calculations provide the optimum value of K (i.e. K_{opt}) and the acceptable range of the switching ratio S. The requirement of speed together with the requirement to achieve K_{opt} limits the freedom of the operating light level I_0 and the freedom of the usable dark resistivity. The light level I_0, the dark resistivity, and the already defined characteristics of the photoconductive layer in turn determine the switching ratio S. If the operating conditions and the photoconductive characteristics so mandated are practical, the dimensions used are acceptable. Otherwise, the thickness t' has to be altered.

After ample calculations have been performed on an OAT combination, a procedure was established[15] to take into account some practical constraints on operating conditions in picking the thickness t' for that OAT, if an acceptable t' is yet to be found. In this procedure, practical limits can be specified on some quantities (e.g. light level, quantum efficiency and lifetime). These specifications in conjunction with K and S lead to an estimate of the thickness t' which would satisfy the practical limits. Focusing calculation is then carried out for verification.

An external matching circuit improves the peak intensity, the conversion efficiency, and the focusing. For P-OAT's, the circuit is matched to the OAT when the entire OAT is activated. For N-OAT's, the circuit is matched to the OAT when no part of the OAT is deactivated. With such an external matching circuit, the OAT usually obtains a significant increase in peak intensity.

The effects of the elemental area, the switching ratio, the figure of merit K on focusing for various OAT configurations were investigated. All the work described above were integrated into a design procedure which lead us to some acceptable candidates for OAT construction.

In all our calculations (of P-OATs and N-OATs), the equivalent circuit consists of a parallel combination of elemental circuits representing elemental areas on the OAT. To facilitate the analysis and characterization, these elements are assumed isolated and independent. In our investigation of both types of OATs, the elemental area segmented for calculation is usually chosen at 0.01 cm x 0.01 cm. The values of K for good focusing change when we change our choice of elemental areas. We found that as long as the elemental area A chosen for calculation is smaller than 0.01 cm x 0.01 cm, the only effect of changing A is to change K, but not to change the other results. When we decrease the elemental area from A to A' in our calculation, the optimum value of K for focusing (K_{opt}) is raised by a factor equal to A/A'. The corresponding peak intensities and lobe ratios around K_{opt} remain virtually unchanged. Such change in K agrees with the inverse proportionality between K and A. In other words, when the elemental area used is 0.01 cm x 0.01 cm or less, consistent conclusion and design is obtained. If the area chosen is too small, however, the assumption of independence and good isolation of these elemental circuits becomes invalid.

When the elemental area used exceeds 0.01 cm x 0.01 cm, inaccurate simulation and inconsistent conclusions result. As the elemental area increases, focusing becomes worse due to insufficient sampling and quantization of the OAT output. The dimension of 0.01 cm is a desirable choice as the elemental area for calculation. Being one eighth of the minimum fringe spacing of the GZP, 0.01 cm is still far from the limit of the photoconductive resolution of thin layers. It insures consistency and sufficient accuracy in results while saving considerable computer time from calculations using much smaller areas.

The lateral coupling between elements is too complicated to model accurately and is neglected in our preliminary calculations. The actual extent and effect of such coupling will await experimental evaluation of the OATs under construction. Although the actual OAT performance will not likely be as good as predicted

from our preliminary calculations, optimistic evidence does exist
in the work of other investigators.[13,14,17,18]

The design and performance of some matched OATs are summarized
in Table 1. Each of these OAT examples provides good focusing
according to our simulation and is selected from the corresponding
category as requiring the lowest light level of operation. The
examples shown here do not reflect their ultimate performance in
other respects. In Table 1, t represents the thickness of the main
piezoelectric layer. The quantity t' represents the thickness of
the CdS layer for P-OAT and the thickness of the CaF_2 layer for
N-OAT. The optimum value of $K(K_{opt})$, the normalized peak intensity
I_p, the first and the second lobe ratios are also tabulated. The
focusing efficiency (F.E.) is defined as the percentage of the
total acoustic power output immediately behind the OAT that is
concentrated in the main lobe at the focal plane. The voltage $V_{0.5}$
is the source voltage required to produce a peak intensity of
0.5mW/cm* in the focus with 3dB/cm propagation attentuation. The
electrical to acoustic power conversion efficiency (C.E.) is also
calculated. For real-time operation, the available light level
incident at the OAT must exceed I_{omin}, and the dark resistivity of
the $Cd_{1-x}Zn_xS$ layer must exceed ρ_{imin}.

In summary, deactivated N-OATs and activated P-OATs can effect
sound focusing if designed with proper materials and dimensions and
operated under certain conditions. The figure of merit K, depending on the characteristic of the photoconductive layer and the
illuminating light intensity, plays an important role in the function
of the OAT. For a fixed acceptable OAT configuration, the light
intensity controls the focusing ability.

Table 1 reveals the minimum incident light level (I_{omin})
carrying the optical focus-inducing pattern to make these OATs
effective. For example, the light intensity I_o to carry the pattern
at the OAT needs to be at least $0.184W/cm^2$ for the CdS + Quartz
P-OAT, and $0.046W/cm^2$ for the CdS + $LiNbO_3$ P-OAT and the CaF_2 + CdS
N-OAT. It is much smaller for the other three P-OATs listed.

The required light source power can be estimated from the area
of the zone-plate and the mechanism for optical projection and
scanning. If the all-electronic holographic approach is used for
projection and scanning, the power of the light source needs to be
two orders of magnitude higher than the power required at the OAT.

---

*Choice of the level of $0.5mW/cm^2$ is arbitrary, although it
coincides with the level recommended by NSF Experiment No. 5 to
stay below in diagnostic instruments.

Table 1

Summary of Dimensions and Performance of Some Matched OAT's

	P-OAT CdS + Quartz	P-OAT CdS + LiNbO$_3$	P-OAT CdS + BaTiO$_3$	P-OAT CdS + PZT 4	P-OAT CdS + PZT 5	N-OAT CaF$_2$ + CdS
t (cm)	0.095	0.1228	0.0913	0.0767	0.0726	0.075
t' (cm)	0.14894	0.07447	5×10^{-4}	1×10^{-3}	1×10^{-3}	0.1197
K$_{opt}$	10^7	10^7	10^5	10^5	10^5	10^7
I$_p$ (W/cm^{2}V^2)	1.24×10^{-2}	1.23×10^{-1}	2.61×10^{-3}	1.66×10^{-3}	3.48×10^{-3}	9.10×10^{-2}
1st l.r.	0.01700	0.01633	0.07113	0.08624	0.05497	0.0155
2nd l.r.	0.00675	0.00445	0.02873	0.04443	0.02246	0.0048
F.E. (%)	1.1	8.02	0.09	0.05	0.12	10.5
V$_{0.5}$	6.4	2.0	13.8	17.4	12.0	2.3
C.E. (%)	10.2	23.5	74.85	77.7	74.2	4.41
I$_o$ min (W/cm^2)	0.13372	0.04593	2.07×10^{-4}	8.3×10^{-4}	8.3×10^{-4}	0.0466
ρ_i min (Ω-cm)	6.7×10^6	1.34×10^7	2.0×10^7	1.0×10^7	1.0×10^7	2.1×10^7

For the CdS + Quartz P-OAT, the CdS + $LiNbO_3$ P-OAT and the N-OAT, the required laser power would be too large to be practical; conventional projection with mechanical scan would be preferable. For the CdS + $BaTiO_3$ P-OAT, the CdS + PZT 4 P-OAT and the CdS + PZT 5 P-OAT, the required laser power would be on the order of a watt and is available. Therefore, these three configurations remain possible candidates for all electronic holographic scanning.

EXPERIMENTAL PROGRESS

Experimental work is underway in separate areas in order to construct prototype OATs. These areas involve the development of a low-resistivity and optically transparent electrode, the development of a procedure for fabrication and deposition of a photoconductive layer with proper parameters and characteristics, and the characterization of the physical properties of the piezoelectric ceramics. Also under experimental investigation are the interfacing and integration of these layers to form an OAT without contact and deterioration problems, the construction and planning of an OAT mosaic for rigidity and economy, and the set up for diagnosis of the OAT transformation and speed responses.

In order to address the photoconductive layer optically, a transparent, conductive electrode layer must be made. The technique developed by McKinney[19] was used to vapor-deposit a layer of In_2O_2 Indium-Tin-Oxide (abbreviated ITO) onto substrates. Measurements were made of the resistivity and light transmission of the films versus film thickenss. The thickness variation is obtained by monitoring the evaporation time from 20 minutes to several hours. Table 2 summarizes the results of these measurements. These results show that an adequate transparent electrode can easily be achieved.

Table 2

Properties of Vapor-Deposited ITO Layers

ITO Thickness (μm)	Resistivity ohm/square	Transmission % at 633 nm
0.03	2260	91
0.05	1202	86
0.43	192	72
0.90	70	50

A more difficult task is the development of a procedure for deposition and fabrication of the photoconductive layer with satisfactory characteristics. The film must have high dark resistivity, fast photoconductive response time and acceptable sensitivity (to achieve significant switching ratio with available light at the OAT). Thick films of CdS with Cu doping have been produced from a wet slurry. These films exhibit high dark resistivities ranging from 10^9 to 10^{10} ohm-cm and achievable switching ratio of 10^7. The photoconductive response times of these films, however, are on the order of 1 m sec. As mentioned earlier, for real-time operation the response time must be no longer than 1 μ sec. Fast response with rise and decay times on this order is deemed feasible[20] with zinc-doped CdS deposited using an R-F sputtering technique.[21] We are experimenting with this technique to produce $Cd_{1-x}Zn_xS$ films for evaluation and to develop and calibrate a procedure for making films suitable for prototype OATs.

The extent of lateral coupling among resolution elements across the OAT sandwich is being evaluated experimentally. Scaled down OATs are fabricated for testing of speed and opto-acoustic pattern transformation. Diagnostic setups are being used to measure the OAT output pattern. Because the major step for optimization of individual layers and integration of individual layers into an OAT device needs further time and efforts toward completion, no result representative of the capability of the final device can be reported at this time. Switching in the OAT acoustic output in response to incident laser beams has been observed with miniaturized OATs even without any attempt to optimize the properties of any of the layers.

There remain problems of combining the layers into an integral OAT device. For example, some of the layers require heat treatment during its fabrication. The ITO layer requires a post-baking procedure at 250° C for 5 hours. The Curie temperature of $BaTiO_3$ is about 120° C; the post-bake will depolarize the piezoelectric. Repoling is not difficult but requires careful planning. Other problems under investigation include various ways of interfacing the layers without creating contact problems and non-linear behavior as well as design and construction of rigid OAT mosaics without causing spurious modes and diffraction.

CONCLUSIONS

A possible transmitter embodiment of a scanning-focused-beam system for orthographic imaging is described in the introduction. The key element is an OAT. An OAT has to be carefully designed and evaluated before construction for experimental testing. Its field transformation behavior and focusing capability are investigated with equivalent circuit model and Fresnel diffraction. Computer simulation takes into consideration the dependence of the OAT performance on its device geometry, amplitude and phase distortions, material properties and operating conditions. It was found from our preliminary considerations and simulations that only OAT configurations with proper materials and dimensions can produce good focusing under certain operating conditions. A procedure is established to design OAT candidates for construction. Some examples of acceptable OAT candidates are given in Table 1. The construction and testing of prototype OATs are initiated.

For acceptable OAT candidates, the light intensity carrying the focus-inducing pattern is an important operating parameter in determining the sound focusing. This light intensity is limited by the available source power and the projection mechanism. For OAT configurations requiring high incident light level, mechanical schemes for projections scanning are contemplated. For OAT configurations requiring low incident light level, all electronic holographic scanning is contemplated.

Although CdS doped with Zn is used as the example for the photoconductive material and zone-plate as the example for the focus-inducing pattern, other materials and patterns are also under consideration.

ACKNOWLEDGEMENT

This work is supported by the National Science Foundation (NSF ENG 74-22184 and NSF ENG 74-22340).

REFERENCES

1. P. Green, L. Schaefer, and A. Macovski, "Considerations for Diagnostic Ultrasonic Imaging," in Acoustical Holography, G. Wade, Ed. (Plenum, New York, 1972), Vol. 4, pp. 97-111.

2. See, for instance, R. Anderson, "Potential Medical Applications for Ultrasonic Holography," in Acoustical Holography, P. Green, Ed. (Plenum, New York, 1974), Vol. 5, pp. 505-513.

3. K. Wang and G. Wade, "Threshold Contrast for Various Acoustic Imaging Systems," in Acoustical Holography (Plenum, New York, 1972), Vol. 4, pp. 431-462.

4. G. Wade, Ed., Acoustical Holography (Plenum, New York, 1972), Vol. 4.

5. P. Green, Ed., Acoustical Holography (Plenum, New York, 1974), Vol. 5.

6. N. Booth, Ed., Acoustical Holography (Plenum, New York, 1975), Vol. 6.

7. G. Wade Ed., Acoustic Imaging (Plenum, New York, 1976).

8. K. Wang and G. Wade, "Comparison of Ideal Performance of Some Real-Time Acoustic Imaging Systems," in Journal of Acoustical Society of America, Vol. 56, No. 3, September 1974, pp. 922-928.

9. B. B. Brendon, "Real Time Acoustical Imaging by Means of Liquid Surface Holography," Ref. 4, pp. 1-9.

10. R. L. Whitman, M. Ahmed, and A. Korpel, "A Progress Report on the Laser Scanned Acoustic Camera," Ref. 4, pp. 11-32.

11. J. Landry, H. Keyani and G. Wade, "Bragg-Diffraction Imaging: A Potential Technique for Medical Diagnosis and Material Inspection," Ref. 4, pp. 127-146.

12. P. Green, L. Schaefer, E. Jones and J. Suarez, "A New, High-Performance Ultrasonic Camera System," Ref. 5, pp. 493-513.

13. C. Sabet and C. W. Turner, "Parametric Transducer for High-Speed Real-Time Acoustic Imaging," Elec. Let., Vol. 12, No. 2, Jan. 1976, pp. 44, 45.

14. C. G. Roberts, "Optically Scanned Acoustic Imaging," ML Report 2361, Stanford University, Calif., 1974.

15. H. Shen, "Some Design Aspects of a High-Sensitivity Ultrasonic Imaging System," Master's Thesis, University of Houston, Dec., 1975, unpublished.

16. K. Wang and H. Shen, "Design of an Optically Controlled Transducer for Ultrasonic Imaging," submitted to Applied Optics.

17. B. A. Auld, R. C. Addison, and D. C. Webb, "Focusing and Scanning of Acoustic Waves in Solids," Acoustical Holography, Vol. 2, Metherell & Larmore Ed., Plenum Press, New York (1970), pp. 117-132.

18. S. A. Farnow and B. A. Auld, "An Acoustic Phase Plate Imaging Device," Acoustical Holography, Vol. 6, N. Booth Ed., Plenum Press, New York, (1975), pp. 259-274.

19. I. D. McKinney, "Technique for Vapor Depositing Thin-Film In_2O_3/SnO_2 Transparent Electrodes," Sandia Labs Report No. Sc-RR-720798, (Nov. 1972).

20. Private communication with David B. Fraser. Also see David B.
     Fraser, "Sputtered Films for Display Devices" Proceedings of
     IEEE (July 1973), pp. 1013-1018.

21. D. B. Fraser and H. D. Cook, "Sputter Deposition of $Cd_{1-x}Zn_xS$
     Photoconductive Films," Journal of Vacuum Science and Technol-
     ogy, Vol. 11, No. 1, pp. 56-59 (1974).

DESIGN CRITERIA FOR MEDICAL BRAGG IMAGING

D. Frieda, P. Spiegler, F. Kearly, M. Greenfield and
R. Stern
University of California, Department of Radiological
Sciences, The Center for the Health Sciences, Los
Angeles, CA   90024

ABSTRACT

We have built a Bragg imaging system using carefully designed
cylindrical and spherical lenses essentially free of spherical
aberrations. We are particularly interested in applying the
system to medical applications and have therefore restricted our-
selves to low frequencies of ultrasound (1-5 MHz). The system has
been used to obtain transmitted and reflected images. In addition
to images, quantitative data on the light intensity in the image
as a function of sound intensity have been obtained. Engineering
data on the optical mounting and transducer are also discussed.

The experience gained so far leads us to believe that a well
designed system will yield useful medical images.

INTRODUCTION

In concept, Bragg-diffraction imaging is a relatively simple
and promising technique for real time ultrasonic imaging. Essentially a sound field is imaged by a light wedge oriented perpendicular to the direction of sound (1, 2). As with other real time
ultrasonic imaging systems, the resolution is proportional to the
wavelength of sound and inversely proportional to the effective
numerical aperture of the system (3). The sensitivity (weakest
acoustical signal that can be imaged) is believed to be comparable
to that of liquid surface holography but poorer than that of the
Stanford Camera. Yet in practice, it has not been possible to

construct a Bragg-diffraction imaging system that yields medical images comparable to those achieved with either of these two systems (4, 5).

We have for the past two years looked into this difficulty and present here our findings.

OPTICS

The optical design of a Bragg imaging system is difficult when using low frequency ultrasound, 5 MHz or less, needed for medical application. Since the wavelength of ultrasound is about 500 to 1000 times greater than that of the light, the Bragg angle is less than a milliradian and the resulting image is highly astigmatic and anamorphic. Further, since the resolution is determined by the wavelength of ultrasound and the effective numerical aperture of the system, much attention must be focused on the optics of the system used to produce and correct the image. The small Bragg angle requires that every effort be made to reduce the forward scatter and spread of unwanted light out of the wedge into the image field.

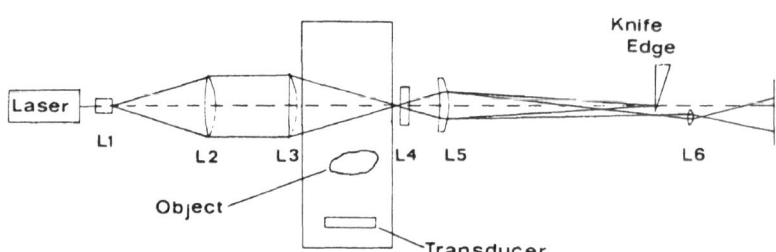

Figure 1.   Schematic diagram of a
             Bragg-diffraction
             imaging system.

The usual system is depicted in Fig. 1. The laser beam is expanded 50 to 100 times in diameter by lenses L1 and L2. The cylindrical lens L3 produced the light wedge which passes through a water-filled acoustic cell for interaction with the sound field. The cylindrical lenses L4 and L5 and the spherical lens L6 are used to remove the distortion from the image.

Our original set-up used single element plano-convex cylindrical lenses. Such lenses are usually free of spherical aberrations to relative apertures of f/12. Using 3 MHz ultrasound, the theoretical resolution was then about six millimeters. Landry, et al (4) had pointed out that lens L3, the wedge forming lens, must be of high quality and designed for minimal spherical aberrations for good vertical resolution. Thus our first attempt at an improvement was to design and replace this lens with a corrected uncemented doublet. The uncemented doublet although more costly and more critical to mount allows for much more flexibility than a cemented one.

The improvement was not as dramatic as expected since now lens L5 was found to be the resolution limiting element. Bragg diffraction imaging is highly astigmatic as first pointed out by Korpel (6). Consider Fig. 2a. "Rays of sound" are emitted at point S. They interact with the light wedge (C'D'O' and CDO) at points A', B', A and B. An image is formed at the line FES (also known as an anamorphic image) and at the line E'F' (also known as an orthoscopic image) to remove this astigmatism, a cylindrical lens, L5, that transfers the first image onto the location of the second image is necessary (see also Fig. 2b). Since the two images are several meters apart, it is easier to use a second cylindrical lens, L4, which is oriented perpendicular to cylindrical lens L5, to bring the orthoscopic image closer to the Bragg cell and to then finally magnify the corrected image with spherical lens L6. Lens L5 must have a relative aperture as good as or better than the wedge forming lens, L3, so as to capture and image the entire wedge of light with minimal spherical aberrations. Our second improvement thus consisted of replacing lens L5 with another corrected uncemented doublet. With this replacement substantial improvement was achieved but still not quite as expected.

Further investigations revealed that the spherical lens L2 was now the resolution limiting element. This lens essentially produced a plane wave of light for lens L3, however, in order to do so effectively its relative aperture must similarly be equal to or better than that of L3 over the area of the useful beam of light. Good images were finally achieved by replacing lens L2 with another appropriately designed spherical doublet.

Our present system has an effective f-number of about six. The experience gained from it suggests that an f/3 system that would yield a 7 cm x 7 cm image is feasible. While we are satisfied with our present components, we often observe ghost images. Their origin is unresolved and currently under investitation. Lens alignment appears to be the critical factor.

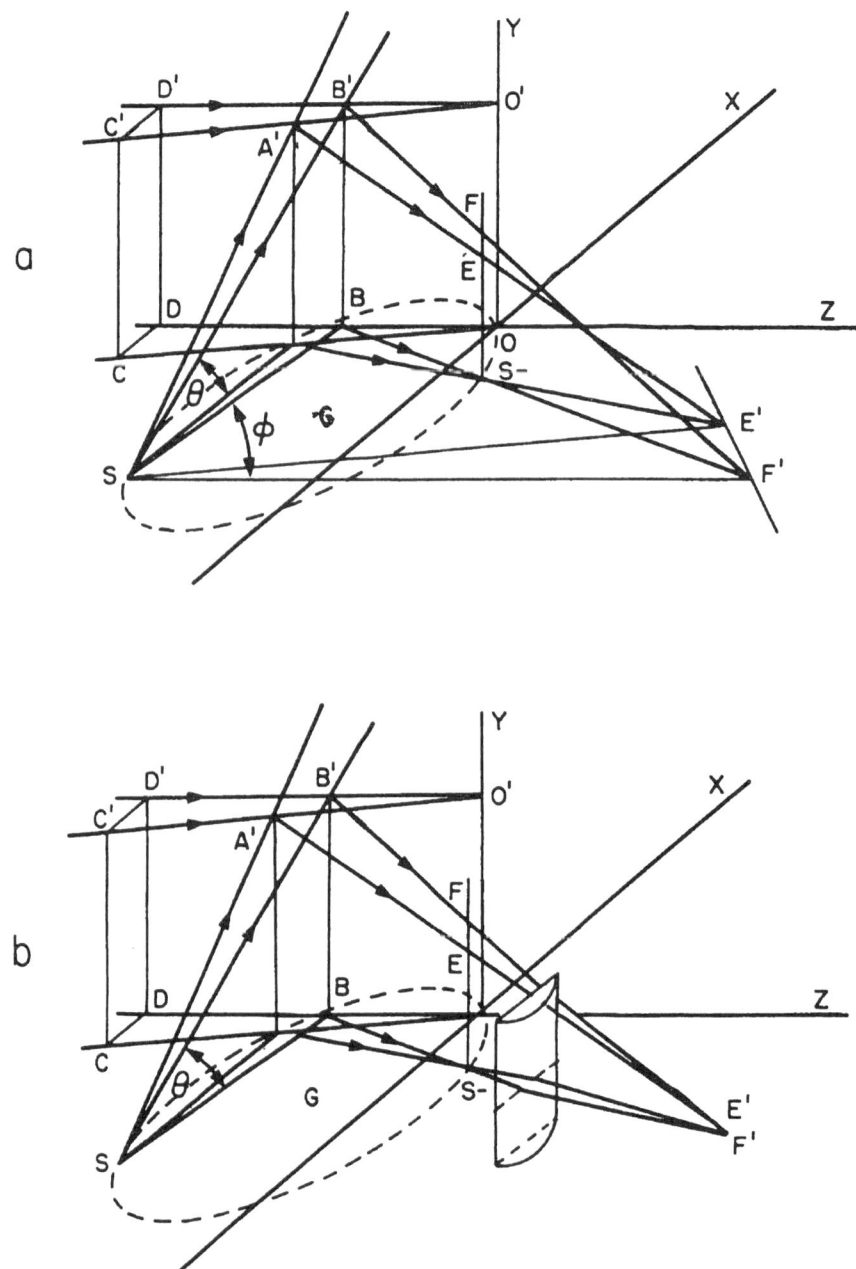

Figures 2a and 2b. Astigmatism and its removal.

ACOUSTICS

We use rectangular PZT-4 plates with fundamental frequencies of 1, 1.5 and 3 MHz, ranging in size from 5 cm x 5 cm to 10 cm x 10 cm. An impedance matching circuit (7) used with radio receivers and also described by Landry, et al (4) was used to match the transducer's impedance with that of the power supply. The complex impedance of the transducer assembly was measured using an HP Vector Voltmeter. These measurements were used to determine the components of the matching network, the resonance frequency of the transducer and the relative efficiencies of the transducer (8). Efficiencies of 50% to 60% were achieved in the fundamental mode and 30% in the first harmonic. The acoustical power radiated per unit area of the transducer was calculated from a knowledge of the efficiencies, the area of the transducer and the power input obtained experimentally with the use of an RF wattmeter.

We used this information to measure the amount of light diffracted as a function of sound intensity. In such experiments, we essentially image a cross-section of the sound beam on a light sensitive diode. The results of such a study are shown in Fig. 3. As can be seen, the amount of diffracted light increases at first with ultrasound power but then saturates. Further increases result in distortions and even a slight decrease in the light intensity of the image of the sound beam. This behavior is further illustrated in Fig. 4. If a flat attenuator is placed in front of the transducer, reducing the sound intensity, then one obtains a graph similar to but shifted to the right as shown in Fig. 4. Similar data are obtained using other transducers at different frequencies. If a more intense laser is used then a similar graph is obtained but with a longer range of image intensities.

These data can be readily explained by the nature of Bragg-diffraction imaging. The transducer produces essentially a plane wave of sound and only those light rays in the wedge that travel at an angle of $\pi/2 \pm \phi B$ (ϕB is the Bragg angle) to the direction of sound propagation can interact and be diffracted. The onset of saturation thus indicates that all available photons are being diffracted. This is easily verified by placing a translucent screen behind lens L5. A dark line can be seen at the center of the light cross-section. This is illustrated in Fig. 5. Further increases in ultrasonic power causes many photons to undergo at least two interactions. This explains both the appearance of distortion and the saturation observed in the image.

From studying the effect of ultrasonic power, transducer size and light source intensity on image brightness, data are being obtained that will be useful in determining what can be imaged.

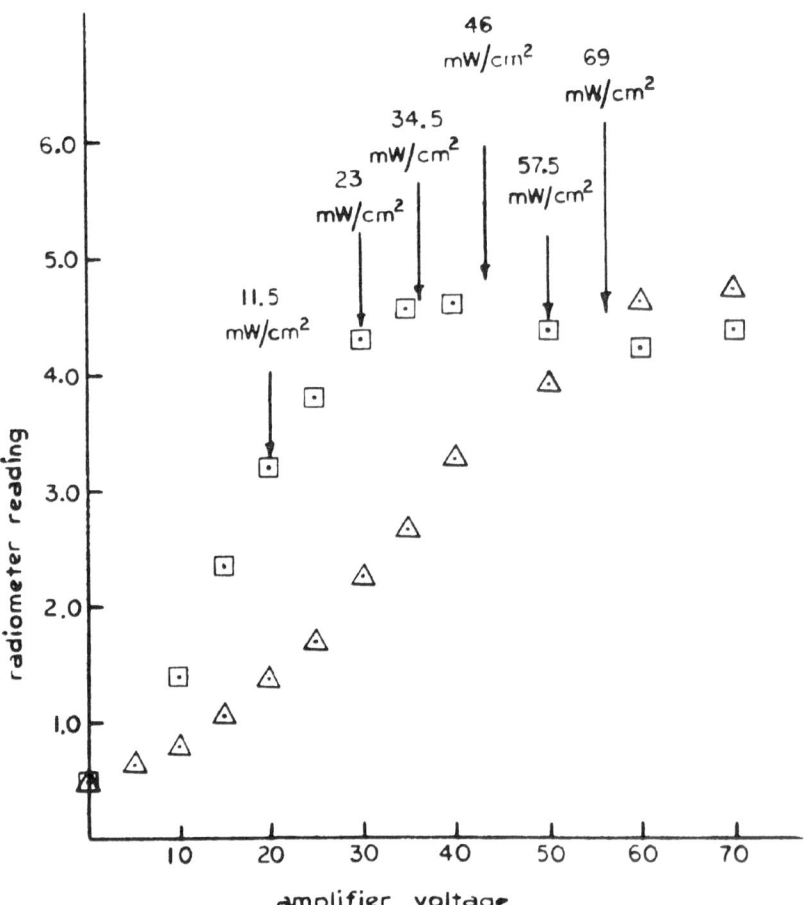

Figure 3. Amount of light diffracted by a plane wave of sound
           versus sound intensity. ⊡ no attenuator in path
           of sound beam. △ polystyrene slab in path of sound
           beam.

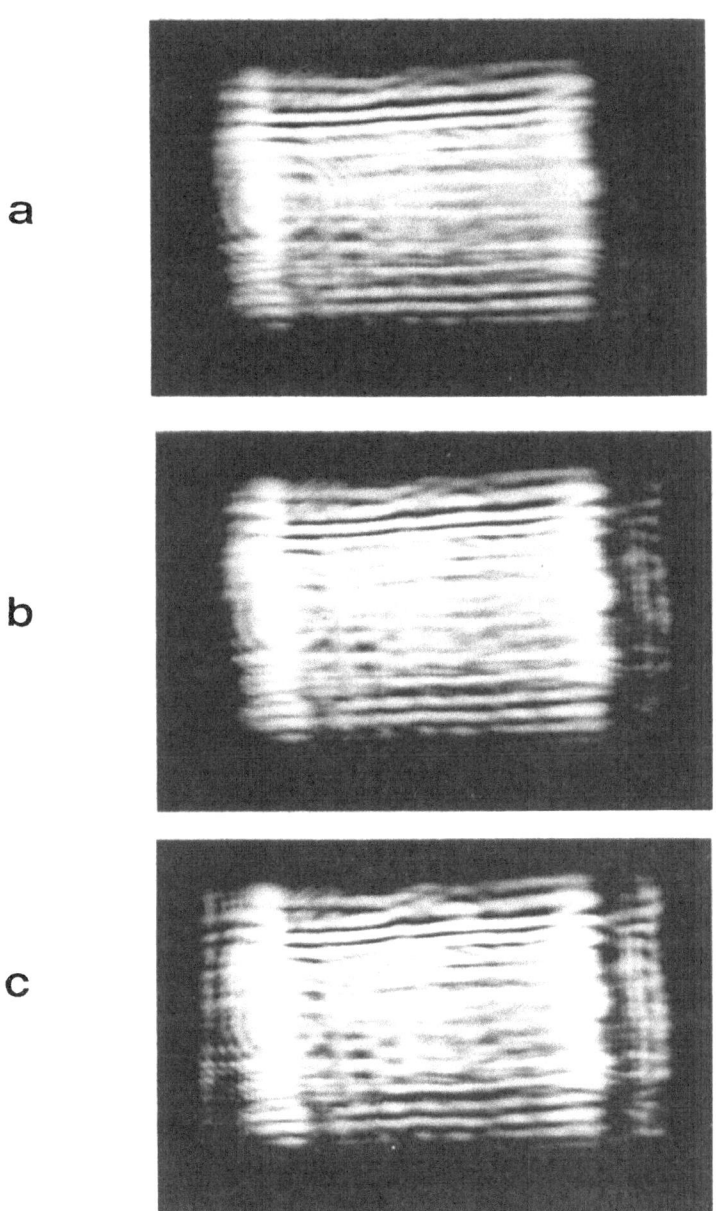

Figure 4.   Images of sound field.   a) 11.5 mW/cm^2; b) 23 mW/cm^2;
c) 34.5 mW/cm^2.

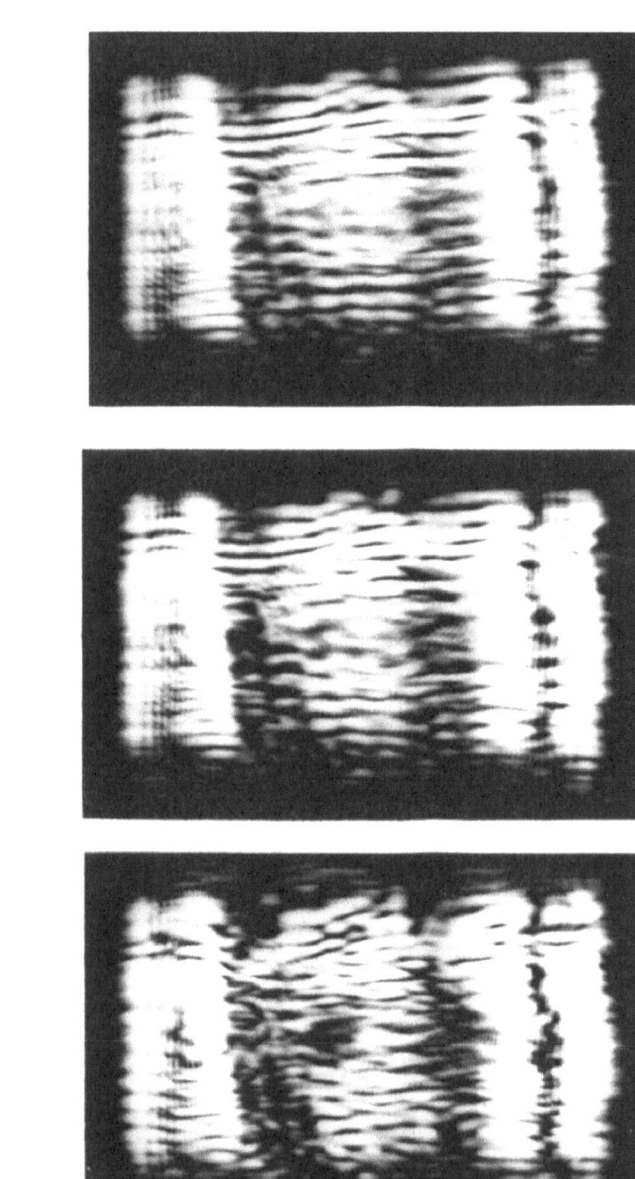

Figure 4 cont'd). Images of sound field. d) 46 mW/cm^2; e) 57.5 mW/cm^2; f) 69 mW/cm^2.

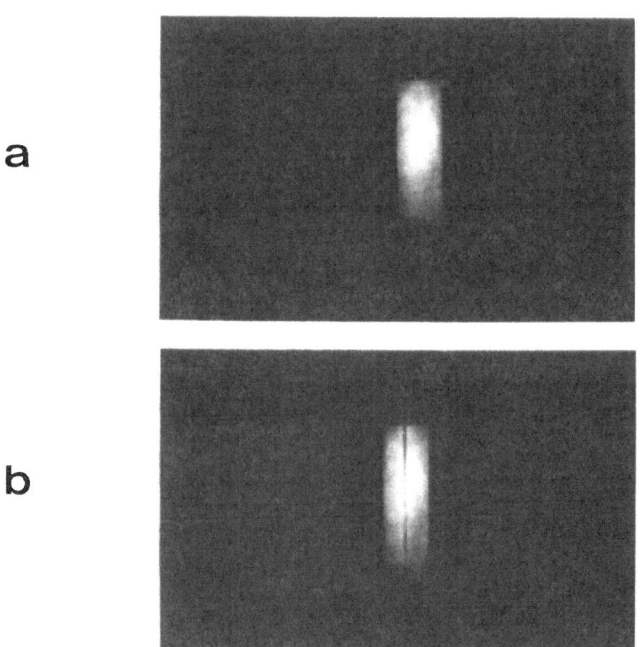

Figure 5.   Removal of light by sound.   a) Cross-section of light
beam behind lens L5 without sound field; b) with sound
field.

The effect of having a saturation level in image brightness and
the size of the range will be to limit the number of resoluble
shades of a gray scale.

IMAGES

In conclusion, we show in Figure 6   some of the latest
images obtained. These results lead us to believe that a well
designed system which yields useful medical images is feasible.

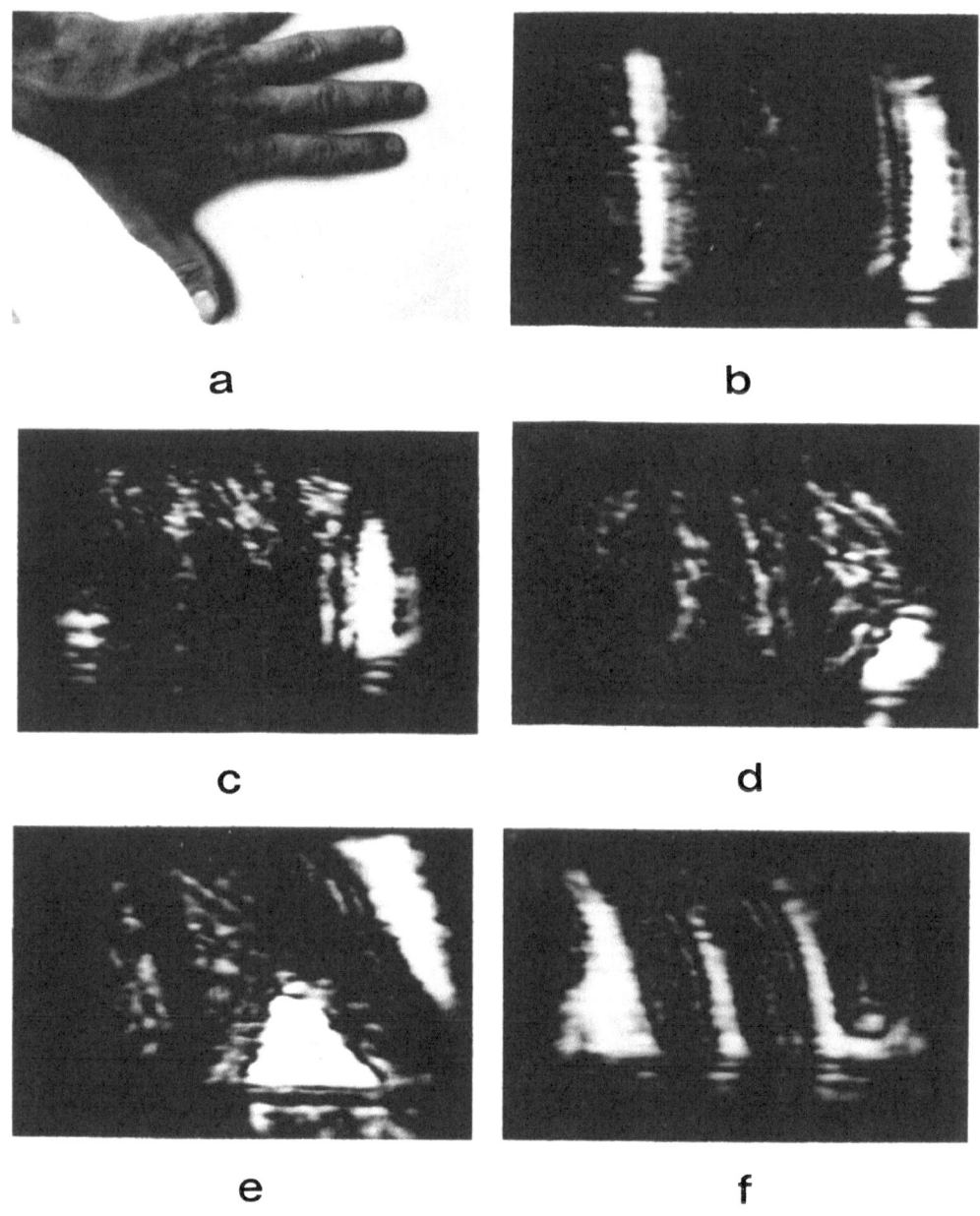

Figure 6.   Images of the hand.   a) light image of left hand;
            b) lower arm above wrist; c) and d) middle of right
            hand; e) thumb and webbing; f) fingers.
            b,c,d,e,f, are ultrasonic images taken with 3 MHz
            ultrasound.

REFERENCES

1. A. Korpel, "Visualization of the cross section of a sound beam by Bragg diffraction of laser light," Appl. Phys. Letters 9(12):425-427 (1966)

2. A. Korpel, "Optical image of ultrasonic fields by acoustic Bragg diffraction," Ph.D. dissertation, University of Delft, Netherlands (1969).

3. R.A. Smith, G. Wade, J. Powers, J. Landry, "Studies of resolution in a Bragg imaging system," J. Acoust. Soc. Amer. 49:1062-1068 (1971)

4. J. Landry, H. Keyani, G. Wade, "Bragg-diffraction imaging: a potential technique for medical diagnosis and material inspection," Acoustical Holography, Vol. IV, Glen Wade (Ed.), New York: Plenum Press (1972).

5. H. Keyani, J. Landry, G. Wade, "Bragg-diffraction imaging: a potential technique for medical diagnosis and material inspection, part II," Acoustical Holography, Vol. V, Philip S. Green (Ed.), New York: Plenum Press (1974).

6. A. Korpel, "Astigmatic imaging properties of Bragg diffraction," J. Acoust. Soc. of Amer. 49:1059-1061 (1971).

7. G. Flesher and G. Wade, Personal communication, University of California, Santa Barbara, November, 1975.

8. American Standard Procedures for Calibration of Electro-acoustic Transducers, American Standards Association, Inc., New York (1958).

ELECTRONICALLY SCANNED AND FOCUSED RECEIVING ARRAY

J. W. YOUNG*

Naval Undersea Center

San Diego, California 92132 U.S.A.

A ninety element electronically scanned and focused receiving array has been built and tested. The discrete transducer elements are formed by partial dicing of PZT-5 ceramic bars. Focused beamforming is accomplished by a hybrid analog-digital electronic system. Correct phasing of the transducer elements is achieved by multiplying the received 500 kHz acoustic signals by a low frequency FM sweep or "chirp" signal. The chirp, whose frequency range determines the focal distance, is generated by a VCO, digitized, and entered into a four-level digital delay line ninety units in length. An output tap on the delay line is provided for each hydrophone. Multiplication of the chirp and the acoustic carrier is accomplished by an integrated circuit digital-to-analog multiplier. The current outputs of the multipliers are added by a network of summing amplifiers. After high pass filtering, an amplitude modulated 500 kHz carrier remains. Single sideband demodulation is used to produce the scanning beam output. This signal is then rectified and low-pass filtered and can be used to intensity modulate a CRT display. The system can focus down to a range of less than one meter and produces a resolution of the order of one centimeter at that distance.

I. INTRODUCTION

The application of electronic phased arrays to underwater acoustics is not new. The pioneering work of Tucker, Welsby, et al.[1,2] at the University of Birmingham in England was performed almost twenty years ago. Their so-called "modulation scanning" technique is very similar to that employed by the system described in this paper. The electronic implementation of the present system

---

*Present address: **AMETEK**, Straza Division,
790 Greenfield Drive, El Cajon, California 92022 U.S.A.

which has been constructed at the Naval Undersea Center (NUC) is, however, significantly different from previous efforts. The Birmingham system was designed for use at moderately long ranges, in most cases in the far field of the array. Thus, initially, the problem of focusing the array was not addressed. More recently, near field beamforming has been considered by the Birmingham group.[3,4,5] However, they have separated the focusing from the scanning operation whereas in the NUC system the two are accomplished together.

The present receiving array is intended to form a portion of an acoustic imaging system which will provide high resolution information on targets at ranges of only 1 to 10 meters. Thus, the ability to conveniently alter the range of focus of the array has been of primary importance in the design of the NUC system.

The work of Havlice, Quate, Kino, et al.[6,7] at Stanford University, forms the immediate technological basis for the present system. Much of the early work by this group was motivated and influenced by Whitehouse of NUC and supported by Navy funding. Recently, the development of an underwater acoustic imaging system has been carried out directly by NUC while the Stanford effort has been oriented toward medical ultrasonics.

II. THEORY

Detailed discussions of focused beamforming have been given by Welsby (Ref. 4) and by Havlice (Ref. 7). For completeness however, a brief survey of the problem will be given here.

Figure 1 shows the geometry of a point source near a linear array of transducers. The source is assumed to radiate an elementary spherical wave,

$$P_s = P_o \, r_o \, \frac{e^{ikR}}{R} \quad , \tag{1}$$

where r_o is a reference distance.

The pressure received at the nth hydrophone is

$$P_s \, (Z_n) = P_o \, r_o \, \frac{e^{ikR_n}}{R_n} \quad , \tag{2}$$

where

$$R_n = \left[r_s^2 - 2 \, Z_n \, r_s \cos \theta_s + Z_n^2 \right]^{1/2} \tag{3}$$

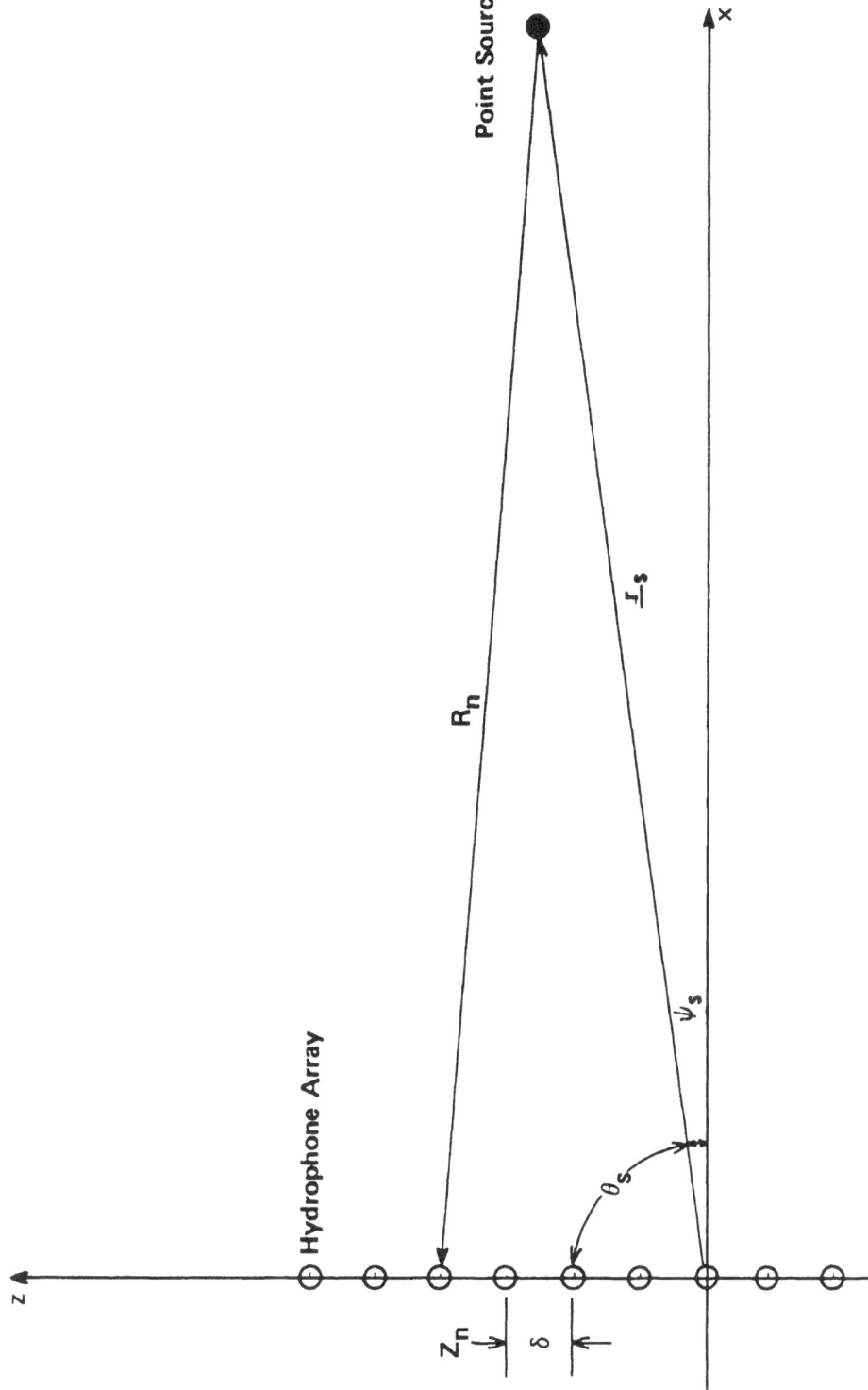

FIG. 1. GEOMETRY OF POINT SOURCE NEAR LINE ARRAY OF HYDROPHONES.

If we make the standard Fresnel zone approximation, we find

$$R_n \approx r_s - Z_n \cos \theta_s + \frac{Z_n^2 \sin^2 \theta_s}{2 r_s} \quad . \tag{4}$$

In terms of the complementary angle, ψ_s, this is

$$R_n \approx r_s - Z_n \sin \psi_s + \frac{Z_n^2}{2 r_s} \cos^2 \psi_s \quad . \tag{5}$$

The expansion of R_n used by Havlice is slightly different and is restricted to small values of ψ_s. However, the two versions are identical to first order in ψ_s and differ only slightly to second order.

The pressure field at the array is approximately given by

$$P_s(Z_n) \approx P_o r_o \frac{e^{ikr_s}}{r_s} e^{-ikZ_n \sin \psi_s} e^{i\frac{kZ_n^2}{2 r_s} \cos^2 \psi_s} \quad . \tag{6}$$

In this form, the pressure has been factored into three terms from left to right which represent, respectively, the spherical wave propagation to the array center, the linear phase shift across the array associated with a source in the far field, and the quadratic phase correction for a near field source. Since the latter two terms vary with position along the array, they must be corrected by the beamformer in order to maximize the array output.

A schematic diagram of a five element segment of the electronic beamformer is shown in Figure 2. The system consists of hydrophones, a delay line with multiple output taps, multipliers and a summation network. A signal, f(t) is entered into the delay line. At the nth hydrophone station the delayed signal, f(t-nτ), is multiplied by the signal from the hydrophone, $h_n(t)$. The outputs of all multipliers are then summed which leads to a system output of the form

$$O(t) = \sum_{n=1}^{N} h_n(t) f(t - n\tau) \quad , \tag{7}$$

where N is the number of hydrophones in the array.

For the previously discussed acoustic point source,

$$h_n(t) = e^{-ikZ_n \sin \psi_s} e^{i\frac{kZ_n^2}{2r_s} \cos^2 \psi_s} e^{-i\omega t} \quad . \tag{8}$$

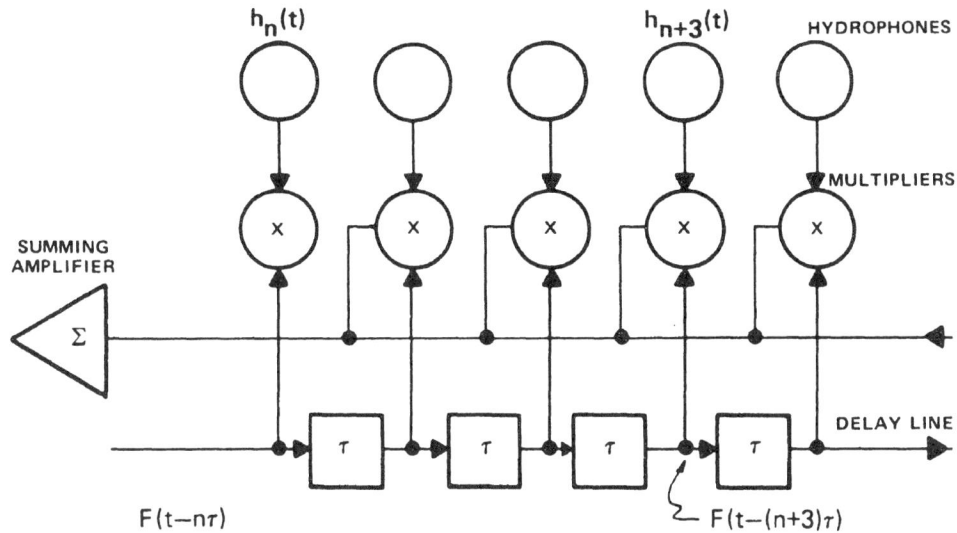

FIG. 2. SCHEMATIC DIAGRAM OF BEAMFORMER.

We shall assume that the hydrophone elements are equally spaced so that

$$Z_n = n\delta \quad . \tag{9}$$

The interrogating signal is chosen to be a linear FM sweep or "chirp" of the form

$$f(t) = e^{-i(\omega_0 t + \frac{a t^2}{2})} \quad . \tag{10}$$

The system output then is given by

$$O(t) = e^{-i(\omega+\omega_0+\frac{a t}{2})t} \sum_n e^{in(a t\tau - ka \sin \psi_s + \omega_0\tau)}$$

$$\times \, e^{i\frac{n^2}{2}\left(\frac{k\delta^2}{r_s} \cos^2 \psi_s - a\tau^2\right)} \tag{11}$$

For suitable choices of parameters such that

$$a\tau^2 = \frac{k\delta^2}{r_s} \cos^2 \psi_s \quad , \tag{12}$$

the n^2 term vanishes. In this case the summation can be performed easily.

The result is

$$O(t) = e^{-i(\omega+\omega_0+\frac{a\tau}{2})t} \; e^{i(\frac{N-1}{2})Q(t)} \left[\frac{\sin\frac{N}{2}Q(t)}{\sin\frac{Q(t)}{2}} \right], \tag{13}$$

where

$$Q(t) = \frac{k\delta^2 \cos^2\psi_s}{\tau r_s} \left(t - \frac{\tau r_s \sin\psi_s}{\delta \cos^2\psi_s}\right) + \omega_0\tau \quad. \tag{14}$$

The signal $O(t)$ consists of a chirp of instantaneous frequency

$$\omega_i = \omega + \omega_0 - (\frac{N-1}{2})a\tau + at \tag{15}$$

which is amplitude modulated by the function

$$A(t) = \frac{\sin\frac{NQ(t)}{2}}{\sin\frac{Q(t)}{2}} \quad. \tag{16}$$

This is identical to the pattern function of an ordinary uniformly weighted line array. It is maximized when $Q(t) = 0$. The presence of the factor $\omega_0\tau$ in Q produces an undesirable offset in the zero position. However, if

$$\omega_0\tau = 2n\pi \quad, \tag{17}$$

this parameter does not affect the amplitude function. In this case, the peak output of the array occurs when

$$t_0 = \frac{\tau r_s \sin\omega_s}{\delta \cos^2\psi_s} = \frac{\tau Z_s}{\delta \cos^2\psi_s} \quad. \tag{18}$$

The factor δ/τ is the effective speed of propagation (v) of the signal $f(t)$ in the delay line.

For small angles such that $\cos^2\psi_s \approx 1$,

$$t_0 = \frac{Z_s}{v} \quad. \tag{19}$$

Thus, the time of occurrence of a peak in the output signal corresponds directly to the lateral position of the acoustic source with respect to the array. In other words, when the chirp pattern propagating through the delay line matches the Fresnel zone pattern of the pressure field on the array both in terms of chirp rate, a, and lateral position, the array response will be maximized. Since the required a depends on r_s, scanning at different ranges can be accomplished by changing the chirp rate. In addition, if a single frequency signal is entered into the delay line, a stable unfocused (far field) beam is formed. By slowly varying the interrogating frequency the beam will be swept through an angular sector. This is the mode of operation which was employed in the original Birmingham system.

As with any array, the resolution of a focused line array is determined by its length in wavelengths. It can be shown that the lateral half-power width of the main lobe of the array pattern is given by

$$\Delta Z_{1/2} = \frac{2.8}{\pi} \frac{\lambda}{L} r_s , \qquad (20)$$

where $L = N \delta$ $\qquad\qquad\qquad\qquad\qquad\qquad\qquad$ (21)

and is the effective length of the array.

The angular width of the pattern is given by

$$\Delta \psi_{1/2} = \frac{\Delta Z_{1/2}}{r_s} = 51° (\frac{\lambda}{L}) , \qquad (22)$$

which is identical to the well known result for a uniform line array in the far field.

Resolution in range is typically much less precise and decreases as range increases toward the far field limit. A detailed discussion of this problem is given by Havlice in Ref. 7. As with conventional arrays, the focused line array with uniform amplitude weighting has side lobes only 13.5 dB below the main lobe. This is undesirable in acoustic imaging due to image dynamic range requirements. However, standard shading techniques can be used to reduce this level although some increase in $\Delta Z_{1/2}$ will result.

III. SYSTEM REQUIREMENTS

The goal of the NUC system is to provide an acoustic imaging capability with lateral resolution of one centimeter at a range of one meter and to be able to operate out to a maximum range of ten meters. Since these distances are within visual observation range in reasonably clear water, it is evident that the acoustic imaging system is intended to operate in extremely turbid water. This situation is commonly encountered in rivers, harbors and in the vicinity of the ocean bottom when the sediment has been stirred up by an undersea vehicle.

The backscattering cross section of the small particles which cause turbidity rises rapidly ($\propto (k\bar{a})^4$, where $k = \omega/c$ and $\bar{a}$ is the mean particle radius) toward the geometrical limit for $k\bar{a} < 1$. For $k\bar{a} > 1$, it is approximately constant. Thus, if acoustic energy is to suffer appreciably less backscattering by the medium than light, it is necessary that $k\bar{a} \ll 1$. If it is assumed that mean particle diameters are of the order of 200 u, $k\bar{a} = 1$ for a frequency of about 2.4 MHz. This is near the operating frequency of the Stanford acoustic imaging array on which the NUC system is modeled.

Based upon these considerations it was apparent that for moderate range operation in highly turbid water it was necessary to significantly lower the acoustic frequency of the system. Unfortunately, this requires an increase in size of the acoustic array in order to preserve the resolution capability. It is also necessary that the acoustic wavelength be smaller than the resolution length. As a reasonable compromise, an operating frequency of 500 kHz, which corresponds to a wavelength of 3 mm, was selected.

The specification of one centimeter resolution at one meter requires an array 90λ in length according to Eq. (20). Since the array was to be scanned over only a small angular range near broadside, an undersampled array of 90 hydrophones spaced at one wavelength intervals was considered acceptable.

IV. THE ELECTRONIC SCANNING SYSTEM

Because of the significantly lower frequency of operation and the larger number of hydrophone outputs to be processed, it was decided that the Stanford surface acoustic wave (SAW) delay line technology was not appropriate to the NUC system. Several factors influenced this decision. First was the significant attenuation of the chirp signal in passing along the delay line experienced in early versions of the Stanford system. This was later reduced greatly, however, by redesign of the output taps. Second was the necessity of operating with a high frequency (50 MHz) carrier and, in particular, the difficulty of achieving efficient multipliers at this frequency. Finally, it was desired to utilize commercially available components as much as possible.

As an alternative to the purely analog Stanford system, a hybrid digital-analog approach was employed. The principal components of the NUC system are shown schematically in Figure 3. The device is based upon a four-bit digital

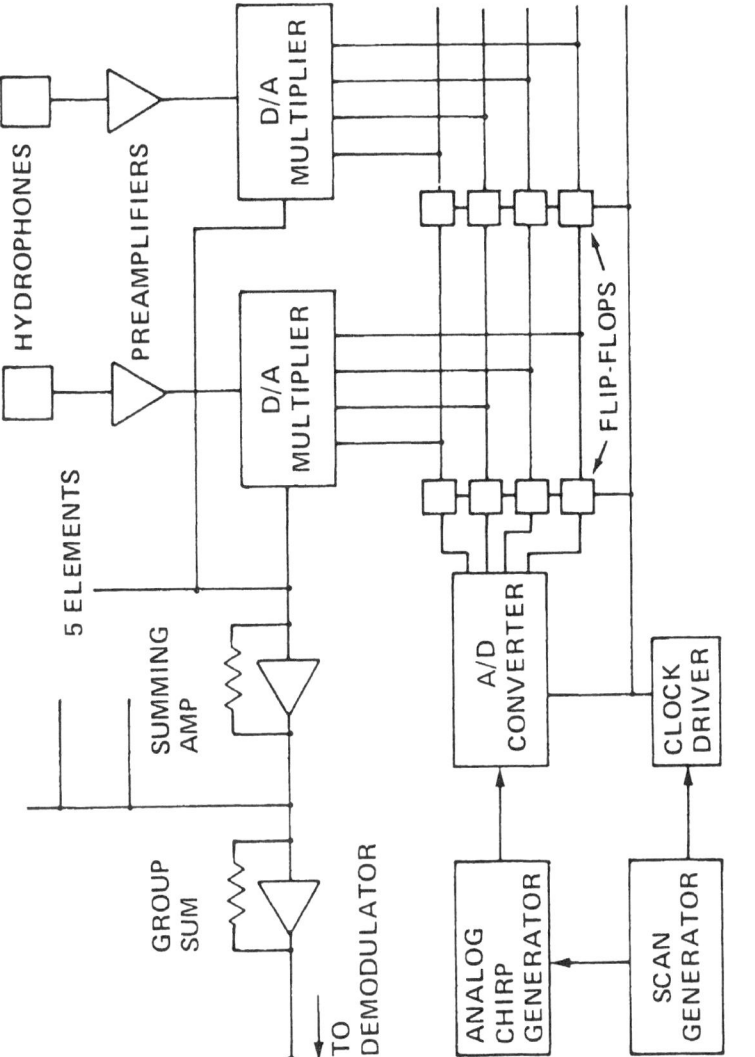

FIG. 3. SCHEMATIC DIAGRAM OF NUC ELECTRONIC SCANNING SYSTEM.

delay line ninety units in length. The other key element is a digital-to-analog (D/A) multiplier which takes a four-bit digital word representing a sample of the chirp and the preamplified signal from a hydrophone and produces an analog current output proportional to their product. The specific integrated circuit (IC) device utilized is a Motorola MC1406 D/A multiplier. This unit can accept a six-bit digital word, but it was felt that four bits provided sufficient accuracy for a demonstration system. The devices cost about $6.00 each at the time the system was constructed. Improved IC's to perform this function are available today for lower cost.

The combination of a multiple bit digital delay line with D/A multipliers as output taps has terminal characteristics similar to the tapped analog delay lines currently being developed using charge coupled devices (CCD's). Unlike CCD's however, there is no signal degradation in passing along the delay line. Another advantage of this technology as compared to conventional analog delay lines is that the clock rate, and hence the effective speed of propagation, is controllable. In the present system the clock frequency is only 40 kHz. However, all the digital components are capable of operating up to several MHz.

The system is constructed in identical modules containing the basic electronics for five hydrophone "channels." Each unit consists of five preamplifiers, five D/A multipliers, twenty delay line flip-flops and a summing amplifier mounted on a single printed circuit (PC) board. Eighteen such boards are required for the present ninety element array.

In addition to the channel boards, one board is required for the system drive electronics and three boards are used for the final summation and demodulation of the array output.

The system control is entirely digital except for the chirp generator. This is a voltage controlled oscillator (VCO) driven by a triangle wave to produce a linear FM sweep. The instantaneous frequency ranges from near zero to a maximum value which depends upon the range of focus of the array. The relationship of these parameters can be determined from the focusing relation of Eq. (12). The instantaneous chirp frequency is

$$f_i(t) = \frac{at}{2\pi} \quad , \tag{23}$$

where $0 \leqslant t \leqslant T/2$

and T is the period of the triangle wave. The array completes a full scan of the field in a time T/2, but the other half cycle is required to refill the delay line for the next scan. Thus, the effective scanning frequency is

$$f_s = 1/T \quad . \tag{24}$$

The maximum frequency of the chirp is given by

$$f_m = \frac{aT}{4\pi} = \frac{a}{4\pi f_s} \quad . \tag{25}$$

The time delay between hydrophones is

$$\tau = 1/f_c \quad , \tag{26}$$

where f_c is the system clock frequency. Making these substitutions in Eq. (12) yields

$$\frac{2 f_m f_s}{f_c^2} = \frac{\delta^2}{\lambda r_s} = \frac{\lambda}{r_s} \quad . \tag{27}$$

since $\delta = \lambda$ for this system.

The relation of f_s and f_c depends upon the field of view to be scanned. In general,

$$f_c \geqslant 2N f_s \quad . \tag{28}$$

At present, the system is configured to scan a field equal to the length of the array, and the equality holds. This leads to the relation

$$\frac{f_m}{f_c} = \frac{L}{r_s} \quad . \tag{29}$$

Thus, for an array of length 270 mm focused over the range interval from one to ten meters,

$$.027 \leqslant f_m/f_c \leqslant .27 \quad . \tag{30}$$

For a 40 kHz clock, the maximum chirp frequency required is 10.8 kHz.

As shown in Eq. (29), the focusing range can be altered by changing either f_m or f_c. In general, however, it is not desirable to change the clock frequency since the scan rate would also be changed. It is expected that future versions of this system will employ direct digital generation of the chirp. This will give greater flexibility to the system and should eliminate certain instability and alignment problems which affect the present VCO.

V. THE DEMODULATOR

A number of simplifications were made for the sake of clarity in discussing the theory of focused modulation scanning in Section II. In order to understand the demodulation process, however, it is necessary to examine some of the practical details. First, the functions $h_n(t)$ and $f(t)$ were taken to be complex exponentials whereas only their real parts have physical significance. If Eq. (7) is corrected such that

$$O(t) = \sum_n \text{Re } h_n(t) \text{ Re } f(t-n\tau) \tag{31}$$

it is found that there are terms for which the quadratic phase factor cannot be made to cancel. Because these terms are incorrectly phased however, they tend to add incoherently and do not appear to significantly degrade system performance.

A second point which has been overlooked is that the MC1406 is a one quadrant multiplier. Therefore, both the chirp and the hydrophone signal must be d.c. offset which results in extraneous terms in the product output. A high pass filter following the summing amplifiers eliminates the low frequency terms. The remainder consists of a carrier at the acoustic frequency, f_a, plus an amplitude modulated chirp. By making the appropriate substitution into Eq. (15) it can be shown that the instantaneous frequency of the chirp is given by

$$f_i = f_a - \frac{f_m}{2} + \frac{2 f_m t}{T} \quad , \tag{32}$$

where $0 \leqslant t \leqslant T/2$. Thus, f_i sweeps from $f_a - f_m/2$ to $f_a + f_m/2$ during the scanning operation. The desired information is contained in the amplitude modulation of the chirp.

In order to extract the modulating signal, a form of single sideband (SSB) demodulator was used. This system is shown in Figure 4. The input signal is split into two channels and mixed with in-phase and quadrature versions of the carrier. The resulting signals are then passed through two wide band 90° phase difference networks. The sum and difference of these two outputs contain only lower sideband (LSB) and upper sideband (USB) information, respectively.

Since the chirp sweeps from the LSB to the USB during the scan, one-half of the field of view is seen on each channel. At this point, the two signals can be recombined or used separately.

The final stage of the demodulator consists of a simple diode detector and low pass filter followed by a variable threshold circuit which removes low level noise.

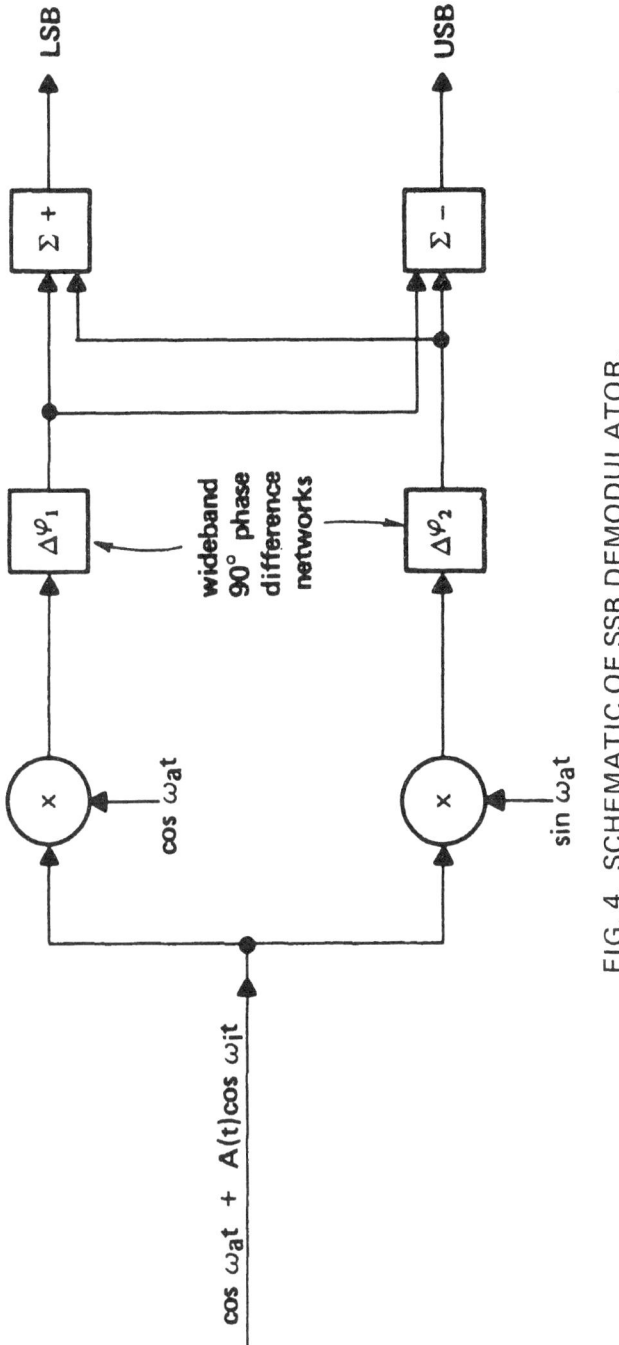

FIG. 4. SCHEMATIC OF SSB DEMODULATOR.

VI. THE TRANSDUCER ARRAY

A one hundred element transducer array was entirely constructed by the NUC Transducer shop. Quasi-independent elements were created by dicing PZT-5 bars with a diamond wire saw. A cut 0.015" in width was made approximately one third of the way into the bar from either side. Since the elements are used in the 33-mode, each hydrophone thus has two independent electrodes.

The complete array consists of three four-inch long bars laid end-to-end and mounted on a one-half inch thick block of butyl rubber. The array is housed in an oil filled aluminum box with a neoprene window on the front. A pair of leads is brought from each hydrophone element through the rubber layer to feed-through connectors on the back wall of the unit. The entire structure is shown in Figure 5.

VII. EXPERIMENTAL RESULTS AND DISCUSSION

For preliminary tests, the array has been mounted in the end wall of a small acoustic tank. The electronic beamforming equipment has not as yet been packaged for submerged operation. It is mounted in a standard 19 inch rack immediately behind the array. Each transducer element is connected to its appropriate electronic channel by a 6 inch twisted pair of wires. The tank is six feet in length and is semicircular in cross section with a diameter of 18 inches. It is lined with horsehair packing material to reduce reflections. However, strong surface reflections are still present and cause serious problems.

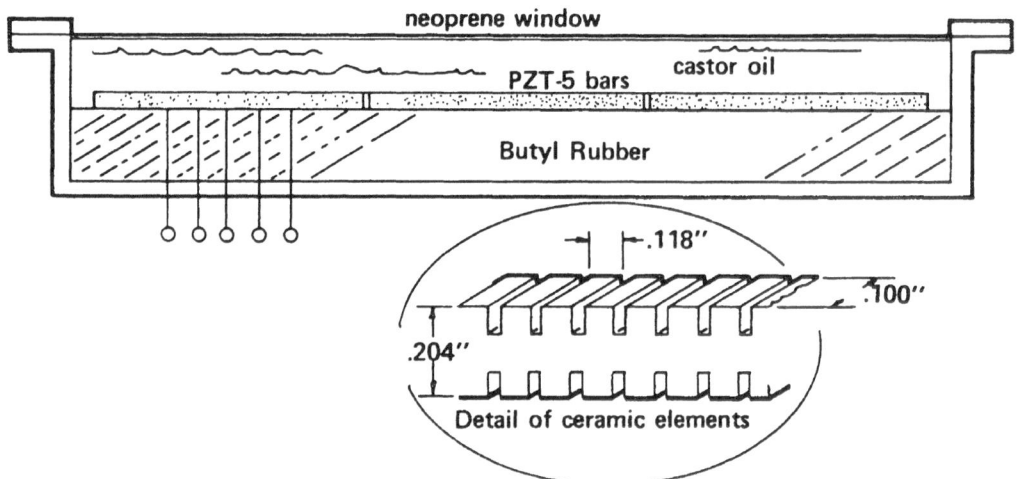

FIG. 5. CONSTRUCTION OF HYDROPHONE ARRAY.

In order to demonstrate the operation of the system, one or two small acoustic projectors have been placed approximately one meter away from the array. The beamformer output is used to either deflection or intensity modulate an oscilloscope. When one transducer is moved laterally across the field, a distinct peak or bright spot can be seen to move across the scope. There typically are numerous spurious responses as well. These appear to be predominantly due to multipath problems in the tank. Until it is possible to test the system in a larger facility, it will be difficult to be sure of the significance of these effects. By careful timing of the position and orientation of the source and proper setting of the threshold level of the detector, however, the very clean response peak shown in Figure 6a can be obtained. When a second projector is placed as close as possible to the first (i.e., 1.5" center-to-center separation), the results shown in Figure 6b are achieved. In this case, there is some evidence of sidelobe response as well. The width of the peaks as compared to their separation indicates that the predicted lateral resolution of about one centimeter has been achieved.

The effect of focusing of the beam has been shown by moving the source transducer to different ranges. There is an appreciable broadening of the response peak when the source is brought closer to the array. Within about 0.5 meters, it is so broad as to be almost lost in the noise background. On the other hand, only a slight degradation of the peak occurs when the source is moved out to the full length of the tank, which is 1.8 meters.

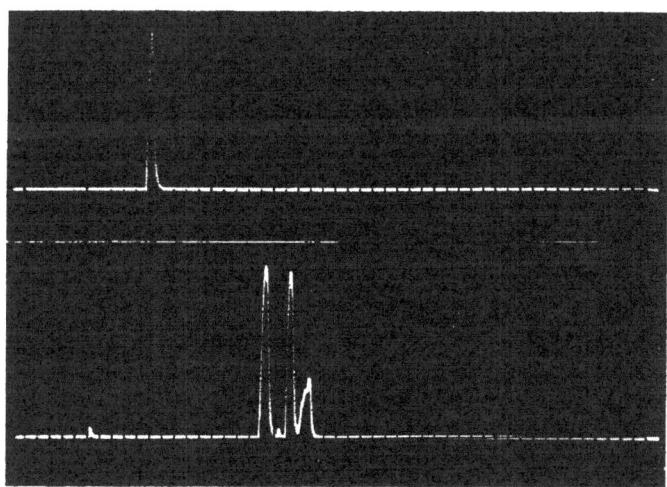

FIG. 6. OSCILLOSCOPE PHOTOGRAPHS SHOWING SYSTEM OUTPUT.
  a. SINGLE SOURCE AT ONE METER,
  b. TWO SOURCES 3.8 CM APART AT ONE METER.

An attempt at reflection imaging has been made by turning the acoustic projectors around and placing them close to the receiving array. Targets such as a steel angle iron and a large spherical transducer were then placed at the focal distance of one meter. It was possible to detect the reflections although the specularity of the targets and the relatively narrow beamwidths of the projectors made adjustment difficult.

Thus far, it has not been possible to make quantitative measurements of the system sensitivity or dynamic range. Until a larger tank facility is constructed or the entire electronic system is packaged for submerged operation, only qualitative indications of the system's performance and limitations can be achieved.

VIII. CONCLUSIONS

It has been shown that the technique of modulation scanning can be successfully applied to the problem of near field beamforming at an acoustic frequency of 500 kHz. A particular implementation using a multi-bit digital delay line and D/A multipliers has been described. Many possible improvements in the system can be suggested. In particular, complete conversion of the scan control circuitry to digital electronics is foreseen. Further tests are required to quantify the system operating characteristics. These will require either an improved acoustic tank facility or complete packaging of the beamforming electronics for underwater operation.

ACKNOWLEDGEMENTS

The author wishes to acknowledge the excellent work of H. Ding, C. Bell, and D. Stephens who designed and constructed most of the electronic system. The printed circuit boards were produced under contract by SDD, Inc. Much of the credit also belongs to L. Reavis, D. Coffer, and R. Dettmer who constructed the transducer array.

REFERENCES

1. D. G. Tucker, V. G. Welsby, and R. Kendall, "Electronic Sector Scanning," Brit. Inst. Radio Eng., 26, 465 (1958).

2. V. G. Welsby and J. R. Dunn, "A High-Resolution Electronic Sector Scanning Sonar," J. Brit. Inst. Radio Eng. 26, 205 (1963).

3. D. G. Tucker, "Near-Field Effects in Electronic Scanning Sonar," J. Sound Vib. 8, 355 (1968).

4. V. G. Welsby, "Electronic Scanning of Focused Arrays," J. Sound Vib. 8, 390 (1968).

5. V. G. Welsby, D. J. Creasey, and N. Barnickle, "Narrow Beam Focused Array for Electronically Scanned Sonar: Some Experimental Results," J. Sound Vib. 30, 237 (1973).

6. J. F. Havlice, G. S. Kino, and C. F. Quate, "Electronically Focused Acoustic Imaging Device," Appl. Phys. Letters 23, 581 (1973).

7. J. F. Havlice, G. S. Kino, J. S. Kofol, and C. F. Quate, "An Electronically Focused Acoustic Imaging Device," Acoustical Holography, Vol. 5, P. S. Green, Ed., New York: Plenum, 1974, pp 317-334.

MONOLITHIC MOSAIC TRANSDUCER UTILIZING TRAPPED ENERGY MODES

H.F. Tiersten, J.F. McDonald, M.F. Tse and P. Das

Rensselaer Polytechnic Institute

Troy, New York 12181 USA

A major difficulty in the fabrication of a large mosaic transducer is the achievement of adequate acoustic isolation of the small transducer elements making up the array. In order to obtain the isolation some workers have combined completely separate individual transducer elements [1] while others have used a large piezoelectric plate with grooves [2, 3]. The latter procedure is somewhat less cumbersome but still difficult for very small element sizes. Recently attention has been directed towards acoustic isolation schemes which do not require grooves. Some of these techniques involve matched terminator backing for the plate [4]. In this way the internal plate reflections that produce coupling are reduced.

In this paper a completely different approach is used. The desired isolation is achieved by employing an effect known as acoustic energy trapping. This effect results from the natural ever-present mass loading and electrical shorting of the metal electrodes. By virtue of this effect sets of electrodes can be placed on a uniform piezoelectric plate a sufficient distance apart for them to act essentially independently.

Basically, the energy trapping is achieved in a frequency range in between the cutoff frequencies of thickness vibration of the electroded and unelectroded regions. Ultimately this constraint sets a limit on the operating bandwidth of the device, which however, can be quite large. However, this type of transducer array has an advantage of small size, less positioning error and smaller fabrication cost because neither plate etching nor bonding are involved, yet the advantages of the well known process of photolithography are retained with regard to the electrode metalization. Another advantage of the trapped energy structure is that it permits a fairly simple theoretical analysis of the wave structure both inside and immediately outside the piezoelectric plate. Finally, the simplicity of the structure enhances the uniformity of response characteristics over a mosaic array of such devices.

THE EXTENSIONAL TRAPPED ENERGY MODE

The phenomena of energy trapping in piezoelectric plates is well known and has been employed for many years in trapped energy resonators [3,5] and monolithic crystal filters. More detailed theoretical analyses have been given for the fundamental modes in the case of thickness-shear [7,8] and thickness-extension [8] using approximate plate equations due primarily to Mindlin [9,10]. More recently detailed analyses have been given for overtone modes of coupled thickness-shear and thickness-twist for the small coupling case [11], [12].

We now consider the partially electroded piezoelectric plate shown in Fig. 1. The piezoelectric plate is assumed to be composed of a hexagonal crystalline material in class C_{6v}[13]. Polarized ceramics are in this symmetry class. The x_3 axis is normal to the plane of the plate and coincides with the hexagonal axis of symmetry.

The pertinent piezoelectric equations for extensional modes in this application are [14]

$$c_{55}u_{3,11} + (c_{55}+c_{13})u_{1,13} + c_{33}u_{3,33}$$
$$+ e_{15}\phi_{,11} + e_{33}\phi_{,33}$$
$$+ (e_{31}+e_{15})\phi_{,13} = \rho\ddot{u}_3 \qquad (1)$$
$$c_{11}u_{1,11} + (c_{13}+c_{55})u_{3,13} + c_{55}u_{1,33}$$
$$- \varepsilon_{11}\phi_{,11} = \rho\ddot{u}_1$$
$$e_{15}u_{3,11} + (e_{11}+e_{31})u_{1,13} + e_{33}u_{3,33}$$
$$- \varepsilon_{33}\phi_{,33} = 0$$

where x_2 dependence is ignored (strip geometry) and $c_{qp}, e_{ip}, \varepsilon_{ij}$, ρ, u_1, u_3, ϕ are the elastic constants, the piezoelectric constants, the dielectric constants, the mass density, the mechanical displace-

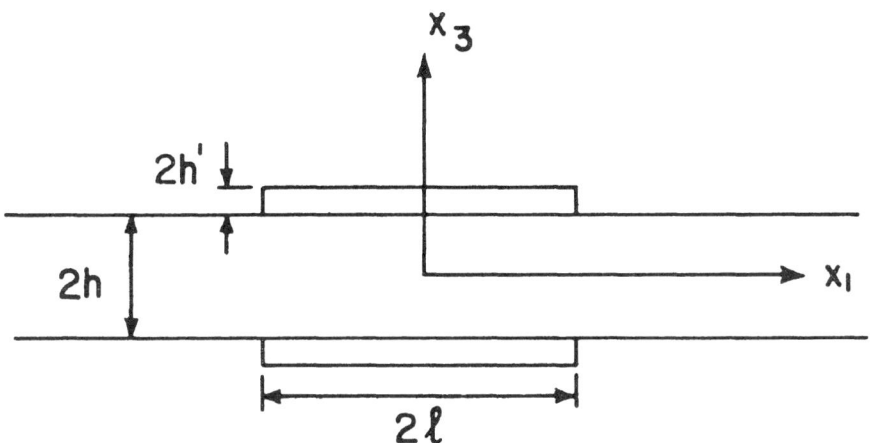

Fig. 1.   Basic trapped energy mode device geometry.

ment components in the x_1 and x_3 directions and the electric potential, respectively. The tensor and compressed matrix notation are
discussed in Chapter 7 of [7].

On the unelectroded surfaces of the plate the pertinent boundary conditions to be satisfied are

$$c_{13}u_{1,1} + c_{33}u_{3,3} + e_{33}\phi_{,3} = 0$$
$$e_{55}(u_{3,1} + u_{1,3}) + e_{15}\phi_{,1} = 0 \tag{2}$$
$$e_{31}u_{1,1} + e_{33}u_{3,3} - \varepsilon_{33}\phi_{,3} = 0$$

at $x_3 = \pm h$. These equations assume that ε_{33} is much larger than
the dielectric constant for the material surrounding the plate. On
the electroded surfaces the boundary conditions are

$$c_{13}u_{1,1} + c_{33}u_{3,3} + e_{33}\phi_{,3} = \mp 2\rho' h' u_3$$
$$c_{55}(u_{3,1} + u_{1,3}) + e_{15}\phi_{,1} = \mp 2\rho' h' u_1 \tag{3}$$
$$\phi = 0$$

at $x_3 = \pm h$ when the electrodes are shorted and where ρ' and $2h'$
are the mass density and thickness of the electrodes.

Following the discussion in Chapter 10 of reference [7] we note
that the plate eigensolutions satisfying the differential equations
(1) and boundary conditions (2) on the unelectroded regions are [15]

$$u_3 = (B_3^{(1)} \sin \eta_1 x_3 + B_3^{(2)} \sin \eta_2 x_3$$
$$+ B_3^{(3)} \sin \eta_3 x_3)\, e^{j\omega t} \cos \xi x_1$$
$$u_1 = (B_1^{(1)} \cos \eta_1 x_3 + B_1^{(2)} \cos \eta_2 x_3 \tag{4}$$
$$+ B_1^{(3)} \cos \eta_3 x_3)\, e^{j\omega t} \sin \xi x_1$$
$$\phi = (B_4^{(1)} \sin \eta_1 x_3 + B_4^{(2)} \sin \eta_2 x_3$$
$$+ B_4^{(3)} \sin \eta_3 x_3)\, e^{j\omega t} \cos \xi x_1$$

where η_1, η_2 and η_3 are the three roots of the bicubic obtained
from the homogeneous differential equations (1) for fixed ξ and ω.

The three sets of amplitude ratios $B_3^{(n)} : B_1^{(n)} : B_4^{(n)}$ are obtained from any two of the three linear equations (1) while the ratios $B_3^{(1)} : B_3^{(2)} : B_3^{(3)}$ are obtained from any two of the three linear homogeneous boundary condition equations (2) for the correct ξ and ω that satisfy the appropriate boundary condition determinantal equation. Typical dispersion curves [16] giving $\omega = \omega(\xi)$ resulting from this procedure are plotted as the dotted curves in Figure 2 in the frequency range of interest.

When a dispersion curve is on the left of the vertical ω - axis the wave number is purely imaginary and the associated solution function decays exponentially with x_1 along the plate.

Plate eigensolutions for (1) and (3) in the electroded regions may also be written as in (4). Using the same techniques as for the unelectroded region one obtains dispersion curves given by the solid curves in Figure 2.

In order to obtain approximate solutions for the problem of the partially electroded plate shown in Figure 1, one must superpose the 4 solutions corresponding to the 4 curves shown in Figure 2 for each region. On account of the symmetry of the structure

$$U_j = \sum_{n=1}^{4} K^{(n)} U_j^{(n)} \quad ; \quad \phi = \sum_{n=1}^{4} K^{(n)} \phi^{(n)}$$

$$\overline{U}_j = \sum_{n=1}^{4} \overline{K}^{(n)} \overline{U}^{(n)} \quad ; \quad \overline{\phi} = \sum_{n=1}^{4} \overline{K}^{(n)} \phi^{(n)}$$

(5)

where the barred quantities are used in the electroded regions. The $U_j^{(n)}$, $\phi^{(n)}$, $\overline{U}_j^{(n)}$ and $\overline{\phi}^{(n)}$ are the previously discussed solution functions for each of the four possible branches of the dispersion curve at the specified value of ω.

The 8 coefficients $K^{(n)}$, $\overline{K}^{(n)}$ $(n=1,\ldots,4)$ are determined by substituting (5) in what remains of the variational principle of linear piezoelectricity without constraints [7] and letting the $K^{(n)}$ and $\overline{K}^{(n)}$ be arbitrary. This procedure is quite lengthy and will not be discussed in detail. However, when executed it produces 8 linear homogeneous equations in these 8 coefficients. The determinantal equation for this homogeneous set determines the resonance frequencies, ω, in terms of ℓ/h for the shorted electrode system.

This paper does not provide the full details of the above procedure and associated numerical calculations, but instead presents the conditions under which the energy trapping effect can be realized practically in a transducer structure. To this end we first note that the critical thickness frequencies are given by

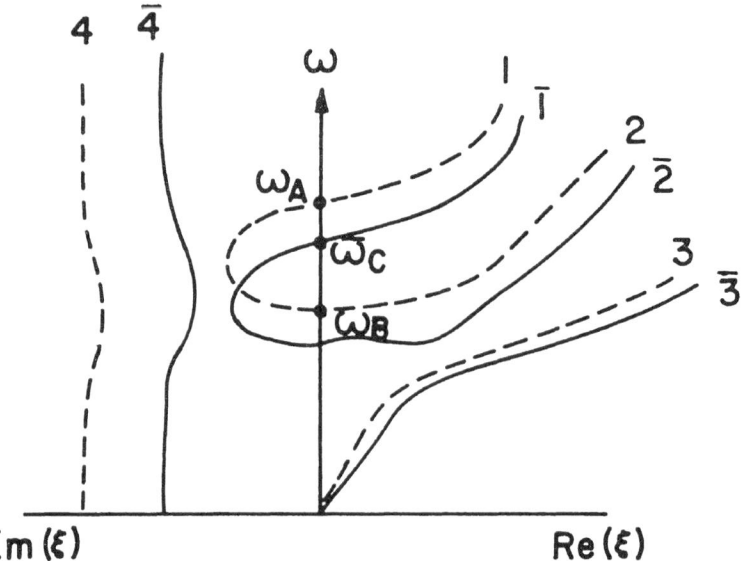

Fig. 2. Dispersion curves for the trapped energy mode device.

$$\omega_A = \frac{\pi}{2h} \left(\frac{\overline{c}_{33}}{\rho} \right)^{1/2}$$

$$\omega_B = \frac{\pi}{h} \left(\frac{c_{55}}{\rho} \right)^{1/2} \tag{6}$$

$$\overline{\omega}_C = \eta_1 \left(\frac{\overline{c}_{33}}{\rho} \right)^{1/2}$$

where $\eta_1 h$ is the lowest root of

$$\text{Tan}\,(\eta_1 h) = \eta_1 h/(k_{33}^2 + R\eta_1^2 h^2) \tag{7}$$

and

$$\overline{c}_{33} = c_{33} + e_{33}^2/\varepsilon_{33}$$

$$k_{33}^2 = e_{33}^2 /(\overline{c}_{33}\,\varepsilon_{33}) \tag{8}$$

$$R = 2\rho' h'/(\rho h)$$

For low modes of large coupling materials the mass loading term $R\eta_1^2 h^2$ is much less than the piezoelectric coupling term k_{33}^2 in (7) and we may write

$$\text{Tan}\,\eta_1 h = \eta_1 h/k_{33}^2 \tag{9}$$

which shows that the piezoelectric effect <u>reduces</u> the thickness frequency for an electroded region from that for an unelectroded region. The greater the reduction, the lower the mode number [18] is. For large coupling materials this reduction is quite large [19]. Thus $\overline{\omega}_C < \omega_A$. If

$$\omega_A > \omega > \overline{\omega}_C \tag{10}$$

the solution for the unelectroded region corresponding to branch 1 in Figure 2 will attenuate along the plate. On the other hand, the solution for the electroded region corresponding to branch $\overline{1}$ will have trigonmetric dependence. Since the curves labeled 1 and $\overline{1}$ are strongly coupled and constitute the dominant portion of the mode in the aforementioned frequency range, this type of mode is referred to as a trapped energy mode. However, it should be remembered that since the mode contains solution functions in the unelectroded region associated with curves 2 and 3, which are propagating, it is not completely trapped. Nevertheless, since the amplitudes of the propagating waves in the modes are relatively small compared to the amplitude of the trapped wave, the mode may be said to be essentially trapped. Since it is imperative that the transducers operate in an extensional mode and the pertinent dispersion curves are almost always of the general shape shown, it is essential that $\omega_A > \omega_B$. From (6) it is clear that the condition for this is that

$$\bar{c}_{33} > 4\, c_{55} \tag{11}$$

In fact, since it is desirable to have the bandwidth as large as possible, it is advantageous to have $\bar{c}_{33}$ as much larger than $4c_{55}$ as possible. Since for $\omega > \omega_A$ the solution function associated with curve 1 in the unelectroded region is not a decaying exponential, there is no trapped energy mode for $\omega > \omega_A$. As a consequence, the bandwidth of the transducer structure is $\omega_A - \bar{\omega}_C$ provided $\bar{\omega}_C > \omega_B$. Since $\omega_A - \bar{\omega}_C$ increases with k_{33}, the larger the piezo-electric coupling, the greater the bandwidth. At this point it should be noted that the unwanted untrapped length-extensional waves in the mode of the trapped energy transducer do not cause much of a limitation on the performance of the device because the motion induced by these waves on the major surface of the transducer is primarily tangential to the fluid and, hence, does not couple strongly.

TRANSDUCER RADIATION FROM A MOSAIC ARRAY

In this section we consider one of the limitations of the device imposed by the finite element sizes and separations required in a mosaic array. These dimensions are dictated by the ratio of the amplitudes (the K's) of the dominant waves in the electroded and unelectroded regions of the plate and by the desired attenuation between adjacent electrodes for the dominant trapped wave. The former consideration is quite complex while the latter is straightforward. However, obviously both dimensions will depend critically on the exact shape of the dispersion curve. Note that the gap $\omega_A - \omega_C$ does not depend on ℓ, the width of the electrode. Even if neither ℓ nor $2\pi/\xi$ is comparable to a wavelength inside the plate material, these dimensions may be comparable to a wavelength outside the plate in the material driven by the transducer [20] because the sound propagation velocity will generally be lower there.

In order to explore the effect these finite sizes have we consider the mosaic array used in a simple phased beam forming application. We will show that the finite element size will not be important in the Fraunhofer (far field) interference limit for the beam. The distance required for this approximation to hold may be shortened with an ultrasonic lens.

We consider the Fresnel interference problem treated by Goodman [21] using the coordinate system shown in Figure 3. The signal impressed on the medium just in front of the plate will be $U(x_1, y_1)$ while the interference pattern in an observation plane a distance z in front of the plate will be denoted by $U(x_0, y_0)$.

Consider the source to be an array of transducers imposing identically shaped spatial excitation patterns $G(x_1, y_1)$ shifted to locations x_{c_i}, y_{c_i} on the plate for $i = 1, \ldots, N$. The superposition of these excitation patterns each driven with phase θ_i and amplitude a_i will produce a source term

$$U(x_1, y_1)$$

$$= \sum_{i=1}^{N} a_i e^{j\theta_i} \; G(x - x_{c_i}, y - y_{c_i}) \tag{12}$$

Since the transducer has a much larger mechanical impedance than the liquid into which it radiates, the transducer effectively imparts a displacement (or velocity) field to the liquid. This means that the boundary value problem for the fluid is a Dirichlet problem for the three dimensional scalar Helmholtz equation, and the relevant Green's function for the plane surface may be obtained using the method of images. If the interference pattern for the beam is near broadside for the array then the Rayleigh-Sommerfeld obliquity factor may be ignored yielding:

$$U(x_0, y_0) = \frac{\exp(jkz)}{j\lambda z} \sum_{i=1}^{N} a_i e^{j\theta_i}$$

$$\exp\left\{\frac{jk}{2z}\left[(x_0 - x_{c_i})^2 + (y_0 - y_{c_i})^2\right]\right\} \tag{13}$$

$$\int_{-\infty}^{\infty}\int G(x_1, y_1) \; \exp\left[\frac{jk}{2z}(x_1^2 + y_1^2)\right]$$

$$\exp\left\{-j\frac{2\pi}{\lambda z}\left[(x_0 - x_{c_i})x_1 + (y_0 - y_{c_i})y_1\right]\right\} dx_1 \; dy_1$$

for the interference pattern in the Fresnel limit, where $k = 2\pi/\lambda$ outside the plate. This result is perfectly general and can be used for any model for the trapped energy mode shape, $G(x, y)$.

For simplicity we represent the distribution of the actual forcing function by a Gaussian shape (which is reasonably accurate):

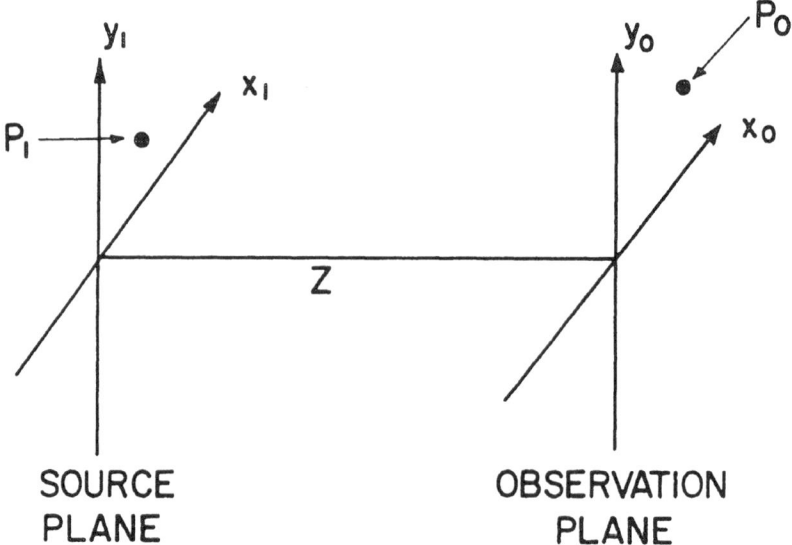

Fig. 3.   Beam forming interference pattern geometry.

$$G(x,y) = \exp\{-x^2/(2D^2)\} \tag{14}$$

It should be stressed that this is only an approximation and that D is chosen to include both the effect of the electrode length, ℓ, and the interelectrode spacing together with its implied attenuation.

We now simply quote the result which is obtained by using simple manipulations:

$$U(x_0,y_0) = \frac{\exp(jkz)}{j\lambda z} \cdot \frac{\pi}{\sqrt{\frac{jk}{2z}\left(\frac{1}{2D^2} - \frac{jk}{2z}\right)}}$$

$$\sum_{i=1}^{N} a_i \, e^{j\theta_i} \tag{15}$$

$$\exp\left\{j\left(\frac{\pi}{\lambda z}\right)\left[\frac{1}{1 - j\frac{kD^2}{z}}\right](x_0-x_{c_i})^2\right\}$$

Now provided kD^2/z is sufficiently small (Fraunhhofer limit) the interference pattern will be independent of D except for scale. Provided D is much less than the length, L, of the array the Fresnel effects may be ignored if kL^2/z is small or for large z. Use of a fixed ultrasonic lens in front of the array can shorten this distance. It should be noted, however, that it is not at all necessary for the mosaic transducer be used in this limit.

EXPERIMENTAL RESULTS

An experiment has been performed to test the effectiveness of energy trapping in practical transducer beam forming device designs. The results, which are still quite preliminary, appear surprisingly successful. The piezoelectric material selected for study was PZT-7A, a polarized ferroelectric ceramic. This material satisfies all of the previously discussed criteria but it does not represent the result of a comprehensive survey of all such materials optimized with respect to desired operating characteristics, nor has the electrode geometry been optimized.

The transducer consists of a one dimensional strip array of electrodes fabricated using photolithography. Very precise control over the thickness and size of the resulting metalization is achievable using this technique. The strips are located on the rear of

the transducer. The front of the transducer is fully electroded
and grounded. Silver paint is used to bind the wires to the
electrodes. (see Figure 4)

The radiation patterns of the transducer array are obtained
using the experimental facility shown in Figure 5. The external
medium in the tank is water. The piezoelectric plate is mounted
in such a way as to prevent any scattering interaction with its
support assembly. This can be pulled vertically using a quiet
motorized drive.

To record the radiation pressure of a transducer it is
excited with a 10 μs RF pulse with center frequency roughly 4.5 MHz.
The ultrasonic radiation is made to interact with light incident
from a He-Ne laser to detect Bragg diffraction [22]. The optical
effect of the ultrasonic wave is to produce a periodic modulation
of the refractive index of the water resembling a dffraction
grating. The strength of the diffracted laser beam is proportional
to pressure intensity for small pressure ranges [23]. The resulting
signal is detected using a photo diode connected to a phase locked
amplifier. The DC output is used to drive the Y input to an XY
plotter. The X plotter input is derived from a ramp waveform whose
period is calibrated to the motorized transducer positioning system.
With the laser beam parallel to the transducer slit direction it is
possible to measure the near field radiation as close as 1 mm or
less from the plate.

Using this technique the near field radiation was measured for
a single 1 mm electrode. The isolation was so complete (Figure 6)
that it was not possible to measure the leakage to propagating modes
using the existing apparatus. The longitudinal radiation from the
ends of the plate where leakage would be strongest was also not
observable. To see this radiation it was necessary to bond the
transducer to some backing material to spoil the trapping and induce
the end leakage. This would place the leakage radiation well below
-40 Db reference the main lobe.

Next the transducer operating bandwidth was measured. This was
found to be about 14% showing reasonable agreement with the calculated value of 12.8%. The resonance is shown in Figure 7, which
shows that the -3 Db bandwidth is 9%.

Finally a 3-element array was excited. The near field is shown
in Figure 8a. Note the uniformity of the responses. Finally the
interference patterns at x = 2 cm and 1 m are shown in Figures 8b
and 8c.

Fig. 4. Strip array showing unbacked plate
with 1mm electrodes with 1mm spacing.

Fig. 5. Experimental facility showing tank, plate, support,
He-Ne laser, photodiode and associated electronics.

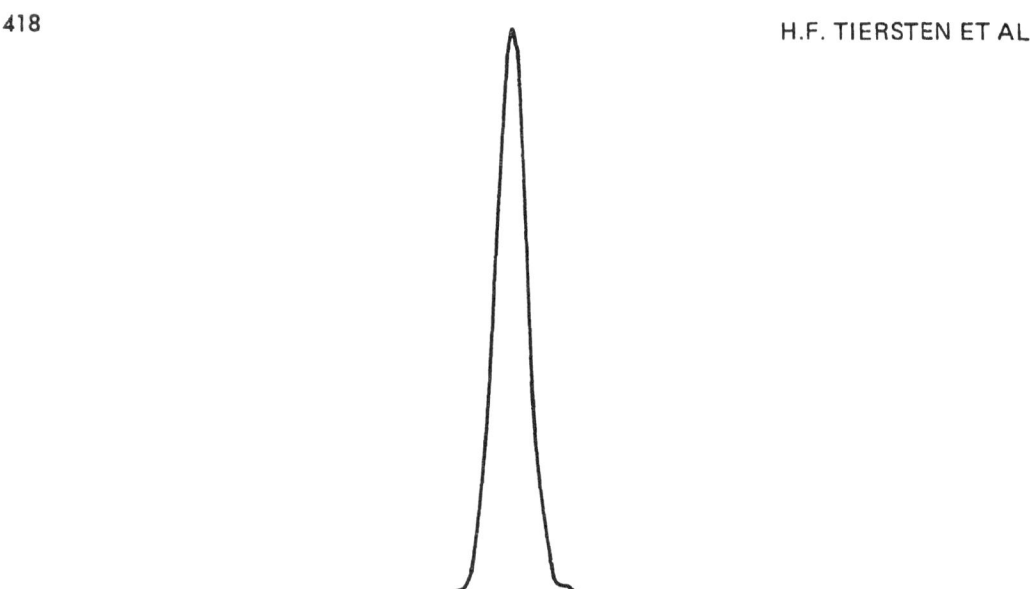

Fig. 6. Near field trapped energy mode radiation for a single
1mm wide strip at z=1mm.

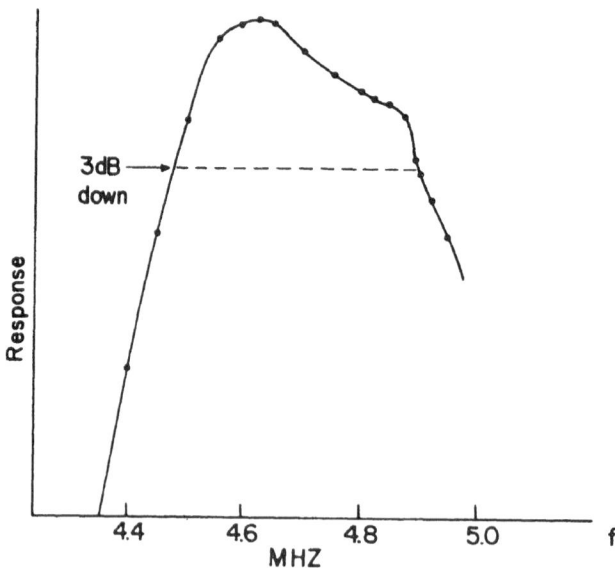

Fig. 7. Measured bandwidth for the basic trapped energy mode
transducer showing about 9% bandwidth at the half power points.
It should be noted that this represents a preliminary result.

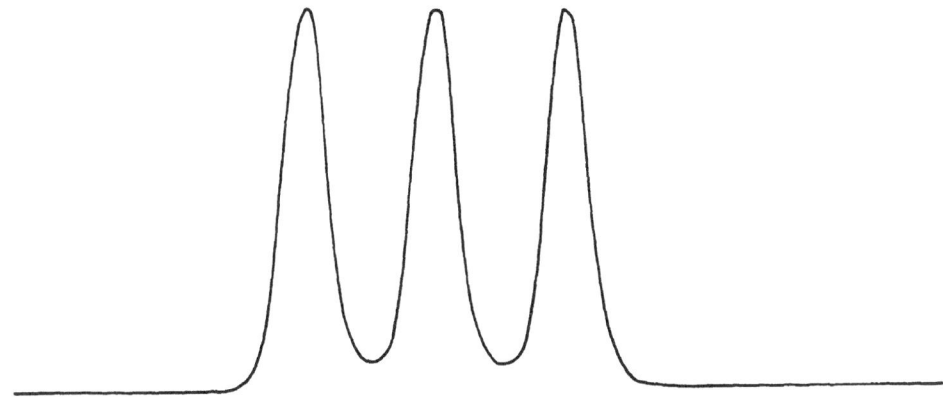

Fig. 8a. Near field for 3 adjacent 1mm strips separated by 1mm at z=1mm.

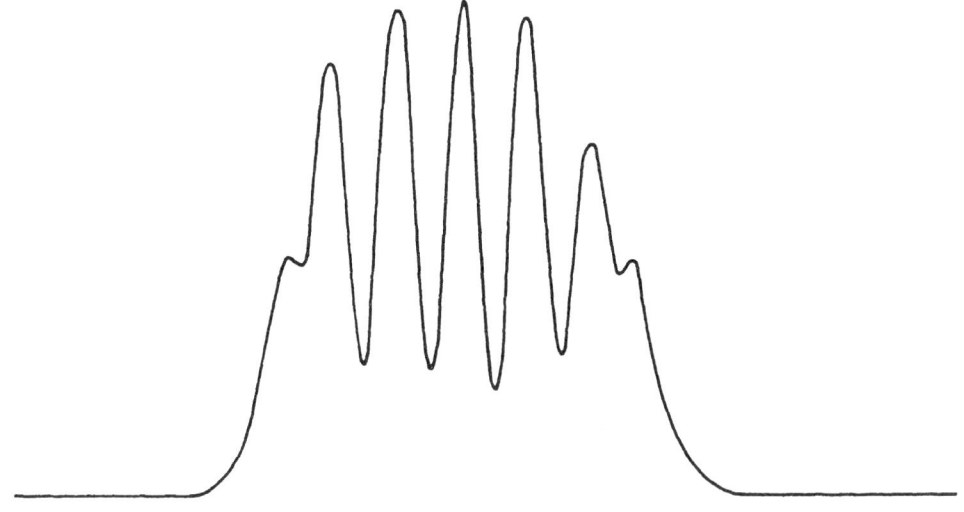

Fig. 8b. Intermediate field corresponding to figure 8a at z=2.5 cm.

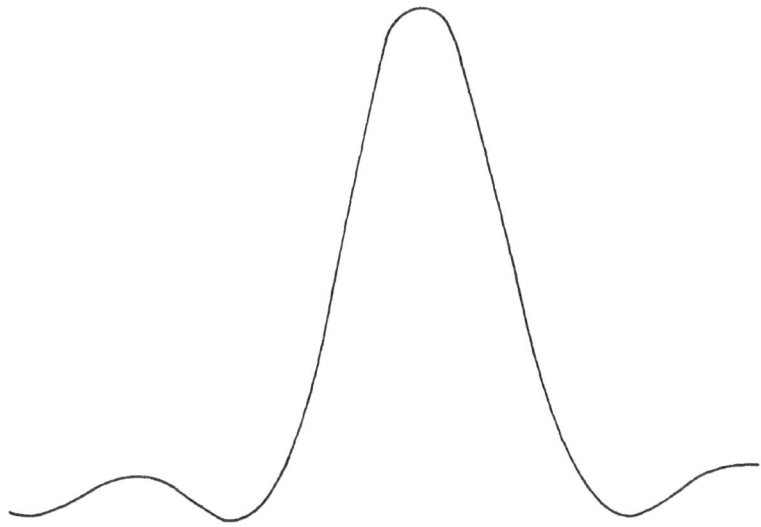

Fig. 8c. Far field (approaching the Fraunhofer limit), z=1m.

CONCLUSIONS

A totally new method of obtaining interelectrode acoustic isolation has been reported for a mosaic array. The method does not depend on grooved or notched plates nor on bonded backing material. The effect is surprisingly strong even in these pre-liminary studies. The mosaic has a uniquely uniform response, usable bandwidth, structural strength and inherent cost advantages. Criteria for materials selection have been given. Finally, it is worth noting that hybrid devices combining the shorted electrode energy trapping effect with other structures (such as grooves, notches, bosses and backing) are possible. Some of these hybrid structures can have a wider bandwidth or other interesting properties.

REFERENCES

1. V.G. Prokhorov, "Piezoelectric Matrices for the Reception of Acoustic Images and Holograms," Sov. Phys. Acoust., Vol. 18 (#3), Jan.-March 1973, pp. 408-410.

2. M.G. Marginness, J.D. Plummer, and J.D. Meindl, "An Acoustic Image Sensor Using a Transmit-Receive Array," in Acoustical Holography, (R.G. Green, Editor) Vol. 5, Plenum Press, New York, 1973, pp. 619-631.

3. N. Takagi, T. Kawaghima, T. Ogura and T. Yamada, "Solid-State Acoustic Image Sensor," in Acoustical Holography, (G. Wade, Editor) Vol. 4, Plenum Press, New York, 1972, pp. 215-236.

4. B.A. Auld, C. DeSilets and G.S. Kino, "A New Acoustic Array for Acoustic Imaging" Ultrasonics Symposium Proceedings, 1974, pp. 24-27.

5. W. Schockley, D.R. Curran and D.J. Koneval, "Energy Trapping and Related Studies of Multiple Electrode Filter Crystals," Proc. 17th Annual Symposium on Frequency Control, 88 (1963). W. Schockley, D.R. Curran and D.J. Koneval, "Trapped Energy Modes in Quartz Crystal Filters," J. Acoust. Soc. Am., $\underline{41}$, 981 (1967).

6. M. Onoe and H. Jumonji, "Analysis of Piezoelectric Resonators Vibrating in Trapped Energy Modes," Electronics and Comm. Eng. (Japan), 84 (1965). M. Onoe, H. Jumonji and N. Kobori, "High Frequency Crystal Filters Employing Multiple Mode Resonators Vibrating in Trapped Energy Modes," Proc. 20th Annual Symposium on Fre-quency Control, 266 (1966).

7. H.F. Tiersten, "Linear Piezoelectric Plate Vibrations,"
    Plenum Press, New York, 1969.

8. R. Holland and E.P. Eer Nisse, "Design of Resonant Piezo-
    electric Devices," MIT Press, No. 56, Cambridge, Mass., 1969.

9. R.D. Mindlin, "High Frequency Vibrations of Crystal Plates,"
    Q. Appl. Math., 19, 51 (1961).

10. R.D. Mindlin and M.A. Medick, "Extensional Vibrations of
    Piezoelectric Crystal Plates," J. App. Mech. Trans. ASME,
    26 pp. 561-569, Dec. 1959.

11. H.F. Tiersten, "Analysis of Intermodulation in Thickness-Shear
    and Trapped Energy Resonators," J. Acoust. Soc. Am., 57,
    667 (1975).

12. H.F. Tiersten, "Analysis of Trapped Energy Resonators Operating
    in Overtones of Coupled Thickness-Shear and Thickness-Twist,"
    J. Acoust. Soc. Am., 59, 879 (1976).

13. G. Weinreich, "Solids," J. Wiley, New York, 1965, Chapter 1.

14. Reference 7, Chapter 7, Equations (7.29).

15. H.F. Tiersten, "Wave Propagation in an Infinite Piezoelectric
    Plate, J. Acoustic Soc. Am. 35, 1963, p. 234.

16. R.D. Mindlin, "Mathematical Theory of Vibrations and Elastic
    Plates," in Proc. 11th Annual Symposium on Frequency Control,
    U.S. Army Signal Engineering Laboratories, Fort Monmouth,
    N.J., 1957, p. 1-40.

17. Reference 7, Eq. (6.44)

18. H.F. Tiersten, "Thickness Vibrations of Piezoelectric Plates,"
    J. Acoust. Soc. Am., 35, 1963, p. 53.

19. M. Onoe, H.F. Tiersten and A.H. Meitzler, "Shift in the
    Location of Resonant Frequencies Causes by Large Electro-
    Mechanical Coupling in Thickness Mode Resonators," J. Acoust.
    Soc. Am., 35, 1963, p. 36.

20. L.A. Harris, "Element Directivity in Ultrasonic Imaging
    Systems," I.E.E.E. Trans. on Sonics and Ultrasonics, SU-22
    (#5) Sept. 1975, pp. 336-340.

21. J.W. Goodman, "Introduction to Fourier Optics", McGraw-Hill, New York, 1968.

22. Yariv, "Introduction to Optical Electronics," Holt, Rinhardt and Winston, New York, 1971, Chapter 12.

23. M.V. Berry, "The Diffraction of Light by Ultra Sound," Academic Press, New York, 1966.

INTEGRATED ACOUSTIC ARRAY

K. R. Erikson
Rohe Scientific Corporation
Santa Ana, California

R. Zuleeg
McDonnell Douglas Astronautics Company
Huntington Beach, California

ABSTRACT

A fully integrated 8 x 8 element acoustic array module has been fabricated and its electro-acoustic performance measured. The matrix of 2 x 2 mm hydrophones is sawcut into a 3 MHz fundamental resonance frequency Y-cut $LiNbO_3$ wafer. Dual-gate deep depletion mode IGFET's are used with each element of the array for addressing and amplification. This integrated circuitry is contained on a single silicon on sapphire wafer and is bonded to the piezoelectric material in one processing operation. Addressing terminals are available along one edge of the array with signal outputs along another edge.

At 1 MHz in a water medium, the sensitivity was measured to be -115 dBV/micro Pascal and the minimum detectable acoustical intensity was measured to be 27 nano Watt/cm^2. The dual gate cascode amplifier provides 15 dB of voltage gain with a 15k ohm load resistor and has a feedback capacitance of 0.05 pf, which provides flat frequency response to 5 MHz.

INTRODUCTION

Acoustical orthographic imaging has great promise in medical ultrasound as well as potential in other areas such as underwater exploration. In such an imaging system whether it is holographic, or uses lenses or computers, a large number of elements is required for good picture quality. Previous orthographic imaging systems have

*This work was done while the author was with the Actron Division of McDonnell Douglas Corporation, Monrovia, CA.

423

used a variety of area detectors such as a liquid surface[1-3], a solid interface[4], a pellicle[5]; an electronically scanned electro-static array[6]; a mechanically[7] or acousto-mechanically[8] scanned piezoelectric linear array; and electronically scanned piezoelectric plates[9] and arrays [10-13].

Non-piezoelectric area detectors have a limited sensitivity and as such are not considered to be viable candidates for medical imaging (at least in the reflection mode[14]). Mechanically scanned piezoelectric systems have excellent sensitivity, but the mechanical scanning implementation leads to bulk and cost in the final system. Whether such systems will be viable remains to be seen.

In considering orthographic imaging we decided to investigate an electronically scanned piezoelectric system which would have no moving parts. We felt such an array ultimately would have the lowest manufacturing cost through the use of large scale integrated (LSI) circuit techniques. A square area array, read out in matrix fashion, was chosen for simplicity and commonality with signal processing and display techniques already in use. Since an insoni-fied piezoelectric element always actively produces an electrical signal, a switch at each element to disconnect the element from the read out scheme is required. In addition, we felt it important to have a stage of gain close to the element for signal-to-noise improve-ment as well as to translate the relativly high impedance of the array to a lower impedance which could drive the matrix read out line.

In the present investigation we chose to construct a receiver array. This was done for simplicity, but also because we believe that the transmitter should be optimized separately from the receiver so that neither transmitter nor receiver is compromised. This is particularly true in the case of systems which might use noncoherent or other designed waveforms with the transmitter. An 8 x 8 array was chosen to demonstrate the principles of the circuitry and read-out. We have succeeded in constructing such an array and present preliminary results of the performance in this paper. Although a considerable amount of work went into the construction of the inte-grated circuit and the production methods, no special optimization of the piezoelectric array itself was done. This awaits further development.

SURVEY OF ARRAY CONSTRUCTION TECHNIQUES

Many different micro-circuit interconnection or bonding tech-niques are in common use and were considered for use in the array construction. Depending on the method used from one (input) to four (input, output, power supply voltage and ground) connections are needed at each element. Thus, in an array of 250 x 250 elements, 62,500 to 250,000 bonds would be required. This number is clearly beyond the capabilities of hand wiring techniques except in a very few specialized cases where the cost is no object. In a production environment, LSI techniques are mandatory. Also the reliability of the bonds must be carefully examined.

Table 1 lists some common techniques for interconnecting elements in a very large array.

Discrete Array

By a discrete array we mean one whose individual elements are assembled by hand or semiautomatically using printed circuits, soldering, wire wrapping, etc., together with individual resistors and capacitors, semiconductors, and integrated circuits. This is the conventional breadboarding approach and is very costly. It would not be considered for a very large number of array elements unless the application demanded it. There are also physical limitations on the space required for the discrete devices. At long acoustical wavelengths where the element size is large ($>$5 cm) this presents no problem. With the 2 x 2 mm size chosen here, this approach has obvious limitations. The advantage is the ability to use state of the art signal processing and devices to optimize each channel.

Hybrid Array

A hybrid array is one in which either a single piezoelectric plate or semiconductor wafer is used as a substrate for many individual elements. This substrate forms the mechanical assembly of the module.

1. Piezoelectric Substrate. The use of a piezoelectric element as a substrate for active device interconnections requires materials with high Curie temperature (the temperature at which piezoelectric action is destroyed). Many of the bonding techniques require heating the substrate to 300°C or higher. Complex circuitry can be used at each element which can provide an extremely versatile array. With this approach at least four bonds per element are needed. Each chip can be tested individually before use, which is a major advantage. It should be noted that the maximum array module size is only limited by the available piezoelectric material sizes, whereas the minimum element size is controlled by the integrated circuit chip size.

a. Wire bonding uses ultrasonic or thermocompression bonds (ball and stitch bonding) resulting in a web of "flying" wires. In either case an integtated circuit in "chip" form is bonded directly to a substrate with solder or epoxy. Individual connections are made from the chip to the substrate using manual or semiautomatic equipment. This is the most widely used method in the semiconductor industry of producing the familiar Dual-Inline Package (DIP). This is a convenient method for interfacing the tiny (3 x 3 mm typical) silicon chips to printed circuits.

TABLE 1   PIEZOELECTRIC ARRAY CONSTRUCTION TECHNIQUES AND THEIR ESTIMATED RELIABILITIES

Construction Technique		Estimated Element Interconnect Failure Rates 15 %/1000 hours
DISCRETE - HAND WIRED, PRINTED CIRCUITS, ETC. (4 BONDS PER ELEMENT)		4.8×10^{-5}
HYBRID -		
PIEZOELECTRIC SUBSTRATE (4 BONDS PER ELEMENT)		
1) WIRE BONDED INTEGRATED CIRCUIT CHIPS	Thermocompression	52×10^{-5}
	Ultrasonic	28×10^{-5}
2) FLIP CHIP BONDED INTEGRATED CIRCUIT CHIPS		4×10^{-5}
3) BEAM LEAD BONDED INTEGRATED CIRCUIT CHIPS		4×10^{-5}
INTEGRATED CIRCUIT SUBSTRATE (1 BOND PER ELEMENT)		
1) WIRE BONDED PIEZOELECTRIC ELEMENTS	Thermocompression	13×10^{-5}
	Ultrasonic	7×10^{-5}
2) FLIP CHIP BONDED PIEZOELECTRIC ELEMENTS		1×10^{-5}
SANDWICH CONSTRUCTION (MACRO-FLIP CHIP) (1 BOND PER ELEMENT)		N.A.
FULLY INTEGRATED		
DEPOSITION OF SEMICONDUCTOR ON PIEZOELECTRIC SUBSTRATE		N.A.
DEPOSITION OF PIEZOELECTRIC ON SEMICONDUCTOR SUBSTRATE		N.A

Virtually all available semiconductors and integrated circuits can be purchased in chip form rather inexpensively, so that there is no limitation other than space and cost in implementing signal processing schemes. Often, however, additional passive components are required, increasing the complexity and therefore cost at the worst possible point - assembly. Most wire bonders are manually operated, and practical human performance factors become very important when large numbers of bond are required. Rework of bad circuits is a problem because of tight spacing and flying leads. Wire bonding is therefore not well suited to automation.

b. Flip-chip bonding is a very promising construction method for a large array. In the controlled collapse system, connections are made with solder bumps which contact plated interconnect wires. The substrate is heated and a heated tool is used to melt the solder. External pressure and capillary attraction seat the chip securely. A robust pillar thus supports the chip as well as replaces the two bond flying lead with only one bond. The tool-chip-substrate combination must reach 180°C to melt the solder. Three seconds is a typical bonding time.

Flip-chip bonding is quite reliable and is suitable for highly automated assembly as IBM demonstrated in their 360 series computers. Only a limited number of integrated circuits are available off the shelf as "flip-chips" so that custom fabrication is usually required. Disadvantages of flip-chip bonding are that the bonds cannot be visually inspected except with infrared viewers and that rework may be difficult.

c. Beam lead bonding is a relatively new process which was developed for ultra-high reliability systems. Beams extending from the chip are formed in the metallization process during chip fabrication rather than attached later. The substrate must be heated to 280°C to 300°C and the bonding tool to about 300°C to 500°C. A wobble bonder is used to produce a thermocompression bond on all leads of the chip in one operation.

Beam lead devices can be replaced as many as six to ten times before any degradation of the substrate metallization is apparent. At present very few integrated circuits are available in a beam lead package. The initial cost of these devices is moderately high, but when reliability and the ease of rework are considered, they become more attractive.

2. Integrated Circuit Substrate. A silicon or sapphire wafer forms the substrate with individual piezoelectric elements individually bonded and electrically connected. This requires high yields for complex circuitry compared to the piezoelectric substrate approach. It has the advantage of very simple bonding requirements since only one bond per element is needed. Ultrasonic, flip-chip and conductive epoxy bonding methods are all useful. For mechanical strength the bond area itself must occupy a relatively large portion of the element area.

All these hybrid methods suffer from the problem that discrete
elements must be processed individually and hence this method
is not truly an LSI circuit technique.

"Sandwich" Construction

In a sandwich construction a large number of circuit elements
on a single wafer are simultaneously bonded to a similar size
piezoelectric wafer. The piezoelectric wafer is diced or made into
a mechanically isolated array using other techniques. A flip-
chip bonding solder bump is placed on each element of the array
with complementary bonding pads on the integrated circuit wafer.
The two large wafers are then bonded together in a vacuum under
heat and pressure. Since this is the method we chose in this
array construction we will defer further discussion until later
in the paper.

Fully Integrated Arrays

By a fully integrated array we mean one in which the entire
module including the piezoelectric material, circuitry and interconnections are made using LSI technology. No mechanical bonding
operations are used. This has the obvious advantage of simplicity
and low cost if the yield is high enough.

1. <u>Direct Deposition of Silicon</u>. An investigation of deposition
of silicon on Lithium Niobate indicated that high temperature
(750°C) vapor phase deposition, i.e., heteroepitaxy, is not
practical with existing equipment and techniques, since a reaction
takes place during the deposition cycle. This reaction produces
a discoloration of the $LiNbO_3$ by hydrogen reduction and leads to
a doping of the silicon film by either Lithium or Niobium. At
lower temperatures film mobility deteriorates and electron beam
evaporated films are preferred. This was demonstrated by deposition of silicon at 500°C with resulting electron mobilities
of 10 cm^2/Vsec. This mobility is low compared to single crystal
values (300 to 600 cm^2/Vsec) for p-type silicon and is in the
range of mobilities obtained with other polycrystalline semiconductors. It is not sufficient for the active devices in the array.

2. <u>Direct Deposition of Piezoelectric Material</u>. Direct
deposition of the piezoelectric material on a silicon or silicon
on sapphire substrate has been achieved by many investigators for
surface acoustic wave (SAW) applications. Successful materials
include Cadmium Sulphide[16] which suffers from stability problems,
Zinc Oxide[17] and Aluminum Nitride[18] which appear to be very suitable
materials. This approach has the advantage of producing an array
using only vacuum sputtering techniques for growing crystals in
addition to conventional diffusion techniques. The piezoelectric
materials deposited to date are very thin (on the order of a few
microns thick at most). The pressure sensitivity of a thickness
mode transducer is a function of the thickness of the

piezoelectric material. Typical material thicknesses for discrete, hybrid or sandwich transducers are on the order of 1 mm. Thus a thin film piezoelectric is 40 to 60 dB less sensitive. This makes such devices impractical for most imaging applications. Perhaps future work in the sputtering technology of piezoelectric crystals will result in the ability to sputter 100-500 micron thickness layers. If this becomes a reality this array construction method would be the one of choice.

Reliability of Interconnection

In assessing various interconnecting technologies, reliability is a major consideration since 62,500 to 250,000 bonds are needed. Published figures[15] (Table 1) can be used as a relative guide to reliability. Note that the flip-chip and beam lead techniques have an estimated reliability slightly better than that of hand wired circuits, whereas the ultrasonic and thermal compression bonding methods are considerably poorer. With a total of 62,500 elements and a reliability of $1 \times 10^{-5}\%$ per 1000 hours a failure rate of under 1% per 1000 hours is achievable. With methods requiring 4 bonds per element, this failure rate is correspondently larger. This is another major factor in the choice of the sandwich or macro flip-chip construction.

SELECTION OF THE PIEZOELECTRIC MATERIAL FOR THE ARRAY

There are many piezoelectric materials available today with different parameters to be considered and traded-off against each other. Table 2 shows the physical properties of some potential transducer materials. The materials other than Lithium Niobate listed in the chart are ceramics and can be expected to exhibit more variation in piezoelectric properties from point to point than a single crystal material such as $LiNbO_3$.

Depending on the construction technique chosen the Curie temperature (the temperature at which piezoelectric activity ceases or the ceramic becomes depolarized) can become the most important single consideration. In flip-chip bonding relatively high temperatures (200°C) must be used for a reliable connection and materials such as PZT5H and PZT5A are not useful. Lead metaniobate ($PbNbO_3$) has a relatively high Curie temperature and has been used by other investigators in construction of an area array[13]. Another very common piezoelectric material for high frequency surface acoustical wave applications is Lithium Niobate ($LiNbO_3$) which has a very high Curie temperature. $LiNbO_3$ has a disadvantage due to its anisotropic expansion which varies 5 to 1 between the X and Z directions. This thermal expansion leads to serious problems in construction.

Table 3 shows the materials parameters applied in figures of merit which relate to performance and other parameters of the array. These arrays are intended to be operated well below the

TABLE 2   PHYSICAL PROPERTIES OF TRANSDUCER MATERIALS

	$PbNb_2O_6$ [1]	$LiNbO_3$ [2]	$PZT5A$ [3]	$PZT5H$ [3]
Elastic Modulus C^D 10^{10} N/m^2	6.8×10^{10}	20	10.57	11.12
Density 10^3 kg/m^3	5.5×10^8	4.64	7.75	7.5
Piezoelectric Constant g 10^{-3} Vm/N	22×10^{-3}	28	24.8	19.7
Coupling Constant k %	35	30	71	75
Relative Dielectric Constant ϵ^T	800	84	1700	3400
Curie Temperature $^\circ$C	550	1210	365	193
Mechanical Q Qm (H_2O load)	15	33	75	65
Velocity of Sound $\sqrt{\dfrac{c^D}{\rho}}$ m/sec	3516	6837	3693	3850
Thermal Expansion 10^{-6} $\Delta l/l$	N.A.	X 16.7 $\times 10^{-6}$ Z 3.2	2.5	2.5

1. Keramos, Inc., Lizton, Indiana 46149
2. Crystal Technology, Mountain View, California 94060
3. Vernitron, Bedford, Ohio 44146

TABLE 3 THICKNESS MODE TRANSDUCER MATERIALS – RELATIVE FIGURES OF MERIT

	$PbNb_2O_6$	$LiNbO_3$	PZT5A	PZT5H
Relative Thickness $$t f_o \propto \sqrt{\frac{C^D}{\varepsilon}}$$	1	1.87	1.05	1.09
Relative Capacitance $$\frac{C}{A f_o} \propto \varepsilon / \sqrt{\frac{C^D}{\varepsilon}}$$	1	0.05	2	3.8
Relative Pressure Sensitivity $$\frac{V f_o}{P} \propto \frac{\sqrt{\frac{C^D}{\varepsilon}}}{1 + C^D \varepsilon \varepsilon_o g^2}$$	1	2.3	0.7	0.5

Where V = Output Voltage   A = Element
       p = Pressure   C = Capacitance
       f_o = Resonance Frequency   t = Element Thickness

fundamental resonance frequency of the piezoelectric plate for
uniform sensitivity and minimum phase shifts across the array.
At high frequencies such as 2-5 MHz, the thickness of the plate can
become very small, leading to mechanical strength or handling difficulties. A relative thickness figure of merit is then expressed
as the thickness times the resonance frequency. $LiNbO_3$ proves to
have a significant advantage over the other piezoelectric materials
because of its high velocity of sound resulting in a thicker plate
for a given resonance frequency.

The capacitance figure of merit shows that PZT5A and PZT5H
have a very high dielectric constant and that $LiNbO_3$ is extremely low
in comparison. In our array this was not a problem since field
effect transistors have an extremely high input impedance. The
pressure sensitivity figure of merit again shows that $LiNbO_3$ has
a significant advantage over the other materials, primarily
because the thickness in the form of velocity of sound appears in
the numerator of the expression. We decided to use $LiNbO_3$ in the
construction of our array despite the problems associated with
thermal expansion which we were able to overcome.

Figure 1 is a photograph and cross-sectional schematic drawing
of the piezoelectric array. An aluminum electrode is vacuum
plated onto both sides. The elements are separated on one side by
precision cutting with an 0.001 inch wide diamond saw. The other
side is continuous and forms the ground electrode. Each transducer
element has a raised mating pad approximately 0.0001 inch high
attached to it. In a vacuum, under heat and pressure, this pad is
brought into contact with a similar pad on the integrated circuit
to make an electrical connection.

The element size (2 x 2 mm) was chosen to provide high spatial
resolution in an image and to have a reasonably large capacitance
($\sim$ 4 pf), yet allow enough area for circuits.

The voltage sensitivity of the element as a receiver at a
frequency less than the resonance frequency is given by[19]

$$\frac{V_1}{P} = \frac{g_{22}t}{1 + C^D \epsilon \epsilon_0 g_{22}^2} \simeq g_{22}t \tag{1}$$

Where V_1 is the output voltage - volts
     P is the acoustic pressure amplitude - N/m^2
     g_{22} is the piezoelectric constant - Vm/N
     t is the element thickness - m
     C^D is the elastic modulus - N/m^2
     ϵ is the relative dielectric constant
     ϵ_0 is the permittivity of free space - Farad/m

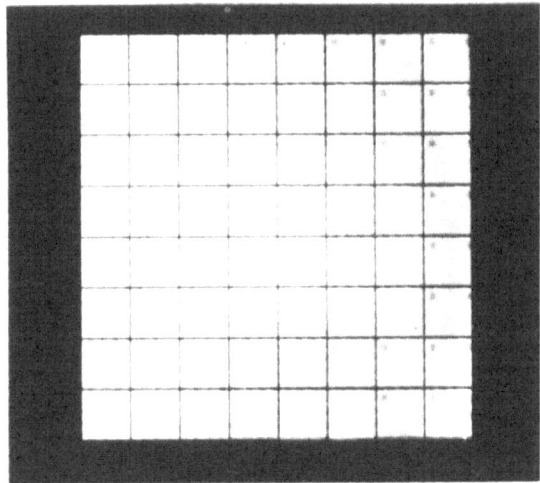

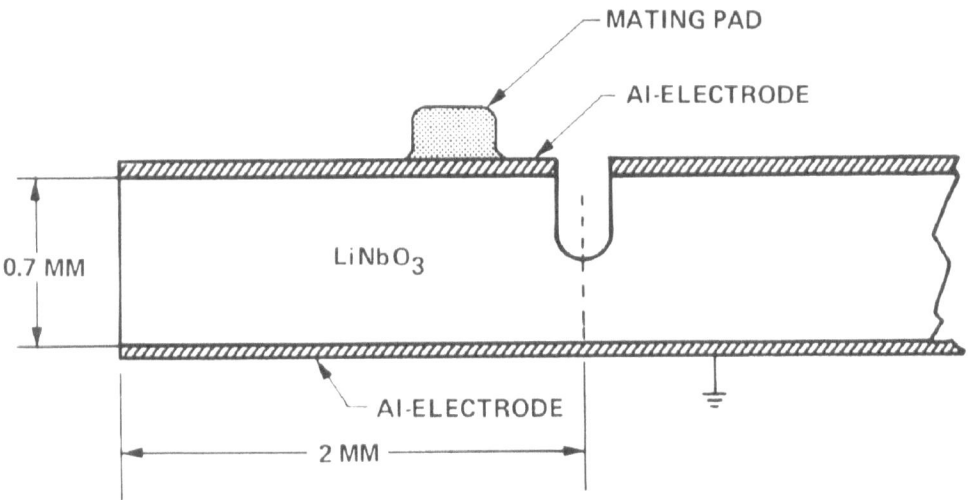

Figure 1. Lithium-Niobate 8 X 8 Transducer Array and Cross-section

The approximation is valid for LiNbO$_3$ since the relative dielectric constant is relatively low and $C^D \epsilon \epsilon_o g^2 \ll 1$. The predicted sensitivity for a thickness of 0.7 mm is -115 dBV/micro Pascal.

INTEGRATED CIRCUIT

Silicon on sapphire is used for the metal-oxide semiconductor transistor (MOST) to avoid capacitive coupling which degrades high frequency response. An N-channel deep depletion mode device (normally on) is used to avoid biasing problems. Figure 2 is a photograph and electrical schematic of the circuit. A dual gate structure is used which operates as a high gain "tetrode" with the gate T$_1$ grounded. With a negative bias applied to this gate, the amplifier is turned off. The function of resistor R ($\sim$1 megohm) is to drain off any charge which may build up on the gate.

The dual gate "tetrode" structure is an essential part of the design since it improves the high frequency response substantially. It provides a 15 dB gain with a 15k ohm load resistor. Since there is no Miller effect capacitance multiplication because of the cascode arrangement, the .05 pf feedback capacitance allows flat frequency response up to 5 MHz. The X-Y readout of each element is readily adapted to external electronics to present an image display. It should be noted that both amplitude and phase are available for all elements of the matrix at the output terminals for further signal processing.

Transducer Amplifier Analysis

A high gain dual gate deep depletion mode insulated gate field effect transistor (IGFET) on silicon on sapphire was selected to give the required high frequency response and for addressing and readout of the array elements. Figure 3 shows the details of the device cross section in schematic form, together with the voltage current characteristics. The "normally-on" deep-depletion mode MOST in saturation has a drain to source current Ids[20].

$$I_{DS} = \frac{\mu W \epsilon_0}{L} \left(\frac{\epsilon_i \epsilon_s}{\epsilon_i a + \epsilon_s t_{ox}} \right) \left[-\frac{q N_D a}{\epsilon_0} \left(\frac{\epsilon_i a + \epsilon_s t_{ox}}{\epsilon_i \epsilon_s} \right) + V_t - V_g \right]^2 \tag{2}$$

Where μ is the effective channel mobility, W,L are the channel width and length, ϵ_o is the permittivity of free space, ϵ_i, ϵ_s are the relative dielectric constant of the oxide insulator and the semiconductor, a is the semiconductor layer thickness, t_{ox} is the oxide thickness, q is the electron charge, N$_D$ is the donor concentration of heteroepitaxial SOS layer, V$_t$ is the threshold voltage due to the surface state layer between oxide layer and silicon, and V$_g$ is the applied external gate voltage.

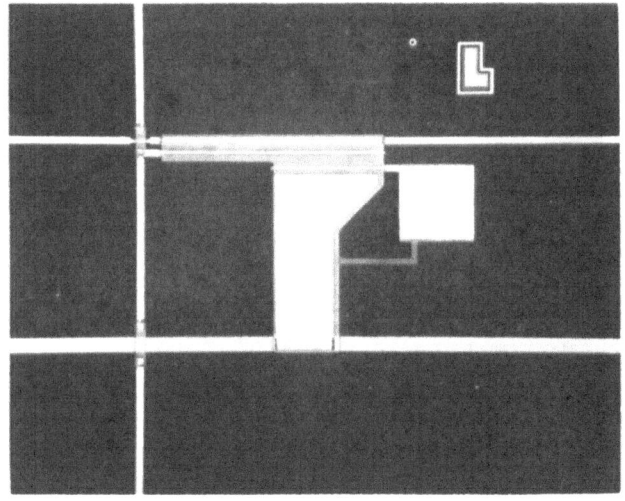

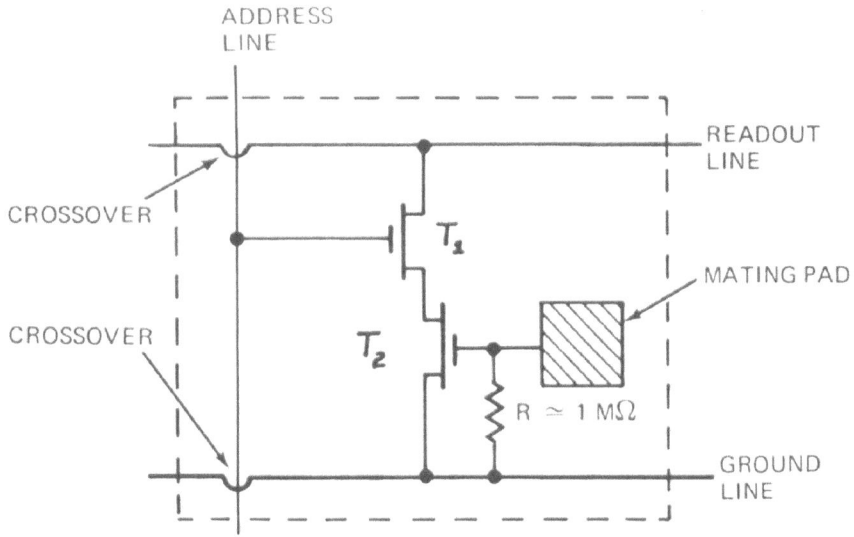

CIRCUIT DIAGRAM OF ONE MATRIX ELEMENT

Figure 2. Array Element Electronics

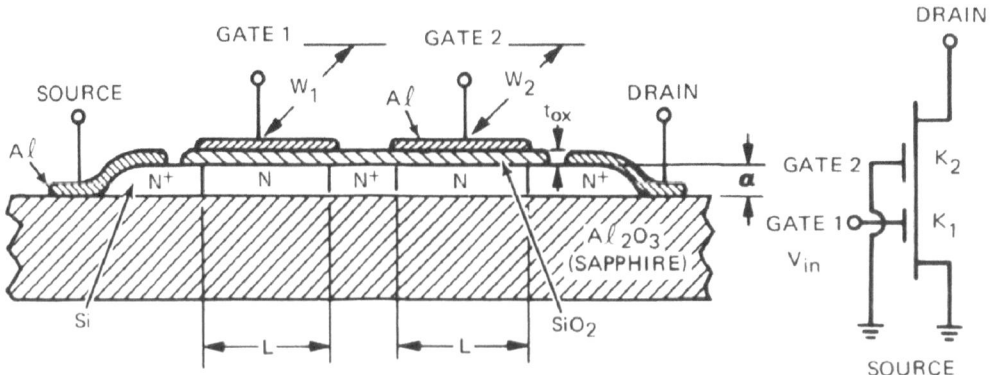

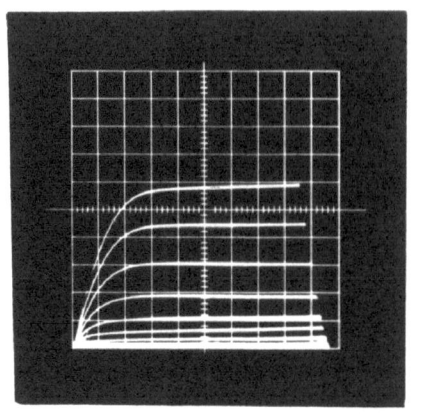

**VOLTAGE-CURRENT CHARACTERISTIC
OF DUAL GATE MOS–SOS**

Figure 3. Details of the Array Element Electronics

Equation 2 can be expressed in the form

$$I_{DS} = \frac{\mu W C_g'}{L}\left[V_0 - V_g\right]^2 = K'\left[V_0 - V_g\right]^2 \tag{3}$$

Where C_g' is the effective gate capacitance per unit area, and V_0 is the built-in voltage taking into account the ionized impurity charges in the n-type substrate material and the surface state charges.

For the dual gate or cascode arrangement the factor $K' = K_1 K_2/(K_1+K_2)$, where K_1 and K_2 are assigned to the geometry of each independent MOST device [21]. This type of device operation is advantageous in the array design because the I_{DS}-value is fixed by geometrical and material parameters. In addition, it does not require an external bias in the quiescent state; i.e., $V_G = 0$. With gate 2 grounded in our design (Figure 3), the I_{DS}-value was selected to be 1 mA at $V_{GS} = 0$. The transconductance of the device is then equal to

$$\tag{4}$$

$$g_m = \frac{dI_{DS}}{dV_{GS}} = -2K'\left[V_0 - V_g\right] \simeq 0.5\, mA/Volt$$

Since the piezoelectric element capacitance C_p is in series with the input capacitance of the MOST, C_p should be much larger than C_g to acheive high charge transfer to the gate capacitance and not impair high frequency response of the amplifier.

This voltage division effect is given by

$$V_{in} = \frac{V_1 C_p}{C_p + C_g} \tag{5}$$

Where V_{in} is the effective input voltage to the MOST and V_1 is the voltage generated by the transducer (Eq. 1).

In our array $C_p \sim 4$ pf and $C_g \sim 1$ pf, so V_{in} is reduced by 2 dB. With acoustical input signals the MOST built-in voltage, V_0, now has an additive term, V_{in}, and the ratio of current change with input signal becomes

$$\frac{\Delta I_{DS}}{I_{DS}} \simeq 2\left(\frac{V_{in}}{V_0}\right) + \left(\frac{V_{in}}{V_0}\right)^2 \tag{6}$$

A linear relation between piezoelectrically induced current and voltage is present for $V_{in} \ll V_0$, which is the case for most practical applications. Similar performance analyses have been applied to polycrystalline piezoelectric field-effect transducers[16,17,22]. With the aid of the transconductance from Eq. 4, the voltage amplification of one stage with a load-resistance of R_L is given by

$$G_V = \frac{V_{OUT}}{V_{in}} = g_m R_L = \frac{\mu W C_g' V_0 R_L}{L} \qquad (7)$$

For typical values in our design $W/L=50$, $V_0=4$ volts, $R_L=15k$ ohm, $\mu =10^3$ cm^2/Vsec, resulting in a gain of 17.5 dB. With the decrease due to input capacitance this is reduced to about 15 dB. The load resistance must be less than 20k ohm for frequency response up to 5 MHz because of capacitive loading of the amplifier stage output from interconnections on the sapphire substrate. For improved circuit performance an impedance converter stage could be added to each readout row on the silicon on sapphire substrate. In addition, the multiplexing (scanning) circuitry can be integrated into the module.

ADDRESSING AND READ-OUT METHOD

Figure 4 is a photograph of the piezoelectric array along with photograph of the complementary silicon on sapphire substrate with completed circuitry ready for mating. Grounding a given column, such as Y_1 and applying a -5V bias (since $V_0=4$ volts) to turn off columns Y_2 - Y_8, makes the output of elements 1 - 8 in column 1 available at the right-hand edge of the array. With the aid of multiplexing circuits, the entire array can be read out.

The completed module is shown schematically in cross-section in Figure 5. The piezoelectric plate is flip-chip bonded to the silicon on sapphire substrate and the device is enclosed in a waterproof container for the experimental verification of its performance. Figure 6 is a photograph of the completed array.

EXPERIMENTAL RESULTS

Figure 7 shows the simple experimental arrangement used to measure some performance characteristics of the array. A standard highly focused 2.25 MHz transducer was driven with a gated sine wave burst at 2.25 MHz and directed at a single element of the array. The response shown in Figure 7 has spurious signals following the incident acoustic pulse. We believe these signals are surface waves travelling across the piezoelectric element. As mentioned earlier, no attempt was made to optimize the performance of the piezoelectric portion of the system. The transducer beam being highly focused had a focal spot size less than the 2 mm square element. Measurements of cross talk were then made by

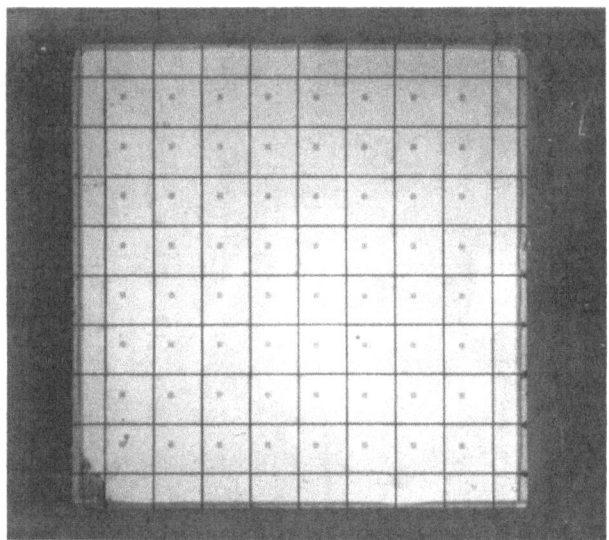

**8 x 8 LiNbO$_3$ ARRAY WITH MATING PADS AND 2 x 2 mm^2 ELEMENT
SEPARATION (3X)**

ADDRESS COLUMNS   Y$_1$ Y$_2$ Y$_3$ Y$_4$ Y$_5$ Y$_6$ Y$_7$ Y$_8$

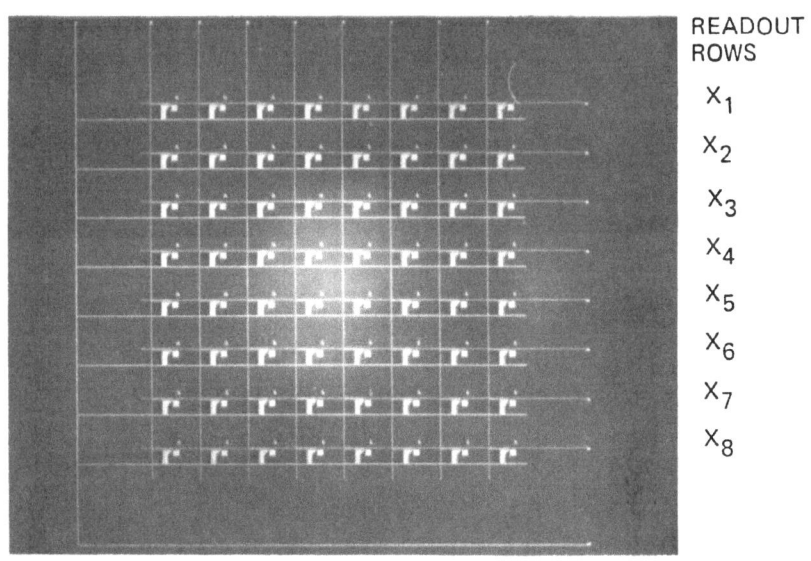

READOUT
ROWS

X$_1$

X$_2$

X$_3$

X$_4$

X$_5$

X$_6$

X$_7$

X$_8$

**8 x 8 ARRAY ELECTRONICS BEFORE MATING WITH
THE LiNbO$_3$ SUBSTRATE (3X)**

Figure 4. Array Wafers Before Bonding

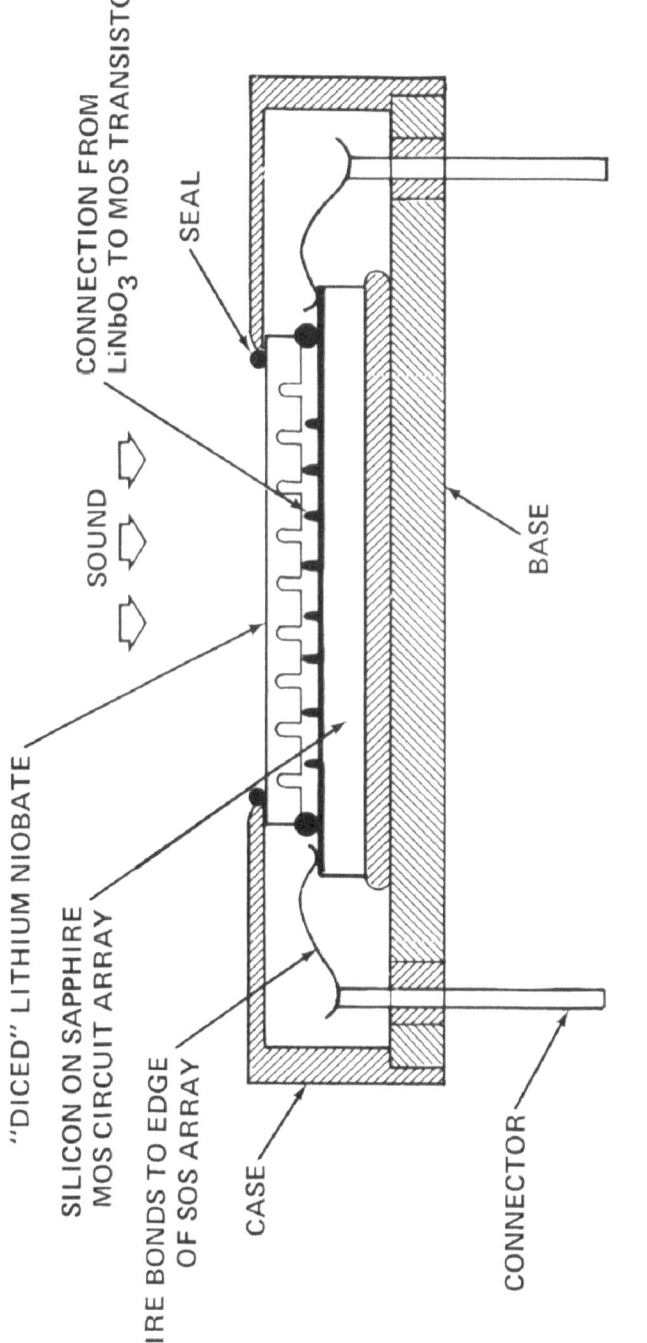

"DICED" LITHIUM NIOBATE

SILICON ON SAPPHIRE
MOS CIRCUIT ARRAY

WIRE BONDS TO EDGE
OF SOS ARRAY

CASE

CONNECTOR

CONNECTION FROM
LiNbO$_3$ TO MOS TRANSISTOR

SEAL

SOUND

BASE

Figure 5. Cross-section of 8 X 8 Element Piezoelectric Array Module-
Laboratory Model

Figure 6. Packaged Array Module

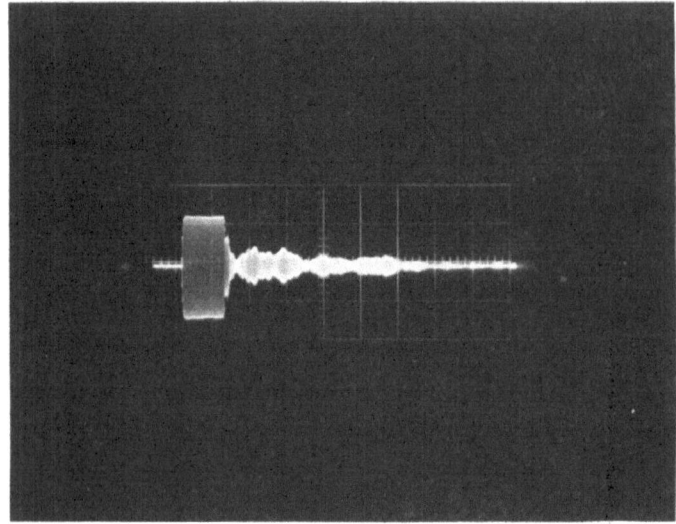

VERTICAL 0.5 V/DIV
HORIZONTAL 20 μSEC/DIV
f $\simeq$ 3 MHz (PULSED)

**SIGNAL RESPONSE OF SINGLE ELEMENT
OF 8 x 8 ARRAY**

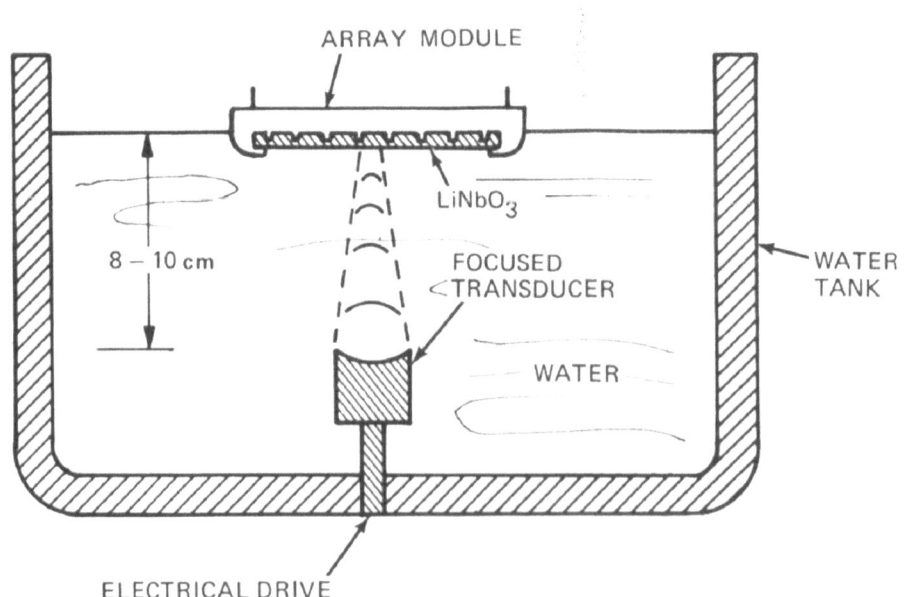

Figure 7. Experimental Arrangement and Test Results

reading out signal levels from adjacent elements. Total cross talk was approximately 20 dB down. This cross talk was a combination of mechanical and electrical coupling.

A second very large (15 x 15 cm) source was then used to measure the uniformity of the array. This source was driven with a gated sine wave at 1 MHz and was placed approximately 10 cm from the array. The array was thus in the extreme near field or uniform plane wave region of the large transducer. This minimized variations in acoustic input due to the large transducer.

The highest sensitivity measured was -115 dBV per micro Pascal in agreement with the predicted value. The variation in sensitivity across the array had a standard deviation of 4 dB from the mean. Sensitivity variations in the array were at worst a factor of about 3 to 1 and 58 of the 64 elements were found to be active. The dynamic range was measured to be approximately 60 dB and the minimum detectable acoustical intensity (due to electronic noise) was measured to be 27 nano Watts per square cm. Electronic read out methods are being developed at this time to allow simulation of an orthographic imaging situation.

SUMMARY

We have demonstrated construction methods which are applicable to a large number of array elements in a production environment. The number of elements in this array can be increased tremendously without a great deal of difficulty with the advent of 3" diameter wafer silicon on sapphire processing techniques in the semiconductor industry. It is now possible to build as many as 50 x 50 elements in a single module. The cost of the modules, of course, depends on the yield in the semiconductor processing. We hope in the near future to be able to demonstrate such a large array module. This work is an ongoing part of the Internal Research and Development (IR&D) activities of the McDonnell Douglas Corporation.

ACKNOWLEDGEMENTS

We would like to thank Mr. John R. Moore, President of the Actron Division of McDonnell Douglas, for his encouragement and foresight in supporting this project. We would also like to thank Paul Friebertshauser, Stan Watanabe and Don Dion for their assistance in the construction of the array.

REFERENCES

1. R. K. Mueller and P.N. Keating, "The Liquid-Gas Interface as a Recording Medium for Acoustical Holography" in Acoustical Holography, Vol. 1, A.F. Metherell et al, Eds., Plenum Press N.Y., 1969, Ch. 3, pp. 49-55.

2. B.B. Brenden, "A Comparison of Acoustical Holography Methods" in Acoustical Holography, Vol. 1, A.F. Metherell et al, Eds., Plenum Press N.Y., 1969 Ch. 4, pp. 57-71.

3. D.R. Holbrook, E.E. McCurry and V. Richards, "Medical Uses of Acoustical Holography", in Acoustical Holography, Vol. 5, P.S. Green, Ed., Plenum Press, N.Y., 1974, pp. 415-451.

4. A.F. Metherell, K.R. Erikson, J.E. Wreede, R.E. Norton, R.E. Greer and R.M. Watts, "A Medical Imaging Acoustical Holography System Using Linearized Subfringe Holographic Interferometry", in Acoustical Holography, Vol. 5, P.S. Green, Ed., Plenum Press, N.Y., 1974, pp. 453-470.

5. R.S. Mezrich, K.R. Etzold and D.H.R. Vilkomerson, "System for Visualizing and Measuring Ultrasonic Wavefronts" RCA Review 35, pp. 483-519 (1974).

6. P.Alais "Acoustical Imaging by Electrostatic Transducers" in Acoustical Holography, Vol. 4, G. Wade, Ed1, 1972, pp. 237-249.

7. W.R.Fenner and G.E. Stewart, "An Ultrasonic Holographic Imaging System for Medical Applications" in Acoustical Holography, Vol. 5, P.S. Green, Ed., Plenum Press, N.Y., 1974, pp. 493-503.

8. P.S. Green, L.S. Schaefer, E.D. Jones and J.R. Suarez, "A New High-Performance Ultrasonic Camera", in Acoustical Holography, Vol. 5, P.S. Green, Ed., Plenum Press, N.Y., 1974, pp. 493-503.

9. J.E. Jacobs, "Ultrasound Image Converter Systems Utilizing Electron-Scanning Techniques", IEEE Trans., Sonics and Ultrasonics, SU-15, pp. 146-152 (1968).

10. G.L. Sackman and R.J. Larkin, "An Electronically Scanned Transducer Array Using Micro Circuit Devices" in Acoustical Holography, Vol. 3, A.F. Metherell, Ed., Plenum Press, N.Y., 1971, pp. 211-223.

11. N.O. Booth and J.L. Sutton, "Holographic Acoustic Imaging" Naval Undersea Center, San Diego, CA, NUC-TP 424.

12. E. Marom, R.K. Mueller, R.F. Koppelmann, G.Z.Zilinskas, "Design and Preliminary Test of an Underwater Viewing System Using Sound Holography" in Acoustical Holography, Vol. 3, A.F. Metherell, Ed., Plenum Press, 1971, pp. 191-209.

13. M.G. Maginness, J.D. Plummer, and J.D. Meindl, "An Acoustic Image Sensor Using a Transmit-Receiver Array" in Acoustical Holography, Vol. 5, P.S. Green, Ed., 1974, pp. 619-631.

14. D.H.R. Vilkomerson, "Analysis of Various Ultrasonic Holographic Imaging Methods for Medical Diagnosis", Acoustical Holography, Vol. 4., G.Wade, ED., 1972, pp. 401-429.

15. J.E. McCormick "On the Reliability of Microconnections" Electronic Packaging and Production, pp. 167-168, June 1968.

16. R.S. Muller and J. Conragan, "Transducer Action in a Metal-Insulator-PIezoelectric-Semiconductor Triode". Appl. Phys. Lett., 20, pp. 156-158 (1972).

17. E.W. Greeneich and R.S. Muller, "Acoustic-Wave Detection via a Piezoelectric Field-Effect Transducer". Appl. Phys. Lett., 20, pp. 156-153, (1972).

18. J.K. Liu, R.B. Stokes and K.M. Lakin, "Evaluation of AIN Films on Sapphire for Surface Acoustic Wave Applications", Proc. IEEE, 1975 Ultrasonics Symposium, 75 CHO 994-450 IEEE, N.Y., pp. 234-237 (1975).

19. T.F. Heuter and R.H. Bolt, Sonics Wiley, N.Y., Ch. 4, 1955.

20. J.S.T. Huang, "Characteristics of a Depletion-Type IGFET", IEEE Trans., ED-20, pp. 513-514 (1973).

21. R.S. Ronen and L. Strauss, "The Silicon on Sapphire MOS Tetrode Source Small Signal Features, LF to UHF", IEEE Trans. on Electron Devices ED-21, 100-109, (1974).

22. R.S. Muller and J. Conragan, "A Metal-Insulator-Piezoelectric Semiconductor Electromechanical Transducer", IEEE Trans. on Elect. Dev. ED-12, pp. 590-595, (1965).

AN EXPERIMENTAL DIGITAL BEAMFORMING SONAR

L.C. Granger, P.L. Greene, G. Zilinskas

Bendix Electrodynamics Division
North Hollywood, California

INTRODUCTION

Previous experimental results using a 250-kHz acoustic imaging system show the potential usefulness of holographic processing in high resolution sonars.[1,2] This paper presents the design and preliminary experimental results of a holographic sonar using a linear array. As presently configured, the system consists of a 128-element line array, a single pre-amplifier/sine-cosine channel, a cylindrical acoustic projector, an Altair 8800 digital microcomputer, and an HP 1335A storage display.

The elements in the hydrophone array are evenly spaced over an 89 cm aperture giving a resolution of 0.54^0 at 178 kHz and a 64^0 azimuth field of view.

The system is currently operated in a serial mode with the hydrophone outputs switched one at a time into the sine/cosine processing channel and stored in the computer memory. Thus, 128 separate T/R cycles are required to accumulate the data for a line image. The stored 128 complex data words are used as the input to a 128 point complex Fast Fourier Transform (FFT) that performs the reconstruction of the image. This FFT is presently done in software. The 128 beam outputs are used to form an intensity modulated image on the storage display. The system is range gated, so that images from increasing ranges are displayed up along the Y-axis of the CRT.

The two main purposes of this work are to explore fast holographic processing algorithms, allowing high range resolutions, and to experiment with complex objects as acoustic targets.

SYSTEM FUNCTION

In normal sonar beamforming, the beampattern for a direction θ is formed by taking the complex signal from each element k, $A_k \exp(j\psi_k)$, and multiplying it by the phase shift representing the correct delay (and gain G_k if desired). This term is $G_k \exp(-j \cdot 2\pi/\lambda \cdot kd \sin \theta)$. All the signals are then summed. (The phase shift is often performed by delay lines.)

The sum is:

$$S(\theta) = \sum_{n=0}^{N-1} G_n A_n \exp \left(j\psi_n - j \frac{2\pi}{\lambda} nd \sin \theta \right) \tag{1}$$

If the look directions "θ_ℓ" are chosen to evenly cover the field of view with N beams, the directions θ_ℓ are defined as:

$$\sin \theta_\ell = \frac{\ell}{N/2} \sin \beta = \frac{\ell}{d} \frac{\lambda}{N} \qquad \text{for} \quad = \frac{-N}{2}, \ldots \frac{N-1}{2} \tag{2}$$

and β is half the field of view.

Thus equation 1 becomes:

$$S(\theta) = \sum_{n=0}^{N-1} (G_n A_n e^{j\psi_n})(e^{-j \frac{2\pi n\ell}{N}}) \tag{3}$$

which is a standard form of a Discrete Fourier Transform applied to the signal vector $(G_n A_n e^{j\psi_n})$.

Thus the holographic processing is equivalent to standard sonar beamforming. The Fourier Transform is performed in the digital domain in this system.

SYSTEM DESCRIPTION

Hydrophone Array

The hydrophone array elements are diced lead zirconate titanate sections. The dicing is required to insure that the elements operate as independent longitudinal vibrators without distorting interactions. This requires a 0.006 inch thick diamond blade for minimum material removal combined with a reduced cutting speed to avoid excessive heat which depoles the ceramic. A thin copper mesh is bonded to the front and back silvered faces of the ceramic to provide electrical contact. The 128 elements are assembled into the 89 cm wide array by bonding a strip of corprene decoupling material to their backs and a neoprene window to the face of the full array mounted in an aluminum frame. The element sizes are chosen so that they have a beampattern of 16^0 by 64^0 to match the system field of view.

Acoustic Projector

The acoustic projector insonifies objects within a 64 degree horizontal by 16 degree vertical sector. The projector consists of a 1-1/16 inch by 7 inch band of 1-1/16 inch by 1/8 inch diced ceramics curved to a radius of 6-5/32 inches.

A computer program for calculating the beampattern function of arc length and radius of a cylindrical section was used to optimize the array pattern for the desired beam width. Putting the projector elements on a cylindrical surface in this manner greatly increases the amount of active surface area and the power handling capability over conventional methods used to produce the same beam widths. The projector beampatterns are shown in Figure 1.

Eight horizontal rows of diced sections have thin silver strips bonded to the front and a thin copper mesh bonded to the back for electrical contact and vertical shading. As in the hydrophone array, the elements are backed by corprene with a neoprene window bonded to the front and mounted in an aluminum frame.

Channel Processor

The system is currently configured with one channel processor unit to which the element outputs are switched serially,

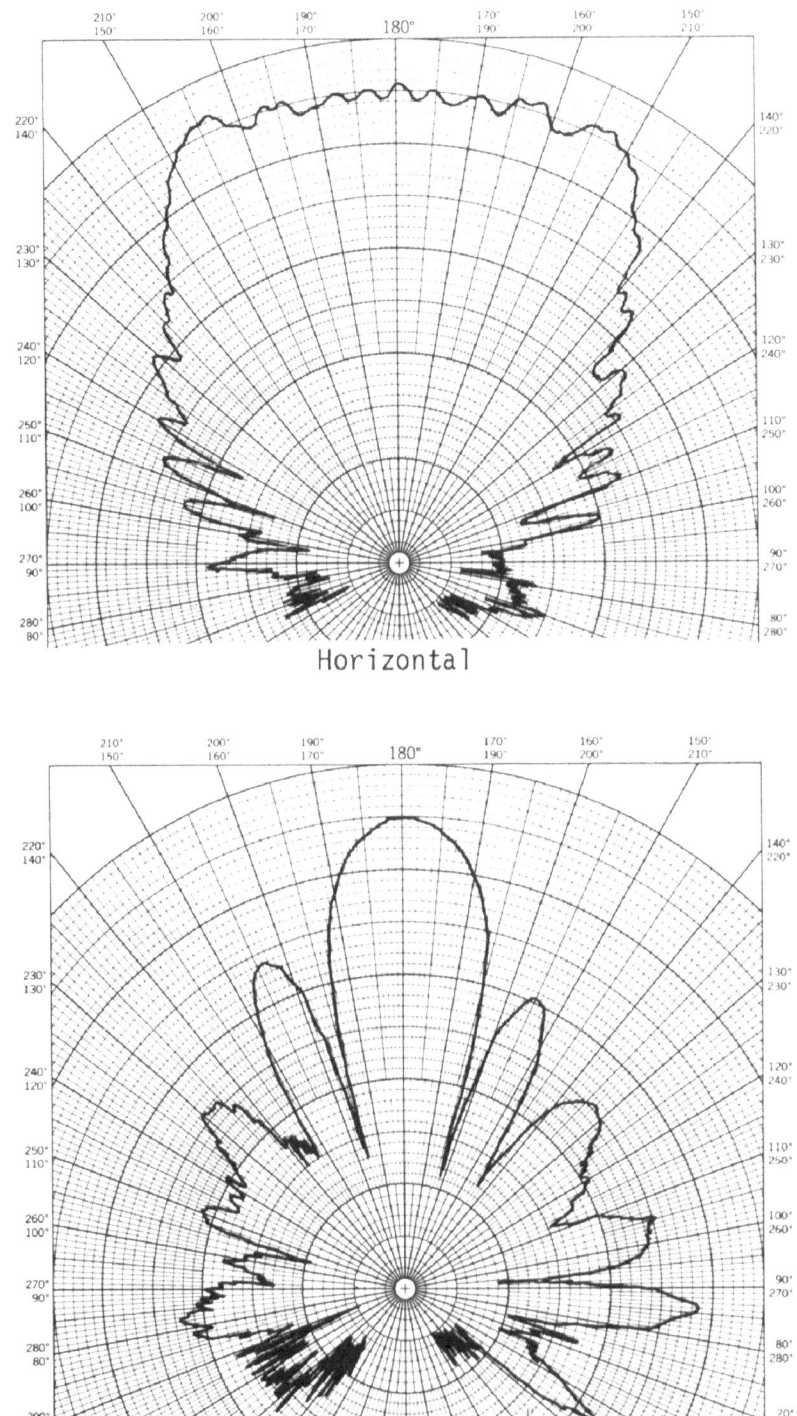

Figure 1. Measured Projector Beampatterns

one per T/R cycle. Using CMOS analog multiplexers, the microcomputer chooses one of the 128 hydrophone elements and transmits
the signal up the cable to the preamp and time variable gain
amplifier (TVG). The TVG amplifier compensates for absorption
and spherical spreading losses as a function of range.

The reference multipliers produce the sine and cosine
functions of the sum and difference frequencies. Upon integration, the sine and cosine of the difference frequency is
obtained, which is then digitized, multiplexed and passed to
the microcomputer. Figure 2 shows a block diagram of the channel processor.

Microprocessor System Controller

An Altair 8800 is used to perform the system controlling
functions. The desired range and the range increments are input
to the computer through a standard teletype. When the system
is started, it erases the display and sets the multiplexer for
the first hydrophone element. Then it outputs a word to the
sonar interface which contains the current range information and
which initiates a transmit/receive (T/R) cycle. When data has
been received and processed by the channel processor, the digital word is read and input to the microprocessor memory. Then
the multiplexer is incremented and the T/R cycle is initiated
for the next element and so on until the 128 elements have each
output some data.

Once the data from a complete line is ready, they are multiplied by the spherical focussing factors for a given range.
The Fast Fourier Transform is then performed and the data output to a display that is also under microprocessor control.

The range is then incremented and the process repeated
for another line. Figure 3 shows a flowchart for the microcomputer program.

Beamformer

The beamforming is performed digitally by the Altair 8800.
Equation 3 shows the basis for using the Discrete Fourier Transform.

An FFT (Fast Fourier Transform) program was written using
an in-place algoritihm with natural order input and scrambled
output. The program follows a flow-chart given by Brigham[3].
The 8-bit data is transformed into 12-bit two's complement form
for ease of computation and the FFT 'butterfly' is performed

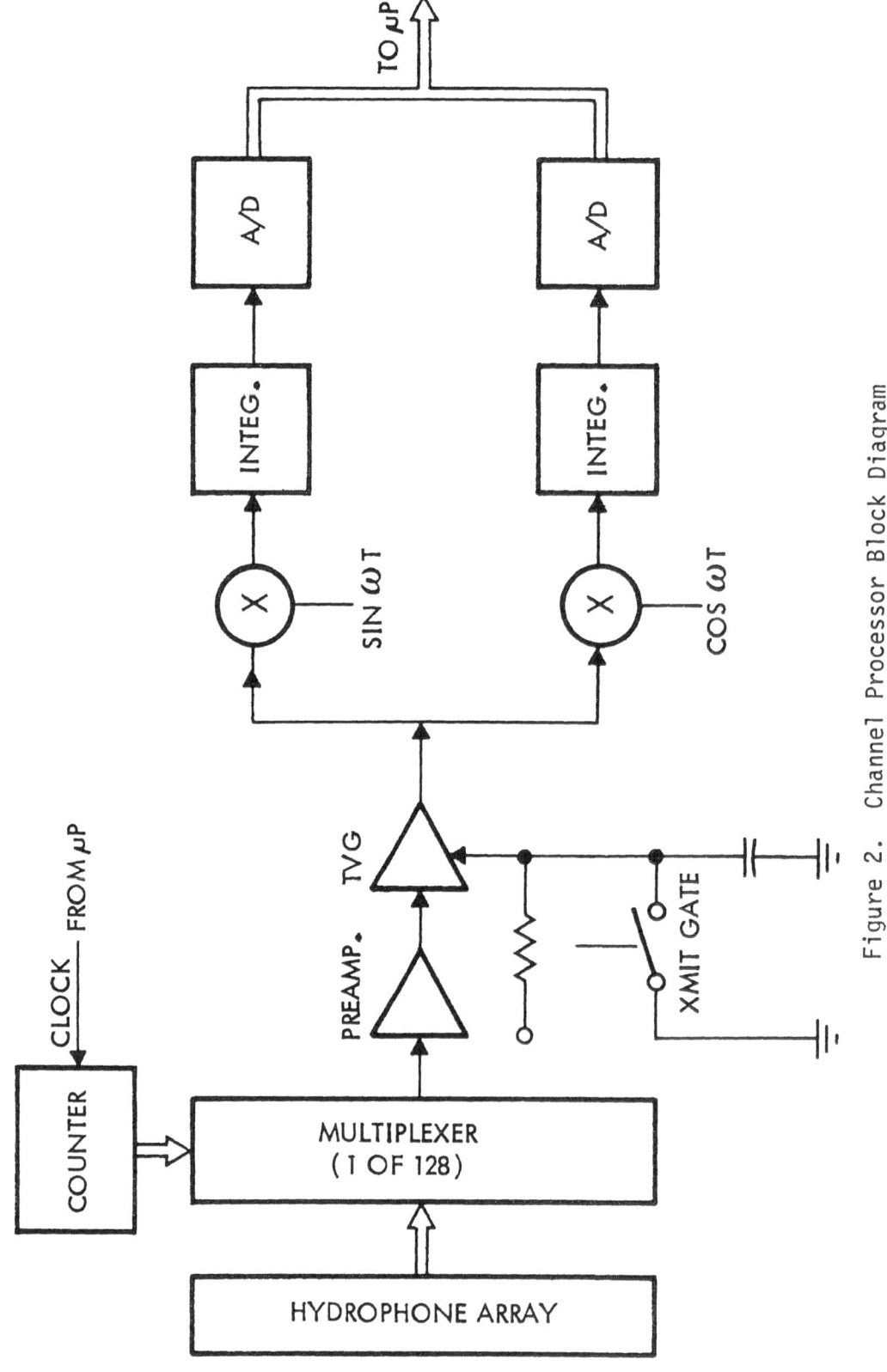

Figure 2.   Channel Processor Block Diagram

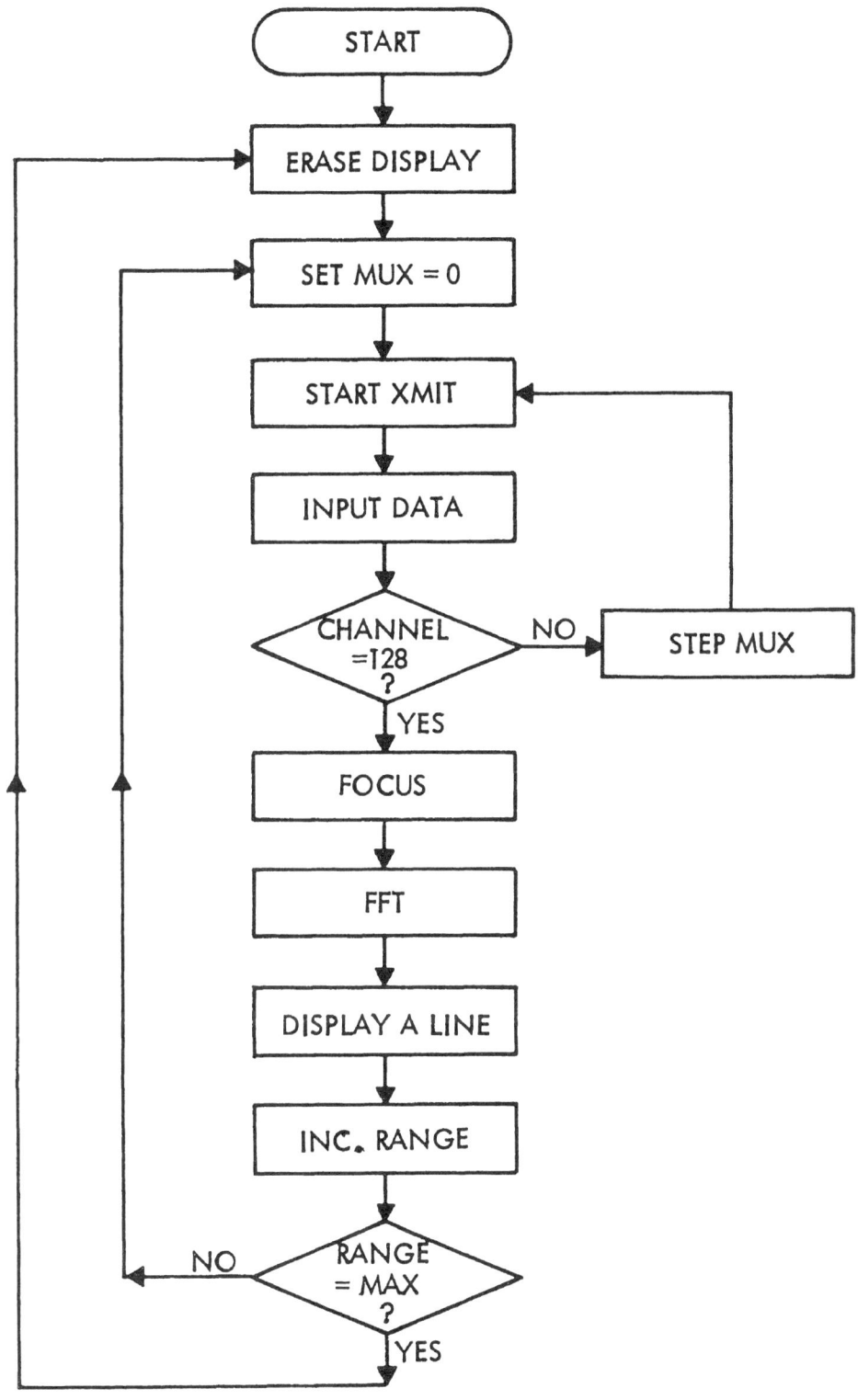

Figure 3. Microprocessor Flow Chart

with 12 x 16 bit multiplies. After the addition is performed, the results are scaled to prevent overflow. This maintains 8-bit accuracy for the final result. The scrambled output is re-ordered before being output to the display.

The time required to perform the 128-point complex FFT is 2.3 seconds of which 1.9 seconds is the software multiplication time. For tank tests, near-field focussing factors are multi-plied with the array data before the digital FFT is performed, as shown in Figure 3. This adds 0.5 seconds to the computation time.

Display

The display used is an HP 1335A storage display which is controlled by the microprocessor. As each line is processed, it is displayed and stored during the time that succeeding lines are being received. The 8-bit data is compressed to fit into the display's dynamic range.

Through input to the microprocessor, the user can select sections of the display to be magnified 1X, 2X, or 4X.

EXPERIMENTAL RESULTS

Azimuth Resolution

To test the basic resolution of the sonar system, images were taken of two point targets (ping-pong balls) at various separations at a range of 5 meters. Spherical focussing was used in the beamforming. Figure 4 shows a single point target, then two targets separated by 1, 2, 4, and 8 times the system Rayleigh resolution angle (ΔS). This is done for targets on-axis and off-axis by 30^0.

The targets are clearly separated at $2\Delta S$ for the on-axis case. The off-axis resolution is slightly poorer due to the effective reduction of the array aperture off-axis and the shading effect of a short pulse across the array.

Though it is somewhat ambiguous to use the Rayleigh criteria for a coherent system where target phase characteristics and exact azimuth position affect the apparent resolution, the exper-iment shows that the criteria is useful as an approximation.

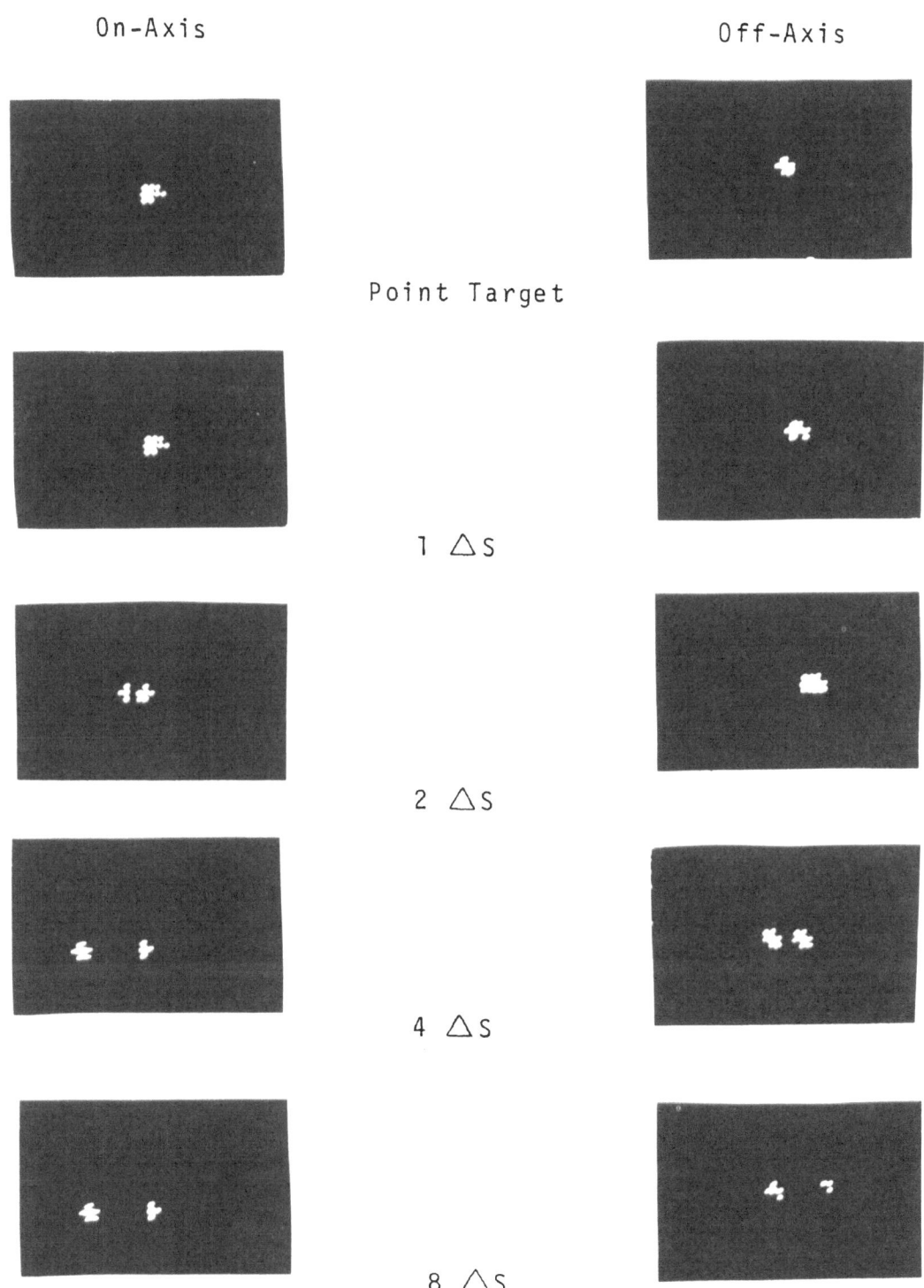

On-Axis   Off-Axis

Point Target

1 △S

2 △S

4 △S

8 △S

Figure 4. Azimuth Resolution

Depth Resolution

Figure 5 shows the display of two on-axis point targets separated in range. The system used a 200 μsec transmitted pulse and a 100 μsec receive gate. The theoretical depth resolution with these parameters is approximately 6 inches. The display clearly distinguishes the targets separated 12 inches but not 6 inches. The reason for the reduced depth resolution appears to be connected with the relation of the channel sampling times to the arrival of the target echoes.

Imaging an Object

Figure 6 shows the acoustic image of a smooth object pictured in Figure 7. This initial experiment of imaging a complex object in B-scan mode shows that much improvement needs to be made before the target can be recognized. A couple of points should be noted, however:

- The acoustic image of the object was remarkably constant, even with minor changes of target/system positional configuration.

- There is a ringing of the target that extends out its apparent return. This is something that we have observed on several images of cylindrical metal objects.

CONCLUSION

The preliminary experimental results reported in this paper are good, which encourages us to make some hardware improvements to allow more advanced experiments. The hardware modifications will allow real-time imaging. These modifications include the construction of 128 parallel channel processors, of a hardwired 128-point FFT, and of an extremely high power acoustic projector. With these changes, the system will be used for bottom imaging experiments in a real ocean environment.

1 Foot Separation

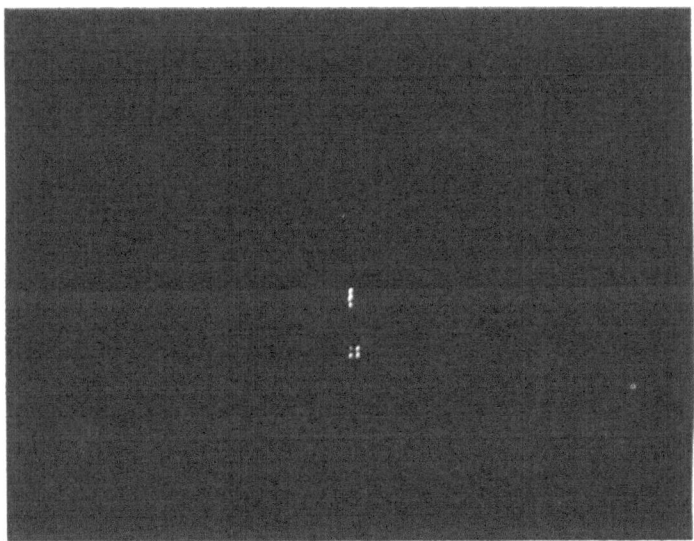

2 Foot Separation

Figure 5. Vertical Resolution

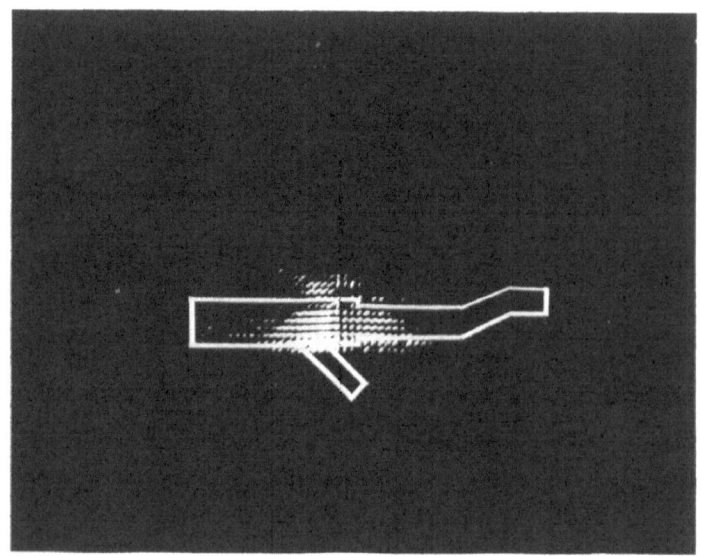

Figure 6.   Display of Acoustic Target

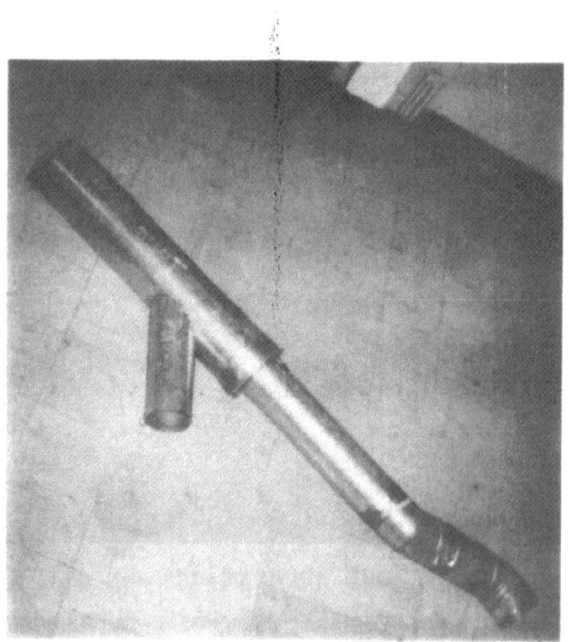

Figure 7.   Photo of Acoustic Target

REFERENCES

1. Marom, et. al., "Design and Preliminary Test of an Underwater Viewing System Using Sound Holography", Third International Symposium on Acoustical Holography, 1970.

2. Keating, et. al., "Complex On-Axis Holograms and Reconstruction Without Conjugate Images", Fifth International Symposium on Acoustical Holography, 1973.

3. Brigham, E. Oran, The Fast Fourier Transform, 1973.

CHARGE COUPLED DEVICES FOR HOLOGRAPHIC AND BEAMSTEERING

SONAR SYSTEMS*

R.D. Melen and J.D. Shott
Stanford University

B.T. Lee and L.C. Granger
Bendix Electrodynamics Division, N. Hollywood, CA

INTRODUCTION

Two different types of charge coupled devices with different applications to acoustic imaging and holography have been developed and are being tested. One device is known as the Charge Coupled Processor (CCP) whose function is to provide signal preconditioning for parallel arrays and output a series of DC signals which can be used to reconstruct the target field. The other device is the Cascade Charge Coupled Device (C3D), an analog lens device which performs beamsteering by changing the CCD clock frequencies to vary the effective delay line lengths of many parallel channels behind the detecting elements.

The basic CCD concept consists of storing electrons (or holes) in potential wells created at the surface of a semiconductor and moving these charges as a packet across the surface by sequential gating of the voltages which created the potential wells. The CCD stores minority carriers (or their absence) under a conductor-insulator-semiconductor capacitor by depleting the area under the conductor of majority carriers. The stored charge will be proportional to the product of the charge injection voltages times the conductor-insulator-semiconductor capacitance.

*Sponsored by Office of Naval Research, Code 212, Contract Number N00014-75-C-0755.

BASIC THEORY OF HIGH RESOLUTION SONAR PROCESSING

The coherent acoustic field from an active target, or the echo from an insonified passive target can be represented by the general solution to the scalar wave equation with appropriate boundary conditions:

$$U(P_0) = \frac{1}{j} \iint \frac{U(P_1) \exp (jkr_{01}) \, ds}{r_{01}} \tag{1}$$

The far-field solution can be written:

$$U(X_0, Y_0) = K \iint U(X_1, Y_1) \exp -j \frac{2}{z} (X_0 X_1 + Y_0 Y_1) dX_1 dY_1 \tag{2}$$

where X_0, Y_0 are the coordinates in the observing plane and X_1, Y_1 are the coordinates in the target plane. In the nearfield, the above expression is modified by spherical focussing factors appearing outside the double integral.

All sonars that seek target information must process the solution to the wave equation. Generally speaking, this can be done in either of the following two mathematically equivalent ways.

1. Measure the field $U(X_0, Y_0)$ in the detection plane and perform inverse Fourier transforms on this field to recover the target. The CCP is designed for use in such a system.

2. Measure the target field $U(X_1, Y_1)$ directly by measuring the outputs of a direction discriminating network. This is traditionally done with delay line summation. The C3D can be applied to this type of beamforming.

THE CHARGE COUPLED PROCESSOR (CCP)

In a holographic system, the field sampled by the array is mixed with a reference signal of the same frequency. This 'freezes' the phase and amplitude distribution of the complex field across the receiving aperture. Fourier transforming this field distribution recovers the image.

To sample the field at an array element, the AC output (at the sonar carrier frequency) is continuously multiplied by a reference signal of the same frequency and integrated over a number of cycles (which is determined by the desired receive gate). A Charge Coupled Processor (CCP) has been developed and tested which will perform this sampling function.

CCD's have three characteristics which are directly applicable to a channel processor element:

1. They can switch or gate a charge which is proportional to an injection voltage at a specific point in time. If the gating frequency is equal to the frequency of the gated input signal, the stored charge will be proportional to the amplitude and phase of the input signal. This is equivalent to multiplying the signal by a reference signal.

2. They can sum or integrate the charges as required by the channel processor by shifting the injected charges into a stationary potential well.

3. They can gate the charge summation to an output register, providing a simultaneous sensing and resetting of the channel processor integrator.

In addition to the above three processor functions, the CCD can form an analog shift register which will multiplex or scan the gated charges stored in the integrators to a single pair of output connections, essentially converting the parallel hydrophone outputs to serial form.

Figure 1 shows a drawing of the CCP component. The received signal is injected into the CCP through the launch diode. The charge that passes through to the launch gate is a function of the bias gate voltage times the signal voltage. The bias gate is driven at the same frequency as the carrier and a charge packet proportional to the product of the bias gate voltage and the signal is accumulated over a number of cycles. This performs the required mixing and integration to form a DC term proportional to the amplitude and cosine of the phase of the complex field at the aperture plane. Another channel with the reference signal driven in quadrature can be used to form the sine term.

This information is then serially multiplexed out of the chip where it can be detected and then used to perform the Fourier transform.

A microphotograph of the CCP is shown in Figure 2. This device was developed to show the feasibility of such a processor and can provide either the sine or cosine components for 4 parallel inputs. The device has been tested and found to perform the following functions: integration, mixing, filtering, and amplification. A scope tracing showing the output of the device is shown in Figure 3. This photo shows the mixing and integration period followed by 4 pulses. After the fourth pulse, the output gate is not reset until the mixing and integration period is complete, which is the reason why the fourth

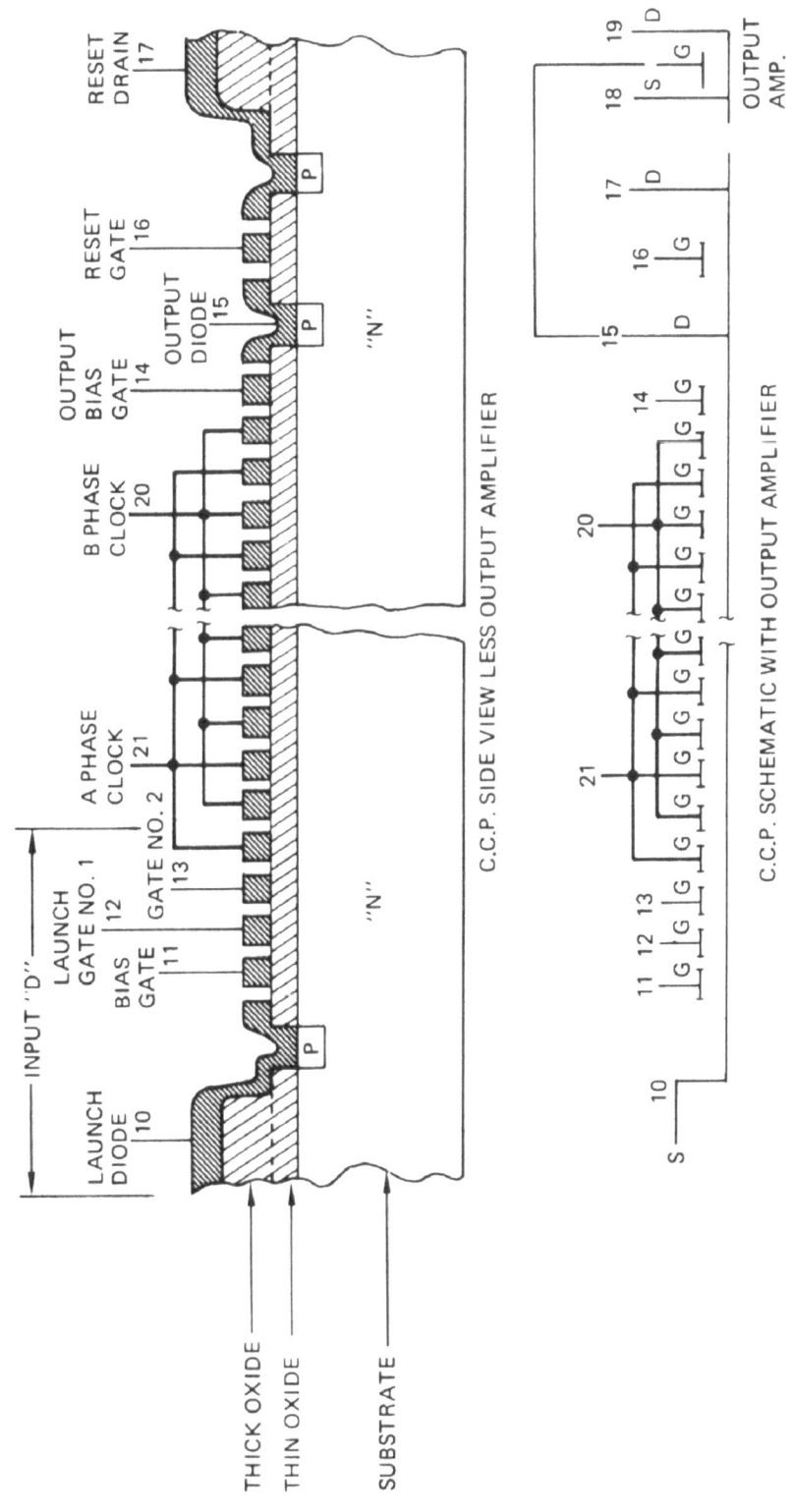

Figure 1.   Charge Coupled Processor (CCP)

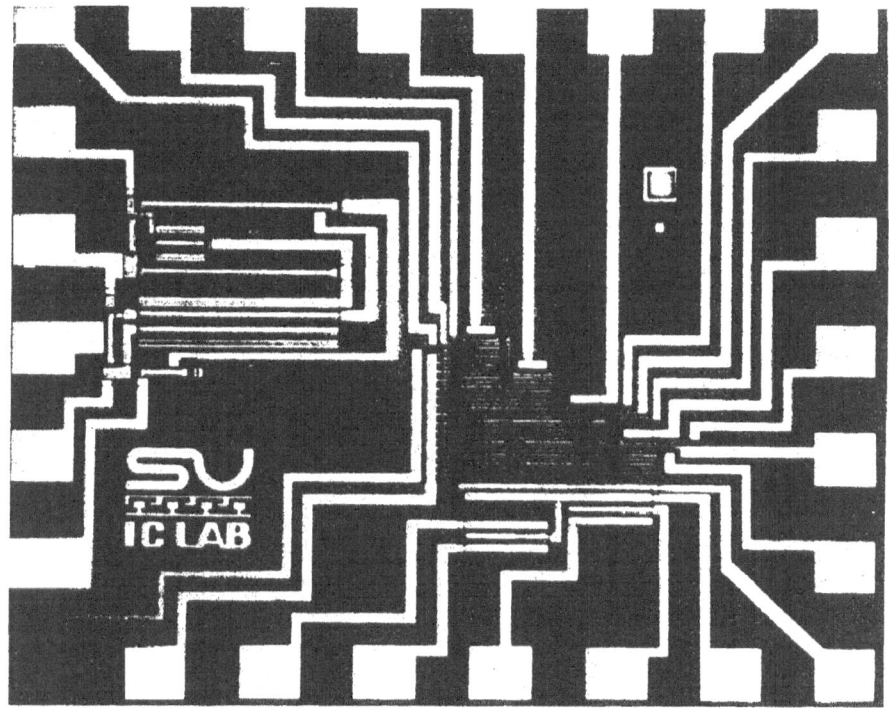

Figure 2. Microphotograph of CCP

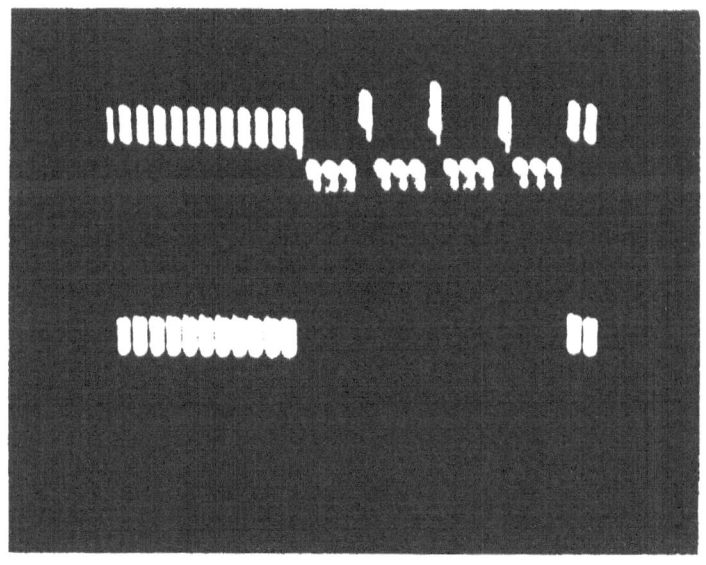

Figure 3. Output of Charge Coupled Processor

output does not appear as a single pulse but as the first of the pulses which continue through the next integration period.

It was expected that this device would have a gain proportional to the square root of the number of cycles in the integration time. Since the mixing and integration function is equivalent to sampling the signal at one point in each cycle, the device gain would be proportional to the number of signal samples.

However, tests of the device showed that the gain leveled off to a constant value. This was the result of the fact that the transconductance of CCD devices is linear with current for low current levels. Due to the mode this device was operated in, for an increasing number of samples, the injection current per sample had to be lowered to prevent saturation. This brought the current levels into the subthreshold region of the curve, resulting in a linear transconductance.

The expected gain of the Charge Coupled Processor was derived in the following manner.

For MOS devices, the transconductance can be described as:

$$g_m = K_1 \sqrt{I_s} \qquad (3)$$

where K_1 is a constant and I_s is the injected current per sample. When one sample is taken, $I = I_s$, the current from a single sample. The total injection current I_s is normally $\frac{1}{2} I_{max}$ (saturation current), thereby operating the device in the center of its linear range. Therefore $I = I_{max}$ (saturation current).

To prevent the charge bucket from overflowing when N samples are collected, the gate bias voltage is reduced so that the injection current for each of the N samples I_s is the total injection current I divided by N.

$$I_s = I / N \qquad (4)$$

Substituting this into equation 3:

$$g_m = K_1 \sqrt{\frac{I}{N}} \qquad (5)$$

The sample gain for N samples is simply:

$$\sum_{j=1}^{j=N} S_j = N \cdot S \qquad \text{(since each sample is the same)}$$

The total gain is equal to the sample gain times the transconductance:

$$G = g_m \times \text{sample gain}$$

$$= K_1 \sqrt{\frac{I}{N}} \cdot N \cdot S \tag{6}$$

$$= K_2 \cdot \frac{N}{\sqrt{N}} \qquad \text{where } K_2 \text{ is a new constant containing } K_1, I, \text{ and } S$$

$$= K_2 \sqrt{N}$$

Thus the device gain was expected to be proportional to the square root of the number of samples.

However, at low currents, the transconductance of a CCD is linear with respect to the injection current. This effect was described by Swanson[1] for MOS devices, but had not been previously demonstrated for CCD's.

As a result:

$$g_m = K_1 I_s \tag{7}$$

and from equation 4, I_s is replaced by I/N. In this case the total gain becomes:

$$G = \text{sample gain} \times \text{transconductance}$$

$$= N \cdot S \cdot K_1 \cdot \left(\frac{I}{N}\right) \tag{8}$$

$$= K_2 \qquad \text{after the N's are cancelled and } S, K_1, \text{ and I are combined into } K_2.$$

Figure 4 is a curve which shows the device gain as a function of the number of samples. In this case K_2 was found experimentally to be approximately 3.

Thus the range of variable gain of this device is only between 0.7 and 3 times amplification and prevents the device from being useful as a processor elements in extended array systems.

This same device was driven as a variable delay line. An off-axis plane wave was simulated at the 4 inputs of the device by applying a phase differential between them. The time delay between the elements was varied by varying the clock driving frequency. As the oscilloscope output of Figure 5 shows, the

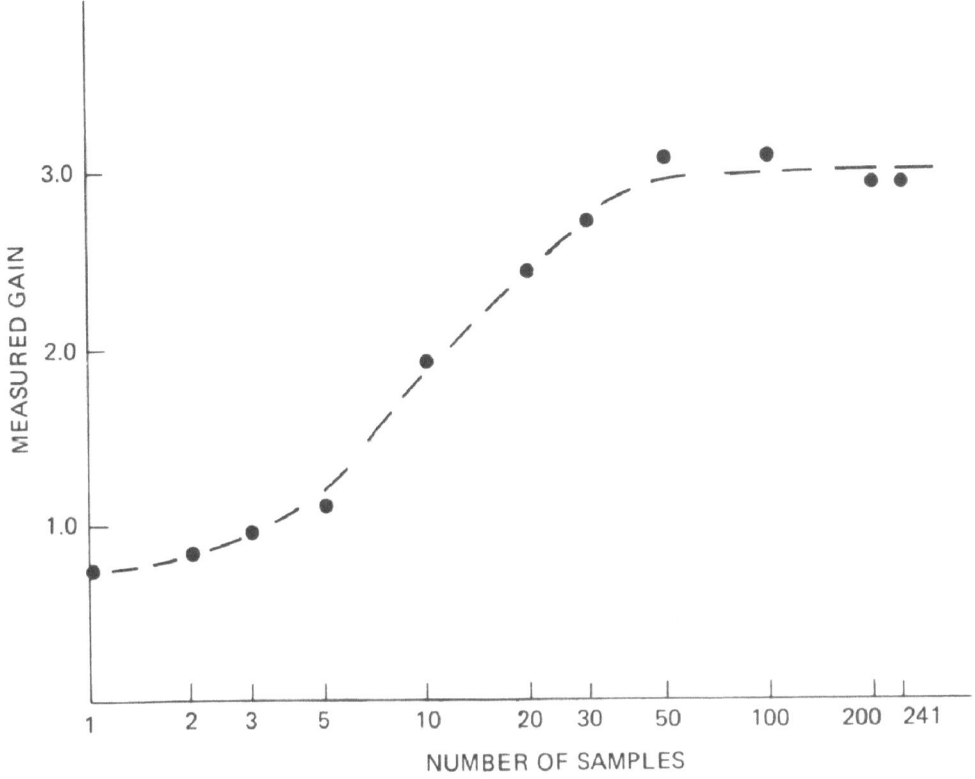

Figure 4.   Measured Gain of CCP

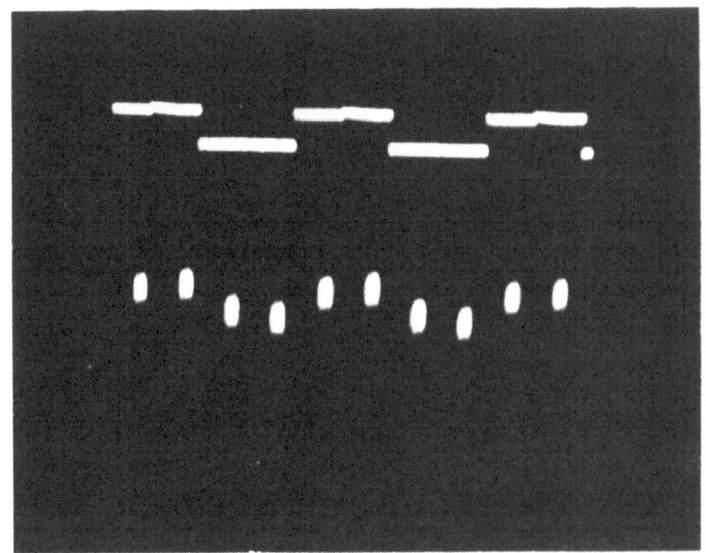

Figure 5.   Output of CCP in Delay Line Mode

output amplitude of the device is at a maximum when the delay
time between elements is the same as the phase delay resulting
from an arriving plane wave.

Since a CCD is essentially a long analog shift-register, such
use as a variable delay line now seems more appropriate than as
a signal preconditioner. The Charge Coupled Processor device
was successfully run as a delay line, but there are only 4 shifts
between each element before the charge from the next input is
dumped into the same bucket. This is equivalent to a single-
element analog lens.

Under a separate program, Stanford had fabricated a 4-element
analog lens known as the Cascade Charge Coupled Device (C3D) [2,3].
This device was developed for use at medical ultrasonic frequencies.
Experiments with the Charge Coupled Processor (as a 1-element
analog lens) showed some of the potential of CCD use as a variable
delay line at typical sonar frequencies. Consequently it was
decided to investigate the 4-element C3D performance at sonar
frequencies.

CASCADE CHARGE COUPLED DEVICE (C3D) ANALOG LENS

Charge coupled devices can be considered as analog shift
registers. One application of CCD's is to use them as electronically adjustable delay lines performing the required delay-sum
operation on broadband acoustic signals. This CCD capability
makes economically feasible an ultrasonic imaging system with
the following features:

1. High resolution
2. Dynamic focussing
3. Adjustable field of view

This device is basically an array of charge coupled delay
lines interconnected on a single integrated circuit substrate and
having multiple sections of delay each clocked independently at
a different frequency. The clock frequency controls the speed
at which a wavefront propagates across the chip. At the interface
between sections or lens elements, the wavefront direction is
changed due to the different velocities in each section. This is
analogous to optical refraction, where Snell's Law ($\sin \theta_1/\sin \theta_2 =
v_1/v_2$) describes the wavefront angle with relation to the relative
velocities. This is the principle by which optical lenses focus.
The output from the parallel channels is summed on the device. Thus
the time delay and summing are performed. The output envelope can
be detected to determine the target field component in one direction. By varying the clock frequencies, beams can be formed in
other directions.

Such a chip could be used singly, with time-varying clock frequencies to perform a 'scan' of the target field. Or parallel chips could be used with different sets of clock frequencies to perform the field sensing for many directions simultaneously.

Figure 6 shows a drawing of such a device. This is a 20-input device which has been fabricated and tested at medical ultrasonic frequencies[2,3]. With further development, NDT frequencies should also be achievable.

Sections 1 and 2 of the device are used for plane wave beam-steering. (The Charge Coupled Processor device, when run as a delay line, is equivalent to one section of this device.) Sections 3 and 4 of the device are used for near-field focussing. Dynamic focussing can be achieved by varying the clock frequencies as returns from increasing ranges are achieved. This feature is not as important for high resolution sonar applications, and could be suppressed by simply driving both these sections at the same frequency. Since their inclusion poses no significant manufacturing problems, this feature can essentially be incorporated 'for free'.

The application of the C3D to high resolution sonar has several attractive features. A block diagram of such a system is shown in Figure 7. In contrast to the traditional LC delay line approach, it does not require careful 'tweaking' in order to form theoretical beams. A 50 dB dynamic range has been measured at medical ultrasonic frequencies. Such a dynamic range would be adequate for high resolution sonar when used in conjunction with TVG. By adding an extra delay section to the C3D device, a block delay can be formed which would enable such a chip to be used as a building block in a much larger system. Plans are being made to fabricate a 32-input device which will require only 40 pins.

The C3D chips are inherently flexible and under microprocessor control (of the clock driving frequencies), individual chips can be switched in or out as required in case a chip should fail. This would provide a measure of redundancy in applications in the field. Also, in cases where oversampling of the beam directions may be desired, the clock frequencies to the chip can be easily altered.

One of the potential problems with applying this to low frequency sonar is the limited length of time that charge can be stored on a chip before the buckets are filled due to thermally generated dark current. This is the limiting factor on the delay length that is achievable. This effect can be greatly reduced by cooling and may not impose severe limitations at low frequencies.

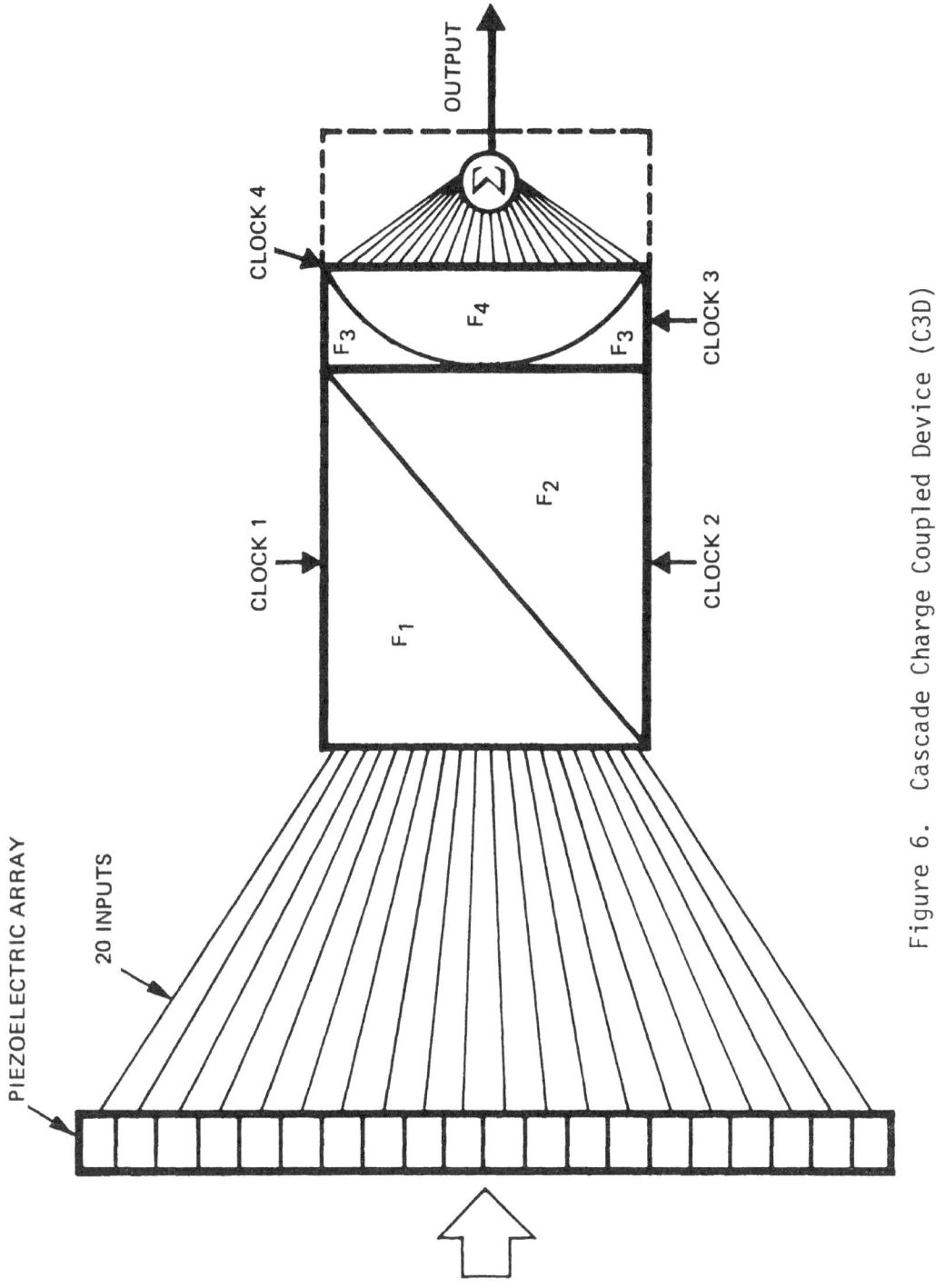

Figure 6.   Cascade Charge Coupled Device (C3D)

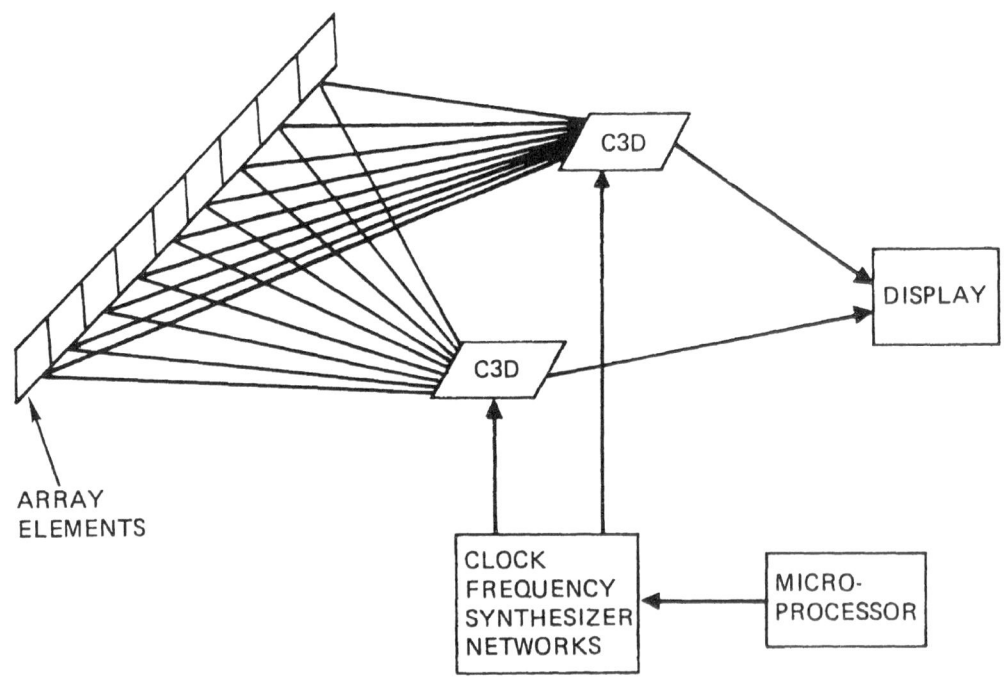

Figure 7. C3D System Block Diagram

Experiments are being undertaken to measure the essential
properties and beamforming capabilities of the C3D at typical
high resolution sonar frequencies of about 200 kHz. CCD performance improves with lower frequencies, so it is expected that the
device will perform well at sonar frequencies since they are lower
than the medical frequencies where the tests have been run to date.

Experiments will be run at 180 kHz. Signals from 20 hydrophones over a 16 cm aperture will be input to the C3D device and
beams will be formed over the entire field of view. Clock frequencies which are multiples of 120 kHz (2/3 the primary frequency)
have been chosen so that any possible combinations of sum and
difference frequencies will not generate noise at the carrier
frequency. Sections 1 and 2 of the device will be operated with
various pairs of c-ock frequencies ranging between 720 kHz and
3 MHz. The lowest frequency of 720 kHz is twice the Nyquist rate.

CONCLUSION

Charge coupled devices are being investigated for application
to acoustic imaging and high resolution sonar. Two types of
devices with different applications have been discussed here.

The Charge Coupled Processor (CCP) has application primarily to narrow-band holographic systems. The Cascade Charge Coupled Device (C3D) has application to broadband time delay systems. Experimental results with such devices indicate that CCD's have the potential for being important components of multi-element acoustic detection systems.

REFERENCES

1. Swanson, R.M. and J.D. Meindl, "Ion-Implanted Complementary MOS Transistors in Low-Voltage Circuits", IEEE Journal of Solid State Circuits, SC7,2, April 1972, pps. 146-153.

2. Schott, J.D., R.D. Melen and J.D. Meindl, "The Cascade Charge Coupled Device: A Single-Chip Lens for Ultrasonic Imaging Systems", 1976, IEEE Solid State Circuits Conference, Philadelphia, February 1976.

3. Melen, R.D., et.al., "CCD Dynamically Focussed Lenses for Ultrasonic Imaging Systems", 1975 International Conference on the Application of Charge-Coupled Devices, San Diego, Octover 1975, pps. 165-171.

CHIRP FOCUSED TRANSMITTER THEORY

J. Souquet, G. S. Kino, and T. Waugh

Stanford University

Stanford, California 94305

I. INTRODUCTION

We have described an electronically focused and scanned acoustic imaging device in earlier papers.[1,2] This device was constructed by using a tapped acoustic surface wave delay line as a phase reference for an array of piezoelectric transducers. By inserting a frequency modulated chirp along the delay line, the device can be used either as a receiver or as a transmitter. In the transmit mode, it acts like a moving lens, emitting a focused beam which scans along a plane parallel to the array at a velocity comparable to the acoustic velocity. The same array can be used in the receive mode. By this means, a B-scan device was constructed which has good transverse and range definition.

We examine here the problem of rapidly moving lenses of this nature, operating in the transmitter mode. We have computed the side lobe levels in this transient mode, and shown that the side lobe levels are slightly worse than in a cw system. By using Hamming weighting, the side lobe level can be made very low. By doubling the number of elements in the array, and connecting them together in pairs, it is also possible to radically reduce the first grating lobe levels. We have carried out experiments to check this theory, which are in good agreement with it.

II. PRINCIPLES OF THE ELECTRONICALLY FOCUSED SYSTEM

It has been shown in previous papers that a linear FM chirp signal can be sent along a surface acoustic wave delay line and used for the purpose of electronic focusing and scanning.

For the transmit mode, we consider an array of transducers as shown in Fig. 1. The chirp signal entering the tapped delay line is taken to be of the form

$$F(t) \;=\; \exp\, j\!\left(\omega_1 t + \mu\, \frac{t^2}{2}\right) \;. \tag{1}$$

At the nth electrode with a coordinate x_n, assuming a velocity v, we obtain

$$F\!\left(t - \frac{x_n}{v}\right) \;=\; \exp\, j\!\left[\omega_1\left(t - \frac{x_n}{v}\right) + \frac{\mu}{2}\left(t - \frac{x_n}{v}\right)^2\right] \;. \tag{2}$$

This signal is then mixed with a signal that varies as $e^{j\omega_2 t}$ to produce a wave of center frequency ω_s. So, after mixing, the output becomes:

$$
\begin{aligned}
F_1\!\left(t - \frac{x_n}{v}\right) \;=\;& F\!\left(t - \frac{x_n}{v}\right) \exp\, j\omega_2 t \\[2mm]
=\;& \exp\, j\left\{\left[\omega_s\!\left(t - \frac{x_n}{v}\right) + \frac{\mu}{2}\left(t - \frac{x_n}{v}\right)^2\right] + \frac{\omega_1 - \omega_s}{v}\, x_n\right\} \\[2mm]
=\;& F_s\!\left(t - \frac{x_n}{v}\right) \exp\, j\, \frac{\omega_1 - \omega_s}{v}\, x_n \;.
\end{aligned}
\tag{3}
$$

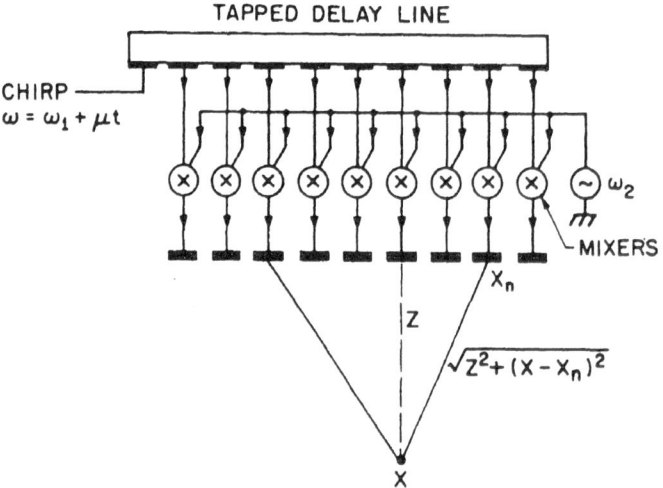

FIG. 1--Illustration of the arrangement transducers-delay line.

If we want a beam with its central ray emitted at right angle to the array at ω_s , we must choose ω_1 such that:

$$\frac{\omega_1 \, x_n}{v} = 2N\pi \quad . \tag{4}$$

The signal reaching the point x,z is then:

$$H(t,x,z) = \sum_{x_n} \exp j \left(-\omega_2 \, \frac{x_n}{v} \right) F_s \left(t - \frac{x_n}{v} - \frac{\sqrt{z^2 + (x_n - x)^2}}{v_w} \right) \tag{5}$$

where v_w is the propagation velocity of the medium.

Taking account of apodization $G(t)$ of the input signal and the angular response $R(\theta)$ of the transducer elements, where θ is a function of the frequency of the signal at the individual element, the output becomes

$$H(t) = \sum_{x_n} G\left(t - \frac{x_n}{v} - \frac{1}{v_w} \sqrt{z^2 + (x_n - x)^2} \right) R\left(\frac{x_n - x}{z} \right)$$

$$\times \; F_s \left(t - \frac{x_n}{v} - \frac{1}{v_w} \sqrt{z^2 + (x_n - x)^2} \right) \exp j \left(\frac{-\omega_2}{v} \, x_n \right) \quad , \tag{6}$$

where $\tan \theta = (x_n - x)/z$. It is convenient to write the time in terms of the time $z/v_w + x/v$ when the ray normal to the array reaches x,z . Thus we write

$$t' = t - \frac{z}{v_w} - \frac{x}{v} \quad . \tag{7}$$

Furthermore it is convenient to define a new variable u which is dependent on the difference in time traveled by the signals reaching x,z via the element at x_n :

$$u = t' - \frac{x_n - x}{v} - \frac{1}{v_w} \left(\sqrt{z^2 + (x_n - x)^2} - z \right) \quad . \tag{8}$$

Thus we can write $H(t)$ in terms of the new variable as

$$H(t') = \sum_{x_n} G(u) \, F_s(u) \, R\left(\frac{x_n - x}{z} \right) \exp j \left(-\omega_2 \, \frac{x_n - x}{v} \right) \quad . \tag{9}$$

This equation includes all the interesting parameters one can use and modify to improve the overall characteristics of the device.

By using a Hamming type of weighting for the function $G(u)$, one can considerably reduce the side lobe level; by a suitable spacing x_n, and connections to the elements of the array, the function $R[(x_n-x)/z]$ can be changed and the grating lobe levels reduced. It is also of interest to investigate what happens when the scanning velocity is changed.

III. THE PARAXIAL APPROXIMATION

Before proceeding to examine the numerical results, it is worthwhile to examine the analytic results which can be obtained by the use of the paraxial approximation.

We will assume for simplicity, that the system is unapodized, so that the chirp has finite amplitude for a time $T = D/v$ where D is the spatial length the chirp occupies on the delay line. Thus we write

$$G(u) = 1 \qquad |u| < T/2$$
$$G(u) = 0 \qquad |u| > T/2 \quad . \tag{10}$$

We will also take $R[(x_n-x)/z] = 1$ and

$$\sqrt{z^2 + (x_n - x)^2} \approx z + (x_n - x)^2/2z \tag{11}$$

and choose

$$\omega_1 x_n/v = 2N\pi \quad . \tag{12}$$

In this case

$$H(t') \approx \exp j\omega_s(t' + \frac{x}{v}) \sum \exp\left[- \frac{j\omega_s(x-x_n)^2}{2v_w z}\right]$$

$$\times \exp\left[j \frac{\mu}{2}\left(t' - \frac{x_n-x}{v} - \frac{(x_n-x)^2}{2zv_w}\right)^2\right]. \tag{13}$$

Neglecting the quadratic term in the last bracket, we see that by choosing the focusing condition $\mu = \omega_s v^2/v_w z$, we can eliminate all the square law phase terms in (x_n-x) and write

$$H(t') = \exp j\left[\omega_s(t' + \frac{x}{v}) + \frac{\mu}{2} t'^2\right] \sum \exp\left\{j\mu t'[(x-x_n)/v]\right\}. \tag{14}$$

We note that at $t' = 0$ all the elements provide contributions in the same phase and so the system is focused on x, z at $t' = 0$.

We must now examine the limits of the summation. These are from

$$-\frac{T}{2} < t' - \frac{(x_n - x)^2}{2zv_w} - \frac{x_n - x}{v} < \frac{T}{2} \quad . \tag{15}$$

It is convenient to write $x_n - x = vn\tau$. Here $\tau = \ell/v$ is the delay time between elements spaced ℓ apart. We see that

$$-\frac{T}{2} < t' - \frac{n^2\tau^2v^2}{2zv_w} - n\tau < \frac{T}{2} \quad . \tag{16}$$

Neglecting squared term in $n\tau$, the limits are:

$$n \approx \frac{t'}{\tau} \pm \frac{T}{2\tau} \quad . \tag{17}$$

This yields the result

$$H(t') = \exp j\omega_s (t' + \frac{x}{v}) \frac{\sin\left(\frac{\pi Dvt'}{z\lambda}\right)}{\sin\left(\frac{\pi \ell vt'}{z\lambda}\right)} \quad . \tag{18}$$

We see that the spot at x, z is illuminated by a focused beam whose amplitude is maximum at $t' = 0$ and illuminates it for a time between 4 dB points of

$$\Delta\left(t - \frac{z}{v_w} - \frac{x}{v}\right) = \Delta t'_s \approx \frac{z\lambda}{Dv} \quad , \tag{19}$$

as we would expect for a lens of width D moving with a velocity v.

We see that the side lobes of the transmitter occur along a line of slope $(dx/dz) = -(v_w/v)$. The definition in the x direction corresponds to

$$\Delta x_s = \frac{z\lambda}{D} \tag{20}$$

and the range definition, i.e., in the z direction for a given excitation time t' corresponds to

$$\Delta z_s = \frac{z\lambda}{D} \frac{v_w}{v} \quad .$$ (21)

Thus the higher the scan velocity, the better the range definition at a given time.

In general, the paraxial approximation will tend to be at its best near the peak of the main lobe, when for all transducer elements $(x - x_n)^2 \ll z^2$. Thus the far out side lobes levels will not be correctly predicted by this formula, and the grating lobe levels will be even more in error.

As a rough estimate we see that the quadratic terms we have neglected contribute a phase error of the order:

$$\Delta\phi_n = \frac{\pi}{\lambda} \left(\frac{x_n - x}{z}\right)^2 \frac{v}{v_w} \left[vt' - (x_n - x) + \frac{v}{2v_w}\left(\frac{x_n - x}{z}\right)^2\right] \quad .$$ (22)

We see that the phase errors are different on both sides of the main lobe and worse for t' positive. They reach a value of $\pi/2$ approximately where

$$(x_n - x)^3 \simeq z^2 \lambda v_w / 2v \quad .$$ (23)

Thus for a chirp of spatial length D, to keep the phase errors from the end elements below this value, we must have

$$D^3 < z^2 \lambda v_w / 16v$$ (24)

This formula is very nearly the same (within a factor of 1.26) to the condition that the time difference between rays reaching the point x, z should be less than the width of the main lobe as given by the paraxial formula.

It should be noted that, in the paraxial formula, grating lobes occur near a time

$$t'_G \approx \pm \frac{z\lambda}{\ell v} \quad .$$ (25)

However, at this point the paraxial theory is badly in error. Instead, if referring to Fig. 2, we take account of the delay time along the delay line and along the ray from the center of the chirp to calculate t'_G, we take the grating lobe center to be where

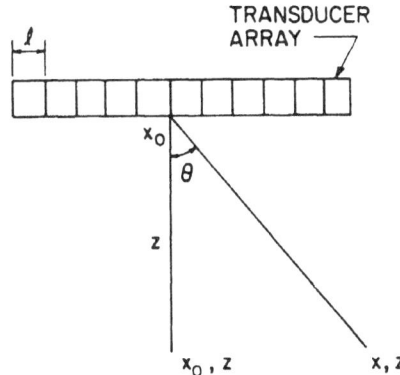

FIG. 2--Diagram showing the coordinates of the system.

$x - x_0 \approx \pm z\lambda/\ell$. So the point x,z is excited at a time

$$t \approx \frac{x_0}{v} - \frac{1}{v_w} \sqrt{(x - x_0)^2 + z^2} \quad , \qquad (26)$$

where x_0 is the position of the center of the chirp. This corresponds to the condition

$$t_G' = \pm \frac{x - x_0}{v} + \frac{z}{v_w} - \frac{1}{v_w} \sqrt{(x - x_0)^2 + z^2}$$

$$\approx \pm \frac{z\lambda}{\ell v} \left[1 \mp \frac{1}{2} \left(\frac{\lambda}{\ell} \right)^2 \frac{v}{v_w} \right] \quad ; \qquad (27)$$

we note that the grating lobes are not equally spaced in time from the main lobe.

IV. NUMERICAL COMPUTATIONS

(a) The Effect of Apodization

We now discuss the more exact results obtained by numerically computing the results without making the paraxial assumption. For the case of the unapodized chirp, we have:

$$\begin{aligned} G(u) &= 1 \qquad \text{for} \quad -\frac{T}{2} < u < \frac{T}{2} \\ G(u) &= 0 \qquad \text{for} \quad |u| > \frac{T}{2} \end{aligned} \qquad (28)$$

where T is the chirp length.

From the paraxial results, we expect the output of the device to yield a sinc-type of focus with a maximum output amplitude at $t' = 0$, the first side lobes being 13.5 dB reduced in amplitude from the main lobe. A numerical computation has been carried out for a 120 element array, where the angular response of the array elements is assumed uniform. The element-to-element spacing is $\ell = 2$ mm , the scan velocity $v = 2000$ m/sec , the velocity in the medium $v_w = 1500$ m/sec , and the frequency $f = 2.5$ MHz . It will be seen in Fig. 3 that the side lobes levels are now not the same on each side of the main lobe. For $t' > 0$ the first side lobe is now only 7 dB down from the main lobe. However, the worst grating lobe response will be seen to be 6 dB down from the main lobe; this is due to the effect of aberrations.

Let us now consider the case of Hamming weighting. $G(u)$ is chosen so that

$$G(u) = k + (1-k) \cos^2 \frac{\pi u}{T} \qquad \text{for} \quad |u| < \frac{T}{2}$$
$$\tag{29}$$
$$G(u) = 0 \qquad \text{for} \quad |u| > \frac{T}{2} \, .$$

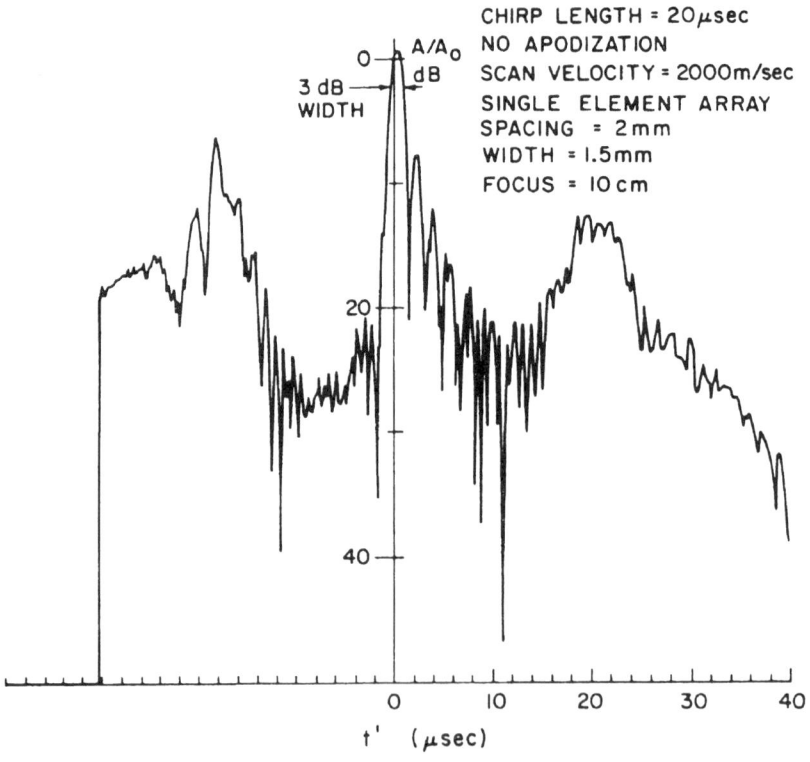

CHIRP LENGTH = 20 μsec
NO APODIZATION
SCAN VELOCITY = 2000 m/sec
SINGLE ELEMENT ARRAY
SPACING = 2 mm
WIDTH = 1.5 mm
FOCUS = 10 cm

FIG. 3--Theoretical focal distribution for the unapodized chirp focused transmitter.

Ideally, for $k = 0.08$, the side lobe level should be reduced by
43 dB from the main lobe. The maximum output occurs at $t' = 0$
and its amplitude is 0.54 times the amplitude of the unapodized
case, and the main lobe is 1.5 times as wide at the 3 dB points
as for the unapodized array. Figure 4 shows an illustration of
Hamming weighting using, for this computation, the same parameters
used for computing the unapodized example. It is clear that, at
the expense of a slight loss in definition, the use of apodization
should be of great help in improving the side lobe level. This
slight loss in definition can also be improved by increasing the
chirp length, as will be seen later in this paper.

(b) Grating Lobe Level Improvement

As can be seen from the previous curves, the grating lobe
levels did not change very much under the influence of Hamming
weighting. These grating lobes are a serious disadvantage in an
imaging system, and we would like to reduce their levels. This
can be done by changing the angular response of an array element.

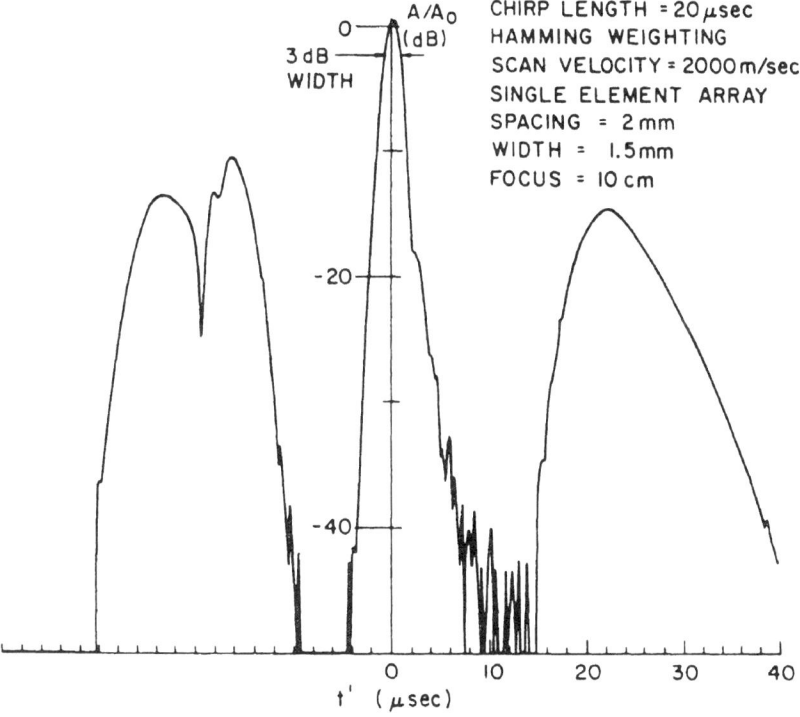

FIG. 4--Theoretical focal distribution for the Hamming weighted
chirp focused transmitter.

For a single transducer element of width d , we have

$$R(\theta) = \frac{\sin(\pi d\ \sin\ \theta/\lambda)}{\pi d\ \sin\ \theta/\lambda} \tag{30}$$

where

$$\theta = \tan^{-1}\left(\frac{x_n - x}{z}\right) . \tag{31}$$

The grating lobe occurs where $\sin\ \theta \approx \ell/\lambda$ and

$$R(\theta) \approx \frac{\sin(\pi d/\ell)}{\pi d/\ell} . \tag{32}$$

So we would expect some reduction in grating lobe level due to the angular response of the array element; but this is not enough. One approach to the problem is to decrease the element-to-element spacing. If done directly, this has the economic disadvantage that the number of electronic circuits must be increased with the number of elements per unit length.

An alternative approach with which we have gained excellent results, is to use double the number of elements per unit length, but connect the elements together in pairs. This yields a transducer response

$$R(\theta) = \frac{\sin(\pi d\ \sin\ \theta/\lambda)}{\pi d\ \sin\ \theta/\lambda}\ \cos\ \left(\frac{\pi\ell}{2\lambda}\ \sin\ \theta\right) , \tag{33}$$

where now there are two connected elements in a distance ℓ .

It will be seen that the response is now exactly zero at the grating lobe where $\sin\ \theta \approx \lambda/\ell$. In fact, by using N elements connected together where before there was one, we can reduce the higher order grating lobe levels. Thus the solution is a powerful one because it does not require the use of extra electronic circuits. The problems with it are: (1) it limits the angular response of the elements, (2) it does not completely eliminate the grating lobes, because in a focused system the individual elements are exciting rays over a range of angles.

Figure 5 shows the improvement over Fig. 4 when a double element transducer is used with the same period $\ell = 2$ mm , with $d = \ell/4$. Now the grating lobe level has been reduced to approximately 15 dB. Figure 6 shows the improvement over Fig. 5 when the spacing between the elments is decreased and double elements are still used. Figure 6 has been obtained for $\ell = 1$ mm .

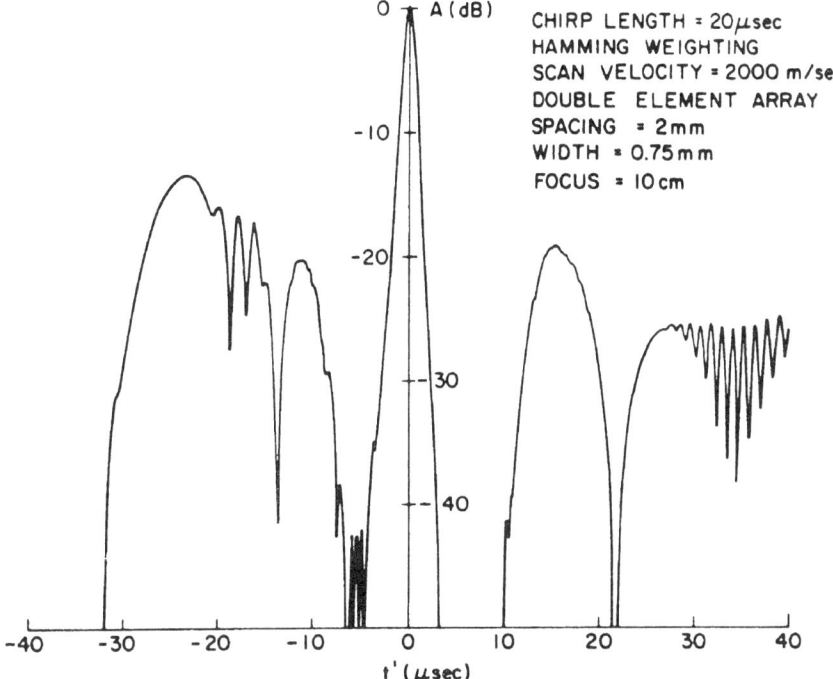

FIG. 5--Theoretical focal distribution for a double element array
     with a 2 mm periodicity.

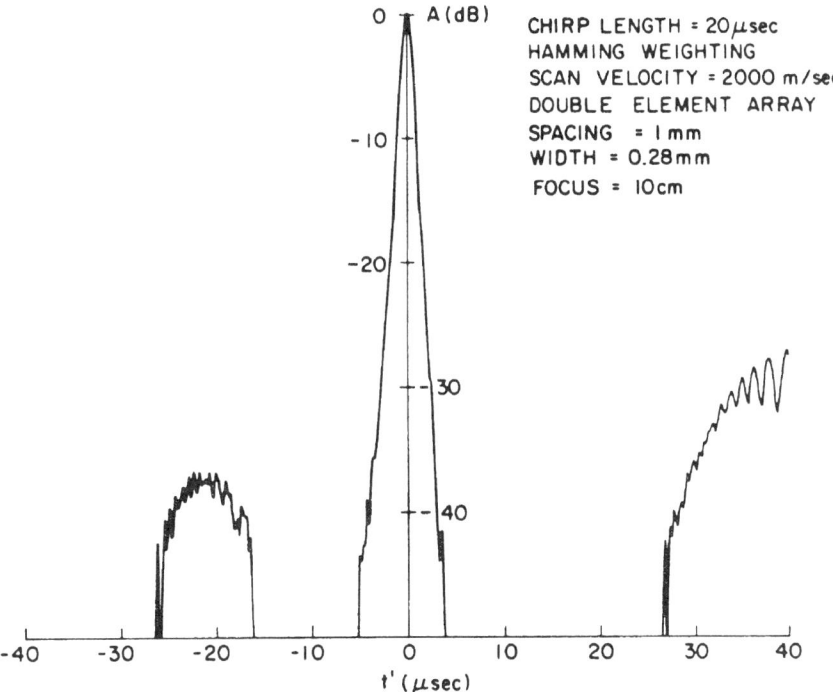

FIG. 6--Theoretical focal distribution for a double element array
     with a 1 mm periodicity.

V. EXPERIMENTAL RESULTS

We have been working at Stanford with an experimental array having 120 elements. The height of the individual elements is 1 cm, their width 0.37 mm. The transducer elements are connected in pairs, the center-to-center spacing between two consecutive pairs is 1.27 mm.

The transducer array is tungsten-epoxy backed and covered on its front side with an epoxy film. The array itself is mounted onto an aluminum housing, as described in a previous paper.

The unapodized chirp output is gated to give a chirp of 60 μsec duration, which corresponds to a beam width D of 4 cm in our system. The center frequency of the system is 2.25 MHz.

A point probe is placed 14 cm away in front of the array, and records the shape of the transmitted focused beam, as shown in Fig. 7. The computed result is shown in Fig. 8.

The comparison between the two results is shown in the following table:

	Experimental Results	Theoretical Results
3 dB width	4 μsec	4 μsec
side lobe level	-10 dB	-11 dB
grating lobe level	-29 dB	-29 dB

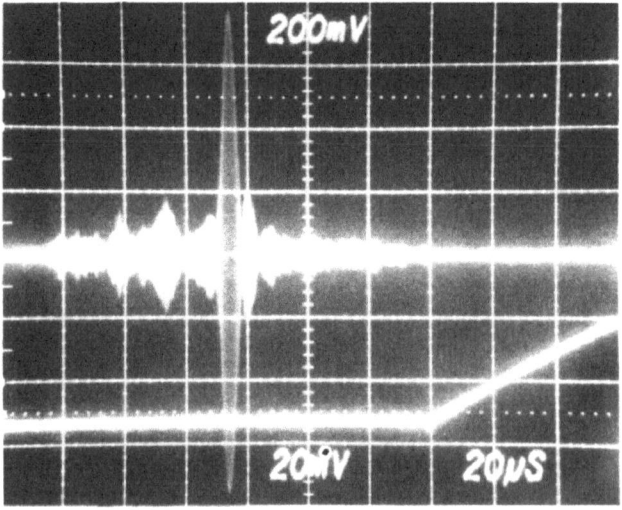

FIG. 7--Transmitted focused beam received at a point probe 14 cm in front of the array. Double element array, elements spaced 1.27 mm apart, chirp length 60 μsec, scan velocity 675 msec, frequency 2.25 MHz.

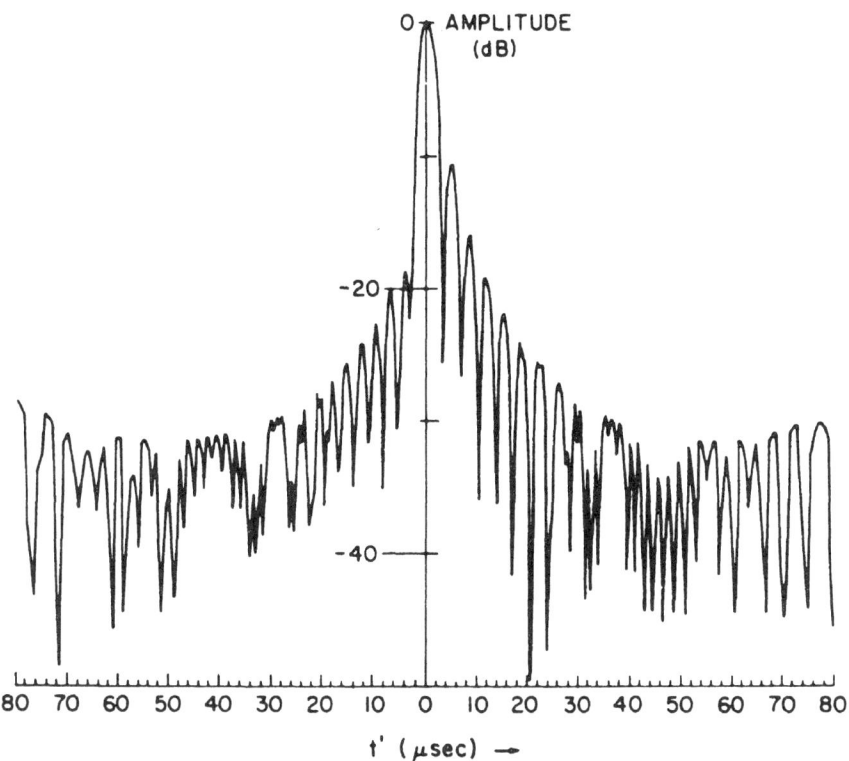

FIG. 8--Theoretical focal distribution corresponding to the experimental result shown in Fig. 7.

Another experiment has been performed on a double element
array with a 1 mm periodicity between each pair. A skewed Hamming
weighted chirp of 40 μsec duration is used; it corresponds to a
beam width D of 4 cm in our system. The center frequency is
2.3 MHz.

Figure 9 shows the transmitted focused beam received at a
point probe placed 10 cm away in front of the array. Side lobe
levels are seen to be -37 dB down from the main lobe, so is the
grating lobe. Figure 10 is the computed result: side lobes are
-40 dB down and grating lobes are -37 dB down from the main lobe.

In both examples the agreement is quite good.

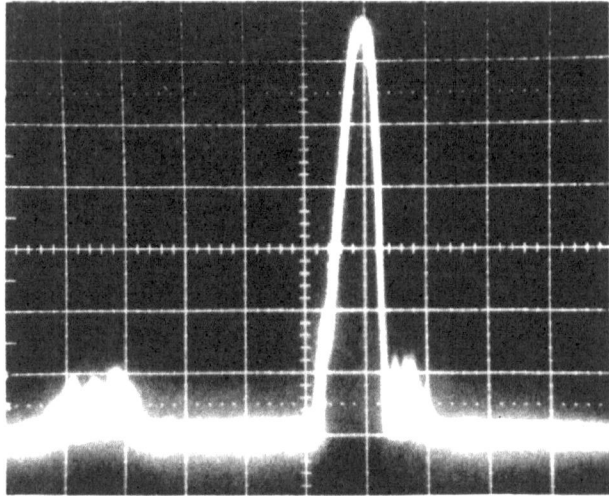

FIG. 9--Transmitted focused beam received at a point probe 10 cm
in front of the array. Double element array, elements
spaced 1 mm apart, chirp length = 40 μsec, scan velocity
1000 msec, frequency 2.3 MHz.

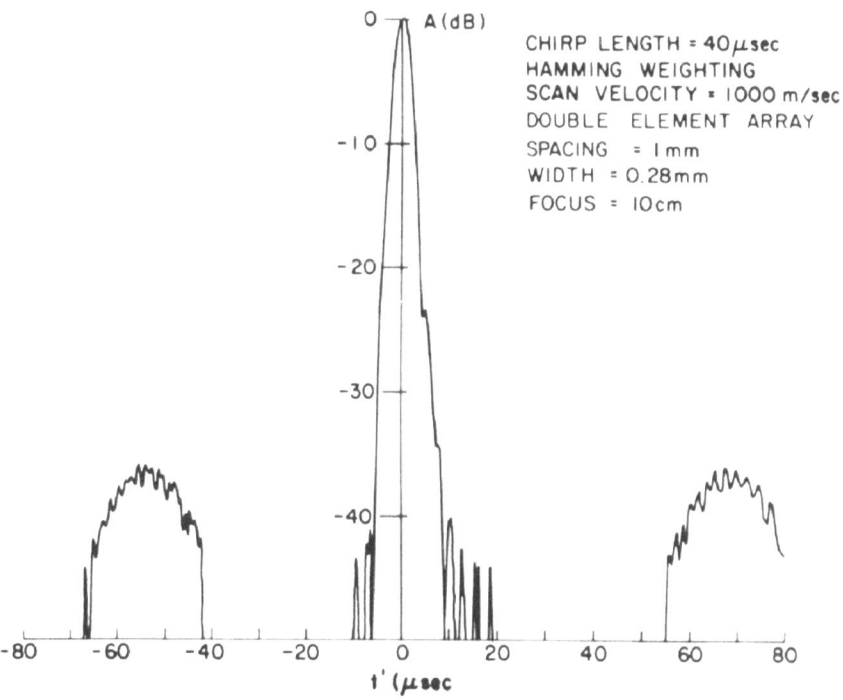

FIG. 10--Theoretical focal distribution corresponding to the experimental result shown in Fig. 9.

VI. NONLINEAR CHIRP

As we have said previously, Hamming weighting deteriorates the resolution, basically because the effective width of the chirp is decreased, as the amplitude of its two ends are reduced. In order to compensate for this effect, the chirp length can be increased. The main drawback in increasing the chirp length is that the linear chirp used in the paraxial approximation theory is no longer valid, it is necessary to use a nonlinear chirp. The equation for the required nonlinear chirp has been derived;[3] however, for simplicity, a quadratic approximation for frequency as a function of time has mainly been used for the computations. Figure 11 shows the frequency versus time characteristic of a linear chirp and a quadratic chirp over 20 μsec. As can be seen from this figure, the frequency error in the chirp introduced by using a linear chirp over 20 μsec is appreciable, but the error in using a quadratic form for the chirp is negligible. In order to illustrate the improvement in using a quadratic chirp, computations have been carried out. We have considered an array of 120 elements connected by pairs. The center-to-center spacing between consecutive pairs is 1.27 mm, the scan velocity is 1500 m/sec, the Hamming weighted chirp length is 40 μsec. Figure 13 shows the theoretical result obtained for a quadratic chirp, compared to Fig. 12 obtained for a linear chirp: the side lobe level has been drastically improved, as well as the resolution.

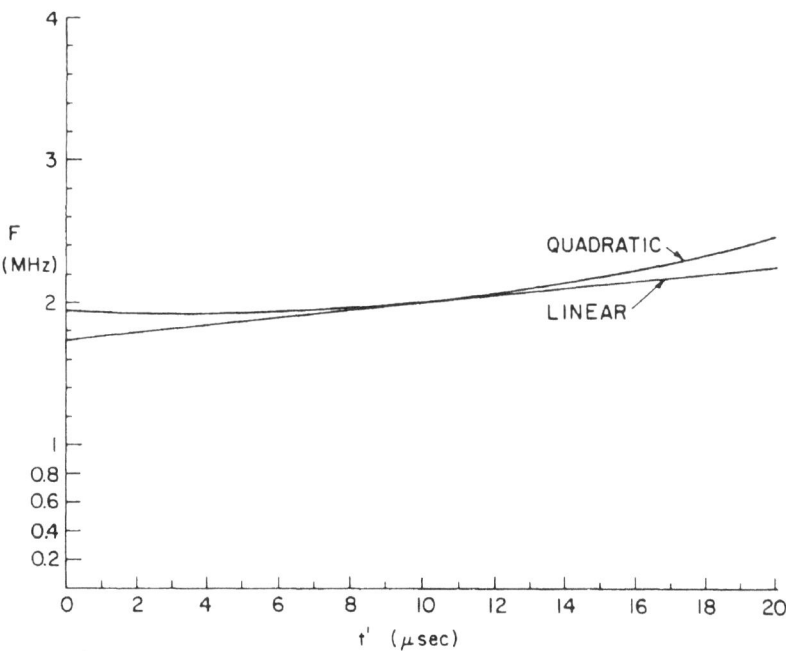

FIG. 11--Frequency vs time characteristic of a linear and a quadratic chirp over 20 μsec.

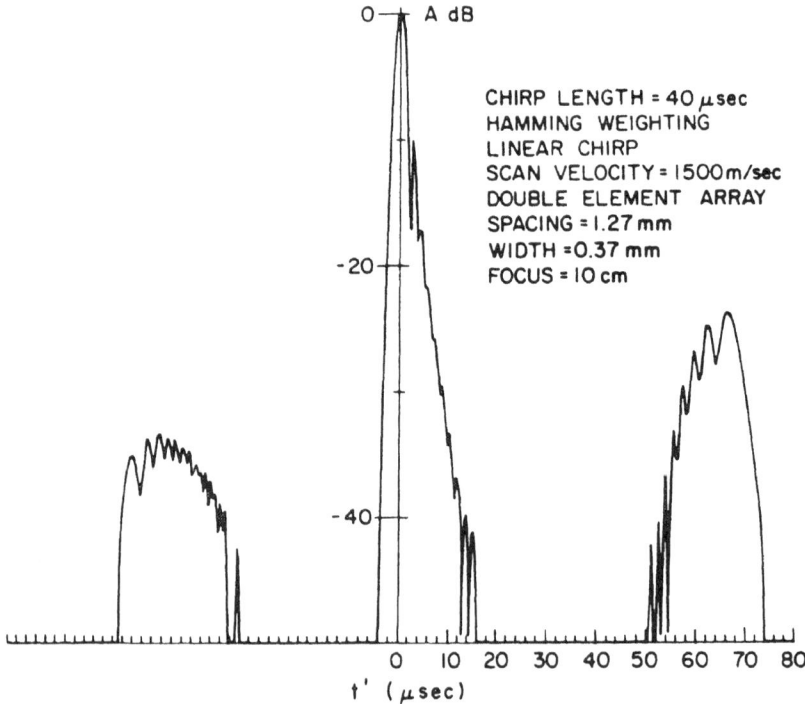

FIG. 12--Theoretical focal distribution for the chirp focused trans-
         mitter - case of linear chirp.

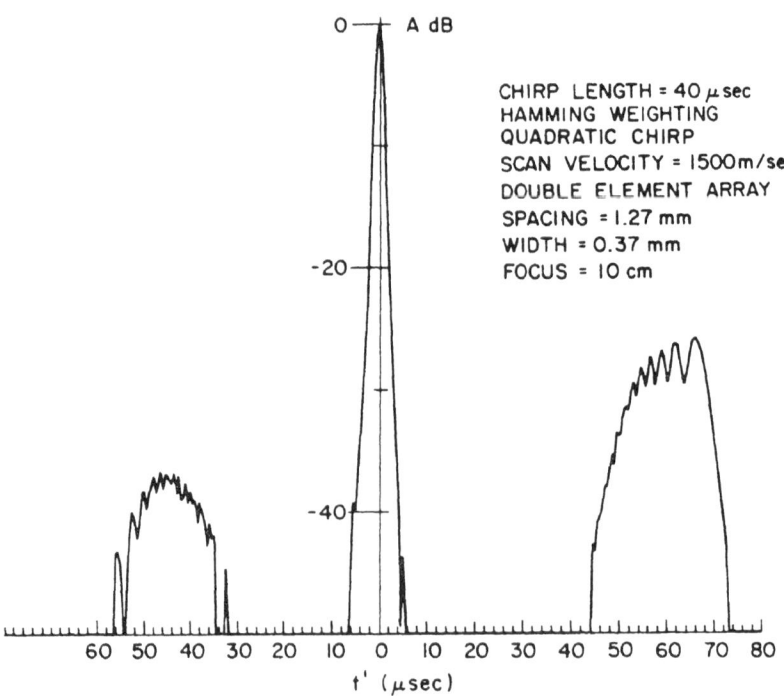

FIG. 13--Theoretical focal distribution for the chirp focused trans-
         mitter - case of quadratic chirp.

VII.   INCREASE IN SCAN VELOCITY

There is still another parameter we can vary, the scan velocity. It would be interesting to see the changes occurring to the device when one increases its value. The main reasons are that it would lead to a better range resolution as well as a physical shortening of the acoustic delay line required.

A computation has been done for a double element array, with the center-to-center spacing between adjacent elements of 1 mm. Computation has been performed for a Hamming weighted chirp of 2 cm length for different scan velocities. Figure 14(a) shows the results obtained with 2000 msec scan velocity and Fig. 14(b) with 6000 msec. As can be seen from those curves, the lateral resolution is identical, 4 mm approximately, and in both cases the grating lobe level is down 30 dB. But the range resolution would be improved from 3 mm in the first case to 1 mm in the second case, with an increase in bandwidth of 3 times. In fact, as can be shown, the range resolution is determined by the bandwidth just as in a normal pulse echo system.

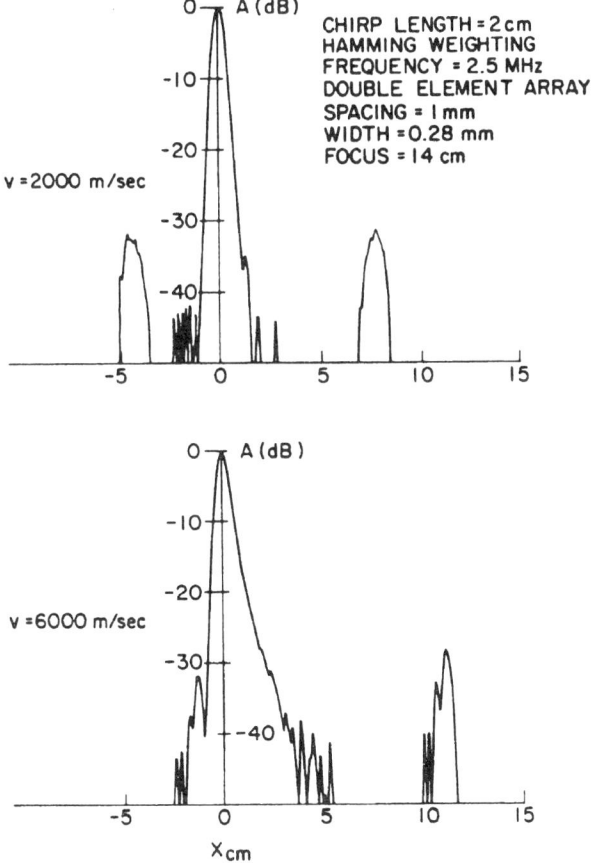

FIG. 14--Comparison of the theoretical focal distribution for two
         different scan velocities.

VIII.   CONCLUSION

We have shown how to design a chirp focused transmitter and improve side lobe and grating lobe levels. Apodization is of great help in side lobe reduction, and we have demonstrated that it is theoretically possible to obtain at least -40 dB side lobe levels and have experimentally observed -30 dB side lobe levels. Element pairing and spacing is also of great help in reducing grating lobe levels and theoretical and experimental results have shown it is possible to achieve a -30 dB grating lobe level. Nonlinear chirp and very fast scan velocity have also been studied in order to improve the system.

The system is normally expected to be used with a receiver, scanned and focused in the same way. To focus on a line z_0 from the array, the receiver is turned on at a time $T_0 = 2z_0/v_w$ after the transmitter. Along the line $z = z_0$ we obtain the product of the receiver and transmitter response from a point x_0, z_0 . However, we can also illuminate a point $x_0, z \neq z_0$ with the main lobe of the transmitter and pick it up at a time $t' - 2(z-z_0)/v_w$ on the receiver, i.e., on a side lobe of the receiver. Thus we obtain from a point two series of side lobes or grating lobes, those illuminated by the main lobe of the transmitter and received by a side lobe of the receiver, and those illuminated by a side lobe of the transmitter and received by a defocused main lobe of the receiver. In a two-dimensional B-scan display, this gives lobes perpendicular to the array and along a line at an angle to the z axis where $\tan \theta = v/v_w$. The response is therefore like that of the receiver or transmitter along these lines, but may be considerably better than this along the x direction. If, however, the system is used in a transmission configuration, the off line side lobes are no longer a problem and in fact the response of the transmitter and receiver are multiplied together which further reduces the side lobe and grating lobes levels.

We conclude that numerical computations show that side lobes are not a severe problem, but grating lobes can provide considerable difficulty particularly with badly defocused beams reflecting from a point $z \ll z_0$.

ACKNOWLEDGEMENTS

This research was sponsored by the National Science Foundation Grant ENG75-18681 and by Rockwell International, for ARPA and AFML under contract RISC 74-20773.

REFERENCES

1. J. Fraser, J. Havlice, G. Kino, W. Leung, H. Shaw, K. Toda,
    T. Waugh, D. Winslow, and L. Zitelli, "A Two-Dimensional
    Electronically Focused Imaging System," presented at the
    IEEE Group on Sonics and Ultrasonics Symposium, November 11-
    14, 1974.

2. J. Fraser, J. Havlice, G. Kino, W. Leung, H. Shaw, K. Toda,
    T. Waugh, D. Winslow, and L. Zitelli, "An Electronically
    Focused Two-Dimensional Acoustic Imaging System," Acous-
    tical Holography, edited by Newell Booth, Vol. 6, 1975,
    pp. 275-304.

3. J. Souquet, Ph.D. Thesis, Stanford University, Stanford,
    California (1976).

FAST BEAMFORMING PROCESSOR

D. A. Gaubatz

Electrical Engineering Department

The Catholic University of America

Washington, D.C.   20064   USA

ABSTRACT

A beamformer implementation is described which combines the computational efficiency of a Fast Fourier Transform algorithm with the speed and economy of analog signal processing hardware. The fast transform algorithm enables a single processor module to provide 32 simultaneous beams when used with a line array of 32 equidistantly-spaced transducers. The signal processing required to implement this function is the equivalent of performing, in real-time, a 32 point Fast Fourier Transform on 32 continuous 100 kilohertz input signals, 100 kilohertz being the isonifier frequency. The fast analog transform technique allows this amount of processing to be performed on a single circuit board, whereas a digital implementation would require a great number of high speed calculations. The operational amplifier circuit configurations which perform the multiplications and summations, and the algorithm characteristics which facilitate analog implementation, are described. The use of multiple Fast Beamforming Processor modules for the real-time generation of two-dimensional images is also discussed.

INTRODUCTION

Two-dimensional image formation can be accomplished by the

This work was supported by the Office of Naval Research under Contract N00014-75-C-0544.

multiple execution of a one-dimensional transform. An acoustic
imaging system has been developed which implements a Fast Fourier
Transform (FFT) algorithm in analog circuitry[1]. The basic one-
dimensional FFT circuit is followed by phase-shifting and combining networks to produce a complete, efficient one-dimensional beamformer. When used with a line array of N equidistantly-spaced
transducers, the beamformer provides N seperate outputs. One output
corresponds to the broadside beam aimed along the normal to the
array. Two groups of (N/2)-1 outputs each correspond to beams aimed
at increasing positive and negative angles of incidence. The one
remaining output responds to wavefronts arriving at both positive
and negative maximum viewing angles. The angle magnitudes are
determined by the transducer spacing, the isonifier operating frequency, and the wavelength of the acoustic radiation in the media.
Beamforming in two dimensions with an orthogonal array of N X N
transducers is provided by 2N Fast Beamforming Processor modules
whose final outputs produce N^2 simultaneous beams. Image processing
for an N X N planar array is accomplished in two stages. Feeding the
N columns of the array transducer outputs to N y-directed beamformer
modules comprises the first stage. The second stage consists in
feeding the N rows of the first stage beamformer outputs to N x-
directed beamformer modules. The second stage outputs represent a
square array of N^2 image points which lead directly to display. The
time required for complete two-dimensional image processing is only
the delay encountered in passing a continuous, narrow-band signal
through two analog beamformer modules. Due to the efficiency of the
FFT circuit, the input-to-output path within a single module passes
through only five operational amplifiers.

Because of the requirement to sample and store a series of
voltages, analog implementation of an FFT for use in the time-fre-
quency domains has not been attractive[2]. In the beamforming application, however, the transform is used to process samples which are
taken spatially as opposed to temporally. The spatially-sampled
voltages are the transducer outputs and are available continuously.
Therefore a transform operating in the space-wave number domains is
not required to perform discrete time sampling, or analog-to-digital
conversion, but can immediately apply analog transform processing to
the transducer outputs. Although the analog approach raises the
standard questions regarding accuracy and repeatability, it does
offer a comparative advantage, at least for the present, in operating
speed and probably also in cost.

The experimental equipment which has been constructed to verify
the analog transform approach to acoustic imaging is based on a 16
point FFT circuit and is used with a 16 X 16 transducer array. The
present discussion concerns the algorithm derivation and implementation details for a FFT circuit which is expanded to 32 points.

Construction of an FFT algorithm in analog circuitry requires,

first of all, that summations are performed by operational amplifiers
and that coefficient multiplications are implemented by amplifier
feedback to input resistor ratios. The real and imaginary parts of a
complex signal must be represented by seperate circuits. Following
this formula, any standard FFT algorithm will lead to a circuit diagram[3]. However, analog circuitry places constraints on transform
algorithms that have no exact counterpart in either hardware or software digital implementations. For example, the distance in address
space that two inputs to a combining operation are removed from each
other is generally immaterial, except for the effort required to generate the addresses. But a great distance in address space usually
translates into a long conductor path which crosses many other paths.
A further example is that the actual amount of computation that needs
to be performed throughout the FFT arrays changes with W, the complex
multiplier. This has the consequence in analog circuitry that some
operations require more amplifiers than others, giving rise to a physical circuit layout of irregular density. Also, each amplifier's
ability to accept and drive a given number of inputs and outputs,
respectively, will not be optimally employed. Another disadvantage
of standard algorithms is that they are formulated for complex inputs
which are unnecessary in the beamforming application. Algorithms
which have been modified for real-valued inputs, however, do not automatically produce a signal flow graph which is optimized for analog
circuitry[4]. The approach taken here is to begin with the matrix-
vector product expression for the 32 DFT's having only real-valued
inputs and to derive an algorithm which satisfies the relevant
criteria.

The relationship of the spatial DFT to beamforming in one dimension is discussed in the next section and the discussion is expanded
to two dimensions in the section that follows. A further two sections
are devoted to the FFT algorithm derivation and implementation. A
final section discusses the design and fabrication details that need
to be addressed for an optimized beamformer module implementation.

ONE-DIMENSIONAL BEAMFORMING AND THE DISCRETE FOURIER TRANSFORM

Beamformer processing requires that a Fourier transform be performed over the pressure field, u(x,t), of the array aperture. This
is given by

$$U[k_x,t] = \int u(x,t)\, \exp[-jk_x x]\, dx \qquad (1)$$

where x is the spatial variable along the aperture and $k_x = k \sin \theta$.
The acoustic wave number, k, is equal to $2\pi/\lambda$, where λ is the wavelength. The angle θ is the angle between the array normal and the
direction to which the beam's maximum sensitivity is steered. Since
the array contains a finite number of transducers, the continuous
integration gives way to a summation over discrete points.

$$U[k_x,t] = \sum_{m=0}^{N-1} u(md,t) \exp[-jk_xmd] \qquad (2)$$

where x is replaced by md, m being the transducer number and d being the distance between transducers. The total number of transducers is N and the (1/N) factor which should precede the summation is temporarily ignored. Evaluating the exponential term in equation (2) yields the Fourier multipliers for the discrete Cosine and Sine transforms. These multipliers function as the transducer output weighting coefficients. The term k_x is not allowed to vary continuously over θ but is restricted to values suitable for FFT processing. The allowed values are given in general by

$$k_x = k \sin\theta = k\, p\, \lambda/(Nd), \quad p = 0,1,\ldots,N-1 \qquad (3)$$

The wave number or spatial frequency term thus becomes $k_x(p)$ and equation (2) is rewritten as

$$U[k_x(p),t] = \sum_{m=0}^{N-1} u(md,t) \exp[-jk_x(p)md], \quad p = 0,1,\ldots,N-1 \qquad (4)$$

Equation (4) makes apparent the similarity between the discrete wave-number variable and W^{-mp}, the standard FFT multiplier.

$$\exp[-jk_x(p)md] = \exp[-j(2\pi mp)/N] = W^{-mp} \qquad (5)$$

Because the spatially sampled transform inputs given by u(md,t) are only real, the complex conjugate relationship applies to the transform outputs. It is therefore sufficient for beamforming purposes to form only the Cosine transforms for $k_x(p)$, $p = 0,1,\ldots,(N/2)$, and the Sine transforms for $k_x(p)$, $p = 1,2,\ldots,(N/2)-1$. The Cosine and Sine transform outputs are in quadrature and supplying an additional 90° phase shift to the Sine transform output produces signals which are combined as shown in equations (6) and (7).

$$V(+p,t) = (+0.5)\{Re\ U[k_x(p),t]+(1/c)\textstyle\int Im\ U[k_x(p),t]dt\} \qquad (6)$$

$$V(-p,t) = (+0.5)\{Re\ U[k_x(p),t]-(1/c)\textstyle\int Im\ U[k_x(p),t]dt\} \qquad (7)$$

The 90° phase shift is provided by an inverting integrator and the constant c is adjusted for unity gain through the integrator circuit at the isonifier frequency. A wavefront incident at $\theta = \arcsin +\lambda(Nd)$, (p = 1), will produce an output only at V(+1,t). The discrete Fourier transformation and half-plane discriminator processing are illustrated in Figure 1 for N = 8 and p = 1. The phase shift and summations following the DFT remain unchanged for $p = 1,2,\ldots,(N/2)-1$. A broad-side-directed beam is represented by p = 0. When p = (N/2), the beam is steered to the limits of the viewing angle and the half-plane

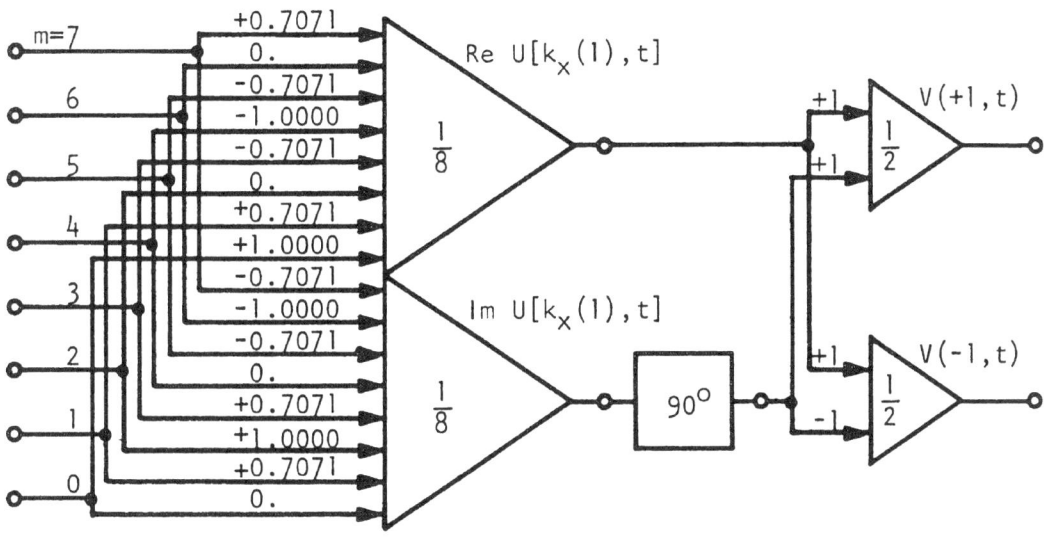

BEAMFORMER PROCESSING

Figure 1

origin can not be determined. Wavefronts incident at greater angles produce aliased images. Equations (8) and (9) complete the beamformer processing for a line array of N transducers.

$$V(\quad 0,t) = (+0.5) \ \text{Re} \ U[k_x(\quad 0),t] \tag{8}$$

$$V(\pm N/2,t) = (+0.5) \ \text{Re} \ U[k_x(N/2),t] \tag{9}$$

TWO-DIMENSIONAL BEAMFORMING

The one-dimensional beamforming described in the previous section can readily be applied to two dimensions. The array becomes a N X N planar array with each transducer output given by $u(x,y,t)$. In terms of the discrete spatial variables, m and n, the outputs become $u(md,nd,t)$. The transducer spacing, d, is the same in both x and y directions. The two-dimensional spatial discrete Fourier transform is given by

$$U[k_x(p),k_y(q),t] = \sum_{m=0}^{N-1} \sum_{n=0}^{N-1} u(md,nd,t) \ \exp\{-j[k_x(p)md+k_y(q)nd]\} \tag{10}$$

$$p,q = 0,1,\dots,N-1$$

Recalling the complex conjugate relationship to reduce redundancy, and seperating the exponential term, the first stage of processing becomes N DFT's per column over the y-directed variable.

$$U[m,k_y(q),t] = \sum_{n=0}^{N-1} u(md,nd,t) \ \exp[-jk_y(q)nd] \tag{11}$$

The second processing stage is N DFT's per row of the first stage outputs.

$$U[k_x(p),k_y(q),t] = \sum_{m=0}^{N-1} U_y[m,k_y(q),t] \exp[-jk_x(p)md] \qquad (12)$$

The requirement for quadrant resolution in two-dimensional beamforming is satisfied by the two successive operations of the half-plane discriminator circuits which are directly attached to the one-dimensional FFT outputs.

FAST ANALOG TRANSFORM DERIVATION

The fast analog transform is optimized for 32 real-valued inputs and analog implementation. The transform employs 96 summing nodes or amplifiers to produce 17 discrete Cosine transforms and 15 discrete Sine transforms, given by equations (13) and (14), respectively.

$$Re\ U[k_x(p),t] = \sum_{m=0}^{31} u(md,t)\ Re\ \exp[-j(2\pi mp)/32],\ p = 0,1,.\ ,16 \quad (13)$$

$$Im\ U[k_x(p),t] = \sum_{m=0}^{31} u(md,t)\ Im\ \exp[-j(2\pi mp)/32],\ p = 1,2,.\ ,15 \quad (14)$$

The 96 summing nodes are evenly distributed in 3 columns of 32 nodes each. The transducer outputs comprise the input array and are expressed as $u_0(m_0)$, $m_0 = 0,1,\ldots ,31$. The time dependency of the input signals is ignored. The input array is followed by 3 computation arrays given by $u_1(m_1)$, $u_2(m_2)$ and $u_3(m_3)$, where m_1, m_2, $m_3 = 0,1,..$. ,31. The derived algorithm will be recognized as a mixed-radix form of 4 X 4 X 2.

The starting point for the algorithm derivation is a 32 X 32 matrix containing the signed-magnitude Fourier multipliers produced by the exponential terms in equations (13) and (14). The multipliers required for the Cosine and Sine transforms for p = 1 are listed in Table 1a, b, respectively. The multiplier magnitudes are rounded to five places and given symbol equivalents which are used later. In Table 1a the magnitude 0.9808 appears four times with the sign progression + - - +. The same magnitude, expressed by B, also appears four times in Table 1b, but with the different sign progression of + + - -. These sign progressions and two others, + + + +, + - + -, are used to form the transform's first computation array. The rows and output vector of the matrix-vector product expression for the 32 DFT's are reordered, grouping together the DFT's in which these sign progressions are evident. This initial reordering is preserved in the output array $\{u_3\}$. The association of sign progressions with output vector groups is as follows: + + + +: $u_3(0)$-$u_3(7)$,

+ - + -: $u_3(8)-u_3(15)$, + - - +: $u_3(16)-u_3(23)$, + + - -: $u_3(24)-u_3(31)$.
The correspondence between the elements of $\{u_3\}$ and the DFT's is
given in Table 3. The reordered 32 X 32 matrix is factored into two
matrices, the second giving the equations for the first computation
array.

$$u_1(m_1)=+u_0(m_1)+u_0(m_1+8)+u_0(m_1+16)+u_0(m_1+24), \; m_1=0,1,.,7 \qquad (15)$$

$$u_1(m_1)=+u_0(m_1-8)-u_0(m_1)+u_0(m_1+8)-u_0(m_1+16), \quad m_1=8,9,.,15 \qquad (16)$$

$$u_1(16)=+u_0(0)-u_0(16) \qquad (17)$$

$$u_1(m_1)=+u_0(m_1-16)-u_0(32-m_1)-u_0(m_1)+u_0(48-m_1), \; m_1=17,18,.,23 \; (18)$$

$$u_1(24)=+u_0(8)-u_0(24) \qquad (19)$$

$$u_1(m_1)=+u_0(32-m_1)+u_0(m_1-16)-u_0(48-m_1)-u_0(m_1), \; m_1=25,26,.,31 \; (20)$$

The first matrix can now be partitioned into sixteen 8 X 8 matrices,
twelve of which contain all zero elements. The four remaining matrices, each of which has eight elements of $\{u_1\}$ as its input vector, can
be factored to produce $\{u_2\}$, the second computation array. The equations for $\{u_2\}$ are listed in Table 2 and they make use of the multiplier symbols assigned in Table 1a. Partitioning the remaining matrices yields sixteen 2 X 2 matrices whose input vectors are elements of
$\{u_2\}$. The equations for the third and final computation array are
listed in Table 3.

Figure 2 contains a signal flow graph representation of the
transform that clearly anticipates analog implementation. Fully one-
fourth of the complete 32 point transform is contained in Figure 2
and the remaining sections follow the same format. All multipliers
having a value between zero and one are contained in the $\{u_2\}$ array.
Output loading on array nodes is equalized in that the $\{u_0\}$ and $\{u_1\}$
nodes drive only four, or fewer, nodes each. The (1/N) factor that
has been ignored up to this point is evident as three seperate factors,
(1/4), (1/4), (1/2), in the signal flow graph. These factors supply
the overall (1/32) factor necessary in the 32 point transform.

FAST ANALOG TRANSFORM IMPLEMENTATION

The signal flow graph of Figure 2 leads directly to the schematic diagram for analog circuitry. Each summer is replaced by an
operational amplifier symbol and the multiplications are implemented
by appropriately scaled resistors. Figure 3 illustrates the equivalent schematic diagram for the summer which produces the $u_2(18)$ output. Figure 3 also introduces the convention of attaching the inputs
multiplied by a positive coefficient to the amplifier's inverting
input. A negative coefficient dictates therfore that connection

Table 1 - a, b

m=	a	b
0	+1.0000=+1	0
1	+0.9808 +B	+H
2	+0.9239 +C	+G
3	+0.8315 +D	+F
4	+0.7071 +E	+E
5	+0.5556 +F	+D
6	+0.3827 +G	+C
7	+0.1951 +H	+B
8	0.	+1
9	-0.1951 -H	+B
10	-0.3827 -G	+C
11	-0.5556 -F	+D
12	-0.7071 -E	+E
13	-0.8315 -D	+F
14	-0.9239 -C	+G
15	-0.9808 -B	+H
16	-1.0000 -1	0
17	-0.9808 -B	-H
18	-0.9239 -C	-G
19	-0.8315 -D	-F
20	-0.7071 -E	-E
21	-0.5556 -F	-D
22	-0.3827 -G	-C
23	-0.1951 -H	-B
24	0. 0	-1
25	+0.1951 +H	-B
26	+0.3827 +G	-C
27	+0.5556 +F	-D
28	+0.7071 +E	-E
29	+0.8315 +D	-F
30	+0.9239 +C	-G
31	+0.9808 +B	-H

Table 2

```
u2( 0)=+ u1( 0)+ u1( 2)+ u1( 4)+ u1( 6)
u2( 1)=+ u1( 1)+ u1( 3)+ u1( 5)+ u1( 7)
u2( 2)=+ u1( 0)        - u1( 4)         1
u2( 3)=+Eu1( 1)-Eu1( 3)+ u1( 5)+ u1( 7)
u2( 4)=+ u1( 0)- u1( 2)+ u1( 4)- u1( 6)
u2( 5)=+ u1( 1)- u1( 3)+ u1( 5)- u1( 7)
u2( 6)=       + u1( 2)        - u1( 6)
u2( 7)=+Eu1( 1)+Eu1( 3)-Eu1( 5)-Eu1( 7)
u2( 8)=+ u1( 8)+Eu1(10)        -Eu1(14)
u2( 9)=+Cu1( 9)+Gu1(11)-Cu1(13)-Gu1(15)
u2(10)=+ u1( 8)-Eu1(10)        +Eu1(14)
u2(11)=+Gu1( 9)-Cu1(11)-Gu1(13)+Cu1(15)
u2(12)=       + u1(10)        + u1(14)
u2(13)=+Gu1( 9)+Cu1(11)+Gu1(13)+Cu1(15)
u2(14)=       + u1(10)        + u1(14)
u2(15)=+Cu1( 9)-Gu1(11)+Cu1(13)-Gu1(15)
u2(16)=+ u1(16)+Eu1(18)+ u1(20)+Gu1(22)
u2(17)=+Bu1(17)+Fu1(19)+Du1(21)+Hu1(23)
u2(18)=+ u1(16)-Eu1(18)+ u1(20)-Cu1(22)
u2(19)=+Du1(17)-Hu1(19)-Bu1(21)-Fu1(23)
u2(20)=+ u1(16)-Gu1(18)+ u1(20)+Cu1(22)
u2(21)=+Fu1(17)-Bu1(19)+Hu1(21)+Du1(23)
u2(22)=+ u1(16)-Cu1(18)+ u1(20)-Gu1(22)
u2(23)=+Hu1(17)-Du1(19)-Fu1(21)-Bu1(23)
u2(24)=+ u1(24)+Cu1(26)+ u1(28)+Gu1(30)
u2(25)=+Bu1(25)+Du1(27)+Fu1(29)+Hu1(31)
u2(26)=- u1(24)-Gu1(26)+ u1(28)+Cu1(30)
u2(27)=-Du1(25)+Fu1(27)+Bu1(29)+Fu1(31)
u2(28)=+ u1(24)-Gu1(26)+ u1(28)+Cu1(30)
u2(29)=+Fu1(25)-Bu1(27)+Hu1(29)+Du1(31)
u2(30)=- u1(24)-Cu1(26)+ u1(28)+Gu1(30)
u2(31)=-Hu1(25)-Fu1(27)-Du1(29)+Bu1(31)
```

Table 3

```
u3( 0)=Re U[kx( 0),t] =+u2( 0)+u2( 1)
u3( 1)=Re U[kx(16),t] =+u2( 0)-u2( 1)
u3( 2)=Re U[kx( 4),t] =+u2( 2)+u2( 3)
u3( 3)=Re U[kx(12),t] =+u2( 2)-u2( 3)
u3( 4)=Re U[kx( 8),t] =+u2( 4)
u3( 5)=Im U[kx( 8),t] =      +u2( 5)
u3( 6)=Im U[kx( 4),t] =+u2( 6)+u2( 7)
u3( 7)=Im U[kx(12),t] =-u2( 6)+u2( 7)
u3( 8)=Re U[kx( 2),t] =+u2( 8)+u2( 9)
u3( 9)=Re U[kx(14),t] =+u2( 8)-u2( 9)
u3(10)=Re U[kx( 6),t] =+u2(10)+u2(11)
u3(11)=Re U[kx(10),t] =+u2(10)-u2(11)
u3(12)=Im U[kx( 2),t] =+u2(12)+u2(13)
u3(13)=Im U[kx(14),t] =-u2(12)+u2(13)
u3(14)=Im U[kx( 6),t] =+u2(14)+u2(15)
u3(15)=Im U[kx(10),t] =-u2(14)+u2(15)
u3(16)=Re U[kx( 1),t] =+u2(16)+u2(17)
u3(17)=Re U[kx(15),t] =+u2(16)-u2(17)
u3(18)=Re U[kx( 3),t] =+u2(18)+u2(19)
u3(19)=Re U[kx(13),t] =+u2(18)-u2(19)
u3(20)=Re U[kx( 5),t] =+u2(20)+u2(21)
u3(21)=Re U[kx(11),t] =+u2(20)-u2(21)
u3(22)=Re U[kx( 7),t] =+u2(22)+u2(23)
u3(23)=Re U[kx( 9),t] =+u2(22)-u2(23)
u3(24)=Im U[kx( 1),t] =+u2(24)+u2(25)
u3(25)=Im U[kx(15),t] =-u2(24)+u2(25)
u3(26)=Im U[kx( 3),t] =+u2(26)+u2(27)
u3(27)=Im U[kx(13),t] =-u2(26)+u2(27)
u3(28)=Im U[kx( 5),t] =+u2(28)+u2(29)
u3(29)=Im U[kx(11),t] =-u2(28)+u2(29)
u3(30)=Im U[kx( 7),t] =+u2(30)+u2(31)
u3(31)=Im U[kx( 9),t] =+u2(30)+u2(31)
```

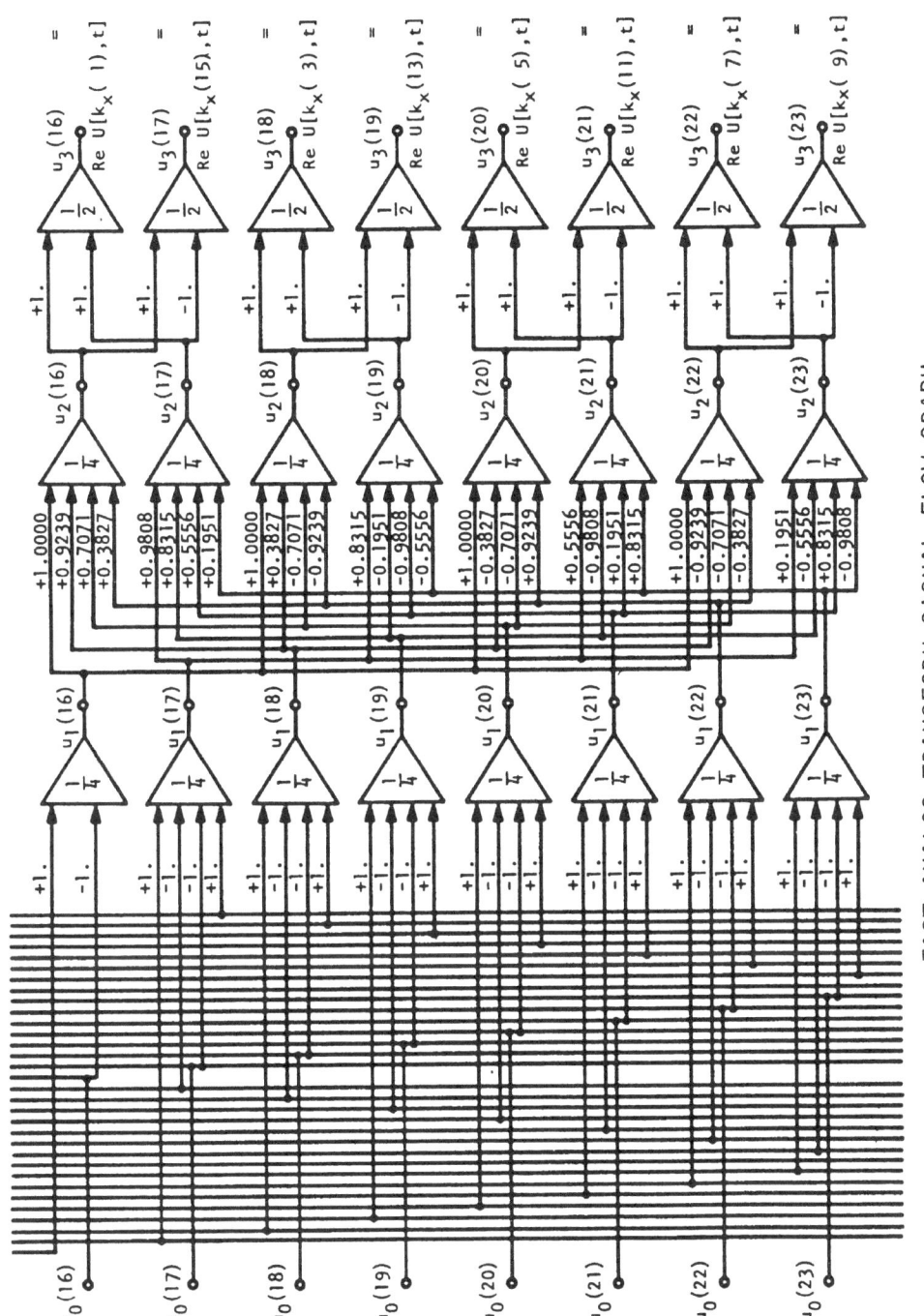

FAST ANALOG TRANSFORM SIGNAL FLOW GRAPH

Figure 2

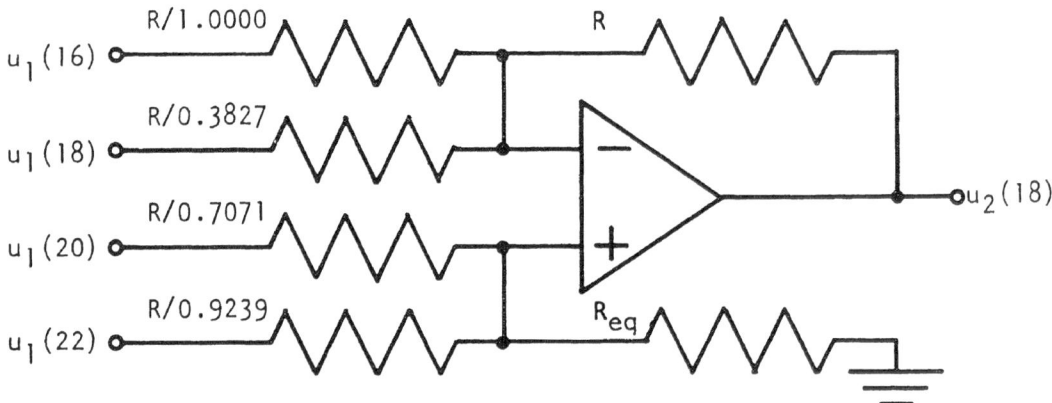

OPERATIONAL AMPLIER CIRCUIT CONFIGURATION

Figure 3

be made to the non-inverting input. This practice maximizes the use
of the amplifiers as inverting summers and simplifies the circuitry.
However, it must be remembered that a complete polarity reversal
occurs at each column of array outputs.

The resistor values are determined by the ratio of the feedback resistor to the relevant multiplier. The one remaining resistor
in Figure 3, R_{eq}, is used to equalize the impedances seen by both
inverting and non-inverting inputs. The value of this resistor must
be calculated individually for each summing node[5]. To implement the
three factors of the overall (1/32) multiplier which are shown in
Figure 2, the feedack resistor, R, is reduced to R/4, R/4 and R/2 for
the factors (1/4), (1/4) and (1/2), respectively. The four input
resistors are left unchanged. The impedance equalizing resistor is
calculated for the reduced feedback resistor. Existing versions of
the transform circuit use an R value of 20,000 ohms.

The operational amplifiers used are of the monolithic integrated
circuit type which are available in single, dual, or quad packaging
configurations. Specifications of importance are sufficient open-loop
gain (about 4 mHz unity gain crossing) and low input bias current to
prevent a large DC voltage from developing through the direct-coupled
circuitry. The amplifiers used are internally compensated, thereby
minimizing the number of external components required. The same amplifier is also used in the integrator circuit which provides the 90°
phase-shift for the half-plane resolution[6].

The existing 16 point Fourier transform circuits were constructed
using standard MIL/EIA 1% tolerance resistors which were chosen to
best match the calculated values. This implementation achieved an
operating system measure of 30% amplitude response for the first side
lobe which compares favorably to the 22% of maximum predicted for an
ideal transducer array-beamformer system[7].

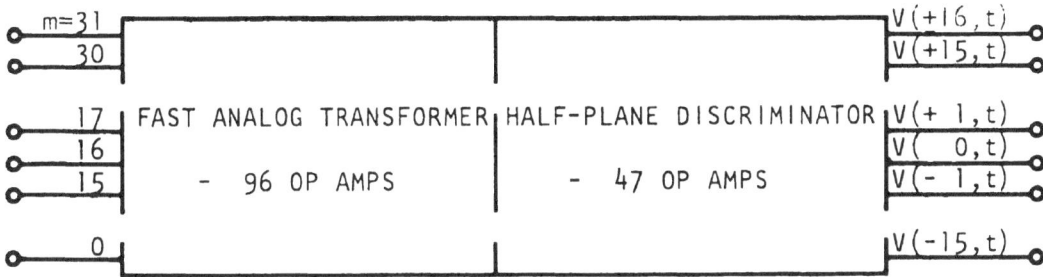

FAST BEAMFORMING PROCESSOR MODULE

Figure 4

The module representation for the 32-transducer beamforming
processor is shown in Figure 4. The fast analog transform and half-
plane discriminator circuits require 96 and 47 amplifiers, respectively. Several amplifiers throughout the circuit can be eliminated
due to localized configurations using less than four inputs, however
this saving is made at the expense of an otherwise highly regular
circuit layout pattern.

DISCUSSION

The Fast Beamforming Processor design presents an alternative to
beamforming via delay element or digital techniques. In implementing
the spatial Fourier transform in analog circuitry, the quantization
error due to analog-to-digital convertors is avoided. However another
form of error is encountered at choosing practical resistor values to
implement the Fourier coefficients. The possibility of custom resistor
fabrication suggests itself, which would also offer decreased component
size as well as improved temperature tracking between resistor values.

Because of the direct-coupled circuitry and internally compensated amplifiers, there is little to inhibit an extensive miniaturization effort. The existing printed circuit cards operate narrow-band
with a nominal isonifier frequency of 100 kHz and little precaution
has been taken regarding the circuit's physical layout. This also
points to the possiblity of extensive miniaturization. The 16 point
cards occupy about one square foot of circuit board area, whereas the
32 point cards using a more compact algorithm and dual amplifiers will
be only slightly larger. Standard printed circuit fabrication techniques necessitate conductor paths having orders of magnitude more
current-carrying capacity than is required in the processor circuitry.
Decreasing the conductor size would have the added benifit of lessen-
ing effects due to distributed capacitance. Capacitance effects would
need to be addressed if the beamformer were to operate at significantly higher frequencies. Faster amplifiers would also be required.

A further degree of design freedom in the construction of the transform lies in the direction of low-power circuitry. The intermediate transform nodes need never make external connection and therefore the only drive requirement to be satisfied is that presented by the following transform circuitry. In addition to the desireable minimization of power consumption, the susceptibility to unwanted cross-coupling through the power supply would be reduced.

Although the use of analog circuitry to perform signal processing functions which are steadily becoming the domain of digital techniques invites problems due to component parameter drift, temperature effects, and so on, it is apparent that further development in the areas discussed above promises to minimize the disadvantages while maintaining the system concept of two-dimensional beamforming via multiple, identical modules. The verification system which has been constructed contains only the thirty-two 16 point transform cards and therefore generates images of objects located in the far-field. Proposals have also been made regarding focusing methods as well as further uses of the system[7]. The 32 point fast analog transform and half-plane discriminator circuitry described here represents about the limit of what can be done on one printed circuit card using standard fabrication and interconnection techniques. A design for a 64 point beamformer which is complete on four circuit cards has been reported elsewhere[8].

CONCLUSION

The beamformer design described here presents an analog approach to the simultaneous formation of 32 beams with a line array of 32 transducers. The use of 64 identical Fast Beamforming Processor modules provides the basis for a two-dimensional imaging system which resolves 1024 points and operates in real-time. The module design presented, as well as hardware which has already been constructed, is compatible with standard printed circuit card fabrication techniques and no special active components are required. The circuit configurations invite extensive optimization, primarily in the module fabrication area. The fast beamformer circuit may be readily adapted to arrays of greater or fewer than 32 transducers or to isonifier frequencies other than 100 kHz. Continued development of the analog FFT-based beamformer module is expected to yield beamformers which are equally useful in one- and two-dimensional applications and which compare favorably with other techniques in the areas of cost, physical size, ruggedness, maintainability, power consumption and speed.

REFERENCES

1. H. F. Harmuth, J. Kamal, S. S. R. Murthy, "Two-Dimensional
    Spatial Hardware Filters for Acoustic Imaging," Applications
    of Walsh Functions and Sequency Theory, H. Schreiber and G. F.
    Sandy, ed., IEEE, New York, 1974, pp. 94-125.

2. G. Ramos, "Analog Computation of the Fast Fourier Transform,"
    IEEE Proceedings, Vol. 58, No. 11, Nov. 1970, pp. 1861-1863.

3. J. W. Cooley, J. W. Tukey, "An Algorithm for the Machine Calcu-
    lation of Complex Fourier Series," Mathematics of Computation,
    Vol. 19, No. 90, Apr. 1965, pp. 297-301.

4. G. D. Bergland, "A Fast Fourier Transform Algorithm for Real-
    Valued Series," Communications of the ACM, Vol. 11, No. 10,
    Oct. 1968, pp. 703-710.

5. R. G. Kostanty, "Doubling Op Amp Summing Power," Electronics,
    Vol. 45, No. 4, Feb. 14, 1972, pp. 73-75.

6. L. Wisseman, J. J. Robertson, "High Performance Integrated
    Operational Amplifiers," Motorola Semiconductor Products, Inc.
    Application Note, AN-204, Phoenix, 1968.

7. H. F. Harmuth, "Generation of Images by Means of Two-Dimensional
    Spatial Hardware Filters," Advances in Electronics and Electron
    Physics, Vol. 40, Academic Press, New York, 1976, pp. 167-248.

8. D. A. Gaubatz, "FFT-Based Analog Beamforming Processor," 1976
    Ultrasonics Symposium Proceedings, Annapolis, Maryland, Sept.
    29-Oct. 1, 1976.

FRESNEL ZONE FOCUSING OF LINEAR ARRAYS

APPLIED TO B AND C ECHOGRAPHY

P. ALAIS, M. FINK

LABORATOIRE D'ACOUSTIQUE PHYSIQUE, INFORMATIQUE
EQUIPE DE RECHERCHE ASSOCIEE AU C.N.R.S.
UNIVERSITE PIERRE ET MARIE CURIE
PARIS , FRANCE

INTRODUCTION

For a long time, physicists have tried to get acoustical pictures like ordinary X rays pictures. The through imaging techniques, holographic or ordinary imaging ones, have not yet proved to be of the same interest for medical use as echographic techniques. On the other hand, these later techniques give essentially cross-sectional B echograms, while the C echographic techniques remain quite limited due to the time required by X Y mechanical scanning. It is possible to run faster using an electronically X scanned one dimensional array, but the lateral resolution of such a device must be higher both in X and Y directions than for B echography. Electronic focusing by delay lines may be combined to an X or Θ scanner (1, 2) this technique requires high performance circuits able to handle the great dynamic range of ultrasonic echoes. We want to present here a relatively simpler technique using holographic focusing useful for both B and C echography.

In coherent optics, a plane monochromatic wave may be focused with an holographic lens which is simply the hologram of a point (Fig. 1) i.e. the classical Fresnel ring configuration. The same focusing may be obtained in acoustics by modulating the amplitude of the emitted signal according to the Fresnel law, i.e. : $\cos(\alpha r^2)$. Both conjugate symmetrical converging and diverging waves are launched. In fact, such a continuous amplitude modulation is not so easy to handle when dealing with an acoustical array and it is much simpler to use a 3 state modulation law $\Pi(\alpha r^2)$ (Fig.2), i.e. to deliver a unique emission signal, either in phase, or in antiphase, or not at all. It is well known that with this 3 state modulation,

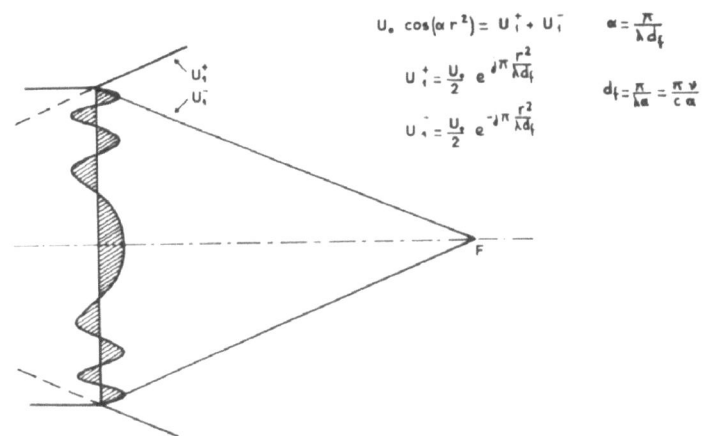

$$U_o \cos(\alpha r^2) = U_1^+ + U_1^- \qquad \alpha = \frac{\pi}{\lambda d_f}$$

$$U_1^+ = \frac{U_o}{2} e^{j\pi \frac{r^2}{\lambda d_f}}$$

$$U_1^- = \frac{U_o}{2} e^{-j\pi \frac{r^2}{\lambda d_f}} \qquad d_f = \frac{\pi}{\lambda \alpha} = \frac{\pi \nu}{c \alpha}$$

Fig 1 — Holographic focalisation by amplitude modulation

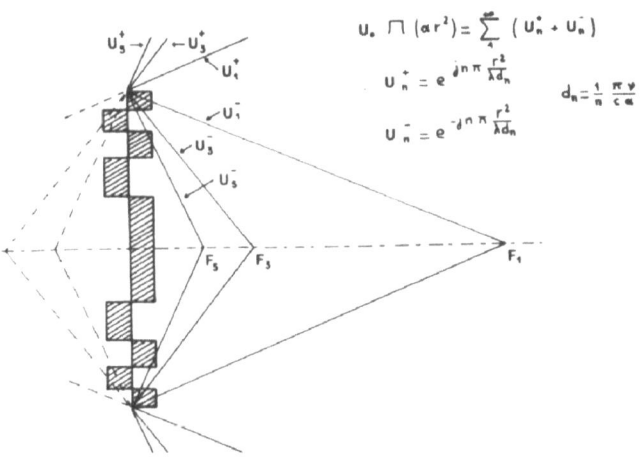

$$U_o \sqcap(\alpha r^2) = \sum_1^\infty (U_n^+ + U_n^-)$$

$$U_n^+ = e^{jn\pi \frac{r^2}{\lambda d_n}}$$

$$U_n^- = e^{-jn\pi \frac{r^2}{\lambda d_n}} \qquad d_n = \frac{1}{n} \frac{\pi \nu}{c \alpha}$$

Fig 2 — Holographic focalisation by 3 states modulation

secondary focuses appear as well as symmetrical diverging conjugate waves. We want to show that when looking for an echographic respon-se, coming from the distance of the main focus, the contribution of all these secondary waves may be not too important for medical uses. This type of spherical focusing has been studied recently at Stan-ford University (3), with a piezoelectric transducer having odd rings and even rings polarized in opposite directions, and in our labora-tory (4) with an electrostatic transducer where the rings are **drawn** on a printed circuit. This paper deals with the analog cylindrical focusing technique, that may be obtained from a one dimensional array (5,6).

THE C ECHOGRAPHIC PROBE

If instead of using a circular Fresnel transducer, we consider a one dimensional X Fresnel transducer with a large aperture in the Y direction, the converging wave will be cylindrically focused on a Y line (Fig.3). Spherical focusing may be obtained by cylindrical focusing in the YOZ plane using an array of cylindrical emitters. It may be checked that there is no problem in associating two different focusing techniques in XOZ and YOZ planes, from the fact that in the Fresnel approximation the X and Y variables are separate in the Fresnel kernel $exp\left[-\frac{j}{}\left(x^2+y^2\right)/\lambda z\right]$ at least for monochromatic radiation.

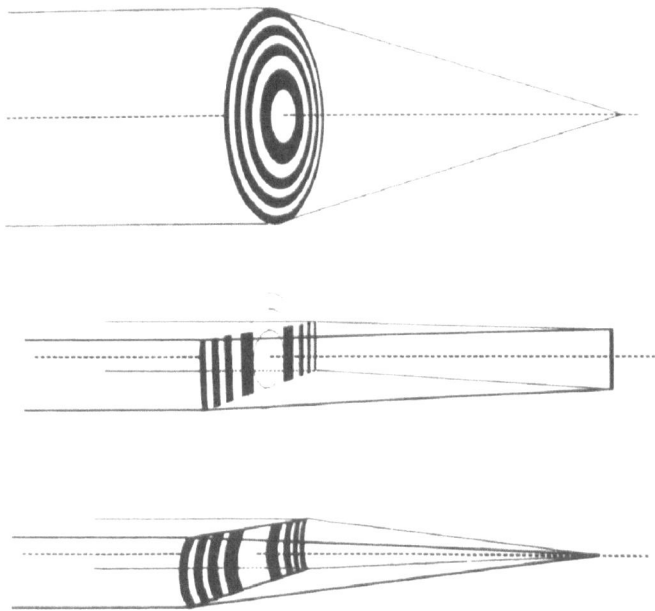

Fig 3 — Various Fresnel focusing techniques

The figure 4 shows how a Fresnel profile $\Pi(\propto r^2)$ may be approached with a linear array of transducers. Each transducer is connected with an electronic inverter to the main emission reception circuitry either in phase or in antiphase through a transformer. Appropriate sampling is determined from the wave length, the aperture, and the focusing distance. We have built a cylindrical probe of 160 transducers, 40 mm wide in the Y direction, 200 mm long in the X direction with a 200 mm radius of curvature, working at 2 MHz. The electronic circuitry may connect the transducers according to Fresnel groups of up to 64 elements and translate the group step by step in the X direction. In the simplest case, the same Fresnel group, i.e. the same focusing is used as well at emission and at reception. The emission signal may have a few oscillations in C echography, and

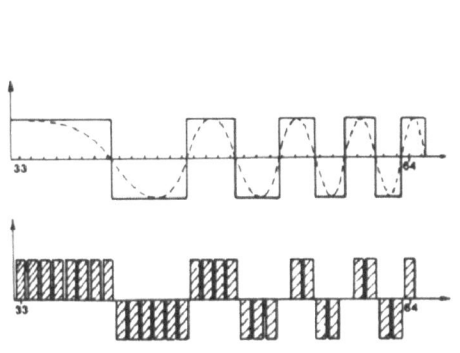

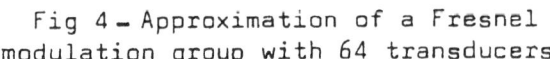

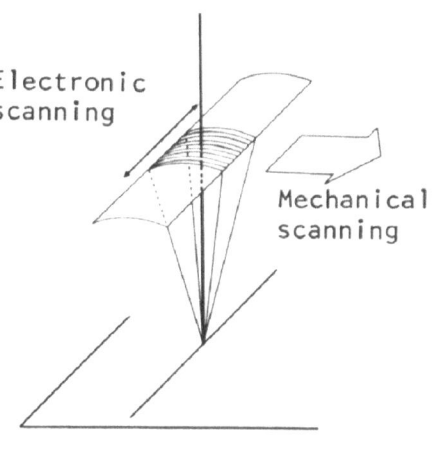

Fig 4 - Approximation of a Fresnel
modulation group with 64 transducers

Fig 5

may be considered as quasi monochromatic. The recurrent frequency of
emission is limited by the time of propagation of ultrasound. Practically in our case the X scanning of the whole array is obtained
with 160 emissions in approximately 0.1 s and a C echogram of 160 by
160 points in 16 seconds using an appropriate mechanical translation
of the probe in the Y direction (Fig. 5).

THE B ECHOGRAPHIC PROBE

The holographic focusing technique may also permit one to get
a great depth of field. For a monochromatic or quasimonochromatic
radiation the focus is located at a distance proportional to the
frequency of the wave. If the emitted signal is no longer monochromatic and has a large frequency bandwidth, a good focusing may be
obtained within a range of distances corresponding to the spectrum
of the signal (Fig. 6) and a single pulse will be focused on a large segment of the OZ axis, giving a large available depth of field.

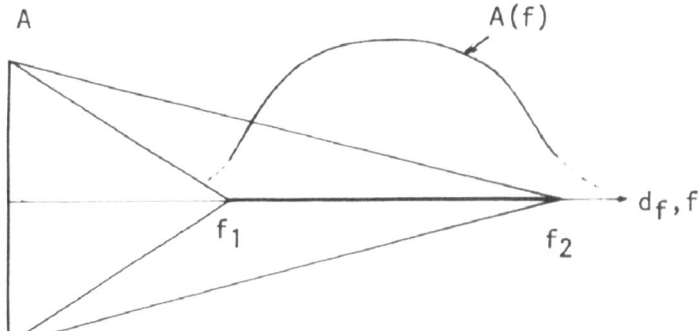

Fig. 6 - Focalisation of a wide band spectrum signal

The figure 7 shows numerical results obtained for a Fresnel group of
64 transducers corresponding to a 80 mm X aperture in our probe. The
amplitude of the acoustical pressure in the XOZ plane is given in
Fig. 7a for a monochromatic signal and the Fig. 7b shows the distribution of the acoustical pressure peak obtained from a signal reduced to 2 oscillations. In this last case the focal spot is much longer and the width is not very wider so that the lateral resolution
is relatively well preserved. In this way the holographic focusing
technique looks like axicon devices and becomes very interesting
for B echography or **3 D** investigation. First, we have checked that
our C echographic probe permitted us to obtain relatively good pictures within a range of distances from 16 to 28 cms from the probe
(Fig.8) with the limitation coming essentially from the cylindrical
focusing. So it is possible to register 3 D information with only
one mechanical scanning.

In B echography, the lateral resolution in the Y direction is
not so strictly required so that it is possible to reduce the Y a-
perture and to obtain a good focusing at all distances by using a
combination of different Fresnel groups. The figure 9 shows that
the whole useful range may be covered with only 3 Fresnel groups.
These groups are stored electronically in PROM devices and are instantaneously available so that the physician may choose any one of
the 3 focusing areas, or come back to the classical operation where
the group is reduced to the central zone and simulates a flat transducer. We have built a B echographic probe well adapted to this new
use. It is flat, 12 mm wide in the Y direction, 120 mm long in the
X direction with 160 elementary transducers working at 3 MHz.

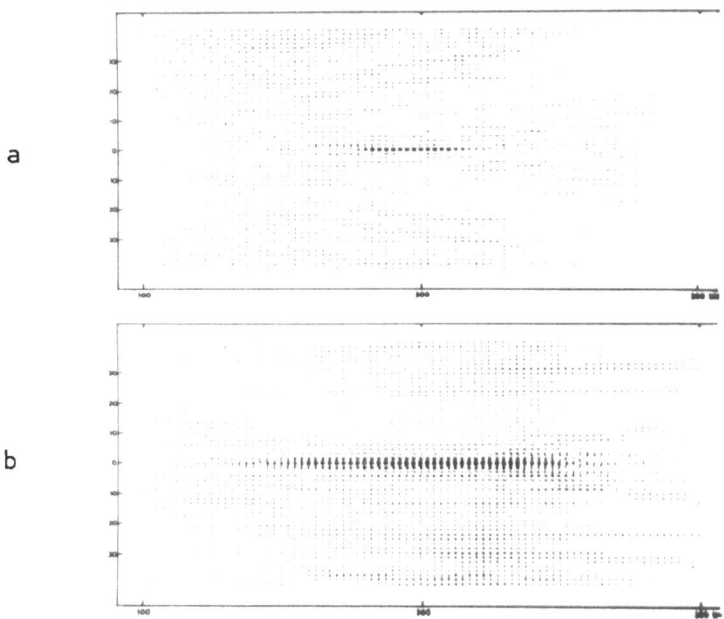

a

b

Fig 7 - Numerical computation of the acoustical field emitted by
64 transducers with a quasimonochromatic signal (a) and a signal
of 2 periods (b)

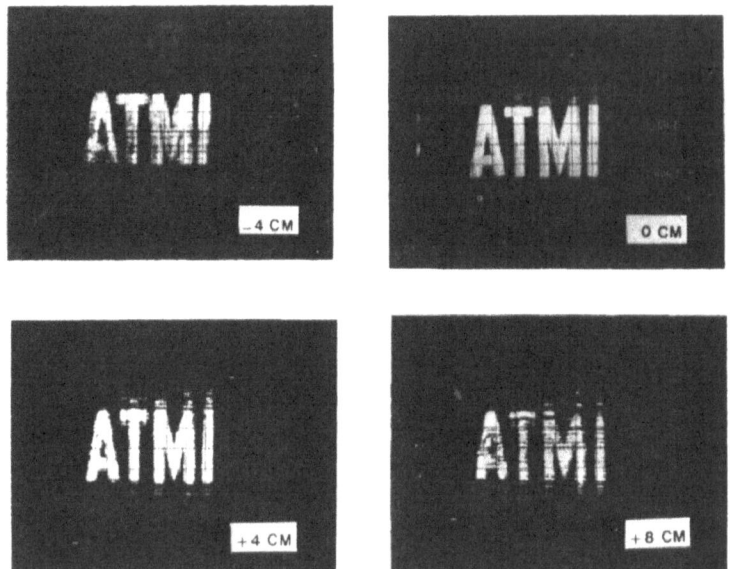

Fig. 8 - C - echograms obtained with letters set, respectively, at
          the focal distance, at - 4 cm and at + 4 and at + 8 cm

When looking for a range of 16 - 17 cm, it is possible to obtain 25
frames per second of 160 lines or 50 frames per second of 80 lines
for fast moving organs like the heart.

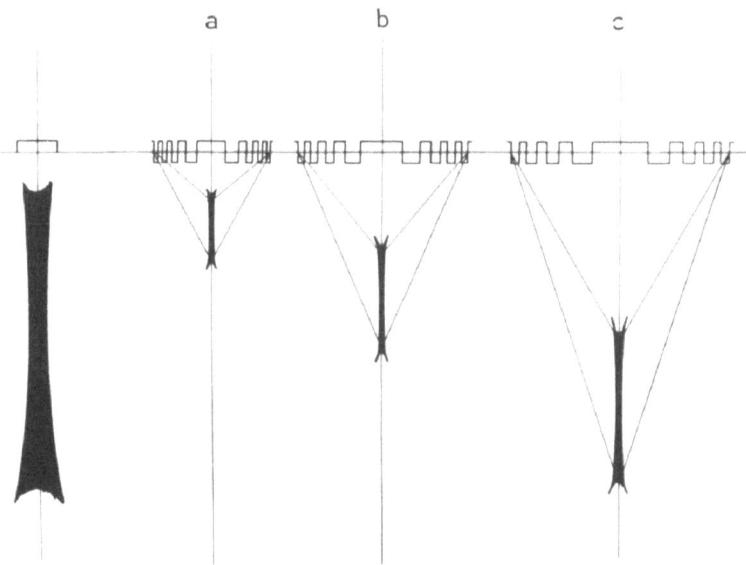

Fig. 9 - Acoustical field focused from 3 differents Fresnel groups
         a,b,c compared to the acoustical fiel emitted from a non
         focusing group.

PHYSICAL CHARACTERISTICS OF 2 DIMENSIONAL HOLOGRAPHIC FOCUSING

It must be recognized that better lateral resolution with higher aperture is obtained in our technique at a little expense of the longitudinal resolution. The figure 10 gives the on axis theoretical response $h^I(t,z)$ for the acoustical pressure observed at a distance z for an emitted signal $\delta(t)$ which may be approached by a brief single pulse. The response $h^I(t,z)$ is hyperbolically modulated from the fact that in 2 dimensional focusing the Fresnel zones have different surface areas in contrast to spherical focusing. The pseudoperiodicity of h^I corresponds to the frequency of the harmonic wave which would be focused at the distance z. The echographic response to a point object located on axis at the distance z is also represented. Obviously, it is the convolution square product of h^I : $h^{II}(t,z) = h^I * h^I$. This last signal has 2n oscillations where n is the number of spatial oscillations retained in the Fresnel group. Practically the echographic response corresponds to an actual emitted signal $s(t)$: $r(t) = s(t) * h^{II}(t,z)$ and has a longer duration than $s(t)$. This is not a strong effect for B echographic examination. Perhaps it could manifest some inconvenients in TM examinations but we have not experienced this situation yet.

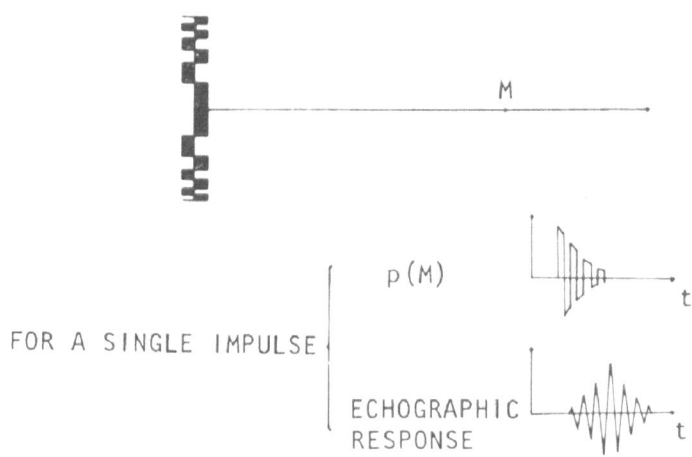

Fig 10 - On axis theoretical response   a) acoustical pressure
                                          b) echographic response

The medical pictures obtained in vivo and in vitro have shown evidence of artefacts due to secondary lobes appearing in the X direction associated with the holographic focusing. We have simulated (Fig.11) numerically holographic focusing with - the ideal Fresnel modulation $\cos(\alpha x^2)$ - the $\sqcap(\alpha x^2)$ modulation - the approaching modulation with a sampled function. It may be checked that the greatest secondary lobes are observed with the same relative maximum amplitude as for the ordinary cylindrical focusing i.e. -13 dB

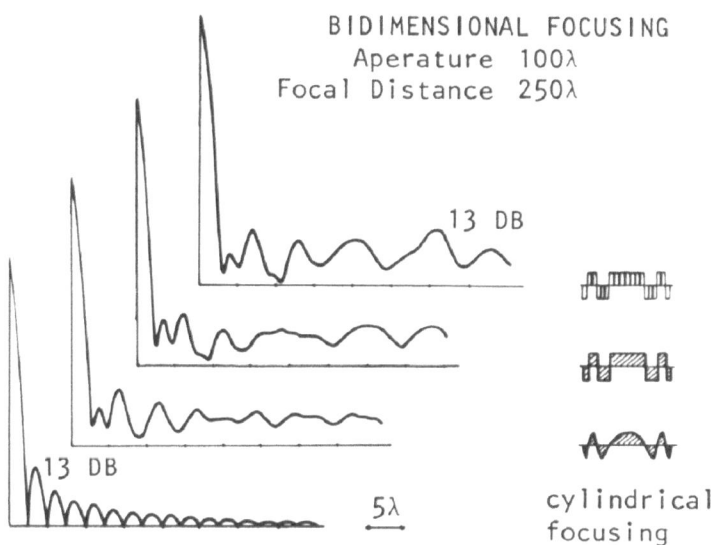

Fig. 11 - Computed pressure at the focal distance

under the main lobe, but in an erratic way and with a location which
may be far from the focus. The figure 12 confirms experimentally these
results and gives the echographic response obtained from nylon wires
located at several distances. Secondary lobes are visible up to distances of the focal spot as large as the used aperture, and their relative
amplitude may be in some cases sufficient to deteriorate echographic
pictures when looking to strong specular reflexions. Our main effort
has been directed to the minimisation of these parasitic effects. In
the following we shall show evidence of the part played by the characteristics of the elementary transducers.

It is well known that focusing from a screened aperture , in
coherent optics as well as in acoustics, gives secondary lobes due

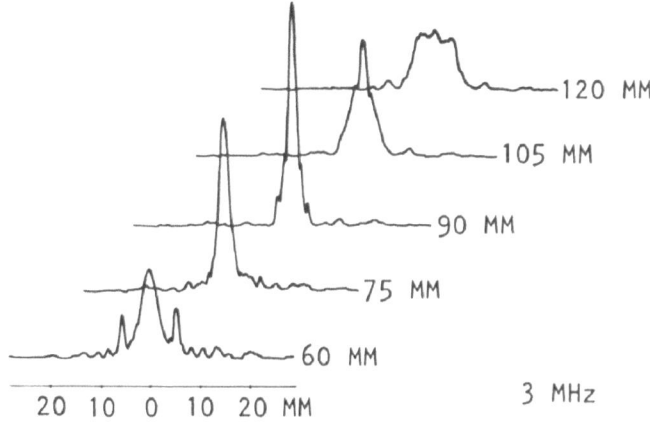

Fig. 12 - Experimental echographic response obtained with the same
          Fresnel profile used at emission and at reception

to the discontinuity of the pupil function which creates oscillations in the focal image determined by its Fourier transform. It is possible to attenuate the secondary lobes by using a progressively attenuating aperture function which is in current use in optics and may be also of great interest in acoustics (7).

Mathematically speaking, it is equivalent to say that the cylindrical focal response at a distance d :

$$(1) \quad h(x) \sim P(x) \, e^{-j\alpha x^2} * e^{j\alpha x^2}, \qquad \alpha = \pi/\lambda d$$

may be written with the oscillating Fourier Transform of P(x) : $\widetilde{\Pi}(f)$ according to :

$$(1') \quad h(x) \sim \Pi\left(\frac{x}{\lambda d}\right)$$

It is apodized when using another pupil function P'(x) =P(x) F(x) where F(x) is an anttenuating function, for example a gaussian function. The modified focal response :

$$(2) \quad h'(x) \sim \frac{1}{\lambda d} \, \Pi\left(\frac{x}{\lambda d}\right) * \mathcal{F}\left(\frac{x}{\lambda d}\right)$$

where $\mathcal{F}(f)$ is the F. T of F(x) may be efficiently apodized by the smoothing effect of the convolution product when $\mathcal{F}(x/\lambda d)$ has the correct shape.

On the other hand, the cylindrical focusing which may be obtained from a linear array of transducers gives a focal response,

$$(3) \quad h''(x) \sim e^{-j\alpha x^2} P(x) \frac{1}{\delta} \, comb\left(\frac{x}{\delta}\right) * f(x) * e^{j\alpha x^2}$$

where the comb function represents the sampling effect with a pitch of length δ and the convolution product ($*$ f(x)) gives account for the aperture of the elementary transducers which is: Rect (x/δ') for truly independent transducers of width δ', but may be notably different when the transducers are coupled. The effect of the finite aperture f(x) is mathematically solved very simply just by commuting the two convolution products. The sampling effect gives rise to secondary Bragg images and it is well known in coherent optics that a correct expression for h''(x) may be written :

$$(3') \quad h''(x) \sim \sum_{-\infty}^{+\infty} \Pi\left(\frac{x-nL}{\lambda d}\right) e^{j\pi n \frac{Lx}{\lambda d}} * f(x)$$

i.e as the sum of images $h''_x(x)$ of order n. $L = \frac{\lambda d}{\delta}$.

The central image $h'_0(x)$:

$$(3'') \quad h''_0(x) \sim \Pi\left(\frac{x}{\lambda d}\right) * f(x)$$

may be apodized as well as h'(x) identifying f(x) to $\frac{1}{\lambda d}\mathcal{F}(x/\lambda d)$.

But an elementary calculation shows that f(x) must have a characteristic width exceeding largely the pitch δ , which gives to each transducer an apparent aperture of about 2 of 3 times the length δ. In this case it may be checked that in the same time the phase factor

$$e^{2j\pi n \frac{Lx}{\lambda d}}$$ attenuates strongly the other Bragg images $h''_n(x)$.

Such an overlapping of each transducer on adjacent ones may be physically obtained from an adequate coupling. When a single transducer is acted, the adjacent ones are also excited and the emission diagram may correspond to an aperture notably greater than the elementary spacing δ . The apparent aperture of each transducer may by easily determined from directivity measurements. The relatively high directivity that the elementary transducers must have for giving a good apodization effect explains why the Bragg focuses are washed out in the same time. The figure 13 shows evidence of this apodizing effect also for holographic focusing through a numerical simulation corresponding to our pratical case. The secondary lobes are reduced by 3 dB and smoothed without appreciable loss in lateral resolution when the optimal directivity is retained for the elementary transducers. In fact, the figure 12 gives the best experimental results we have obtained, taking this effect in account when we conceived the probe.

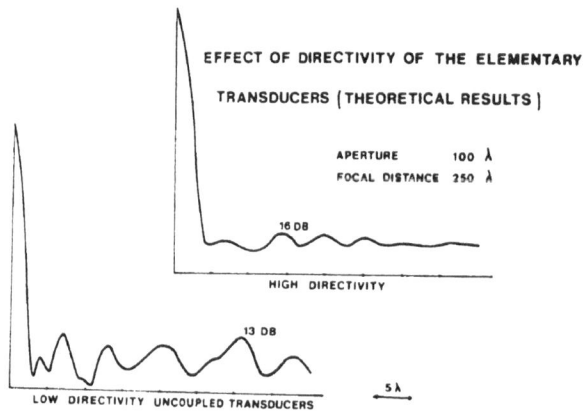

Figure   13

The most important secondary lobes attain -25 dB under the main lobe and the difference with the predicted theoretical results (-32 dB) in the optimal case must be imputed to the insufficient directivity of the elementary transducers and also to the anavoidable dispersion encountered in the characteristics of the transducers which create an asymmetrical echographic response and may enhance some secondary lobes.

AN APODIZING TECHNIQUE MINIMIZING SPECULAR REFLECTIONS AND PERMITTING
TRACKING FOCUSING.

A simple solution for reducing the effects of secondary lobes is
to associate the emission of a flat ordinary transducer simulated by
a group of transducers acted in parallel, as it is classically done
in other arrays, and the holographic focusing at reception. The figure
14 shows the theoretical echographic responses (from a thin wire) when
using the flat transducer (or the holographic focusing) both at emission and at reception : 1x1 (or 2x2) and when using the mixed combination 1x2. An obvious apodization comes from the fact that the echographic response is the simple product of the transfer function at emission by the transfer function at reception. The transfer function for
a classical transducer is not good for lateral resolution, but it is
well apodized so the result 1x2 loses a little in lateral resolution
compared to 2x2 but it is much better apodized. The figure 15 shows
experimental evidence of this apodizing effect. Other advantages come
from this special technique :

Using a thin beam for illuminating and a large aperture for receiving, minimizes strongly the importance of specular reflections
of interfaces parallel to the probe, that may be easily understood
from the figure 16 and that property permits one to follow the contour
of organs much better.

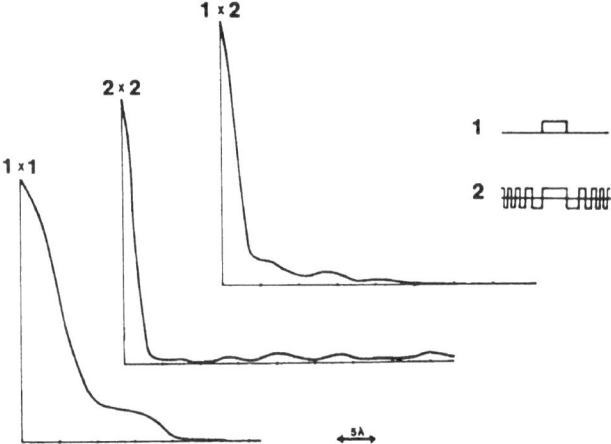

Fig.14 — Theoretical echographic response obtained with different
Fresnel profiles used at emission and at reception

A non focalised beam at emission may be combined with different
Fresnel groups at reception associated with echoes coming from different distances. The figure 9 shows that with 3 different focusing
groups, it is possible to cover the whole range of interest. We have
built an electronic circuit able to modify the Fresnel group in a few
microseconds so that we may practice some kind of tracking focusing.

This work is too recent for presenting experimental results in this
paper.

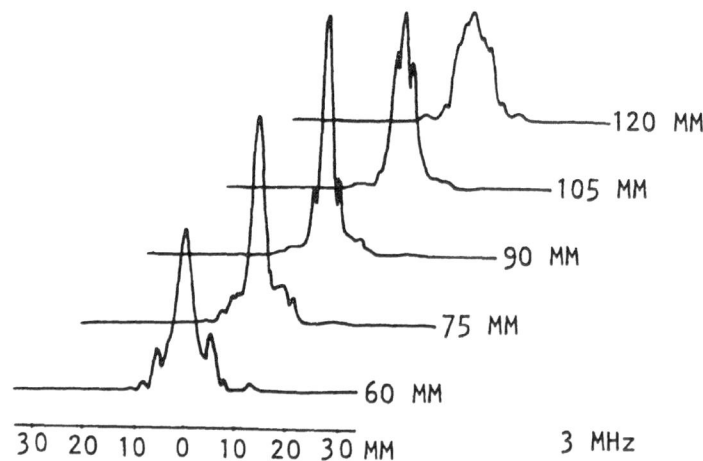

Fig. 15 - Experimental echographic response obtained with two
different Fresnel profiles (1x2) used at emission and at reception

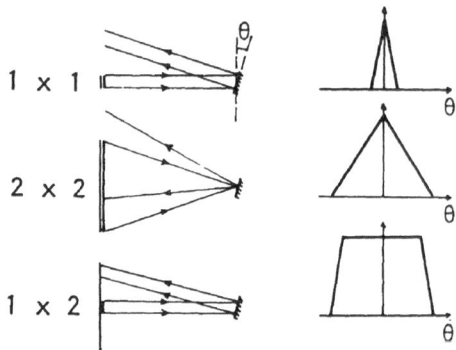

Fig. 16 - Echographic response from a specular target versus angle

MEDICAL PICTURES

We have experimented with C echography of organs in vitro and
real time B echography in vivo. The figures 17 and 18 show C echograms
of a 7 months old foetus, where it may be checked that the lateral
resolution approaches 2mm in X and Y directions for a field of
200x200mm. We are looking now for a special setting appropriate for
medical use. The figures 19 and 20 give obstetrical results obtained
from pregnancies at different stages where the beneficial effects of
increased lateral resolution may be recognized. The figure 21 (a,b)
gives 2 pictures of the heart obtained without focusing i.e according
to the classical technique 1x1 and with combined technique 1x2.

The contours of the different organs are better described in this last case.

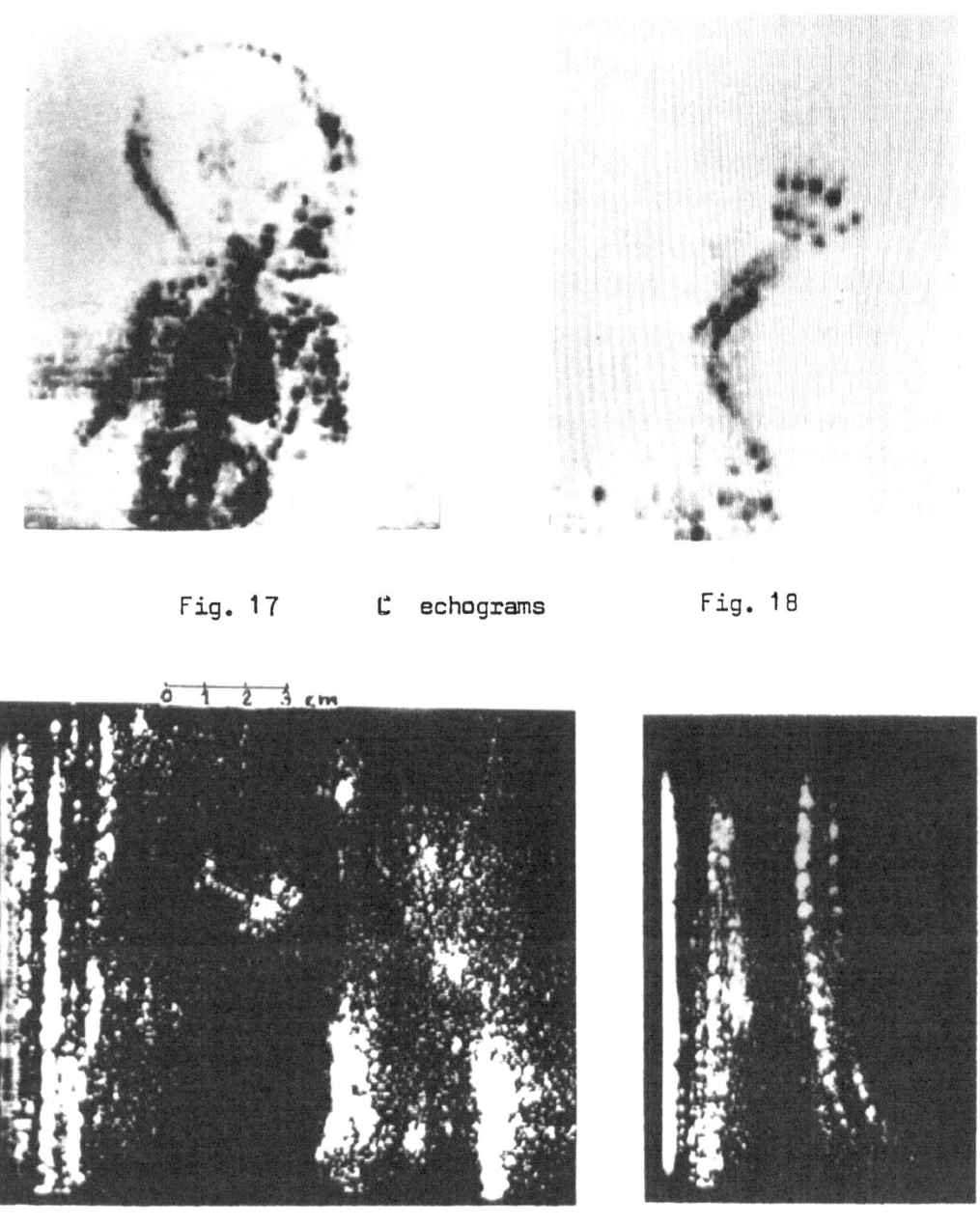

Fig. 17   C echograms   Fig. 18

Fig. 19   B   echograms   Fig. 20
3 months fetus   6 months fetus

Vertebral columns

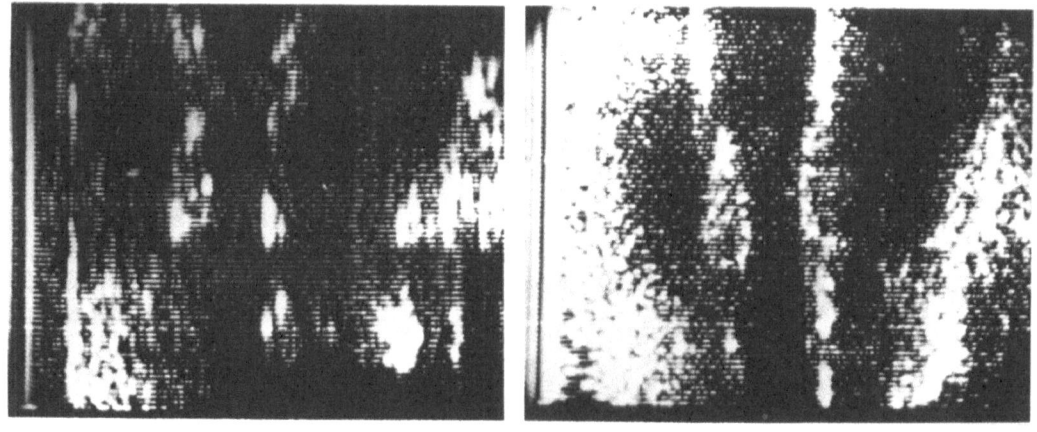

a   (1x1)   b (2x2)

Fig. 21 - B echograms of the heart

CONCLUSION

The holographic focusing technique which requires only simple commutating electronic devices has been proved to be of immediate interest for real time B echography when using special apodizing procedures. Besides, this technique has led to C echograms obtained in 15 s showing appreciable quality for excised organs. The medical interest of such C echograms remains to be proved by operating in vivo with a new special setting.

ACKNOWLEDGEMENTS

The authors wish to thank R. LALIMAN and C. SASSIER for technical assistance given in this work.

BIBLIOGRAPHY

1. F.L. THURSTONE, O.T. RAMM, Acoustical holography, Vol. 5, (1974) pp. 249-259.
2. L. POURCELOT, J.M. POTTIER, M. BERSON, Th. PLANIOL, Proceeding of the 2nd European Congress on Ultrasonics in Medicine, Munich, May 1975.
3. B.A. AULD and S.A. FARNOW, Acoustical Holography, Vol. 6, (1975) pp. 259-274.
4. M. LAGREVE, 1976 Thesis, Université Pierre et Marie Curie, Paris, France.
5. P. ALAIS, M. FINK, Proceedings of the 2nd European Congress on Ultrasonics in Medicine, Munich, May 1975.
6. P. ALAIS, M. FINK, B. RICHARD, J. PERRIN, 1st meeting of the world federation olf Ultrasound in Medicine and Biology, San Francisco (August 1976).
7. A. NIGAM, Acoustical Holography, Vol. 6, (1975), pp. 689-710.

DIFFERENTIAL PHASE CONTRAST IMAGING IN THE ELECTRONICALLY FOCUSED ACOUSTIC SYSTEM

G. Kino, W. Leung, H. Shaw, D. Winslow, and L. Zitelli

Stanford University

Stanford, California 94305

ABSTRACT

Phase contrast imaging, which has been used in optical imaging for years, has recently been introduced into acoustic imaging, in systems using mechanical scanning. In this paper, a phase contrast imaging system is introduced which employs electronic scanning and focusing. Differential phase contrast imaging is used, which compares the phases at adjacent points on the object which are separated by a constant distance along the scan line. The advantage of differential phase contrast imaging, as compared to fixed reference phase contrast imaging, is that the reference and signal beams travel almost identical paths, suppressing the effects of vibration, temperature, and other external influences. The present system consists of an acoustic receiver using a 100 element PZT array, which has two identical main beams separated slightly in space, focused at the same distance and scanned simultaneously. The relative phases between the two beams can be varied, and for "dark field" imaging they are set 180° out of phase. The output of the receiver then reproduces either the phase distribution across the object, or its spatial derivative. Two dimensional differential phase contrast images are obtained by adding a mechanical frame scan. The peripheral sensitivity of this system is presently 12° phase difference, and the ultimate objective is a few degrees.

INTRODUCTION

Phase contrast techniques are commonly used in optical microscopes to help visualize transparent objects. In this approach,

differences in optical velocity or path length are converted into
differences in intensity, allowing properties of highly transparent objects to be detected and displayed. These same techniques have been applied to acoustic imaging in visualizing objects
that have negligible acoustic absorption but which show spatial
variation in acoustic velocity or path length. One of the important potential applications of phase contrast imaging is in detecting stress regions developed around cracks. The acoustic
velocity is altered in such regions, leading to different acoustic
phase shifts. By measuring the phase profile, information concerning regions of residual stress and plastic flow, such as occur
near the tip of a crack, may be mapped out. Another important
area of applications is in the visualization of biological tissues.
Independent measurements of both amplitude and phase can also be
very important. For example, in a tissue section which has approximately uniform thickness, it is important to know how much
of the attenuation is due to absorption and how much to impedance
mismatch. Since there is very little variation in the density of
biological material, large variations in acoustic impedance must
be due to changes in acoustic velocity. For samples without a
uniform thickness, comparison of the phase and amplitude images
should also help to determine how much of the observed attenuation
is due to the absorption constant of the material and how much to
the local thickness of the specimen.

The ordinary phase contrast method measures the absolute
phase profile of objects. The acoustic system introduced here
measures the first derivative of the phase profile, i.e., the
intensity of the image is proportional to the square of the changes
in phase shift as a function of position in the sample. Thus,
abrupt changes of path length or velocity in objects are made to
stand out clearly with high contrast. This technique has been used
in optical imaging and is referred to as the Nomarski Differential
Interference Method.[1] The basic arrangement of an optical system
of this type is shown in Fig. 1. A polarized light beam is split
into two mutually coherent components by a suitable arrangement of
optical prisms (1) and these two components are displaced trans-
versely from one another by approximately one micron. After passing through the specimen, the two components are recombined at
prism (2), and the interference image produced above the analyzer
is magnified and viewed through the eyepiece. Thus the intensity
of the image is related to the phase difference between the two
beams.

In the acoustic differential phase contrast imaging system
introduced here, the acoustic receiver contains a phased array
which has two identical major lobes separated slightly in space,
focused at the same depth and scanned at the same speed. Therefore, this system is an acoustic analog of the Nomarski imaging
system.

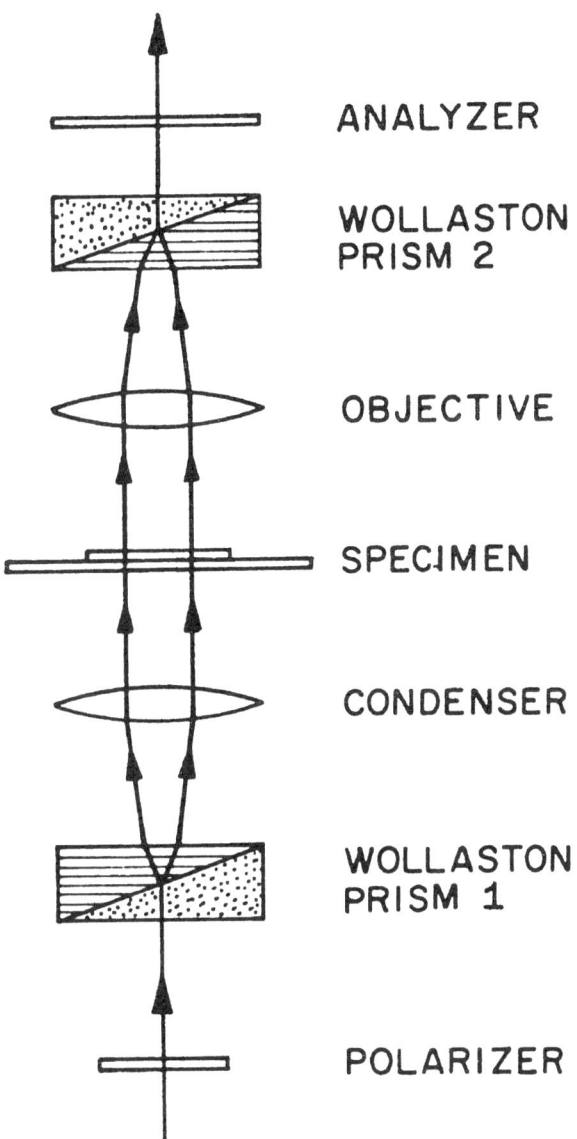

FIG. 1--Schematic diagram of a transmission Nomarski differential
phase contrast system. (The separation between the two
beams is exaggerated.)

PRINCIPLE OF OPERATION

The most important part of the acoustic differential phase contrast imaging system is the acoustic receiver, which employs electronic scanning and focusing. The basic receiver has been described in several papers,[2,3] and only a very brief account will be given here.

The acoustic receiver consists of a PZT array of 100 elements and a surface acoustic wave delay line with 100 taps, interconnected one-to-one by means of 100 double-balanced mixers. The surface wave delay line provides the necessary and phase delay references for the array. Consider a point source in front of the receiver. Its wavefronts excite the receiver elements with a phase distribution which is parabolic as a function of distance along the array. If an FM chirp is launched on the surface wave delay line, the frequency will vary linearly and hence the phase will vary quadratically along the delay line. If the center frequency of the chirp is correctly chosen, the phase distribution along the delay line will be the same as that along the array, and the outputs of all the mixers will be in phase and can be added together. Thus a strong signal is obtained and the receiver is said to be focused at the point source. Since the FM chirp is traveling along the delay line, the receiver is also scanning in the same direction.

In differential phase contrast imaging, the receiver is focused at two adjacent points in the object and scanned at the same time. This can be achieved in several different ways. The approach used here is to launch two FM chirps in the delay line, which have identical slopes but differ slightly in center frequency. The focal length of the two receiving beams is determined by the chirp slope, and their separation is determined by the difference in their center frequencies. These two chirps are obtained by mixing a single chirp with a low frequency signal in a doubly balanced mixer. The carrier is suppressed and only the sidebands (which are chirps) remain. These two chirps have the same chirp rate and are separated by $2f_c$, where f_c is the frequency of the low frequency signal. Furthermore, the relative phase between the two output chirps can be varied by changing the starting phase of the low frequency signal. As a result, the relative phase between the two main beams in the receiver can be adjusted at will. The output of the receiver is the vector sum of the contributions from these two beams. A schematic diagram of the entire setup is shown in Fig. 2.

In addition to phase variations, amplitude variations produced by the sample are also contained in the differential phase contrast images. In most applications, it is desirable to separate the phase and amplitude information. This can be achieved by using the saturation characteristics of the mixers. For the mixers used here, an

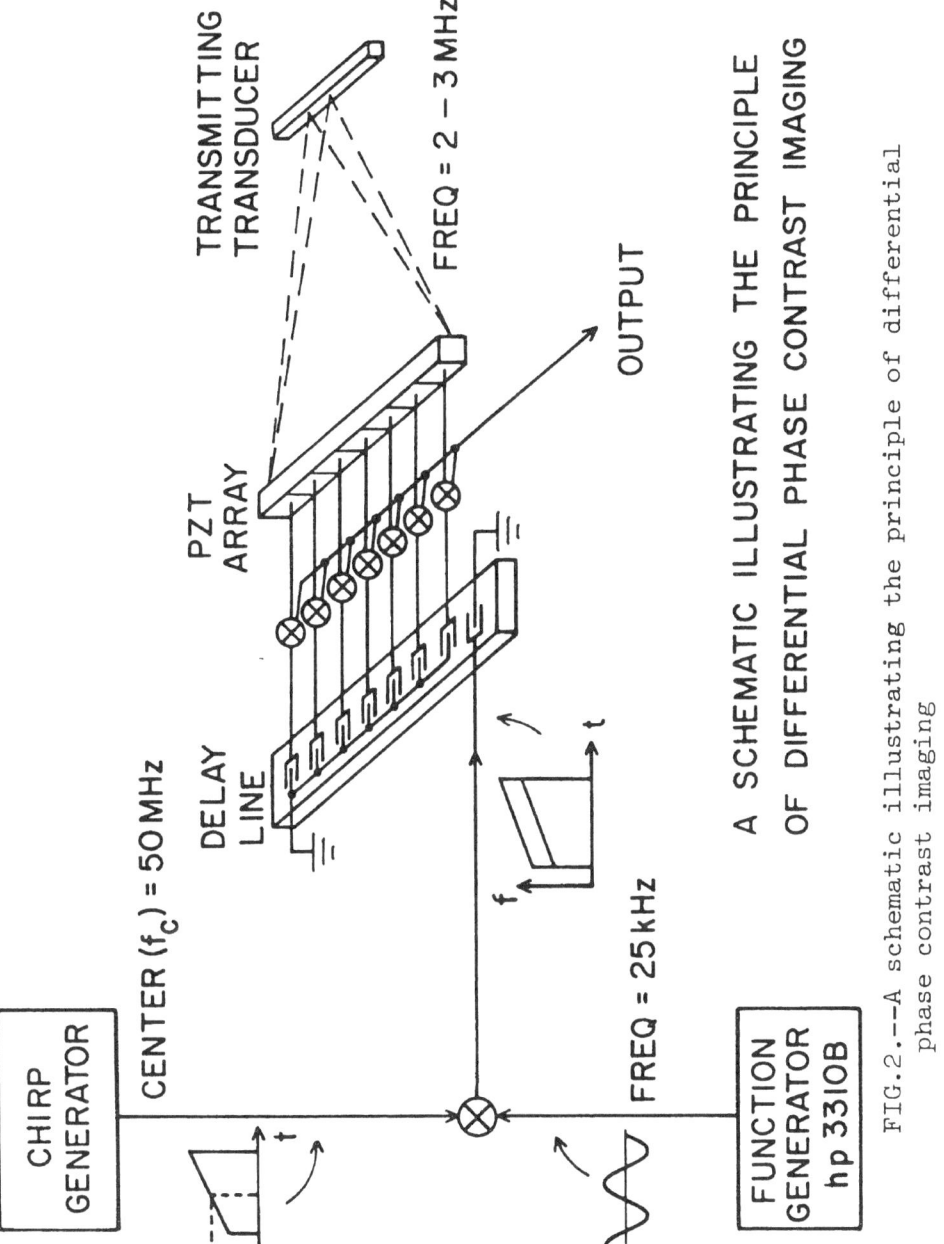

FIG. 2.—A schematic illustrating the principle of differential phase contrast imaging

rf signal of 50 mV peak-to-peak from the PZT transducers is sufficient to cause saturation. As a result, the system is insensitive
to amplitude variations in the specimen under examination.

EXPERIMENTAL RESULTS

The acoustic receiver used in this system can function as
either an ordinary amplitude scanner (with one single chirp in the
delay line) or as a differential phase contrast scanner (with double chirp). For amplitude scan, the point response of the receiver
is shown in Fig. 3(a). The position and magnitude of the signal
correspond to the position and magnitude of the point source respectively. The chirp in the delay line is apodized to reduce sidelobe levels. The point responses of the receiver in the differential phase contrast mode are shown in Fig. 3(b) and (c). In Fig.
3(b), the two beams of the receiver are 180° out of phase so that
the overlapping portions of the two pulses cancel each other. In
Fig. 3(c), the two beams are in phase so that the overlapping portions of the two pulses add. The two pulses may not be distinguished
if they are too close together.

Since the receiver is a scanning device, two dimensional images
can be obtained by using the setup shown in Fig. 4. A horizontal
strip transducer serves as an illuminator. The line scan is provided electronically by the receiver and the frame scan by mechanical motion. Several types of objects have been used to test this
system. In order to get the best dynamic range, the receiver is
usually operated in the "dark field" mode. In this mode of operation, the two main beams are 180° out of phase. Therefore, there
will be no output from the receiver if the two beams are looking at
two identical sources. To understand the operational principles
more clearly, assuming that there is no object under examination
and that the horizontal strip illuminator has a uniform phase along
its length, the signal from the receiver will be two pulses corresponding to the two ends of the strip illuminator as shown in Fig. 5.
When the beams are inside the strip illuminator, there is no output
because the two beams are 180° out of phase. Outputs occur only
when one of the beams is inside the illuminator while the other is
outside.

The first test object used was a 0.005 inch-thick, 8 mm-wide
scotch tape. The response of the system is shown in Fig. 6. An
additional phase shift is introduced by the scotch tape. Therefore, the two beams are no longer 180° out of phase when one of
the beams is inside the scotch tape and the other is outside. The
edges of the scotch tape are then detected. The magnitude of the
output from the receiver is proportional to the amount of phase
shift introduced by the scotch tape.

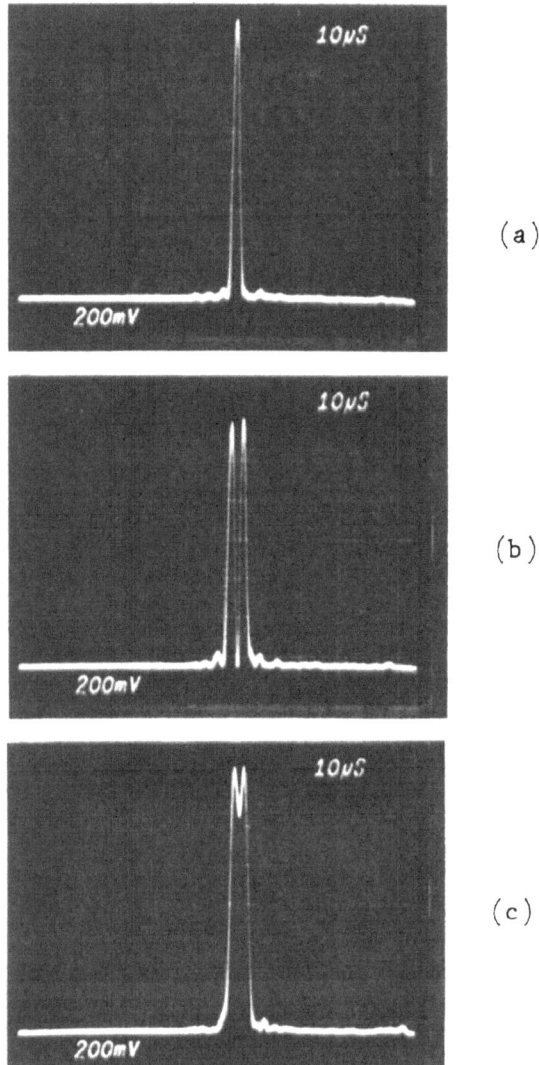

FIG. 3--Point responses of the receiver. (a) Amplitude scan, single chirp; (b) with two chirps which are 180° out of phase; (c) with two chirps which are in phase.

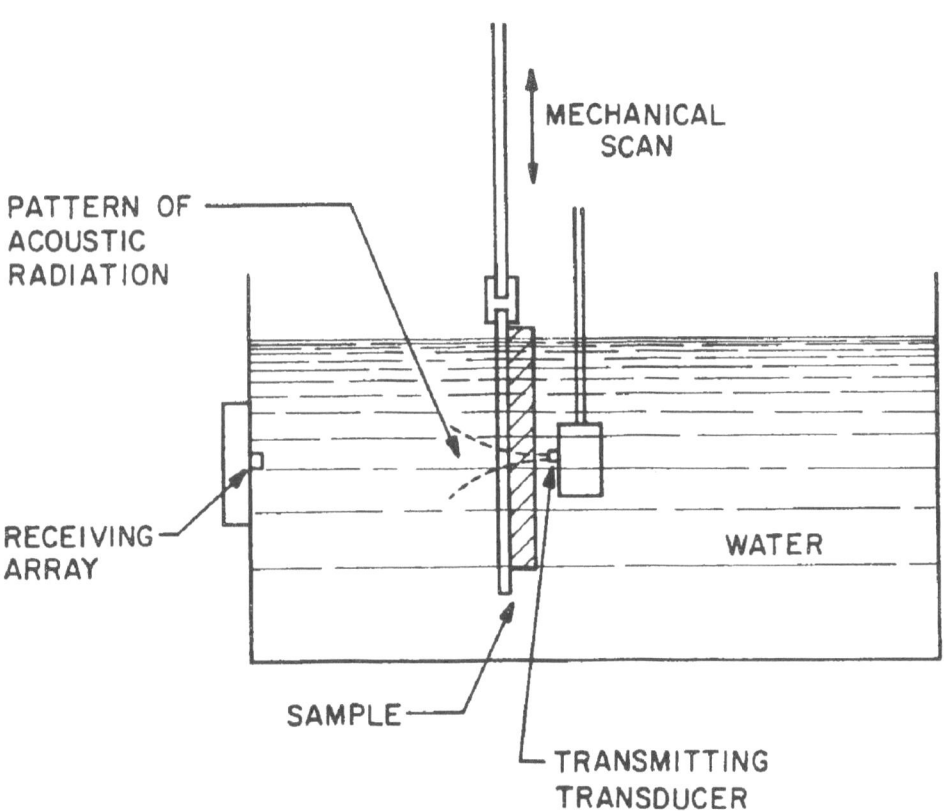

FIG. 4--The experimental arrangement of the setup to obtain two-
dimensional acoustic images of samples.

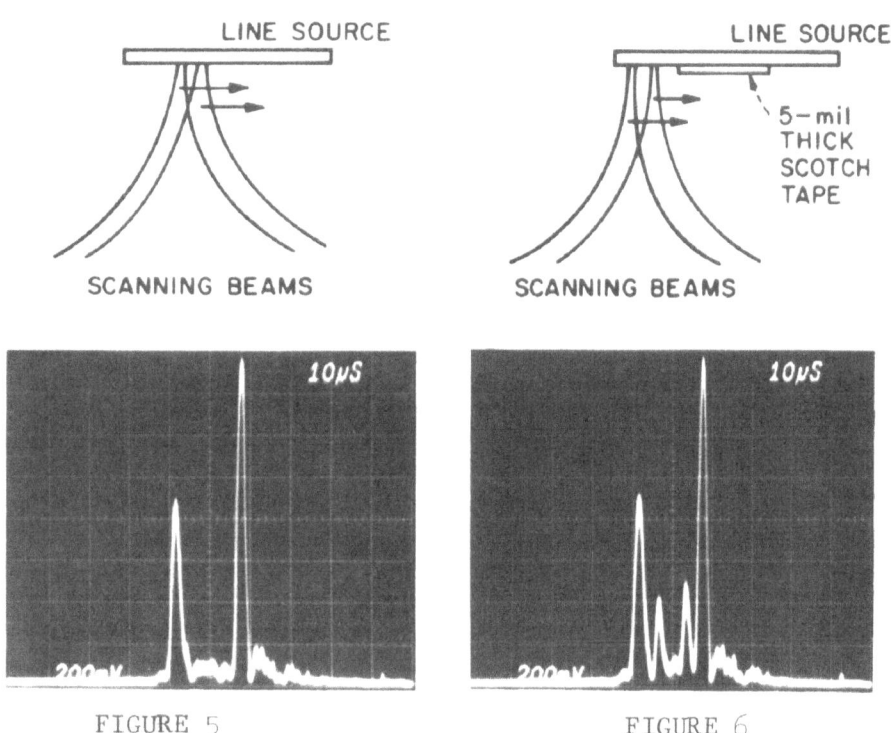

FIGURE 5   FIGURE 6

FIG. 5--Receiver is in "dark field" mode, showing the edges of
      the line source.

FIG. 6--Receiver is in "dark field" mode, showing the edge of
      the line source and scotch tape. The signals produced
      by the two edges of the tape is 6 dB above the noise.

The second test object was a 1/8 inch-thick plexiglas plate with one oval shaped and one circular depression on its surface. The depths of the oval and circular depressions were 0.02 and 0.03 inches respectively. The frequency used was 2.3 MHz and the separation of the two beams was approximately 0.6 cm. The relative phase shifts are approximately 120° and 180° respectively. A two-dimensional image of this object obtained with this system is shown in Fig. 7. Note that the two edges of the holding rod can also be seen in the picture.

This system is capable of measuring much smaller phase shifts. In Fig. 8, two-dimensional images of another object, which consisted of a piece of plexiglas with a 0.005 inch oval shaped depression are shown. The relative phase shift corresponding to the depression is about 40°. The image on the right is an amplitude-only image, and that on the left is a differential phase contrast image. Because of the negligible difference in acoustic absorption, the oval shaped depression cannot be seen in the amplitude-only image, while it stands out clearly in the differential phase contrast mode. The holder of the object can be seen in both images.

The degree of cancellation between the receiving beams determines the phase sensitivity. Since the output is the vector sum of the two beams, the magnitude of the output signal will then be proportional to $\sin \theta/2$, where θ is the relative phase between the two beams. In Fig. 6, a signal corresponding to a phase shift of 20°, produced by a piece of scotch tape 0.005 inch thick, at 2.8 MHz, is shown. This signal is 6 dB above the noise, which is produced mainly by imperfect cancellation of the receiving beams. From these results, the peripheral sensitivity of the system is about 12°.

CONCLUSION

The basic principles of a new acoustic differential phase contrast imaging system with electronic focusing and scanning have been demonstrated. The system is very versatile, because the phase difference and spacing between the receiving beams can be arbitrarily adjusted. Amplitude-only images can be obtained simply by putting a dc voltage instead of a low frequency signal into the mixer which produces the double chirps. Because both beams travel across the same environment, the entire system is quite insensitive to external disturbances resulting from temperature variations and mechanical vibration. The phase sensitivity is good and is expected to be improved, down to the range of a few degrees. An ultimate objective is to use the system for detecting small imperfections, and to measure stress and the presence of plasticity in materials.

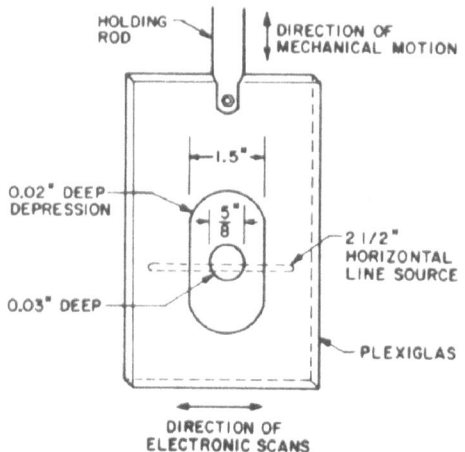

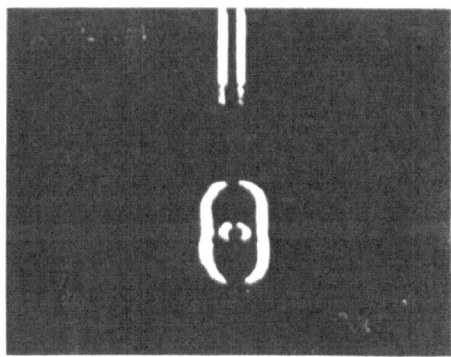

FIG. 7--Schematic and acoustic differential phase contrast image of a plexiglas sample.

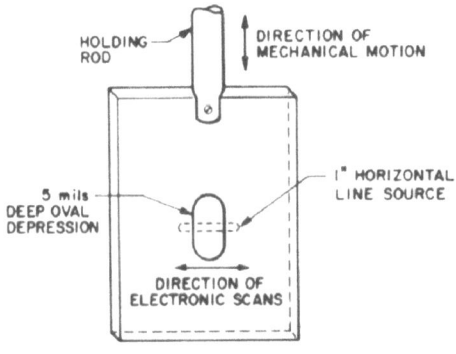

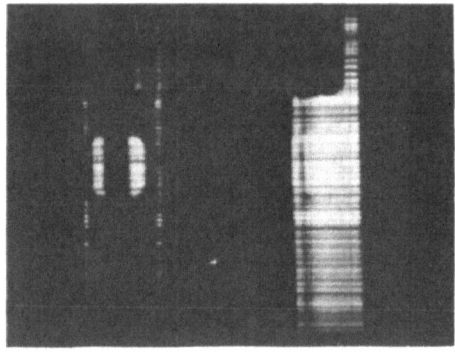

FIG. 8--Schematic and acoustic images showing a 5 mils deep
         depression on a piece of plexiglas. The phase shift
         introduced is about 40°.

REFERENCES

The work reported in this paper was supported by the Electric Power Research Institute under Contract RP609-1.

1. M. Francon, "Isotropic and Anisotropic Media. Application of Anisotropic Materials to Interferometry," in Advanced Optical Techniques, Van Heel, Editor, (North-Holland Publishing Co., Amsterdam, 1967), Chapter 2.

2. J. Fraser, J. Havlice, G. Kino, W. Leung, H. Shaw, K. Toda, T. Waugh, D. Winslow, and L. Zitelli, "An Electronically Focused Two-Dimensional Acoustic Imaging System," in Acoustical Holography, Vol. 6, edited by Newell Booth, Plenum Press, New York, 1975, p. 275.

3. J. F. Havlice, G. S. Kino, and C. F. Quate, "A New Acoustic Imaging Device," 1973 IEEE Ultrasonics Symposium, Monterey, California, December 1973.

HOLOGRAPHIC ADAPTIVE PROCESSING - A COMPARISON WITH LMS ADAPTIVE PROCESSING

P. N. Keating and T. Sawatari

Bendix Research Laboratories

Southfield, Michigan 48076 U.S.A.

ABSTRACT

A holographic adaptive processing technique which is an advanced version of one described earlier,[1,2] has been applied to a passive underwater acoustics scenario used by Frost[3] to test the well-known "Constrained Least-Mean-Squares (LMS) Algorithm" for adaptive array processing. This application allows the holographic approach to be functionally compared with the LMS approach. Results are presented which show that the holographic technique works better on the Frost Scenario (consisting of partly coherent signal, interferences, and white noise) than the LMS technique.

I. INTRODUCTION

The major purpose of this paper is to compare holographic adaptive processing[1,2] with Least-Mean-Square (LMS) adaptive processing.[3,4] In order to perform this comparison, we describe (a) a modified and improved version of the adaptive null processing technique which can be applied to an input signal situation essentially the same as that used by Frost for the application of his LMS algorithm,[3] and (b) the results obtained by applying the null processing to this target scenario. Frost's algorithm is a typical LMS adaptive processing method for underwater acoustic arrays.

The null processing method is rather different in philosophy and approach from many other adaptive beam-forming approaches[3-7] such as those similar to Frost's algorithm. As a result, it is difficult to compare this nulling technique with many of the other approaches.

One way to carry out this comparison is to do it on the basis of functional performance. In line with this, we apply the nulling approach to a target 'scenario' already used for the simulated application of the Frost algorithm.[3]

The simple adaptive holographic null processing, which was compared with DICANNE[8] and a holographic enhancement technique[9,10] in a recent article, is not well adapted to the Frost scenario. Frost had only four sensors in his array, and thus the different directional signals are not well-resolved. If two or more signals of not too dissimilar strength are not resolved during conventional beam-forming, then the simple adaptive routine will experience serious difficulty in finding the correct null positions, because of interference. Consequently, the efficiency of the processing is significantly decreased, as described in the first part of Section II of this paper.

In the second part of Section II, we describe a higher-level approach which utilizes differences in spectral characteristics in a more complex manner than originally proposed.[2] This more-recent approach falls more or less midway between the original method (where filtering is carried out and then all the processing is spatial processing) and "more conventional" adaptive beam-forming (ABF) methods (e.g. Frost's; where the spectral and directional responses of the processor are optimized together - at considerable cost).

In the first two parts of Section III, we briefly review the LMS algorithm and the underwater acoustic scenario which Frost used, in preparation for the numerical comparison. In the last part of the same section, the test results are described where it is shown that the present nulling approach works better on this particular target scenario.

II. HIGHER-LEVEL HOLOGRAPHIC PROCESSING

A. Multi-Null Processing

The iteration method of holographic null processing, the principle of which was described in the previous paper,[2] is now discussed in a more general sense.

Assume that $(M + 1)$ discrete spatially-locallized sources are distributed in the far field, and a linear array is used. The signal of frequency ω, (the holographic information) detected by the linear array is

$$\tilde{A}_\ell = \sum_{m=0}^{M} A_m e^{-ix_m a_\ell} \qquad \ell = -n, -(n-1), \ldots 0, \ldots n-1, n \text{ and } N = 2n+1 \tag{1}$$

where $\tilde{A}_\ell$ is the data obtained by the ℓth sensor, $x_m = (2\pi/\lambda) \sin\theta_m$, θ_m is the mth target direction, λ is the acoustic wave length corresponding to the frequency ω, a is the array element separation, A_m is the amplitude and phase of the mth spatially-locallized noise, and A_0 is the signal of interest coming from a direction θ_0.

The reconstructed image is then given by

$$\tilde{G}_k = \frac{1}{N} \sum_\ell^M \tilde{A}_\ell e^{i2\pi k\ell/K} \tag{2}$$

$$= \sum_{m=0}^M A_m f(k-\nu_m) \tag{3}$$

where

$$f(\mu) = \frac{1}{N} \sum_\ell^N e^{i2\pi\mu\ell/K} \tag{4}$$

and

$$\nu_m = \frac{Ka}{2\pi} x_m = \frac{Ka}{\lambda} \sin\theta_m \tag{5}$$

Note that $f(0) = 1$.
The assumption here is that the power $|\tilde{G}_{k_0}|^2$ in the look direction does not represent $|A_0|^2$ very well, even though we chose $k_0 = \nu_0$, because the contributions from the strong, locallized noise sources to the look direction $(|A_m f(k_0 - \nu_m)|)$ are comparable to or sometimes greater than $(|A_0|)$.

In order to remove these noise contributions from the look direction, we locate N nulls sequentially on each noise direction. The resultant image is expressed as

$$\tilde{G}_k' = \sum_{m=1}^M \hat{A}_m f(k-\nu_m) + \sum_{m=1}^M \hat{A}_m' f'(k-\nu_m)$$

$$+ A_0 f(k-\nu_0) \tag{6}$$

where $f'(\mu) = \dfrac{d}{d\mu} f(\mu) \tag{7}$

$$\hat{A}_m = \sum_{n>m}^M A_n f(\nu_n - \nu_m) \tag{8}$$

$$\hat{A}_m' = -\Delta\nu_m A_m \tag{9}$$

and $\Delta\nu_m$ is the error in the location of the null due to the side lobe contributions from other noise sources. This expression is obtained by abbreviating all higher-order terms than those of $f(\nu_n - \nu_m)$ and $\Delta\nu_m$. If $\Delta\nu_m \simeq 0$; i.e., the location of its coherent source is accurately identified, G_k' has an expression similar to G_k of Eq. (2), except that A_m is changed to $\hat{A}_m$. If N, which is the number of array elements, is larger than M, which is the number of noise sources, and if each source is separated more than the distance of the resolution limit of the array, then $\hat{A}_m$ is always less than A_m, and therefore the noise is reduced.

It is obvious in this case that the noise in Eq. (6) can be re-duced further by successively applying nulls at the peaks of the remaining noises. Needless to say, the peaks of the residue appear in the same locations (ν_m) as those of the original strong sources.

However, if $\Delta\nu_m$ is not very small $(\Delta\nu_m \simeq 1)$, then the second term of Eq. (6) is dominant and the situation is rather complicated. It is not impossible to locate many more nulls around each ν_m and reduce the contribution in the look direction from the spatially localized noise. However, in practice, if one cannot make $\Delta\nu_m$ much smaller than unity, the present multi-null processing is not very effective. In order to implement the multi-null method, it is therefore nec-essary to establish an accurate null location.

B. Frequency Analysis for Determination of Null Location

We have seen in the previous subsection that the efficiency of the holographic null processing is dependent upon the effectiveness of the null-location search. In other words, minimizing $\Delta\nu_m$ is very important. The straightforward null searching procedure described in the previous paper[1] is not very effective if the different directional signals, which have different spectrum distributions, are not resolved by the directivity of the array. This is often the case; Frost's scenario[3], which he used for the application of his LMS algorithms and which we are going to use for a comparison test of the present holographic processing method in a later section, is a typical example.

A method is described in this section to find the location of the null more accurately. The principle we propose to use is to start by working in the frequency band where the strongest direct-ional noise source is a maximum so that we can find its null position with a minimum of interference. This is in contrast with the prev-ious approaches where we would concentrate on the frequency band where the desired signal power was greatest. A basic assumption that is made in our approach is that there is no more than one significant directional signal in the same two-dimensional resolu-tion cell in both bearing and frequency space. (If this assumption is invalid, the Frost algorithm, and similar approaches, will not work.) The higher-level approach involves the following steps:

1. Carry out an appropriate spectrum analysis.
2. Determine the frequency band in which the total power (summed over the sensors) is a maximum, and carry out conventional reconstruction (or beam-forming) on this data.
3. Assuming this does not give the signal of interest clearly observable and resolved, next use this beam-forming data to find the "best null" location (e.g. find the bearing angle of θ_m, at which the reconstructed image has a peak).
4. Find the next-biggest spectral peak in total power and repeat the above steps to find the "best null" position for the second strongest source.
5. Continue until the frequency band selected is that associated with the desired signal.
6. Carry out null processing on the data from this frequency band using the null positions determined in steps 3 and 4 above.

III. COMPARISON TEST

Prior to comparing the performance of the above higher-level holographic processing technique with that of the conventional LMS adaptive processing, we here briefly review the LMS adaptive processing and describe the underwater acoustic scenario which Frost used for testing of his LMS algorithm.

A. LMS Adaptive Processing

Detailed discussions of the LMS filter or LMS adaptive processing are extensively given in the literature.[3-8],[11] However, the following brief review is given for a completeness in the comparison of the higher-level null processing (HLNP) method with LMS processing.

The LMS filter is a filter which minimizes the difference between the output of the filter and an ideal output, in a statistical sense. The filter is given as a solution of the well-known Wiener-Hopf equation and is expressed in terms of power spectra:

$$F(\omega) = \phi_{1d}(\omega)/\phi_{11}(\omega) \tag{10}$$

where $\phi_{11}(\omega)$ is the power spectrum of the input signal and $\phi_{1d}(\omega)$ is a cross power spectrum between the input signal and desired response. If the information is gathered in a discrete manner such as in an array processor, it is more convenient to use vector-matrix notation for the Wiener filter:

$$\underset{\sim}{W} = R_{xx}^{-1} \underset{\sim}{P} \tag{11}$$

where $\underset{\sim}{W}$ is the optimum weight vector such that the optimum output of the filter is $\underset{\sim}{Y} = \underset{\sim}{W}^{\dagger}\underset{\sim}{X}$, $\underset{\sim}{X}$ is the input vector and $\dagger$ indicates a transposition of the vector, R_{xx} is the covariance matrix of the input ($R_{xx} = E[\underset{\sim}{X}\cdot\underset{\sim}{X}^{\dagger}]$, E indicates a statistical average), and the vector $\underset{\sim}{P}$ is the cross correlation between the input vector $\underset{\sim}{X}$ and the desired response, which is the scaler quantity d_i, ($\underset{\sim}{P} = E[d_i\underset{\sim}{X}]$).

The Frost algorithm is a practical approach to finding an approximate solution. The method of steepest descent is used and, in an adaptive manner, the weight vector $\underset{\sim}{W}_{j+1}$ for the next input vector $\underset{\sim}{X}_{j+1}$ is estimated from the present weight vector and the gradient of the mean square error between the desired and the present input. As an approximation, the gradient is given as a quantity proportional to the present input vector $\underset{\sim}{X}_j$. As a result, this algorithm does not require an exact measure of the covariance matrix nor a matrix inversion operation. Frost has developed an algorithm where each $\underset{\sim}{W}_j$ is constrained so that the filter maintains a chosen frequency characteristic for the look direction, while rejecting noises from other directions.

B. Frost's Scenario

The Frost scenario consists of three spatially localliced signals which are partly incoherent, together with different white-noise signals injected into each of the four channels used in the Frost simulation.[3] The four sensors are assumed to be uniformly spaced at τ-second intervals on a line. The signal of interest is represented by a pseudo-Gaussian sequence with a bandwidth $\Delta F = 0.1/\tau$, centered at frequency $f_o = 0.3/\tau$, and is taken to be arriving at $0°$ (i.e., normal to the array). The directional noise sources A, B have bandwidths $0.05/\tau$ and $0.07/\tau$, respectively, center-frequencies $0.2/\tau$ and $0.4/\tau$, respectively, and bearings of $45°$ and $60°$, respectively. The total power in each of the sources A, B is 10 times that of the signal of interest, while the total power in each of white-noise signals is equal to the signal power.

The signals to be used by the numerical comparison described in the following section must be discrete time-series. Because it is desirable to have an integral number of samples in the sensor-to-sensor delays, we have used slightly different beam directions from Frost's, namely, $45.1°$ and $61.0°$. These correspond to a sampling rate of 24 samples per τ-second interval, and 17- and 21-sample delays for the two beam-directions. The target scenario is summarized in Table I.

In the present simulation, the three directional, partly-coherent

signals were obtained by (a) generating random phases with a proba-
bility density which is uniform between $-\pi$ to π and otherwise zero,
(b) calculating the amplitude from the expected frequency spectrum
of a bandpass single-pole recursive digital filter corresponding
to each signal of a given central frequency and bandwidth, (c)
superimposing these spectral components at a frequency interval
determined by the data acquisition time of the present systems, and
(d) multiplying the sequences by appropriate factors to obtain the
correct power-levels given above. Thus, there are 3 partly coherent
sequences representing the incoming beams and 4 white-noise sequences,
one for each channel.

Table I   Signals and Noises in the Simulation

SOURCE	POWER	DIRECTION (0° IS NORMAL TO ARRAY)	CENTER FREQUENCY (1.0 IS $1/\tau$)	BANDWIDTH
LOOK-DIRECTION SIGNAL	0.1	0°	0.3	0.1
NOISE A	1.0	45° (45.1°)*	0.2	0.05
NOISE B	1.0	60° (61.0°)*	0.4	0.07
WHITE NOISE (PER TAP)	0.1			

*( ) IS BRL'S NOISE DIRECTION

The total signal provided by each sensor is then obtained by
adding the three partly-coherent sequences with appropriate delays
to simulate the directional signals, and then adding a white-noise
sequence to each.

C. Results of Numerical Comparison

The scenario we simulated was not exactly identical with that of
Frost's (see Table I and also Table I in Ref. 3). Fig. 1 is the
power spectrum of our input signal in comparison with that shown in
Frost's paper (Fig. 10 in Ref. 3). Our power spectrum is more
smoothly spread than his.

The frequency response of Frost's processor is shown in Fig. 2
in comparison with that used in the present processor. It is to be
noted that in the holographic processing, the input information

falling in the spectrum other than that shown in Fig. 2 has been
utilized to identify null locations accurately. This is very important in order to increase the efficiency of the nulling process,
especially when a low-resolution array, such as in the present
case is considered.

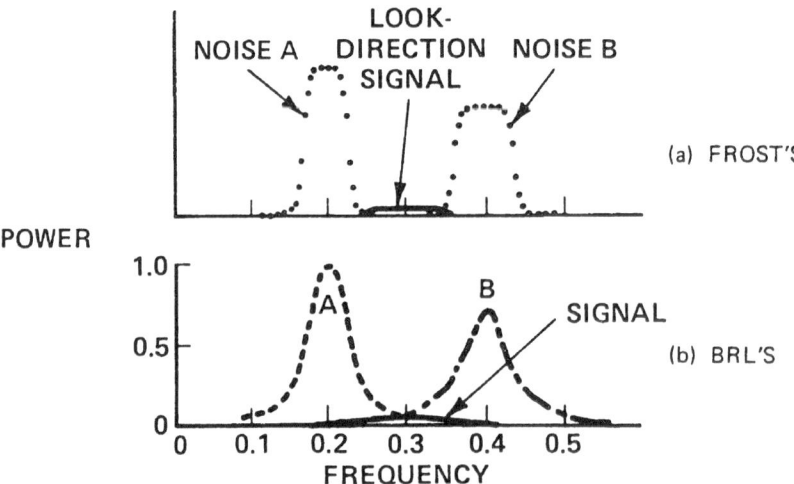

Figure 1 - Power Spectral Density of Input Signal

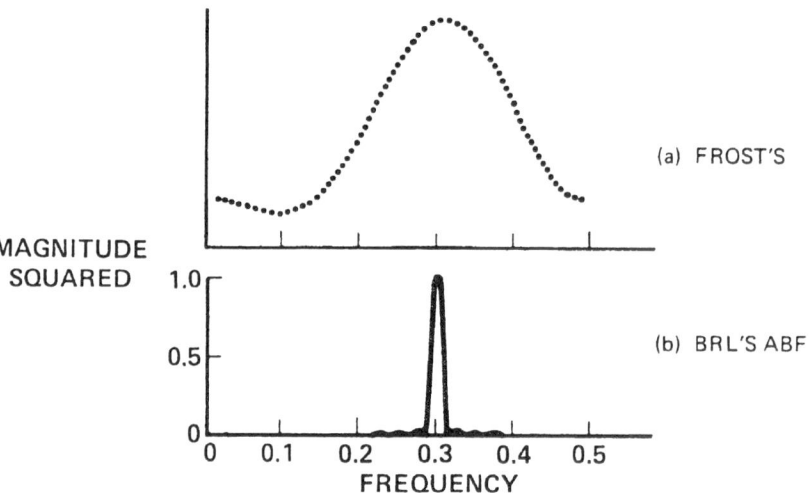

Figure 2 - Frequency Response of the Processors in Look Direction

With these differences, we can compare the performance of the two methods in the following manner. The quantity we are interested in is the power in the look direction. The power before processing and the optimum output power after processing are listed in Table II. These differences are due to the differences mentioned above, i.e., the minor input spectrum differences and (primarily) the difference in the frequency response of the processors.

Table II – Power in Look Direction

	T.P. OF IDEAL OUTPUT SIGNAL (OPTIMUM OUTPUT POWER) B	T.P. OF UNPROCESSED OUTPUT WITH ALL NOISES C	T.P. OF PROCESSED OUTPUT WITH ALL NOISES D	RELATIVE ERROR OF THE PROCESSORS E = (D−B)/B
FROST'S	0.09	0.252	0.117	30%
BRL'S	0.023	0.034	0.023*	0%*

*EXPECTION OF THE PROCESSED OUTPUT IS THE SAME AS THAT OF THE DESIRED SIGNAL. THEREFORE, THE RELATIVE ERROR IS ZERO. HOWEVER, OUR COMPUTER RESULTS INDICATE THE OUTPUT FLUCTUATION ABOUT 20% AROUND THE VALUE LISTED.

Frost's simulation data (Fig. 11 in Ref. 3) show that after 30 iterations of the adaptive process, the output data converged to about 3.3, which is about 32% off from the desired value. The time required to carry out 30 iterations is $58 \times 30 \simeq 30$ minutes (assuming $\tau = 1$ second). On the other hand, in our method, deviation of the output power from the optimum power is less than 20%. The expectation value of the deviation is zero; however, our data for different input-randomness fluctuates about 20%. It should be noted that this fluctuation is due to "coherent" superposition of all three signals due to residual correlations, in contrast to Frost's signal which has zero cross-correlation from the time difference greater than 25τ. This coherent superposition produces rather unpredictable fluctuations on the spectrum (speckle pattern) in that peaks of the spectrum vary due to the randomness of the phase of the signals. This caused the fluctuation of the final output value. However, if we impose the condition of zero cross-correlation betwen the signals, as Frost did, we expect a consistent result with zero deviation.

Typical results from our processor in the look frequency-band are shown in Fig. 3, in comparison with the associated data. The data is presented in Fig. 3 in the form of logarithmic intensity for each of 16 beams formed from the four detector signals. Fig. 3 (a)

is a baseline result which is the result of reconstructing, at
the center-frequency of the desired signal $(0 \cdot 3/\tau)$, the ideal
signal, i.e., the input data does not contain any noise or interference. Fig. 3 (b) is the adaptively-processed and reconstructed
result, at the center frequency of the desired signal, which
contains the desired signal and all the noise and interference
specified in Table I. Fig. 3 (c) is the conventional reconstructed
result from the same input data as that used in (b). The improvement resulting from our processing is clearly seen in these data,
and in Fig. 3 (b) in particular. It is interesting to note that,
in Fig. 3 (c), where substantial interference is present, the
power in the look direction is less than that in Fig. 3 (b). This
is due to a coherent anti-phase superposition of signal and interference, i.e., residual correlations.

Fig. 4 is the reconstruction of the input data within the frequency bands at the center frequencies of the respective locallized
interferences. These data were utilized to find the location of
nulls in the manner described in Section II B.

It can be seen from these results that, even using the Frost
scenario, the present holographic adaptive processing shows much
faster and more accurate convergence to the desired signal than
Frost's.

Even if Frost were to use FFT temporal processing to obtain
the same frequency response as we did (and this would be a natural
modification), he would still be left with an average 8% error in
output power.

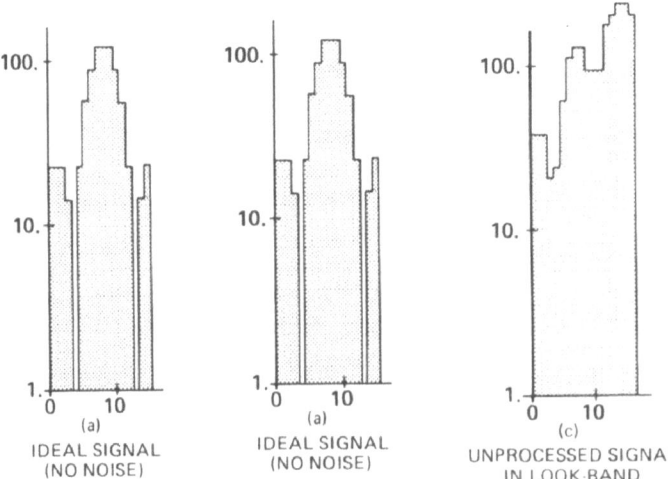

Figure 3 - Beam Forming on Desired Frequency Component

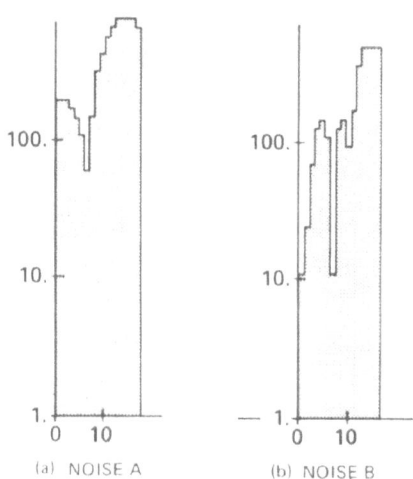

Figure 4 – Beam Forming on Noise-Frequency Bands

REFERENCES

1. R. F. Steinberg, P. N. Keating, and R. F. Koppelmann, "Acoustical Holography", Vol. 6, Ed. by N. Booth, 539. (Plenum Press,
   New York, 1976).

2. P. N. Keating, R. F. Koppelmann, R. K. Mueller, and R. F.
   Steinberg, J. Acoust. Soc. Am. 59, 106 (1976).

3. O. L. Frost, Proc. IEEE 60, 926, (1972).

4. N. L. Owersley, "Signal Processing", Proc. of the NATO Advanced
   Study Institute on Signal Processing with Particular Reference
   to Underwater Acoustics, Loughborough, England, 1973, Ed. by
   J. W. R. Griffithes, et al, 591, (1973).

5. H. Cox, J. Acoust. Soc. Am., 54, 3, 771 (1973).

6. W. F. Gabriel, IEEE 64, 239, (1976).

7. B. Widrow, J. R. Glover, Jr., S. M. McCool, J. Kaunitz, C. S.
   Williams, C. H. Hearns, J. R. Zeider, E. Dong, Jr., and R. C.
   Goodlin, IEEE 63, 1692 (1975).

8. V. C. Anderson, J. Acoust. Soc. Am., 45, 2, 398, (1969).

9. R. K. Mueller, R. R. Gupta, and P. N. Keating, J. Appl. Phys.,
   43, 457 (1972).

10. P. N. Keating, R. R. Gupta, and R. K. Mueller, J. Appl. Phys.,
    43, 1198 (1972); also in "Acoustical Holography", Ed. by Glen
    Wade, 4, 251, (Plenum, New York, 1972).

11. Extensive bibliography will be found, for example, in H. L.
    Van Trees, "Detection, Estimation, and Modulation Theory",
    Parts I and III, John Wiley and Sons, New York (1971).

A SCANNED CURVILINEAR ACOUSTICAL HOLOGRAPHY SYSTEM FOR GEOLOGICAL INSPECTION OF INTERNAL STRUCTURES IN ROCK

H. D. Collins, J. Spalek, W. M. Lechelt
and R. P. Gribble

HOLOSONICS, INC.
2950 George Washington Way
Richland, Washington 99352

This paper describes a curvilinear scanned acoustical holographic device for imaging various internal objects in geological structures. The rapid-scanning system was developed to provide near real-time holography without the stringent requirements of the usual rectilinear scanned systems. The system has the capability of generating a circular scanned hologram in 15 seconds and this could be reduced with the addition of multiple receivers. The acoustic reference beam is simulated electronically with a selectable inclination angle to provide spatial separation of the images.

Using this system, various internal voids in large marble samples have been imaged with λ (i.e., 6mm) lateral resolution at depths greater than 25cm.

CONCLUSIONS

The curvilinear scanned acoustical holographic technique employing simultaneous source-receiver configuration has successfully imaged internal defects in geological structures. We have imaged 1cm diameter defects in marble at depths greater than 20cm with 6mm lateral resolution. This system has the following unique characteristics:

- Optimum resolution and efficient illumination of the objects or defects

- Extremely rapid scanning (near real-time) with the minimum complexity of electronic instrumentation

Ability to image objects directly in the projected
scanned aperture.

The use of the focused transducer simulating a point source
and receiver provides the necessary requirements (i.e., large
aperture, high power capabilility, etc.), but has the inherent restriction of operating pulse echo. The pulse repetition rate or
frequency determines the scan rate, hologram generation time and
the maximum acceptable phase shift frequency of the simulated off-
axis reference. The pulse operation is essentialy a sampling technique and the sampling rate must be compatible with the phase
shift frequency if adequate grating information is to be recovered
in the recorded hologram. The sampling frequency (i.e., pulse frequency) must be at least twice the maximum phase shift frequency to
insure the minimum sampling rate. This may severely increase the
hologram generation time if multiple transducers or arrays are not
used for imaging objects deep in geological structures which require
extremely low pulse repetition rates and thus low scanning speeds.

INTRODUCTION

The rapid curvilinear scanning system was initially developed
in 1968 at Battelle Northwest for biological and ocean bottom topography imaging[1]. In 1970 the system was modified to provide under-
sodium viewing capabilities for the Fast Flux Test Reactor at
Richland, Washington[2].

The circular scanning system was redesigned in 1975 to provide
rapid holographic imaging techniques for the inspection of geological structures under an NSF grant[3] (Scanned Acoustical Holography
for Geological Prediction in Advance of Rapid Underground Excavation). This paper describes the latest system developed to image
large internal voids, cracks, etc. in rock from one exposed surface utilizing the near real-time inspection capabilities of the
curvilinear scanning system. The system has two modes of operation: (1) variable circular path scanning and (2) translational
constant circular path scanning. With the translational mode, the
length of the aperture is limited only by the length of the translation table.

Basically, the circular scanner operates the same as the usual
rectilinear scanned holographic system,[4,5,6,7] except the source-
receiver transducer scans a circular path as it is being translated the length of the aperture. The system consists of three
sections: the rotational disk, the translation platform table,
and the electronic instrumentation unit. The rotational disk on
which the transducers are located is approximately 61cm in diameter
and has rotational capabilities from 0 to 500 r.p.m. The translation platform on which the rotation system is located has various
scanning speeds from 0 to 25cm/min. The hologram generation time

and the translational velocity determines the aperture length. The angular velocity of the disk, the number of transducers and the translational velocity determine the number of scan lines generated per second and the sampling density. For example, if we employ a single transducer and require 150 scan lines in the hologram to be generated in one second with a sample density of 1mm, the disk must rotate at 150 r.p.s. and the translation speed must be 15cm/sec. If two transducers are placed 180 degrees apart on the disk, two lines are scanned per revolution and the generated time is reduced by a factor of two. If N transducers are used, the generation time is reduced by a factor of N; thus real-time imaging capabilities are available at low scanning speeds with the use of arrays on the rotational disk.

The signal conditioning features are the same as the rectilinear system with the exception of electronic simulated acoustic off-axis reference signal. The reference requires a variable frequency control voltage to generate the linear diffraction grating on the curvilinear scanning raster. The basic difference between the two systems is the raster pattern imposed on the hologram. The rectilinear scanner imposes a straight line diffraction grating which produces the usual multiple images in the reconstruction. The circular scanner imposes a curved line grating which does not produce multiple images but disperses the information over the entire reconstruction plane. Hildebrand and Brenden[8] describe the relationships between the minimum sampling density and the scanning configuration to prevent aliasing in the reconstruction.

SIMULATED ACOUSTIC OFF-AXIS REFERENCE BEAMS

If we assume an acoustic plane wave reference beam, then the reference beam can be expressed as

$$S_{R(x,y)} = P_{R(x,y)} \cos\left[\omega t - \phi_{RS}(x,y)\right] \tag{1}$$

where

$$\phi_{RS}(x,y) = \frac{2\pi}{\lambda_S} x \sin \alpha_{RS}.$$

No loss of generality results in only considering two dimensions with the object located on the (x-z) plane. The signal contribution to the acoustic receiver by the reference beam is

$$P_{R(x,y)} \cos\left(\omega t - \frac{2\pi}{\lambda_S} \sin \alpha_{RS} V_x t\right) \tag{2}$$

where V_x is the scanning velocity of the acoustic receiver and

$$\omega_{RS} = \frac{2\pi}{\lambda_S} V_x \sin \alpha_{RS} \tag{3}$$

Equation (2) represents a sinusoidal wave whose phase is a function of the scanning velocity (V_x) and the inclination angle (α_{RS}).
Thus, the acoustic plane wave reference beam can be simulated
with an electrical signal of this form and combined with the object
signal in a balanced mixer or multiplier. If the direction of propagation of the plane wave reference is perpendicular to the x
axis (i.e., $\alpha_{RS} = 0$), then $\omega_{RS} = 0$ and the reference signal is
simply $P_{R(x,y)}\cos \omega t$. This electrical signal when combined with
the object signal in a mixer would simulate an on-axis plane wave
acoustic reference beam. Employing an on-axis reference beam requires imaging objects outside of the projected scanning aperture
to provide separation of nondiffracted light and the images in the
reconstruction. Now if the inclination angle is not zero, then the
electrical reference signal must be phase shifted to simulate an
off-axis acoustic reference beam. The simulated inclination angle
is a function of the scanning velocity, wavelength of sound, and
the phase shifter control voltage frequency. Naturally, in three
dimensions, skewed beams are possible with phase shifting and time
delay circuits. The phase of the electronic reference can be
shifted with respect to time by the following voltage waveform
shown in Figure 1.

The control voltage radian frequency (ω_p) can be expressed
in terms of ω_{RS}:

$$\omega_p = 2\pi f_p = \frac{2\pi}{\lambda_S} V_x \sin \alpha_{RS} \tag{4}$$

The velocity component in the "X" direction is

$$V_x = \omega r \sin\omega t = V\sin\omega t \tag{5}$$

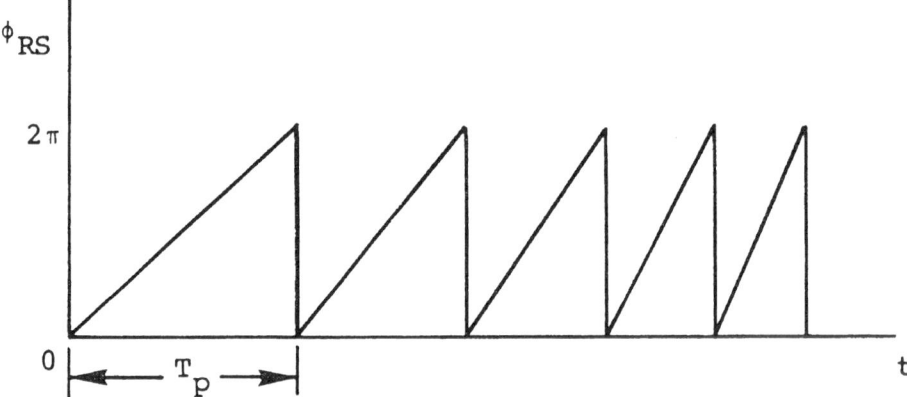

FIGURE 1.   PHASE SHIFTER VOLTAGE WAVEFORM

when ω is the radian frequency of the disk and r the radius. The control voltage frequency (f_p) can be expressed in terms of f_{RS}:

$$f_p = \frac{V_x}{\lambda_s} \sin \alpha_{RS} = \left| \frac{V}{\lambda_s} \sin \omega t \sin \alpha_{RS} \right| \tag{6}$$

If the inclination angle and the acoustic frequency are constant, then the control frequency is a function of the disk velocity and transducer position. The control frequency is maximum at $n\pi/2$ where $n = 1, 2, \ldots$ and minimum at $n\pi$ where $n = 0, 1, 2 \ldots$.

The phase shifter voltage waveform is shown in Figure 1.

If the simulated plane wave is inclined with respect to the hologram plane (i.e., $\alpha_{RS} > 0$), then a linear diffraction grating will be imposed on the hologram. The grating spacing is a function of the scanning velocity (V_x), phase shifter frequency (f_p), and the hologram magnification (m). The grating spacing on the hologram is

$$d = \frac{V_x T_p}{m} = \frac{V_x}{m f_p} = \frac{\lambda_s}{\sin \alpha_{RS}} \tag{7}$$

where

$$\alpha_{RS} = \sin^{-1} \left(\frac{\lambda_s m f_p}{V_x} \right).$$

If $\alpha_{RS} = \pi/2$, then the finest grating imposed on the hologram has a spacing of λ_S and the maximum frequency is

$$f_{p_{max}} = \frac{V}{m\lambda_s} \ldots \text{ where } \omega t = \frac{n\pi}{2} \text{ for } n = 1, 2, \ldots \tag{8}$$

It should be obvious that any grating spacing can be imposed on the hologram by proper adjustment of the phase shift control voltage frequency. The grating spacing can be less than a wavelength which indicates that electronic simulation is more versatile than using an acoustic reference beam. The grating lines are usually constructed perpendicular to the X axis in curvilinear scanning. This results in two sets of gratings. Figure 2a is a typical grating imposed on the curvilinear scanned hologram. Figure 2b is the acoustic hologram and the reconstruction of a point source located directly in the center of the projected scanning aperture. The hologram shows the familiar zone plate located directly in the center of the aperture and also shows the diffraction grating as a result of phase shifting the reference. Figure 2c shows the reconstructed image of the point source.

The image was reconstructed using a helium-neon laser (6328Å) at 10m and the image was diffracted approximately 1cm from the undiffracted light. The lateral image displacement from the

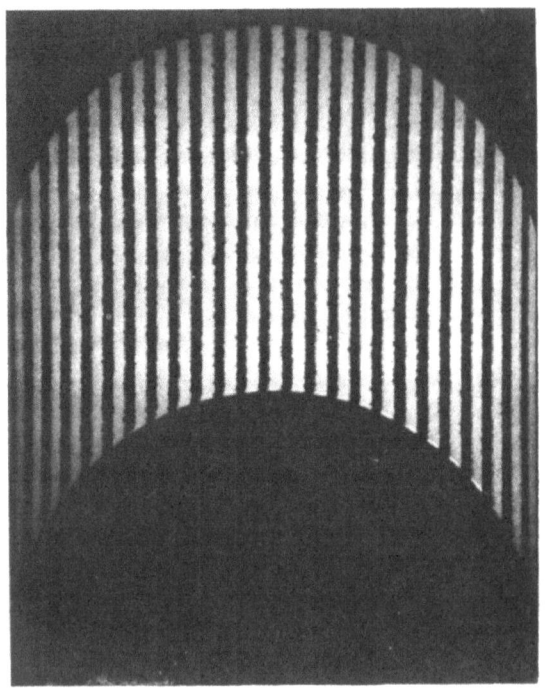

(a) LINEAR GRATING IMPOSED ON THE CURVILINEAR SCANNED HOLOGRAM

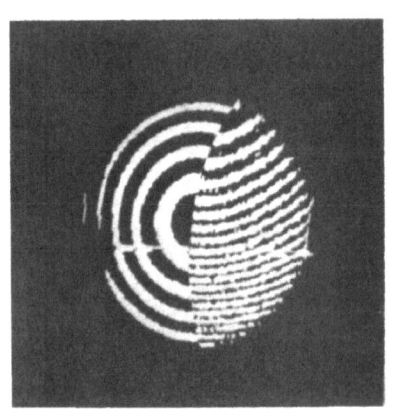

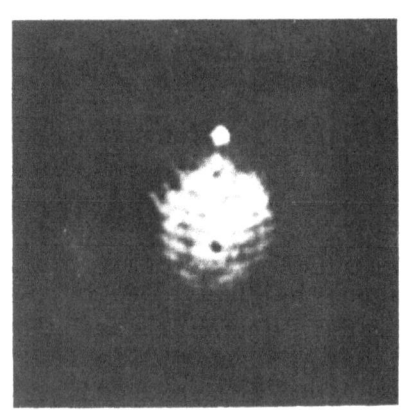

(b) EFFECTS OF OFF-AXIS
    REFERENCE ON THE ACOUSTIC
    HOLOGRAM OF THE POINT OBJECT

(c) OFF-AXIS RECONSTRUCTED
    IMAGE OF A POINT OBJECT

FIGURE 2. CURVILINEAR SCANNED HOLOGRAMS AND RECONSTRUCTION

undiffracted light was calculated using either of the following equations:

$$\delta \simeq \frac{\lambda_L}{\lambda_S} \; mnr_b \; \sin \alpha_{RS} \tag{9}$$

or

$$\delta \simeq \frac{\lambda_L \; mnf_p \; r_b}{V_x} \tag{10}$$

where λ_L = reconstruction wavelength
        λ_S = construction wavelength
        m = hologram magnification
        n = order of diffracted light
        r_b = hologram to image distance
     α_{RS} = inclination angle (with respect to the horizontal) of the simulated reference beam
        f_p = phase shift control voltage frequency
        V_x = scan velocity of the acoustic receiver.

The results were in excellent agreement with the theoretical calculations. This technique of electronic phase shifting the reference beam provides the necessary conditions for imaging directly in the projected scanning aperture and for separation of the images with the undiffracted light. The ability to image directly in the scanning aperture increases the lateral and radial resolution by increasing the effective aperture. Use of this technique and point source-receiver scanning provides an additional increase in the resolution by a factor of two and enhances the object illumination.

RESOLUTION

The most convenient parameter for evaluating the quality of the holographic imaging system is resolution. The resolving power of an acoustical imaging system is by definition the smallest lateral or longitudinal separation in which two object points can be distinguished in the image. A corollary to this is using only one point object and defining the minimum distance it can be moved and detected in the image. The single point criterion is often used in holography to avoid the phase interference problem associated with two points as a result of coherent illumination[9].

The lateral and longitudinal resolutions using the Rayleigh criterion for a circular focused transducer are

$$\Delta X_t \simeq 1.22 \; \frac{\lambda s f}{a} \text{ and } \Delta r_t \equiv 2\lambda_s \left(\frac{f}{a}\right)^2 \tag{11}$$

where λs is the sound wavelength, f the focal length of the transducer and a the diameter of the transducer lens.

The lateral and longitudinal holographic resolution for the simultaneous source-receiver configuration are defined as

$$\Delta X_h \simeq \frac{\lambda_s r_1}{2L} \quad \text{and} \quad \Delta r_h \simeq \frac{\lambda s}{2} \left(\frac{r_1}{L}\right)^2 \qquad (12)$$

where r_1 is the object to hologram distance and L the effective hologram aperture.

These expressions are quite similar if we assume r_1 and L are equal to the focal length (f) and the lens diameter (a). Thus, we have an imaging system with a different criterion for resolution. Usually the transducer resolution dominates unless the object distances are great, compared with the focal length.

The crossover point can be calculated by equating the lateral resolutions. The depth at which crossover occurs in rock with this system is approximately

$$r_1 \simeq \frac{2fL}{a} \simeq 100\text{cm}. \qquad (13)$$

Figure 3a shows the typical geometry used in verifying the lateral resolution of the system experimentally. Two point objects are located within the cross-over region where the dominating resolution is determined by the focused transducer. They are separated at the calculated Rayleigh distance (i.e., 6mm) assuming a 15cm focal length, 3.8cm diameter and 1MHz frequency.

The hologram using an on-axis electronic simulated acoustic plane wave reference is shown in Figure 3b. It is the usual composition of two familiar zone plates where the point objects are sufficiently close together. The composite diffraction lens or hologram shows the two points will be easily resolved in the reconstruction. Figure 3c is the hologram constructed using an off-axis simulated plane wave reference. The important difference to note between the two is the higher spatial frequency content of the on-axis hologram. The spatial frequency content of the hologram is directly dependent on the object and the inclination angle between the object and reference beams. The off-axis reference essentially imposes a linear spatial frequency modulation on the object frequency spectrum. Usually the recording display is resolution limited and it is desirable to minimize the spatial frequencies by providing the minimum acceptable inclination angle between the object and reference beams. Naturally, this angle is dependent on the spatial separation of images in the reconstruction.

Figure 3d shows the reconstructed image of the two point objects. The objects are distinctly separated in the image verifying the theoretical Rayleigh resolution criterion for the focused

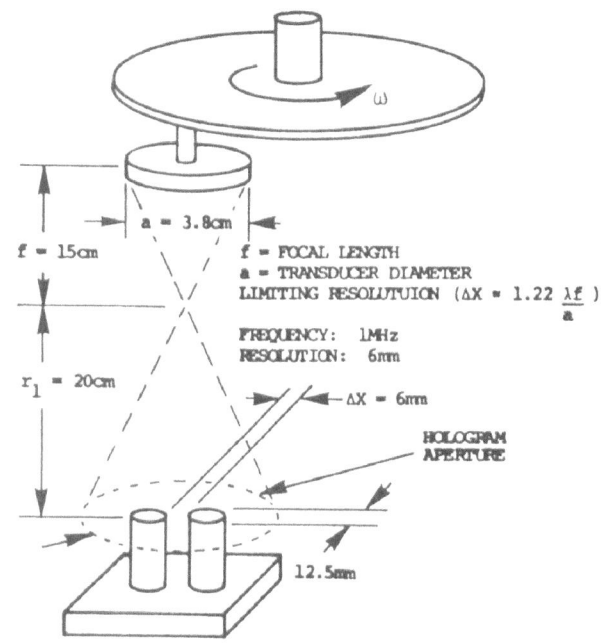

f = FOCAL LENGTH
a = TRANSDUCER DIAMETER
LIMITING RESOLUTUION ($\Delta X = 1.22 \frac{\lambda f}{a}$)

FREQUENCY: 1MHz
RESOLUTION: 6mm

(a) RESOLUTION MEASUREMENT GEOMETRY

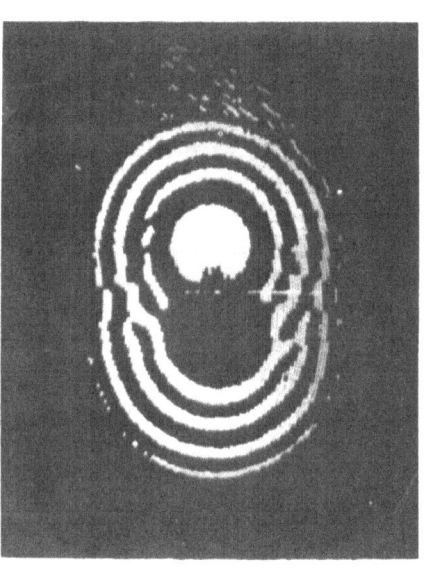

(b) HOLOGRAM OF THE TWO
     OBJECTS WITH AN
     ON-AXIS REFERENCE

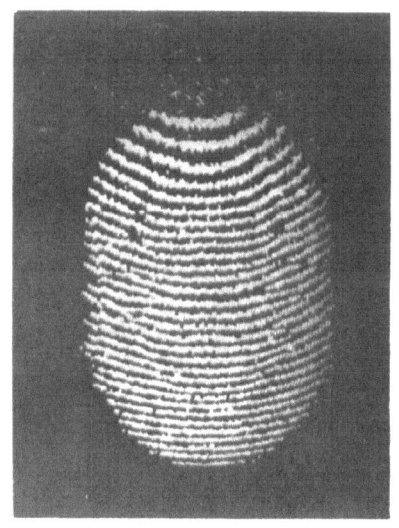

(c) HOLOGRAM OF THE TWO OBJECTS
     WITH AN OFF-AXIS REFERENCE

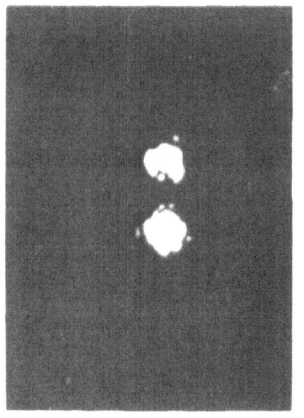

(d) RECONSTRUCTED
     HOLOGRAPHIC IMAGE

FIGURE 3.   CIRCULAR SCANNER RESOLUTION MEASUREMENTS

transducer. The holographic resolution of the system is approximately 1.33λs or 2mm employing this experimental arrangement.

CURVILINEAR ACOUSTICAL HOLOGRAPHY SCANNER: SYSTEM DESCRIPTION

A 61cm diameter curvilinear scanner with associated electronics and interface controls was designed, fabricated, and used for large-aperture imaging of test objects in rock. Photographs of the curvilinear holographic system are shown in Figures 4a and 4b. A block diagram of the curvilinear acoustical holography system in a translational axis configuration is shown in Figure 5. The curvilinear acoustical holography system consists of the following components:

- 61cm diameter circular scanning disk with transducer translation mechanisms

- Rotational vertical drive (shaft, gearboxes, motor)

- Electrical control and interface apparatus (slip rings, stepping motor switches and control circuits, shaft encoders, etc.)

- Acoustical holography signal processor (HolScan System 200)

- Optical processor/reconstructor.

A close-up view of the 61cm diameter circular scanning disk is shown in Figure 4b photograph. The disk has four slots with worm and gear rail mechanism for radially translating the acoustic transducers and counterweights. A block diagram of the rotation and translation synchronization electronics and controls is shown in Figure 6.

Some of the operational characteristics of the laboratory model curvilinear scanner system are:

- Rotational scan speed range: 10 to 500 r.p.m., continuously variable

- Translational axis speed range: 0 to 25cm/min

- Max. diameter of circular scan aperture: 56cm

- Radial-helical scan speed range: 0 to 0.46 cm/sec.

- Circular-scan radial increments ΔR/rev = .05cm/rev to 1.4cm/rev.

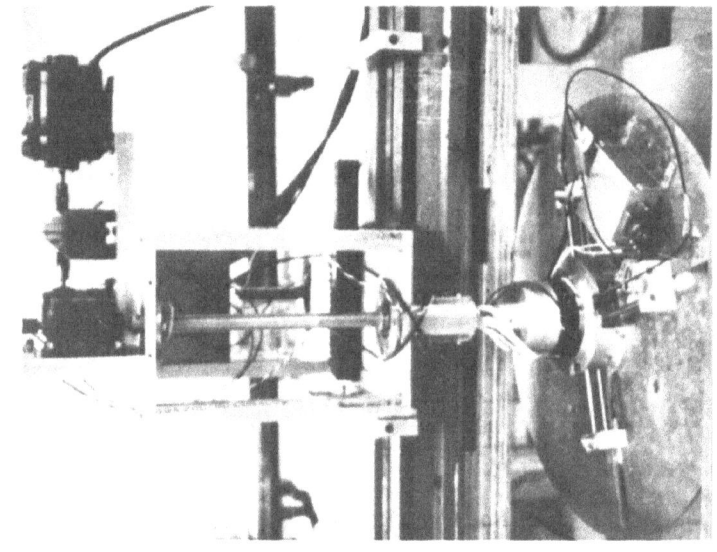

(b) CLOSEUP OF 61cm DISC
CIRCULAR SCANNER

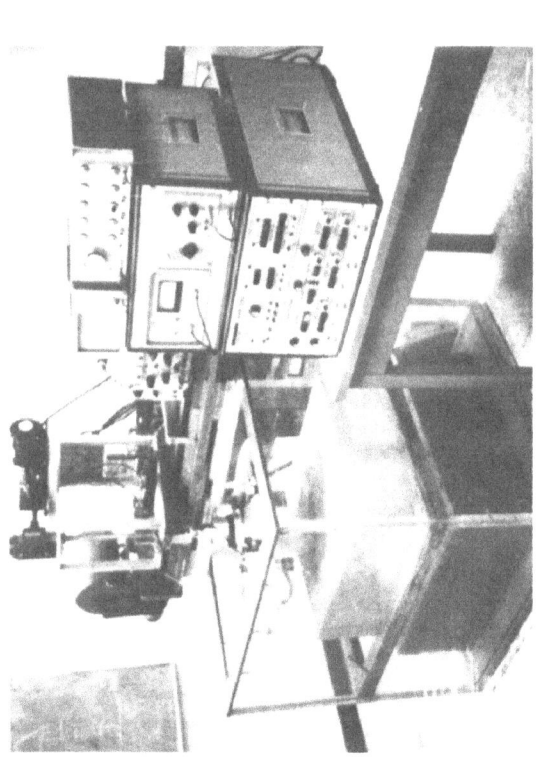

(a) CURVILINEAR SCANNER ABOVE A LARGE BLOCK
OF MARBLE IN LABORATORY SETUP

FIGURE 4. CURVILINEAR SCANNER IN LABORATORY TESTS CONFIGURATION

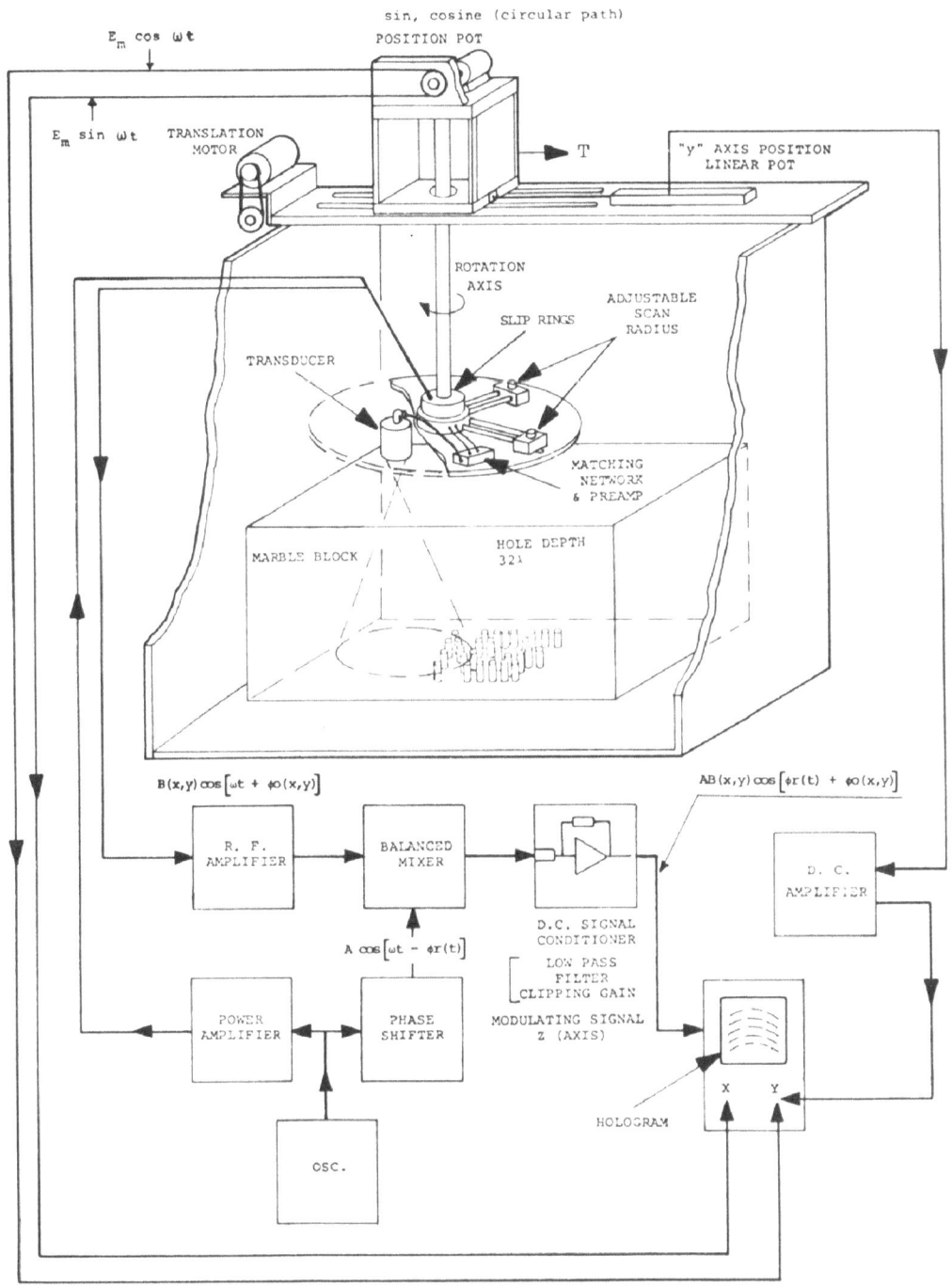

FIGURE 5. CURVILINEAR ACOUSTICAL HOLOGRAPHY SYSTEM

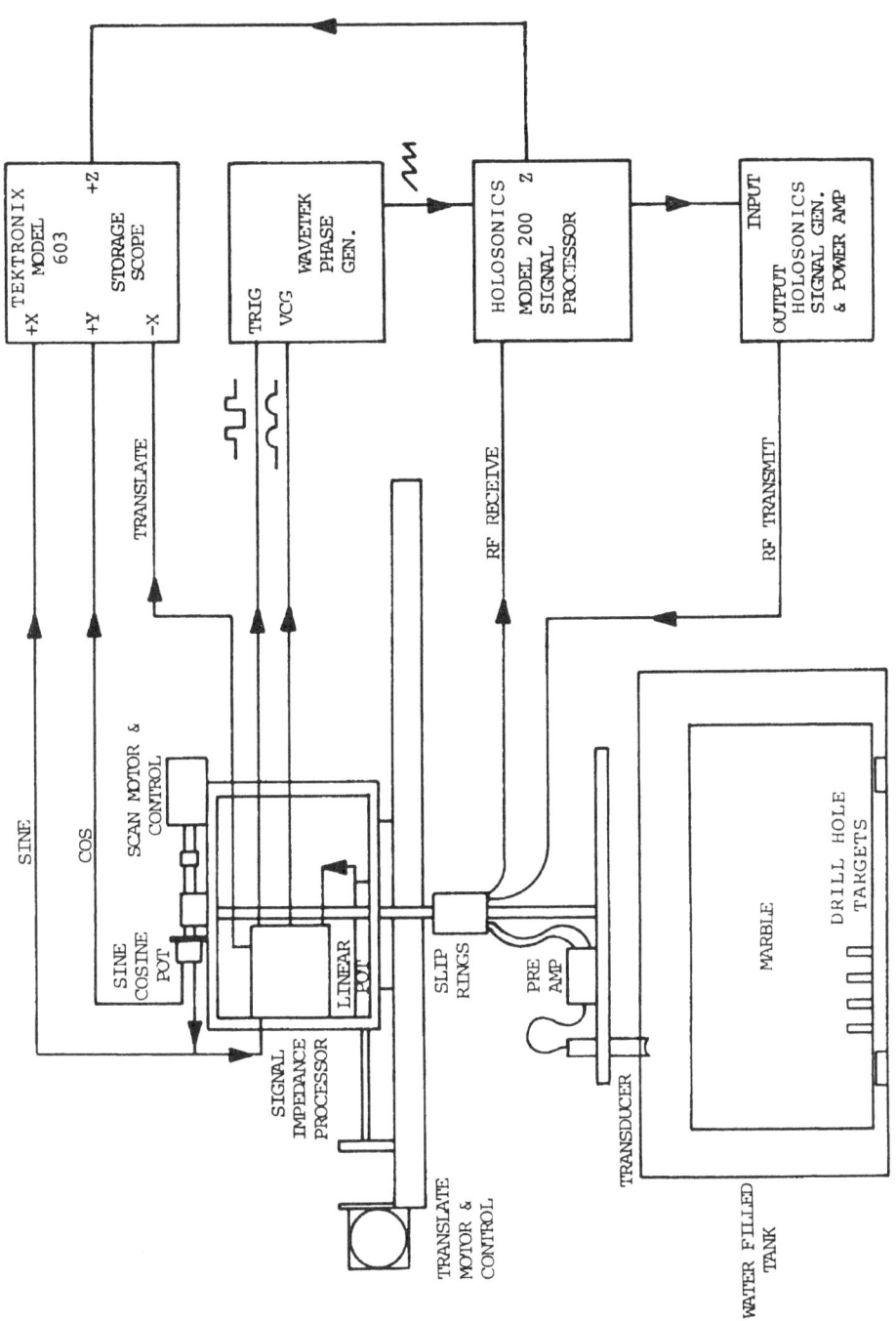

FIGURE 6. ROTATION AND TRANSLATION SYNCHRONIZATION
DEVICES AND CONTROLS

The variable radius scan capabilities will be used for generating scanned holograms in spaces where physical constraints do not permit axis translation, e.g., such as on the face of an underground bore. The variable radius scanned holograms require a more complex off-axis reference (phase-shift) electronics than the constant radius/axis translation holograms.

The analog phase shift electronics used in the initial laboratory experiments had voltage inability problems which would negate their use in subsequent field tests. To avoid the instability problems, digital phase shifter controlled by microprocessor was designed and constructed. Holographic imaging using the digital phase-shifter electronics with radial scans is in progress.

EXPERIMENTAL RESULTS

The results of experiments conducted with the curvilinear acoustical holography system operated in the constant radius/axis translation mode are presented in the following figures. Figure 7 shows a simplified block diagram of the curvilinear system and a preliminary test pattern and image of an object (H) submerged in water. The next figure, Figure 8 shows an acoustical hologram and the reconstructed image of a large (7.6cm X 10cm) letter "H" which was also submerged in water. A larger aperture reconstructed holographic image of an "NSF" letter pattern is shown in Figure 9. The "NSF" letters were constructed from styrofoam covered aluminum rods arranged in a circular pattern. The geometry of the letter pattern is shown in Figure 9a. The approximate size of the letter pattern was 9cm x 30cm. Again the target pattern was submerged in water, at a distance equivalent to the expected distance of targets in the test rock sample. Figure 9b shows the circular hologram geometry in a test tank. The reconstructed image of the large aperture "NSF" pattern in water is shown in Figure 9d.

The next figure, Figure 10 shows a large circular aperture holographic image of an "NSF" pattern of holes in a large rock sample. The geometry of the curvilinearly scanned hologram and of the holes drilled in a large block of marble are shown in Figure 10a and b. Figure 10c shows the acoustical hologram of the hole pattern in marble, located at a 32λ depth below the rock surface. The reconstructed image of the large aperture hologram of the hole pattern in marble is shown in Figure 10d.

The tests using the curvilinear acoustical holography scanner in the axis translation mode were concluded with the above presented results. Preliminary test raster patterns and focused images obtained with the curvilinear system operated in the variable radial scan mode are shown in Figure 11. The variable radial scan mode will be used for generating holograms of objects in spaces where

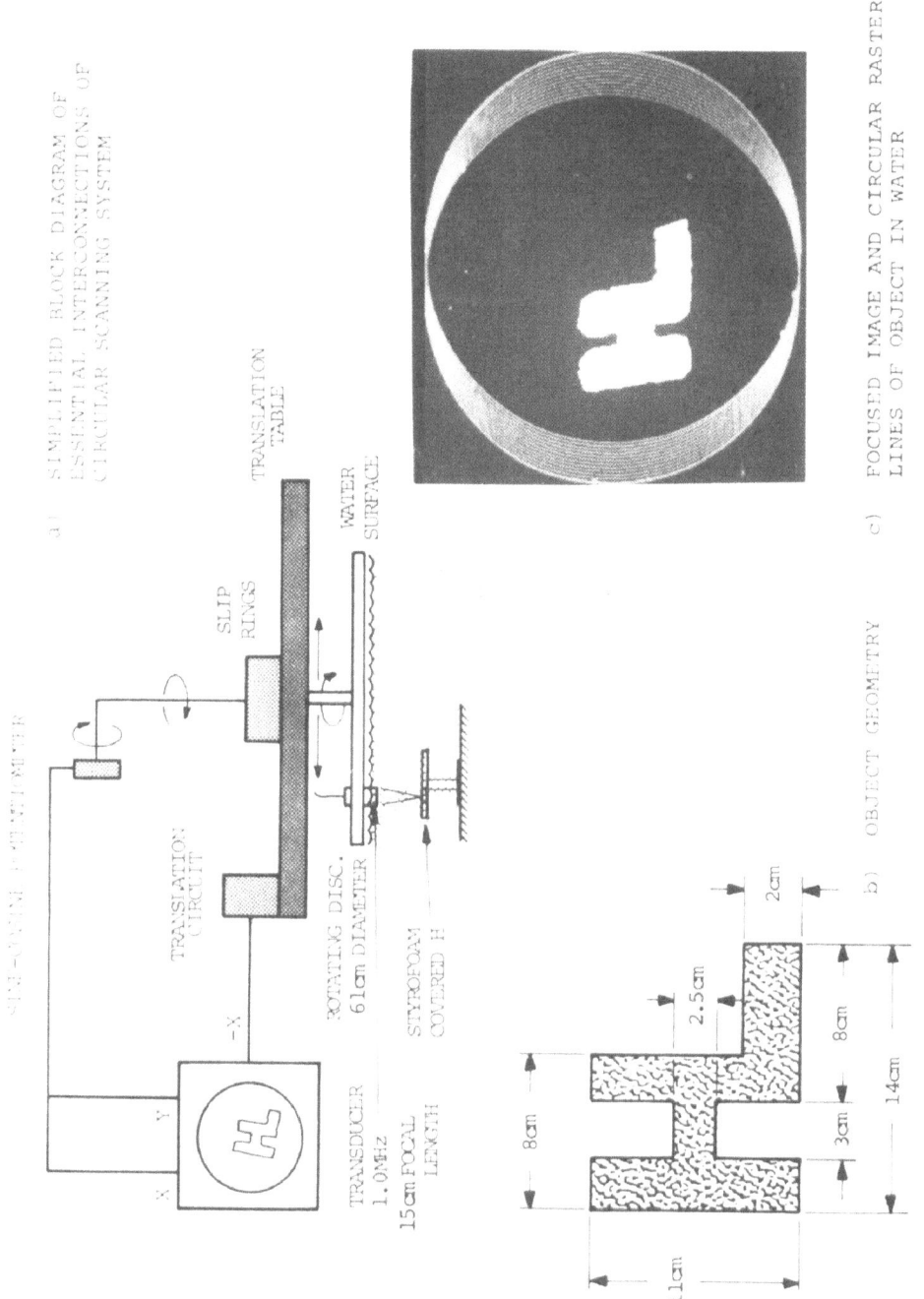

a) SIMPLIFIED BLOCK DIAGRAM OF ESSENTIAL INTERCONNECTIONS OF CIRCULAR SCANNING SYSTEM

b) OBJECT GEOMETRY

c) FOCUSED IMAGE AND CIRCULAR RASTER LINES OF OBJECT IN WATER

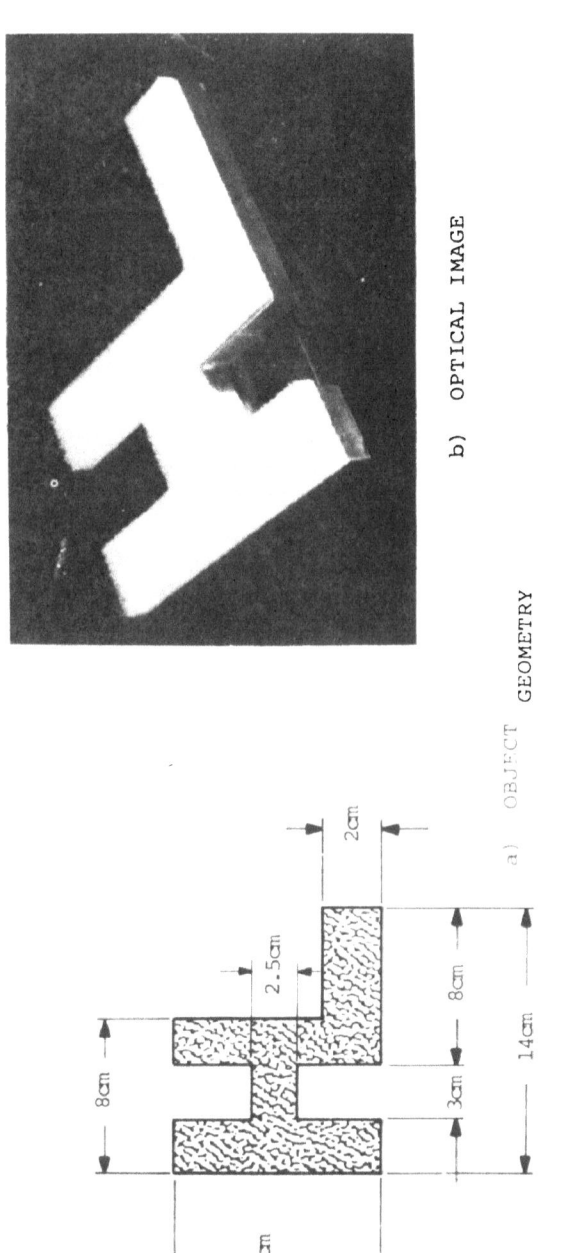

b) OPTICAL IMAGE

a) OBJECT GEOMETRY

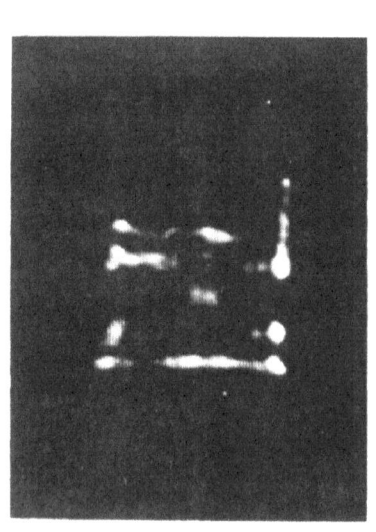

d) RECONSTRUCTED HOLOGRAPHIC IMAGE

f = 1MHz
MAGNIFICATION
m = .216

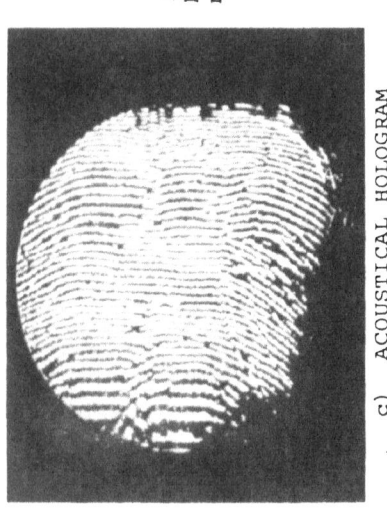

c) ACOUSTICAL HOLOGRAM

FIGURE 8. CIRCULARLY SCANNED HOLOGRAPHIC IMAGE OF THE LETTER "H"

1.3cm DIAMETER X 10cm ALUMINUM RODS
1.3cm SEPARATION EDGE TO EDGE

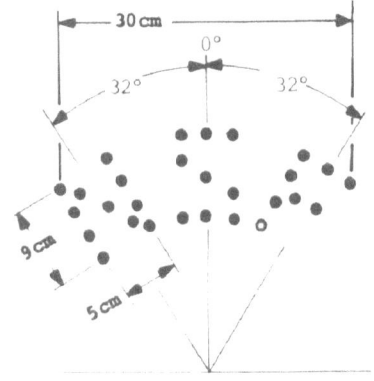

a) NSF PATTERN FOR RESOLUTION
   TESTS

TEST PARAMETERS:

f = 1MHz; 3.8cm DIAMETER TRANSDUCER,
15cm FOCAL LENGTH, MAGNIFICATION: .227

NSF PATTERN

b) CIRCULAR HOLOGRAM GEOMETRY
   IN LABORATORY TANK

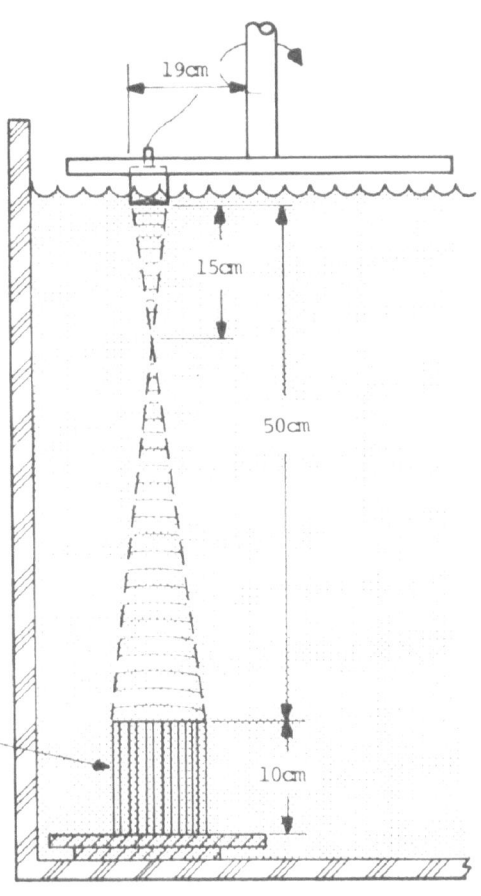

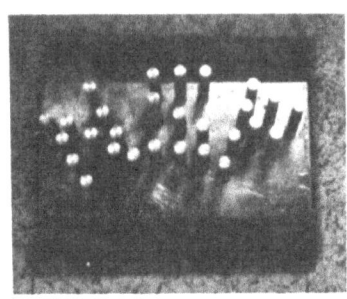

c) PICTURE OF 'NSF' PATTERN

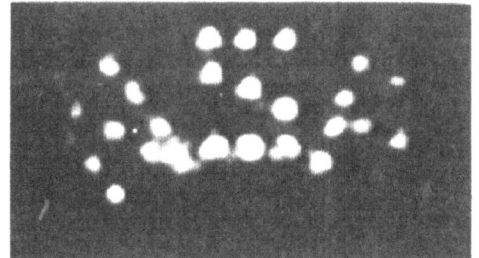

d) RECONSTRUCTED HOLOGRAPHIC
   IMAGE

FIGURE 9. LARGE APERTURE CIRCULAR-SCANNED HOLOGRAM
          OF THE NSF PATTERN

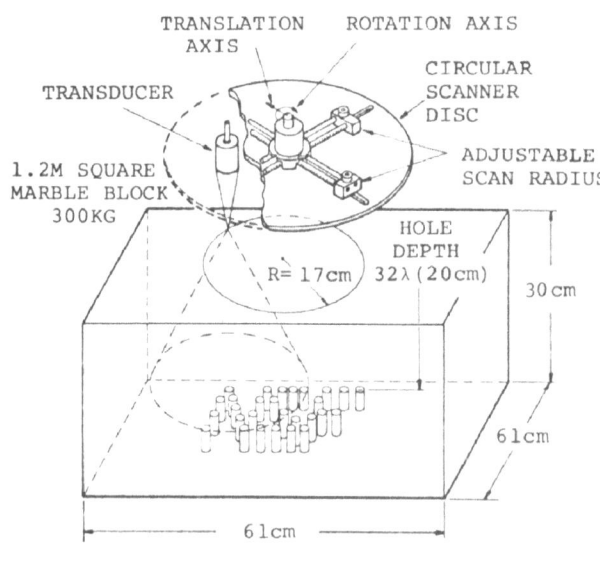

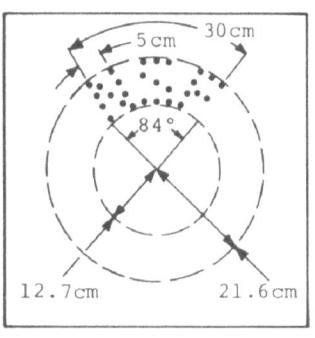

GEOMETRY FOR SCANNED HOLOGRAM
IN MARBLE BLOCK

NSF PATTERN IN LARGE
ROCK SAMPLE

TEST PARAMETERS: f = 0.81MHz; 3.8cm DIA. TRANSDUCER, 15cm FOCAL LENGTH;
                   v MARBLE = 6.4mm/µsec; MAGNIFICATION = .19

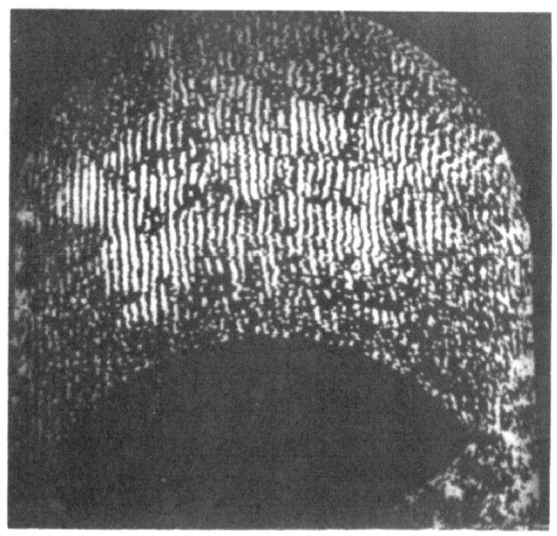

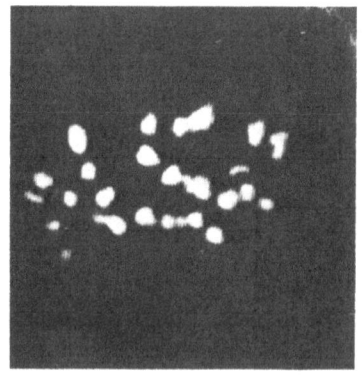

HOLOGRAM OF HOLES AT 32λ
DEPTH IN MARBLE

RECONSTRUCTED IMAGE

FIGURE 10. LARGE CIRCULAR APERTURE HOLOGRAPHIC IMAGE
           OF NSF PATTERN HOLES IN A LARGE ROCK SAMPLE

(a) SIGNAL GATED ON TOP SURFACE ON TARGET SUPPORT BLOCK

(b) SIGNAL GATED ON TOP SURFACE OF TARGET

FIGURE 11.   ACOUSTICAL IMAGING WITH VARIABLE RADIUS CURVILINEAR
              SCAN SYSTEM

physical constraints do not permit axis translation, such as on the face of an underground tunnel.

A full scale, prototype system of the variable radius curvilinear scanner, which has a large 2.4 meter diameter aperture and digital phase shifters controlled by a microprocessor was designed and constructed at Holosonics under NSF sponsorship. Tests with the large aperture prototype system are presently in progress.

REFERENCES

1. Battelle Northwest Research Report on "Ultrasonic Holography Using a Circular Scanning Technique" to Holotron Corporation, Wilmington, Delaware; October 1968.

2. H. Dale Collins, "Various Holographic Scanning Configurations for Under-Sodium Viewing", BNWL-1558, March 1971.

3. T. O. Price, B. B. Brenden, H. D. Collins, J. Spalek, "Scanned Acoustical Holography for Geological Prediction in Advance of Rapid Underground Excavation"; Phase I Technical Report submitted by Holosonics to National Science Foundation; NSF Grant No. GI43686, May 30, 1975.

4. H. D. Collins, B. P. Hildebrand, "The Effects of Scanning Position and Motion Errors on Hologram Resolution", Acoustical Holography Volume 4, Plenum Press, New York, 1972, pp. 467-500.

5. H. D. Collins and R. P. Gribble, "Acoustic Holographic Scanning Techniques for Imaging Flaws in Thick Metal Sections" Seminar-in-Depth on Imaging Techniques for Testing and Inspection (SPIE), February 14-15, 1972, Los Angeles, California.

6. H. D. Collins and B. P. Hildebrand; "Evaluation of Acoustical Holography for Inspection of Pressure Vessel Sections" Paper presented to the Joint ASME (USA) Institution of Mechanical Engineers, London, England, May, 1972.

7. H. Dale Collins and B. B. Brenden, "Acoustical Holographic Transverse Wave Scanning Technique for Imaging Flaws in Thick-Walled Pressure Vessels", Acoustical Holography Volume 5, Plenum Press, New York, pp. 175-136 (1974).

8. B. P. Hildebrand and B. B. Brenden, An Introduction to Acoustical Holography, Plenum Press, New York 1972.

9. J. W. Goodman, Introduction to Fourier Optics, McGraw-Hill, New York, 1968.

A FEW EFFECTIVE PREPROCESSINGS IN SYNTHETIC APERTURE SONAR SYSTEM

Takuso Sato, Osamu Ikeda, Hajime Ohshima, and
Hiroyuki Fujikura

Tokyo Institute of Technology
4259 Nagatsuda, Midori-ku, Yokohama-shi, Japan

ABSTRACT

Two effective preprocessings of the detected signals in
synthetic aperture sonar system are presented.

One of them is the preprocessing to eliminate the effect of
the phase turbulence of the medium. The fundamental idea is the
utilization of the reference signal detected at a fixed point on
the scanning plane and to take the correlation between it and the
signal detected by a scanning detector to cancel the effect of the
turbulence. The theoretical considerations are assured by computer
simulation. Also a complete system is constructed and a typical
experimental result is presented which shows the effect
dramatically.

The other preprocessing is based on the utilization of
superposition of complex images obtained by multiple frequencies of
waves, and it is divided further into two categories; i)
superposition after linear signal processing and ii) superposition
after nonlinear heterodyne processing. As for the former technique,
the effect of the spectral synthesis in frequency domain to
increase the range resolution is discussed theoretically. As for
the latter technique, discussion made is on the effect of changing
the frequencies of the reference waves nonlinearly in synthetic
aperture sonar systems. Basic experimental results are also shown.

Although most of the results have been or are to be submitted
as papers in a few journals(Refs. 2 and 5-8), comprehensive
discussions are presented in this paper.

I. INTRODUCTION

The synthetic aperture technique has been used in sonar systems as an effective and proper means to get high azimuth resolutions. In these synthetic aperture sonar systems the image reconstruction is carried out mostly by using computer data manipulation. While in radar systems optical reconstruction is used. In sonar systems the speed of the waves is rather slow and consequently the accumulation of the data can be carried out almost in real time through the A/D conversion and digital computer techniques. This fact suggests also that fairly complicated signal processings or preprocessings can be inserted as the preprocessor for the synthetic aperture sonar system without disturbing the reconstruction process or without severly affecting the speed of the image reconstruction.

In this paper, two effective preprocessings of the detected signal before displaying the final images are presented. Although these methods can be applied directly to optical reconstruction systems without losing their special features, the computer image reconstruction is expected in these preprocessings and all experimental results are obtained in such a system.

The first one is the preprocessing to eliminate the effect of the phase turbulence of the medium. As is well known, one of the most troublesome problems of the synthetic aperture sonar system or holographic imaging system is the fluctuation of the phase of the detected signals, which is resulted mainly from the turbulence of the medium. And when the fluctuation of waves larger than one half of π radian exists the image reconstruction becomes almost impossible. To overcome this difficulty J. W. Goodman et al. proposed an effective means(Ref. 1). This is based on the cancellation of the fluctuation by the correlation analysis between the object and reference signals. However, Goodman's discussions are limited in optics, where continuous waves and holographic optical reconstruction are used.

In this paper, the above method is applied to synthetic aperture sonar system, where pulsed waves are used and synthetic aperture processing is applied. In a concrete system two point receivers are used, i.e., one is fixed and the other is scanned. The signal from the former receiver is used as the reference signal and the signal from the latter receiver is used as the object signal. Then, by taking the correlation between them, the effect of the turbulence is cancelled.

Theoretical developments are presented by taking account of the special features of the sonar systems, such as the relation between the scanning speed and the rate of the fluctuations. An experimental result obtained by using our prototype of Sequential Synthetic Aperture Sonar System (Ref. 2) is also presented. The experimental result shows clearly the effect of the preprocessing. Almost all of these results have been reported in Refs. 5 and 6.

The second preprocessing is concerned with the improvement of the range resolution. It is based on the utilization of multiple frequencies of waves.

When the synthetic aperture technique is applied, the azimuth resolution is improved as much as we desire by increasing the aperture to be synthesized. On the other hand, the resolution in range direction is left to the utilization of very short pulsed waves. When the pulse width is reduced extremely to a width as narrow as several wave lengths, the detection of the phase and amplitude of the reflected waves becomes very difficult. This is the result of the decrease of the resistance to noises, besides the difficulty of generation of complete signals of such a short duration.

In conventional systems two effective methods to get high range resolution without decreasing the pulse width have been used; i) the technique of linear FM and pulse compression(Ref. 3) and ii) interferometrical holography(Ref. 4).

In this paper, these methods are reconsidered from a more general point of view and they are developed so that they can be applied effectively in synthetic aperture sonar systems. Firstly a theoretical development of the general idea of the utilization of multiple frequencies of waves is presented. This is based on the linear superposition of complex images obtained by multiple frequencies of waves. And it is shown that by synthesizing the spectram of multiple frequencies of waves, the images which might be obtained when a very short duration of pulses are used can be obtained by using actual pulses with a relatively long duration. This is the outline of Ref. 7.

The other technique is based on the superposition after the nonlinear heterodyne operation of the detected signals. Discussion made on this method is the possibility and limitation for obtaining any desired interval of the beating on the reconstructed images by changing the frequencies of the electric reference waves nonlinearly. Some theoretical discussions and experimental results are shown. For further details see Ref. 8.

II. PREPROCESSING TO ELIMINATE THE EFFECT OF TURBULENCES
(Refs. 5 and 6)

A. Principle

Turbulence of the medium is one of the most serious factors which disturb synthetic aperture imaging. In this chapter, a preprocessing technique of the received signals for eliminating the effect of the turbulence is described. The basic idea is to use the signal received at a fixed point on a scanning plane as the reference signal, and to obtain the stable images within the scanning time of the conventional synthetic aperture sonar systems. For this purpose the spatial averaging of the effect of the turbulences is accomplished through the process of aperture synthesis. First, the principle and the effectiveness and limitations of the method are discussed theoretically. Then, the discussions are assured by computer simulation. Moreover a typical experimental result is given to show the effectiveness.

If a pulsed signal $S_t(t)=P_T(t)\exp[j\omega t]$, where $P_T(t)$ is the pulsed waveform with the duration T, is transmitted from the fixed transmitter-receiver positioned at r_f, as shown in Fig. 1, then the following signals will be received at r_{si} and r_f, respectively:

$$S_r(t,r_{si})=\iiint dr\rho(r)S_t(t-\frac{|r_f-r|+|r-r_{si}|}{v})e^{j\Psi(r_f,r,r_{si},t)} \qquad (1)$$

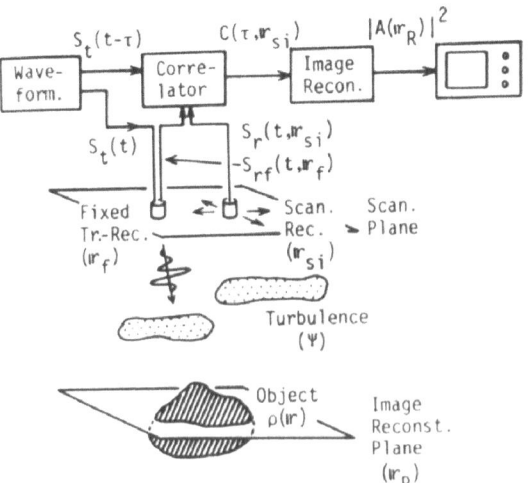

Fig. 1. Schematic diagram of synthetic aperture sonar system with the correlation preprocessing part.

$$S_{rf}(t,\mathbf{r}_f)=\iiint d\mathbf{r}\rho(\mathbf{r})S_t(t-\frac{2|\mathbf{r}_f-\mathbf{r}|}{v})e^{j\Psi'}(\mathbf{r}_f,\mathbf{r},t). \qquad (2)$$

Where ρ is the reflective index of the object, Ψ is the phase fluctuation due to the turbulence of the medium, and v is the sound velocity.

Now, let us use the following signal, which is the correlation function between Eqs. (1) and (2), as the hologram signal:

$$C(\tau,\mathbf{r}_{si})=< S_r(t,\mathbf{r}_{si})S_t^*(t-\tau) >_T < S_{rf}(t,\mathbf{r}_f)S_t^*(t-\tau_f) >_T, \qquad (3)$$

where $<>_T$ represents the integration of the duration T and τ_f is a fixed delay time. Then, let us obtain the reconstructed image by the following relation:

$$A(\mathbf{r}_R)= \frac{1}{N_s} \sum_{i=1}^{N_s} C(\frac{|\mathbf{r}_f-\mathbf{r}_R|+|\mathbf{r}_R-\mathbf{r}_{si}|}{v},\mathbf{r}_{si}), \qquad (4)$$

where $\mathbf{r}_R$ is the image point and N_s is the number of sampling points.

In a case where the turbulence is uniform, that is, if the spatial correlation length is much larger than the dimension of the scanning plane, the effect of the turbulence on the image is completely cancelled. If the turbulence has a spatial correlation length whose projected dimension(X_e) on the scanning plane is smaller than the dimension(X_s) of the plane, and if the temporal correlation between the neighbouring sampling points can not be neglected, then the variance of the image of Eq. (4) results in

$$Var\{A\}= \frac{Q^2}{\mu N_s}(1- \frac{X_e}{X_s}), \qquad (5)$$

where Q^2 is the order of the intensity of C of Eq. (3) and μ is given by

$$\mu= \frac{\Delta_s}{\tau_0 v_s} \ (< 1), \qquad (6)$$

where Δ_s is the sampling spacing, τ_0 is the temporal correlation length, and v_s is the scanning speed.

Eqs. (5) and (6) show that the variance decreases with the increase of the number of sampling points and increases with the decrease of the spatial correlation length or with the increase of the temporal one. This means that the effect of the spatial averaging through the synthesis of the aperture is reduced when the temporal correlation length of the turbulence increases. In this case, the azimuth resolution of the image is reduced by the factor $X_S/(X_S-X_e)$, which compares with the case of no turbulence, since the effective aperture for synthesizing the image is reduced to X_e. (For further details see Refs. 5 and 6.)

B. Computer Simulation

The model used for the computer simulation is shown in Fig. 2. Where the scanning is carried out every Δ_S in one dimension over the range($-X_S/2,X_S/2$). The object is composed of two point targets seperated by $2\lambda z/X_S$ at the depth of z below the scanning line. The thin layer of the turbulence is placed at the depth of z_T. The phase shift of the waves passing through the layer is given by Gaussian random numbers whose expectation is zero and standard deviation is $\pi/2$.

The typical result of the simulation is shown in Fig. 3, where Fig. 3(a) is the case of the spatially uniform turbulence and (b) and (c) are the cases of the turbulence whose spatial correlation length is smaller than the scanning range. It is seen from Fig. 3 (a) that the effect of the turbulence is eliminated completely by the preprocessing, while the image(dashed line) obtained without using the preprocessing is completely blurred out. It is seen from (b) and (c) that the resolution and stability of the images obtained by the preprocessing is reduced with the decrease of the correlation length, in agreement with the theoretical considerations. (See Ref. 5.)

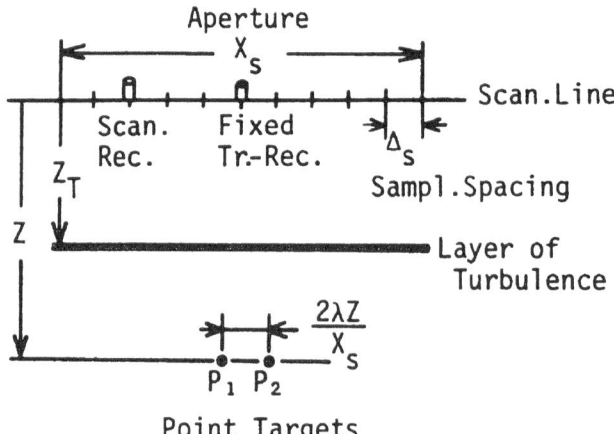

Fig. 2. Model used for computer simulation.
        X_S= 50 mm, Δ_S= 0.5 mm, z_T= 100 mm, z= 200 mm, λ= 1.5 mm.

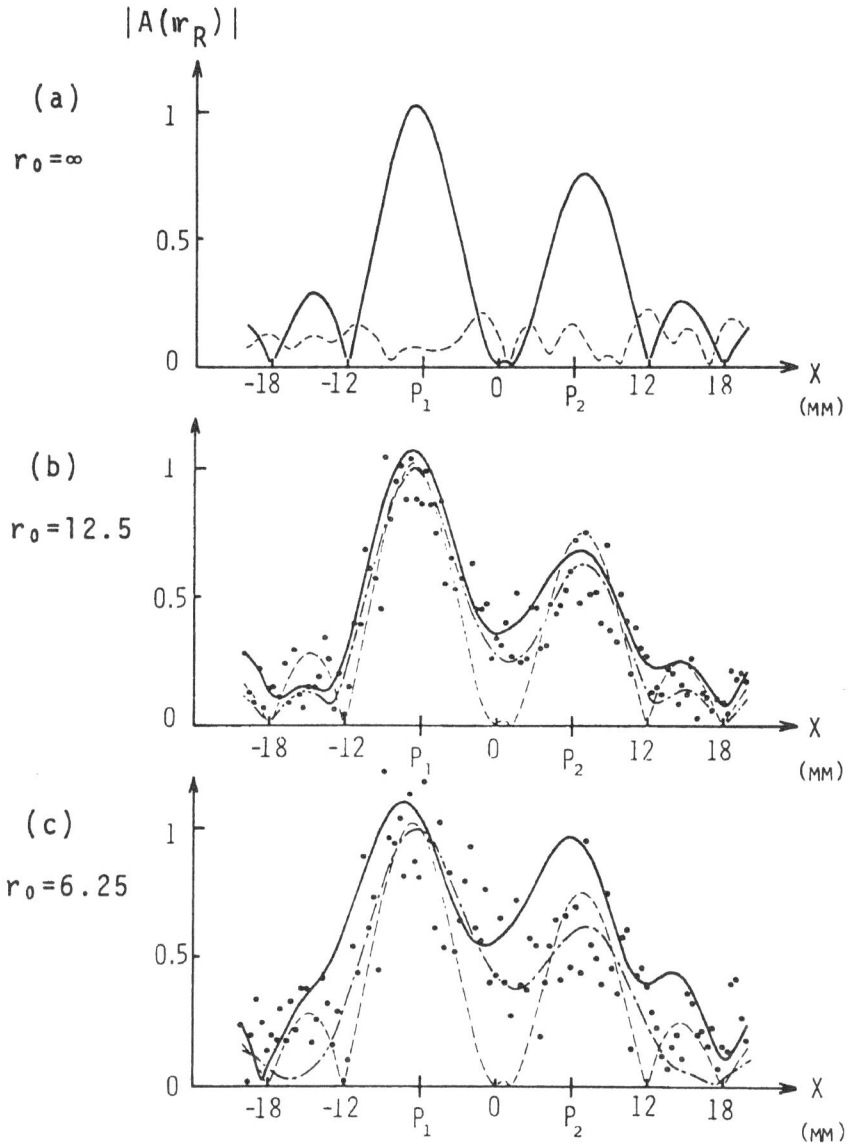

Fig. 3. Result of computer simulation, where $\rho_1=1.0$, $\rho_2=\sqrt{0.5}$, and r_0 is the spatial correlation length of the turbulence. In (a), the solid curve shows the image obtained by using the preprocessing and the dashed one shows that without using it. In (b) or (c), the solid curve shows a typical image obtained by a single scanning, the dots show the image for the case where a different sample of the turbulences is used for a different image point, the chained curve shows the expectation of the image, and the dashed curve shows the image for the case of no turbulence, all of which are concerned with the effect of the preprocessing.

C. Experimental Result

The experimental system is constructed by combining the prototype of synthetic aperture sonar system(Ref. 2) and a specially designed preprocessing part, whose block diagram is shown in Fig. 4. The system displays the computationally reconstructed images on a CRT according to the two-dimensional scanning along a circular line. The turbulence is produced by inserting an oscillating acrylic sheet between the scanning plane and the object.

A typical experimental result for the object which consists of three steel balls is shown in Fig. 5. Where (a) shows the image obtained by using the electric reference signal for the case when the turbulence is absent; (b), the case when the electric reference signal is used with the turbulence included; and (c),the case of the received reference signal also with the turbulence included. It is seen from the result that the image of (b) is completely destroyed by the effect of the turbulence, whereas the image of (c) is almost free from its effect. This dramatically shows the effect of the preprocessing. (See Ref. 6.)

D. Discussion

A signal preprocessing method for eliminating the effect of the turbulence of the medium is presented in this chapter. The effects of other deteriorating factors, such as random motion of a transmitter or receiver, or curved paths of ultrasonic waves due to the inhomogeneity of the medium may also be compensated for by means of computational preprocessing of the received signals and/or in the process of the reconstruction of the images.

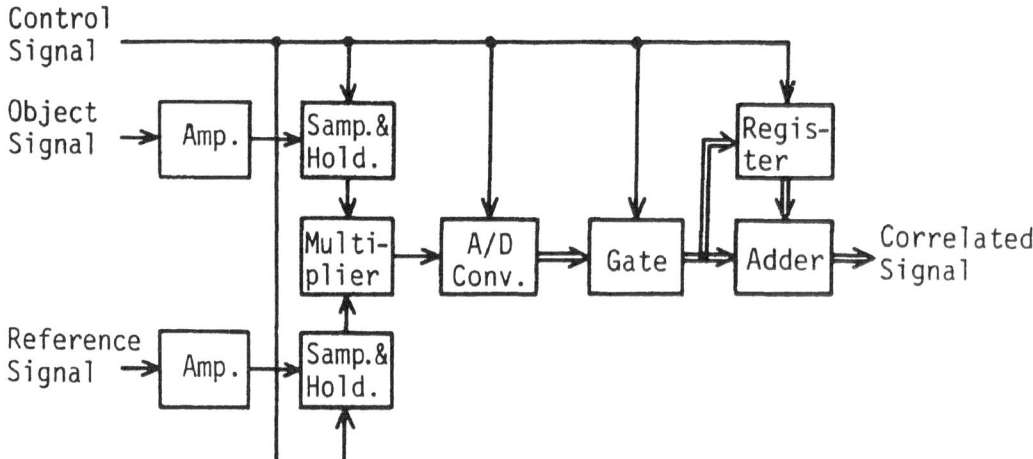

Fig. 4. Block diagram of the correlation preprocessing part of the experimental system.

(a)

Image obtained by using the
electric reference signal
for the case of no turbulence

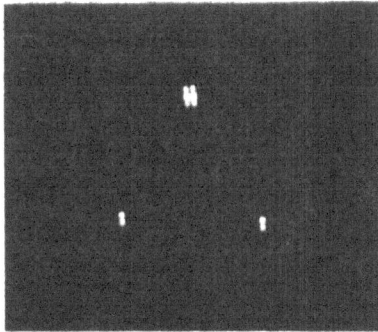

(b)

Image obtained by using the
electric reference signal
for the case of the turbulence
included

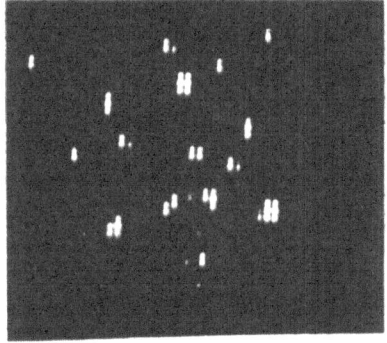

(c)

Image obtained by using the
received reference signal
for the case of the turbulence
included

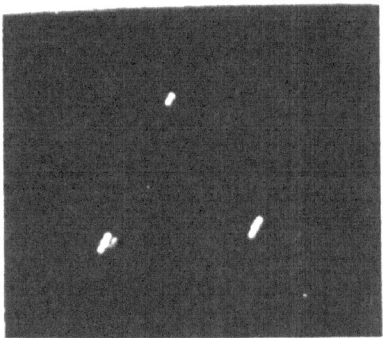

Fig. 5. Experimental result.

III. PREPROCESSINGS TO INCREASE THE RANGE RESOLUTION

A. Linear Superposition of Complex Images
        Obtained by Multiple Frequencies of Waves(Ref. 7)

In conventional Radar or Sonar systems, linear FM or phase-
coded pulsed signals have been employed to increase the range
resolution(Refs. 3 and 4). In this section, by developing the
essential meaning of those methods, a formulation is presented to
the high range resolution through the synthesis of informations for

multiple frequencies of continuous waves in a synthetic aperture
sonar system. The fundamental idea is the synthesis of the Fourier
transformation of an equivalently short pulse by scanning the
frequency used for synthetic aperture sonar imaging. (The details
are shown in Ref. 7.) The mathematical formulation is given below,
referring to the fundamental system shown in Fig. 6.

For transmitted continuous signal $S_t(t,\omega)= \exp[j\omega t]$, the
received signal at $\mathbf{r}_s$ may be given by

$$S_r(t,\mathbf{r}_s,\omega)= \frac{K_1}{\lambda} \iiint d\mathbf{r} \rho(\mathbf{r},\omega) S_t(t- \frac{2|\mathbf{r}_s-\mathbf{r}|}{v},\omega), \qquad (7)$$

where K_1 is a constant and λ is the wave length. When the
heterodyne detection of the received signal is performed, the
resulting signal is expressed by

$$C(\tau,\mathbf{r}_s,\omega)= < S_r(t,\mathbf{r}_s,\omega) S_t^*(t-\tau,\omega) >_{T_c}, \qquad (8)$$

where T_c is the integrating time. After registering $C(\tau,\mathbf{r}_s,\omega)$ for
all sampling points and for every ω in the interval(ω_1,ω_2), the
reconstructed image is given according to the following operation:

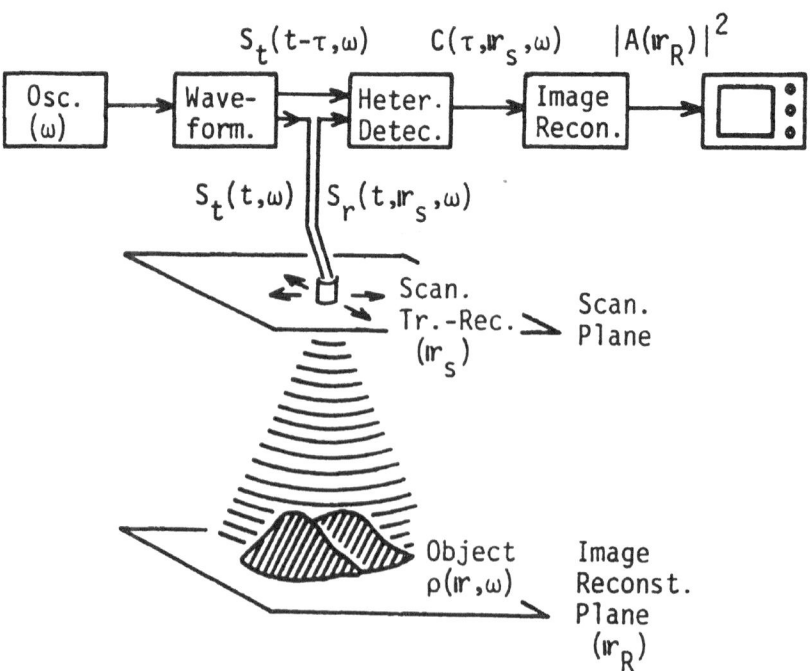

Fig. 6. Schematic diagram of the superresolution ultrasonic imaging
by combining spectral synthesis with aperture synthesis.

$$A(\mathbf{r}_R)= \frac{K_2}{WS} \iint_S d\mathbf{r}_S \int_W d\omega\ C(\frac{2|\mathbf{r}_S-\mathbf{r}_R|}{v},\mathbf{r}_S,\omega)f(\omega),\qquad (9)$$

where $\mathbf{r}_R$ is the coordinate of the image point, $W=\omega_2-\omega_1$, S is the scanning area, $f(\omega)$ is the compensating factor for ω, K_2 is a constant, and both summations with respect to the sampling point and to the frequency scanning are replaced by integrals.

Now, if $\rho(\mathbf{r},\omega)=\rho_1(\mathbf{r})g(\omega)$ is assumed and if $f(\omega)$ is chosen as $\lambda/g(\omega)$, then Eq. (9) becomes as follows:

$$A(\mathbf{r}_R)= \frac{K_1K_2}{X_SY_S} \iiint d\mathbf{r}\ \rho_1(\mathbf{r})\iint d\mathbf{r}_S \operatorname{sinc}[\ \frac{W}{v}(\ |\mathbf{r}_S-\mathbf{r}_R|-|\mathbf{r}_S-\mathbf{r}|\ )]$$

$$\times\ e^{j\ \frac{\omega_1+\omega_2}{v}(\ |\mathbf{r}_S-\mathbf{r}_R|-|\mathbf{r}_S-\mathbf{r}|\ )},\qquad (10)$$

where the rectangular area of $X_S\times Y_S$ is assumed as the scanning area. It is seen from the comparison between Eq. (10) and that of conventional synthetic aperture sonar systems(Ref. 2) that a pulsed waveform, e.g., $P_T(t)$ in the previous chapter, is replaced by the sinc function in this case. From Eq. (10) the resolution limits in azimuth and range directions are given by

$$\Delta x= \lambda_m z/2X_S,$$

$$\Delta y= \lambda_m z/2Y_S,$$

$$\Delta z= v/2F,\qquad (11)$$

where $\lambda_m= 4\pi v/(\omega_1+\omega_2)$ and $F= W/2\pi$. Thus, the first and second results in Eq. (11) agree with the resolution limits of conventional synthetic aperture sonar systems, and the last result is the newly added one, the consequence of the spectral synthesis. (The details and further discussions are given in Ref. 7.)

B. Nonlinear Beating(Ref. 8)

Interferometric fringing methods have been studied in Synthetic Aperture Radar and Acoustical Holography to obtain the three-dimensional information in the images. For this purpose two seperate receivers(Ref. 9) or multiple frequencies of waves(Ref. 4) are used. In this section, a new preprocessing method of the received signals in synthetic aperture sonar is presented as an auxiliary means to the interferometric method using two

frequencies for the purpose of increasing the resolution in range
direction.

The schematic diagram of the nonlinear beating synthetic
aperture sonar system is shown in Fig. 7, where the nonlinear
heterodyne detection is carried out in the preprocessing part. By
referring to the figure and, in the same way as in the previous
section, the hologram signal is given by

$$C = \sum_{k=1}^{2} \rho_k P_T(t - \frac{2R_k}{v}) \cos[\frac{(\omega_1 - \omega_2)R_k}{v} - \theta(t)] e^{-j\frac{(\omega_1 + \omega_2)R_k}{v}}. \tag{12}$$

The cosine term gives the interferometric   fringes, which are
determined by the first term resulted from the actual distance
between the object and the transmitter-receiver and by the
artificially added term $\theta(t)$. The case, where $\theta(t)=0$ and continuous
waves($T \to \infty$) are used, corresponds to the conventional
interferometric holography. In some cases, we can see more clearly
or easily the relative positions, e.g., of two neighbouring point-
like targets by giving an appropriate form of $\theta(t)$. It becomes
possible also to control the interval of the interferometric
beating in the image by controlling $\theta(t)$. (For further details
see Ref. 8.)

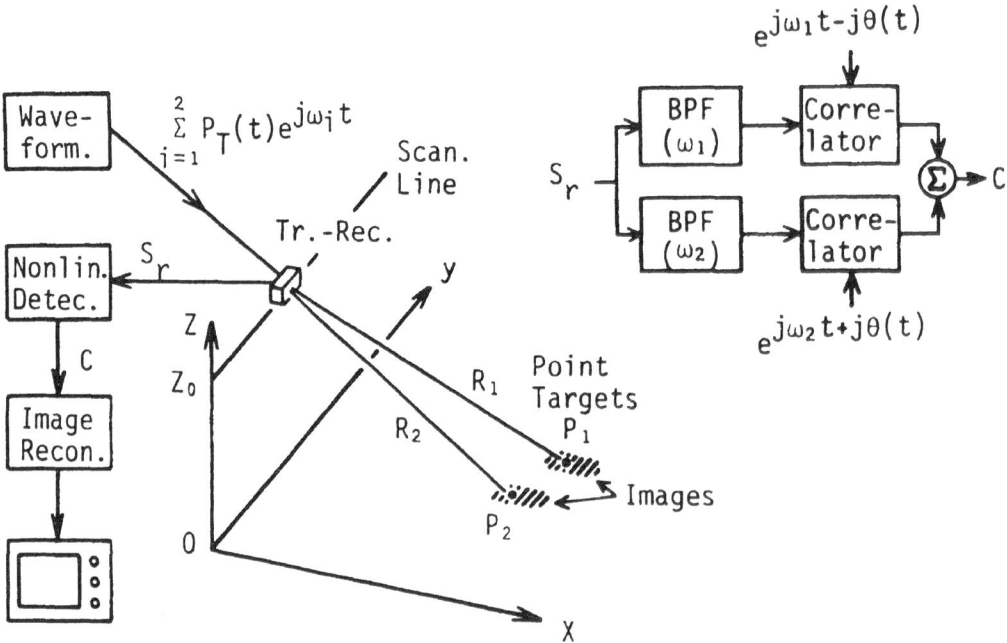

Fig. 7. Schematic diagram of the nonlinear beating synthetic
        aperture sonar system.

(a)

Linear beating

$\theta(t) = \alpha_0 t$

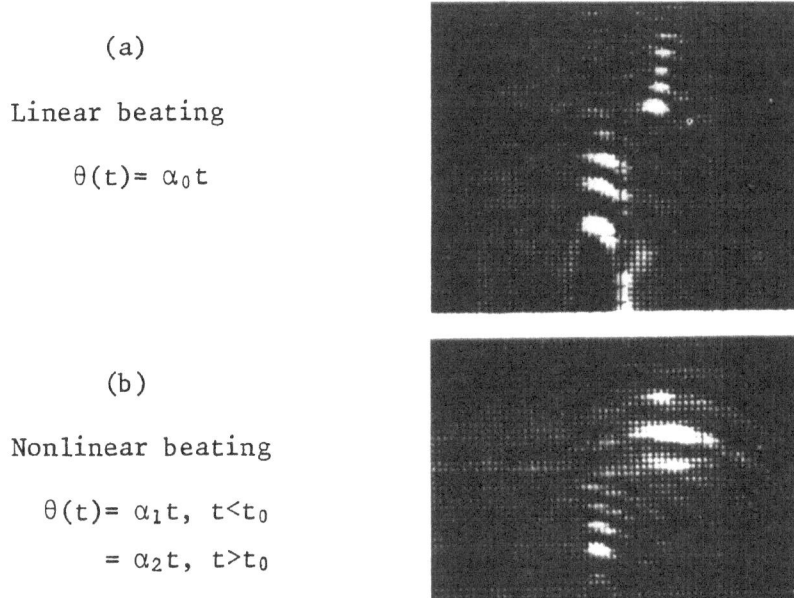

(b)

Nonlinear beating

$\theta(t) = \alpha_1 t, \quad t < t_0$

$ = \alpha_2 t, \quad t > t_0$

$\alpha_1 > \alpha_2$

Fig. 8. Basic experimental results.

Basic experimental results are shown in Fig. 8. Two steel balls are used as the object. The images are prolonged due to the wide duration of the pulse used. Fig. 8(a) shows the linear beating, where $\theta(t) = \alpha_0 t$ for a constant α_0. Fig. 8(b) shows the nonlinear beating, where $\theta(t) = \alpha_1 t$, $t < t_0$ and $\theta(t) = \alpha_2 t$, $t > t_0$. They show clearly the effect of the nonlinear beating preprocessing.

REFERENCES

1. J. W. Goodman, W. H. Huntley, Jr., D. W. Jackson, and M. Lehmann, Appl. Phys. Let., Vol. 8, pp. 311-313 (1966).

2. T. Sato and O. Ikeda, " Sequential Synthetic Aperture Sonar System ", submitted to IEEE Trans. on Sonics and Ultrasonics.

3. L. J. Cutrona, " Synthetic Aperture Radar " in Radar Handbook, M. Skolnik, Ed. (McGraw-Hill, New York, 1970), Chap. 23.

4. K. Suzuki and B. P. Hildebrand, " Holographic Interferometry with Acoustic Waves " in Acoustical Holography, N. Booth, Ed. (Plenum Press, New York, 1975), Vol. 6, pp. 577-595.

5. O. Ikeda, T. Sato, and H. Ohshima, " Synthetic aperture sonar
    in turbulent media ", accepted for publication in J. Acoust.
    Soc. Am.

6. O. Ikeda, T. Sato, and H. Fujikura, " Synthetic aperture sonar
    in turbulent media(II) ", to be submitted to J. Acoust. Soc. Am.

7. T. Sato and O. Ikeda, " Superresolution ultrasonic imaging
    method by combining spectral synthesis with aperture synthesis",
    submitted to J. Acoust. Soc. Am.

8. T. Sato, O. Ikeda, and H. Fujikura, " Nonlinear beating in
    synthetic aperture sonar ", to be submitted to J. Acoust. Soc.
    Am.

9. L. C. Graham, " Synthetic Interferometer Radar for Topographic
    Mapping ", Proc. IEEE, Vol. 62, No. 6, pp. 763-768 (1974).

10. P. N. Keating, R. F. Koppelmann, and R. K. Mueller,
    " Maximization of Resolution in Three Dimensions " in
    Acoustical Holography, N. Booth, Ed. (Plenum Press, New York,
    1975), Vol. 6, pp. 577-595.

APPLICATION OF SHEAR WAVE FOCUSED IMAGE HOLOGRAPHY
TO NONDESTRUCTIVE TESTING

Katsumichi Suzuki, Fuminobu Takahashi and
Yoshihiro Michiguchi

Atomic Energy Research Laboratory, Hitachi Ltd.

Ozenji, Tamaku, Kawasaki, Kanagawa, Japan

ABSTRACT

This paper describes a method for sizing vertical and oblique
flaws or defects within metal structures by the use of scanned acoustical holographic interferometry in the focused image hologram mode.
The image of an internal flaw obtained from focused image holography
consists of interference fringes or contour lines across the flaw's
surface. The fringe separation represents half wavelength deviations in depth from the scanning plane and the length of the fringes
corresponds to the width of the flaw. Thus, it is possible to evaluate
flaw size by measuring the number and length of fringes appearing on
the focused image hologram.

The theory developed in the present paper gives the relation
between the flaw size and the number and separation of fringes on
the hologram. Measurements resulting from the application of the
theory to artificial flaws in metal test pieces are examined in the
experiments in this paper. The deviations of the obtained heights
from the true values measured by physical means are 10 % for the
4.0 mm height flaws at 3.0 MHz 5 % at 5.6 MHz. Natural vertical
flaws produced at the electron beam welded interface on a steel test
piece are also imaged and the size of flaws is measured by acoustical
interferometry.

INTRODUCTION

Nondestructive size evaluation of flaws or cracks in a metal structure which has been put into service is very important from the viewpoint of safety in order to estimate the strength of the structure. A conventional pulse-echo method has been widely used to detect flaws which exist in the structure and evaluate flaw size. It is rather difficult, however, to evaluate the flaw size by the conventional ultrasonic pulse-echo method, because retrieval of flaw size information is dependent only upon the intensity of the reflected ultrasonic waves.

On the other hand, application of acoustical holographic interferometry to the field of nondestructive testing has attracted special interest recently because of its capability for quantitative deformation analysis[1], contour generation[1,2], and aberration correction[1,3]. Theoretical works[4,5] have also been written in this field.

Acoustical holographic interferometry has at least two advantages over the conventional ultrasonic pulse-echo method. One is quantitative flaw size evaluation and the other is an imaging capability for the flaws. These two features are very attractive for applying acoustical holographic interferometry to quantitative nondestructive testing.

This paper describes a method for sizing vertical and oblique flaws in metal structures using scanned acoustical holographic interferometry in the shear wave focused image hologram mode. Size measurements of known artificial flaws by physical means and natural flaws by X-ray techniques, were compared to the results of measuring these same flaws by an acoustical holographic interferometry process based upon theoretical considerations, and the results were presented in this paper.

THEORETICAL ANALYSIS

Acoustical focused image holography is able to retrieve information about shape and size of a flaw from fringe structure of an interferogram. The method described in this paper employs a simultaneous source-receiver scanning technique with a pulse-echo mode focused transducer. The reference wave is simulated electronically with a selectable inclination angle.[2]

Let the x axis be in the scanning direction of the transducer. The reflected wave S_d from the object at the transducer and the reference wave S_r are given by:

$$S_d(x) = A_d(x)\cos \omega t + \phi_d(x)$$
$$S_r(x) = A_r(x)\cos \omega t + \phi_r(x)$$

$$(1)$$

respectively, where $A_d(x)$ and $A_r(x)$ are the amplitude of the reflected and reference waves, $\phi_d(x)$ and $\phi_r(x)$ are the phase of the two waves. An interference term S of the two waves given by Eq.(1) is then:

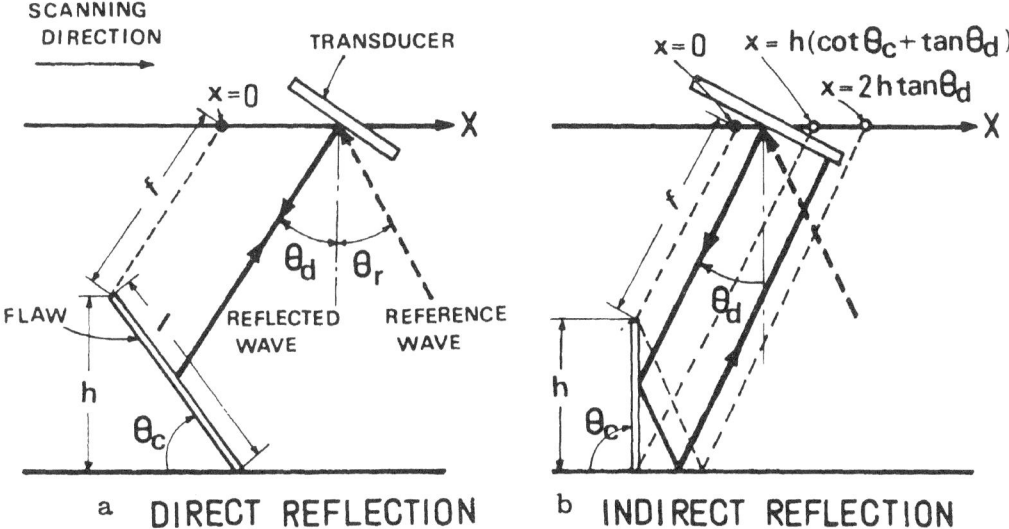

Fig. 1   Simulated image hologram construction geometry in water:
         (a) direct reflection   (b) indirect reflection.

$$S(x) = \frac{A_d(x)A_r(x)}{2} \cos\left[\phi_d(x) - \phi_r(x) \right] \tag{2}$$

which describes the periodic spatial variation of intensity in the
interference fringes.   The periodicity of the fringes is governed by
the phase difference $\phi_d(x) - \phi_r(x)$, and the signal S takes maximum
value at $\phi_d(x) - \phi_r(x) = 2n\pi (n=0, \pm1, \pm2, \ldots)$.

The phase $\phi_d(x)$ depends on the propagation path of the reflected
wave from the flaw.   Here, we assume that the object to be imaged is
a flat flaw in a metal plate.   Furthermore, the following three kinds
of propagation paths are assumed to calculate the phase $\phi_d(x)$:

Path 1 (Direct reflection):
    transducer → flaw → transducer

Path 2 (Indirect reflection):
    transducer → flaw → bottom surface of the metal plate →
    transducer

Path 3 (Indirect reflection):
    transducer → bottom surface of the metal plate → flaw →
    transducer.

Fringes Formed by the Direct Reflection

Figure la shows the simulated image hologram construction geometry in water for the direct reflection. As stated before, the x axis is in the scanning direction of the transducer and the origin of the x axis is taken as the position of the transducer where the acoustic wave emitted from the transducer crosses at the top of the flaw.

The phase ϕ_{d1} at the position x of the transducer is given by:

$$\phi_{d1} = k\left\{ 2(f + r) \right\} \tag{3}$$

where k is the wave number, f is the distance between the origin and the top of the flaw, and $(f + r)$ is the distance between the transducer and the flaw. Geometrical relation between x and r is expressed as:

$$x = r \sin\theta_d(1 + \cot\theta_d \cot\theta_c)$$

where θ_d is the angle of the incident wave and θ_c is the angle of flaw inclination. Equation (3) is then expressed as:

$$\phi_{d1} = 2k\left\{ f + \frac{x}{\sin\theta_d(1 + \cot\theta_d\cot\theta_c)} \right\}. \tag{4}$$

The phase, ϕ_r for the inclined reference wave is denoted as:

$$\phi_r = kx\sin\theta_r \tag{5}$$

where θ_r is the inclination angle of the reference wave. From Eqs. (4) and (5) we can derive the phase difference:

$$\phi_{d1} - \phi_r = k[2f + x\{\frac{2\sin\theta_c}{\cos(\theta_d - \theta_r)} - \sin\theta_r \}] \tag{6}$$

which gives the fringe spacing, Δx_1, on the x axis in the scanning plane as:

$$\Delta x_1 = \frac{\lambda\cos(\theta_d - \theta_c)}{2\sin\theta_c - \sin\theta_r\cos(\theta_d - \theta_c).} \tag{7}$$

By measuring the fringe spacing, Δx_1, we can obtain the inclination angle θ_c of the flaw from Eq. (7).

Geometrical relation between the length of flaw and its projection x_L to the x axis is given by:

$$x_L = \frac{L\cos(\theta_d - \theta_c)}{\cos\theta_d}. \tag{8}$$

Therefore the number of fringes, n_1, on the x axis is expressed as:

$$n_1 = \frac{x_L}{x_1} = \frac{L\{2\sin\Theta_c - \sin\Theta_r\cos(\Theta_d - \Theta_c)}{\cos\Theta_d} . \tag{9}$$

This equation can be rewritten by using the relation between L and h: $h = L\sin\Theta_c$, so that

$$n_1 = \frac{h\{2\sin\Theta_c - \sin\Theta_r\cos(\Theta_d - \Theta_c)}{\sin\Theta_c \cos\Theta_d} \tag{10}$$

Note that the height or length of the flaw is obtained by measuring the flaw inclination angle, Θ_c, from Eq. (7) and the number of fringes, n_1, from Eq. (9) or (10). Width of the flaw is easily obtained by measuring the length of each fringe.

Fringes Formed by the Indirect Reflection

When the angle of flaw inclination angle Θ_c is close to 90°, the intensity of directly reflected waves from the flaw surface becomes weak, while the indirect reflection is dominant. Figure 1b shows the first type of the indirect reflection path (Path-2). The second type of the indirect reflection path (Path-3) is the reverse path of the first type.

We assume in this section that the direction of the emitted acoustic wave from the transducer and the direction of the received acoustic wave by the transducer are parallel.

To receive these two kinds of the indirectly reflected waves by the transducer, the following condition between the height of the flaw, h, and the radius of the transducer, a, is necessary:

$$h \leqq a\cos\Theta_d . \tag{11}$$

When the transducer is in the range of $0 \leqq x \leqq h(\cot\Theta_c + \tan\Theta_d)$, the phase of the reflected wave is given by:

$$\phi_{d2} = k(\frac{2d}{\cos\Theta_d} - 2h\sin\Theta_d + \frac{2x\sin\Theta_d}{\cos\Theta_c + \tan\Theta_d}) \tag{12}$$

where d is the distance between the transducer and the bottom surface of the metal plate. The fringe spacing along the x axis in the scanning plane is given by:

$$\Delta x_2 = \frac{(1 + \cot\Theta_d \cot\Theta_c)}{\sin\Theta_r(1 + \cot\Theta_d \cot\Theta_c) + 2\sin\Theta_d} . \tag{13}$$

Then the number of the fringes, n_2, is expresses as:

$$n_2 = \frac{h\tan\theta_d}{\lambda} [\sin\theta_r(1 + \cot\theta_c \cot\theta_d) + 2\sin\theta_d]. \tag{14}$$

When the transducer position lies in the range of $h(\cot\theta_c + \tan\theta_d)$ $\leq x \leq 2h\tan\theta_d$, the same calculation procedure gives the fringe spacing Δx_3 and the number of fringes, n_3, for the second type of the propagation path (Path-3) of the indirectly reflected wave:

$$\Delta x_3 = \frac{\lambda (1 - \cot\theta_d \cot\theta_c)}{\sin\theta_r(1 - \cot\theta_d\cot\theta_c) + 2\sin\theta_d} \tag{15}$$

$$n_3 = \frac{h\tan\theta_d}{\lambda}[\sin\theta_r(1 - \cot\theta_d \cot\theta_c) + 2\sin\theta_d]. \tag{16}$$

The total number of fringes, n, formed by the above mentioned two kinds of the indirect reflection is given by:

$$n = n_2 + n_3 = \frac{2h\tan\theta_d}{\lambda} (\sin\theta_r + 2\sin\theta_d). \tag{17}$$

Fringe Characteristics for Special Angles

Each fringe spacing and number of fringes corresponding to the propagation paths are listed in Table 1 for special angles.

In the case of $\theta_r = 0°$, note that the number of fringes is independent of the flaw inclination angle, θ_c. Furthermore, if we take as $\theta_r = 0°$ and $\theta_d = 45°$, the number of fringes is independent of the flaw inclination angle, θ_c, and the propagation paths which are assumed in the previous section, and depends only on the height of the flaw and the wavelength. This independence of the propagation paths is advantageous for nondestructive testing, since it is not necessary to identify the propagation path to measure the height of the flaw from the number of fringes appearing in the image hologram.

In the case of $\theta_r = 0°$ and $\theta_c = 90°$, expressions for the fringe spacing are extremely simple as given in the table.

Table 1. Fringe spacing Δx and number of fringes n for special angles

			Fringe spacing (Δx)	Number of fringes (n)
$\theta_r = 0°$	Direct reflection	(Path-1)	$\dfrac{\lambda \cos(\theta_d - \theta_c)}{2\sin\theta_c}$	$\dfrac{2h}{\lambda \cos\theta_d}$
	Indirect reflection	(Path-2)	$\dfrac{\lambda(1 + \cot\theta_d\cot\theta_c)}{2\sin\theta_d}$	$\dfrac{2h\tan\theta_d\sin\theta_d}{\lambda}$
		(Path-3)	$\dfrac{\lambda(1 - \cot\theta_d\cot\theta_c)}{2\sin\theta_d}$	$\dfrac{2h\tan\theta_d\sin\theta_d}{\lambda}$
$\theta_r = 0°$ and $\theta_d = 45°$	Direct reflection	(Path-1)	$\dfrac{\lambda(1 + \cot\theta_c)}{2}$	$\dfrac{2\sqrt{2}\,h}{\lambda}$
	Indirect reflection	(Path-2)	$\dfrac{\sqrt{2}\,\lambda(1 + \cot\theta_c)}{2}$	$\dfrac{\sqrt{2}\,h}{\lambda}$
		(Path-3)	$\dfrac{\sqrt{2}\,\lambda(1 - \cot\theta_c)}{2}$	$\dfrac{\sqrt{2}\,h}{\lambda}$
$\theta_r = 0°$ and $\theta_c = 90°$	Direct reflection	(Path-1)	$\dfrac{\lambda\sin\theta_d}{2}$	$\dfrac{2h}{\lambda\cos\theta_d}$
	Indirect reflection	(Path-2)	$\dfrac{\lambda}{2\sin\theta_d}$	$\dfrac{2h\tan\theta_d\sin\theta_d}{\lambda}$
		(Path-3)	$\dfrac{\lambda}{2\sin_d}$	$\dfrac{2h\tan\theta_d\sin\theta_d}{\lambda}$
$\theta_d = 0°$	Direct reflection	(Path-1)	$\dfrac{\lambda}{2\tan\theta_c - \sin\theta_r}$	$\dfrac{h}{\lambda}(2 - \sin\theta_r\cot\theta_c)$

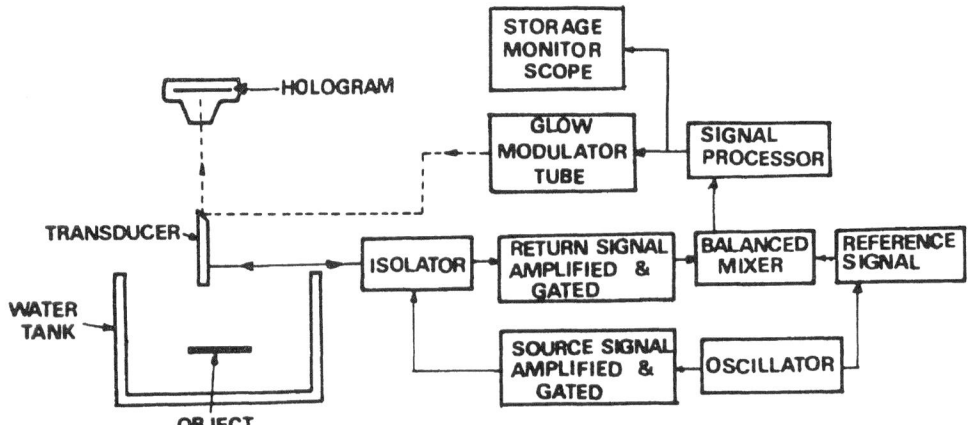

Fig. 2   Simplified block diagram of the acoustic image holography
           system.

EXPERIMENTAL RESULTS

Present experiments were performed on the scanning type acoustical holography system (Canon Holosonics, Inc. , Model-200). The
system is shown in block diagram form in Fig. 2. One of the most
remarkable point of the system is that the inclination angle of the
electronically simulated reference wave can be chosen at will[2].
The maximum scanning area is 15.2 cm x 15.2 cm. The scanning line
density is chosen to be 33 lines/cm in the present experiment.

Two kinds of transducers used in the present experiment are both
the focused type, 2.54 cm in effective diameter and have 10.1 cm focal
lengths in water. Their resonant frequencies are 3.0 and 5.6 MHz.
All of the experiments are performed by shear wave mode insonification.
The incident angle of the wave is 45° ($\theta_d = 45^\circ$), and the inclination
angle of the reference wave is chosen to be 0° ($\theta_r = 0^\circ$) for all
experiments in the present work. Artificial planar flaws in two kinds
of stainless steel test pieces and natural flaws in a steel test piece
are imaged by the image holography technique. Details of the test
pieces are explained in the corresponding section.

Artificial Vertical Flaws with Various Heights

The lower part of Fig. 3 shows the cross section of five artificial
vertical flaws produced by the discharge method in a 10 mm thick stainless steel plate. Height and width of flaws are also shown in the figure.

These flaws are insonified by 3.0 and 5.6 MHz, 45° acoustic shear
waves. The hologram construction geometry and the obtained image

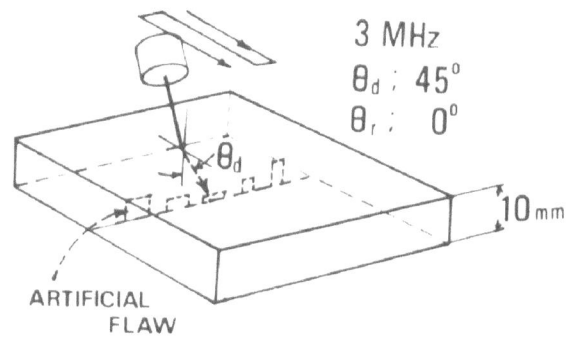

3 MHz
θ_d : 45°
θ_r : 0°

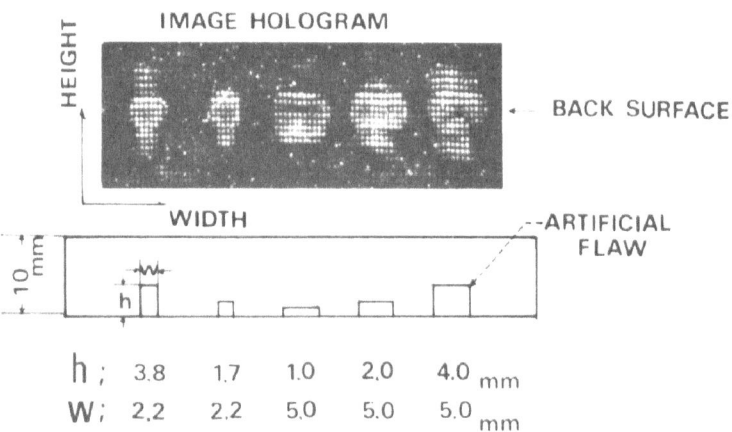

Fig. 3   Cross section of artificial vertical flaws, their image
hologram and hologram construction geometry.   An
arrow beside the hologram indicates a position corresponding to the back surface of the stainless steel plate.

hologram with 3 MHz insonification is shown in Fig. 3. The position
corresponding to the back surface of the metal plate is shown by an
arrow. Note that fringes shown in the hologram are formed by the
indirect reflection, since the measured fringe spacing (Δx = 0.75 mm)
agrees well with the theoretical fringe spacing (Δx_2 = 0.73 mm) for
the indirect reflection.

As seen from the image hologram shown in Fig. 3, the number of
fringes are not proportional to the height of flaws. Because the image

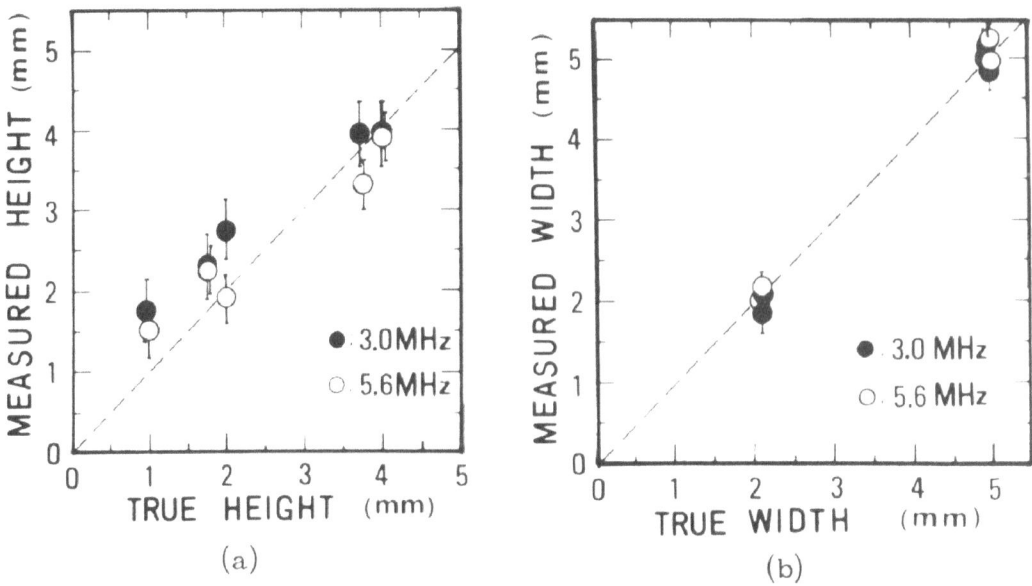

Fig. 4. Comparison of the measured size of vertical artificial flaws
       with the true size measured by physical means; (a) is for the
       height of flaws, and (b) is for the width of flaws.

size of a flaw differs from the actual size of the flaw mainly due to the
spread of the acoustic beam at the focal point in image holography,
when the flaw size is close to or less than the beam diameter of the
acoustic wave. The sizing error due to the spread of the acoustic beam
is corrected by numerical calculation.

The height of the artificial flaws are obtained from the measured
number of fringes by the equation listed in Table 1. The width of the
flaws obtained from the maximum length of the fringes. Then, the
sizing error caused by the spread of the acoustic beam is corrected
by numerical calculation. The result is shown in Fig. 4. The
deviations of the obtained heights of flaws from the true values measured
by physical means are 10 % for the 4.0 mm height flaws at 3.0 MHz
and 5 % at 5.6 MHz. The corresponding values for widths of flaws are
less than 6 % for both 3.0 and 5.6 MHz.

Artificial Flaws with Various Inclination Angles

Figure 5 shows the image hologram construction geometry of
artificial flaws made in a 10 mm thick stainless steel plate by the discharge method. The inclination angles of these internal flaws are 30°,
45°, 60°, and 90°. An edge of the one side of the plate (not shown in
the figure) is cut at angles varying from 45° to 90° in 5° steps. These
internal and external artificial flaws are insonified by 3.0 and 5.6 MHz
acoustic waves. An image hologram of the eight internal flaws is also
shown in Fig. 5.

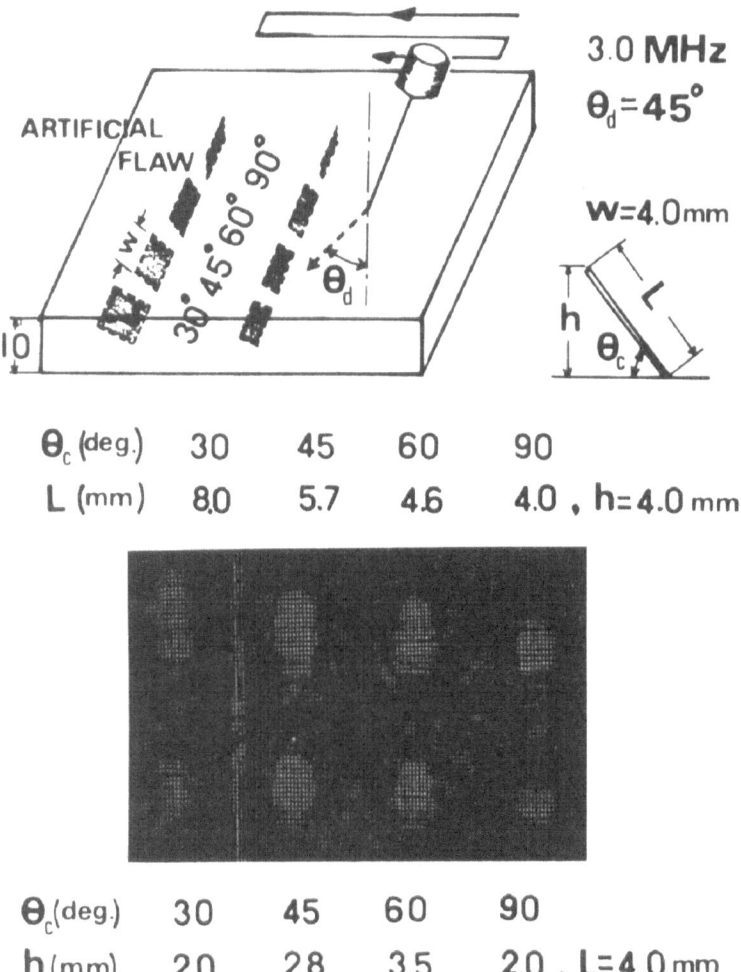

Fig. 5   Image hologram construction geometry of artificial oblique
         flaws in a stainless steel plate and the image hologram of
         the internal flaws.

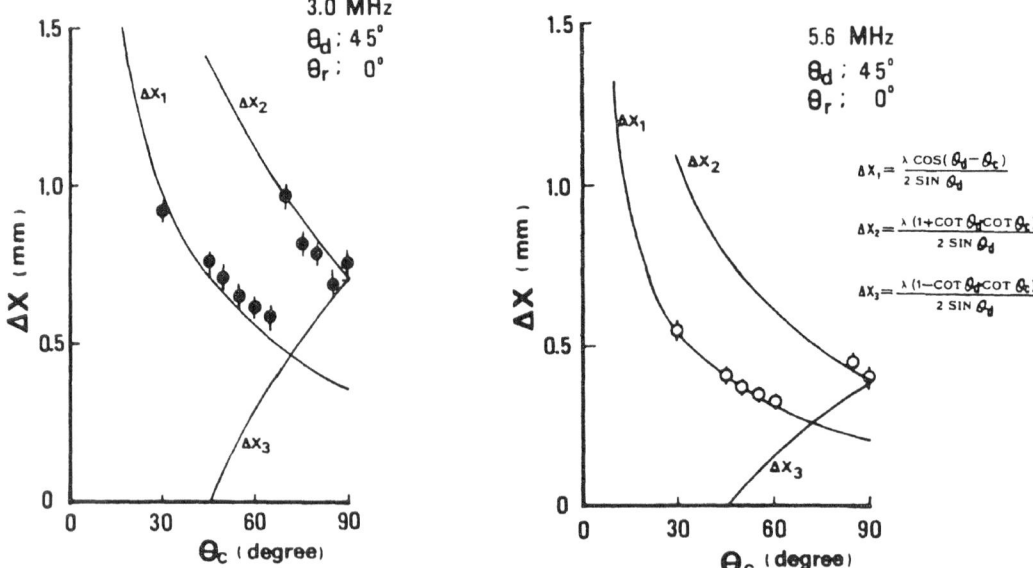

Fig. 6.   Experimental fringe spacings for oblique internal and ex-
          ternal flaws versus the inclination angle of flaws.   The
          fringe spacings predicted by the theory are also shown.

As stated in the theoretical analysis section, the fringe spacing
depends on the propagation path, the flaw inclination angle, Θ_c, and
the incident angle of the acoustic wave, Θ_d . Figure 6 shows experimental results for the relation between the fringe spacing, Δx, and
the inclination angle of the flaw , Θ_c, with theoretical predictions. It
is obvious from Fig. 6 that the direct reflection is dominant for flaws
where the inclination angle, Θ_c, is less than 60° for $\Theta_d = 45°$. On
the other hand, the indirect reflection (defined as Path -2 in the
theoretical analysis section) is dominant for flaws which have an
inclination angle over 65°. Intensity of the indirect reflection for
Path-3 is very weak and the corresponding fringe spacing, Δx_3, is not
obtained in the present experiment.

In the case of $\Theta_r = 0°$ and $\Theta_d = 45°$, the number of fringes is
independent of the type of the propagation path and of the inclination
angle, Θ_c, of the flaw as stated in the teoretical analysis section.
The experimental results for the verification of the teoretical prediction
are shown in Fig. 7. The figure represents the relation between the
height of the internal artificial flaws with four kinds of inclination
angles ($\Theta_c = 30°$, 45°, 60°, and 90°) which are shown in Fig. 5   and
the measured number of fringes. Evidently, the experimentally
measured number of fringes agrees well with the predicted number of
fringes from the theory.

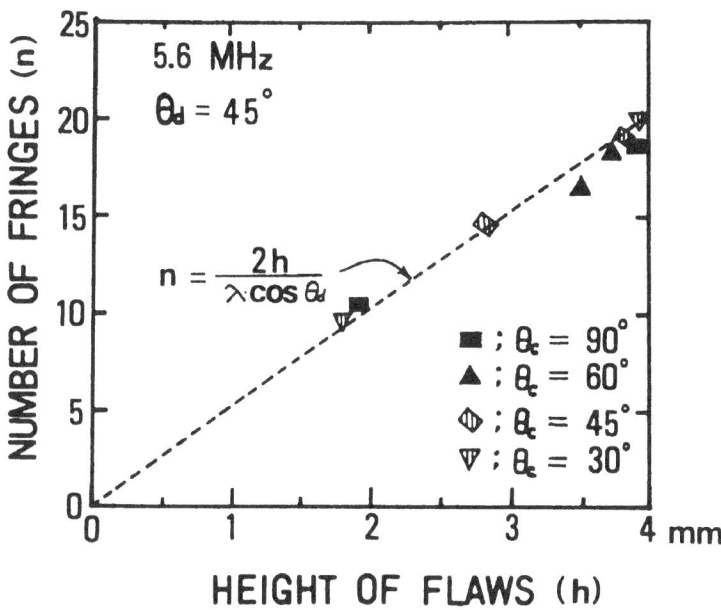

Fig. 7   Theoretical and experimental relations between the number
          of fringes and the height of oblique artificial flaws.

Natural Vertical Flaws in the Electron Beam Welded Interface

Figure 8 shows the geometry of the image hologram construction
for flaws at the electron beam welded interface in a 70 mm thick steel
test piece. The special electron beam welding conditions are adopted
to produce vertical flaws mainly in the welded interface. An example
of the obtained vertical flaw is shown in the photograph in Fig. 8. An
image hologram is also shown in the figure. Twenty three vertical
flaws are imaged.

The type of the reflection is considered as the direct reflection
by the following two reasons. The first reason is that the measured
fringe spacing 0.37 mm is very close to the theoretical fringe spacing
$(\triangle x_1 = 0.35$ mm) which is calculated as the direct reflection. The
second reason is that the propagation time measurement of the reflected waves from flaws indicates that flaws are located at about 25 mm
in depth from the surface of the test piece. This means that the indirectly reflected waves cannot be received by the pulse-echo mode
transducer.

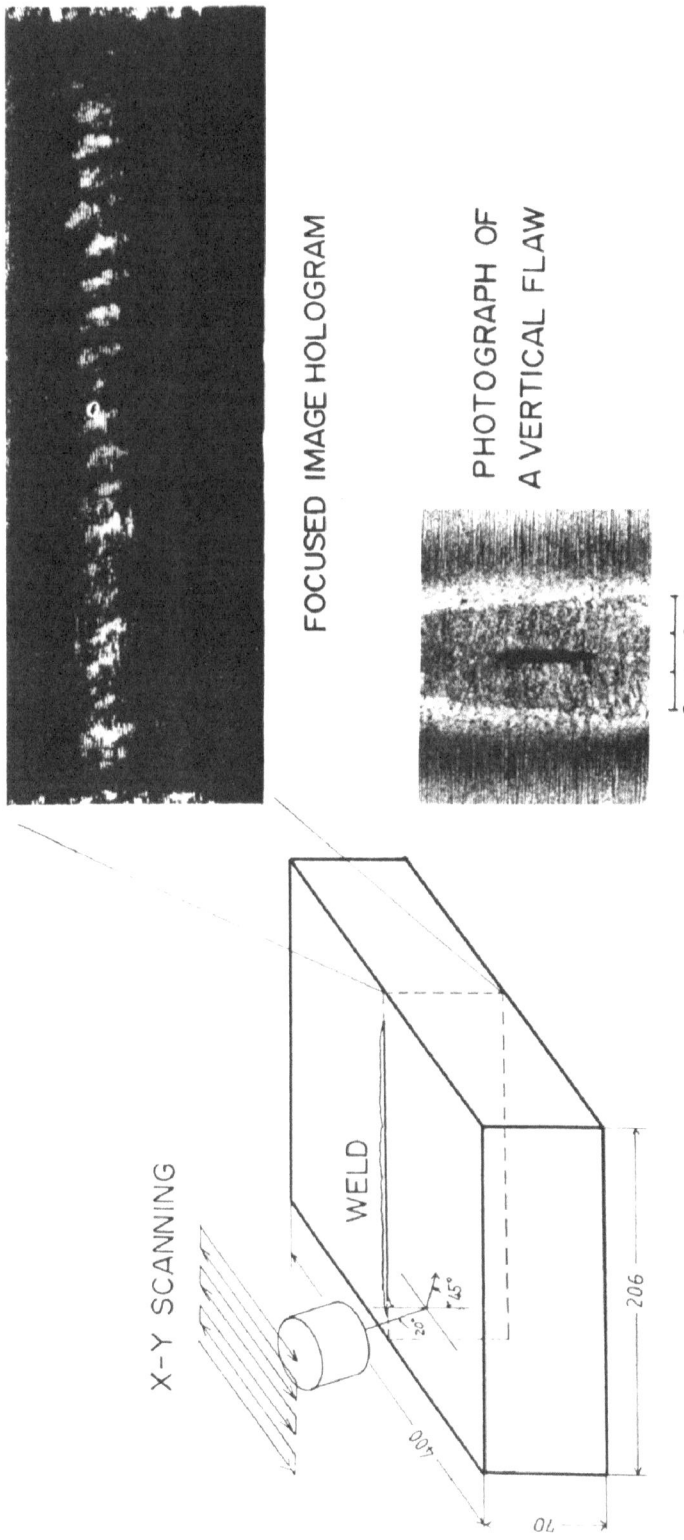

Fig. 8.   Image hologram construction geometry of natural flaws produced at the electron beam welded interface in the 70 mm thick steel test piece, and the image hologram of flaws. The photograph of a vertical flaw is also shown.

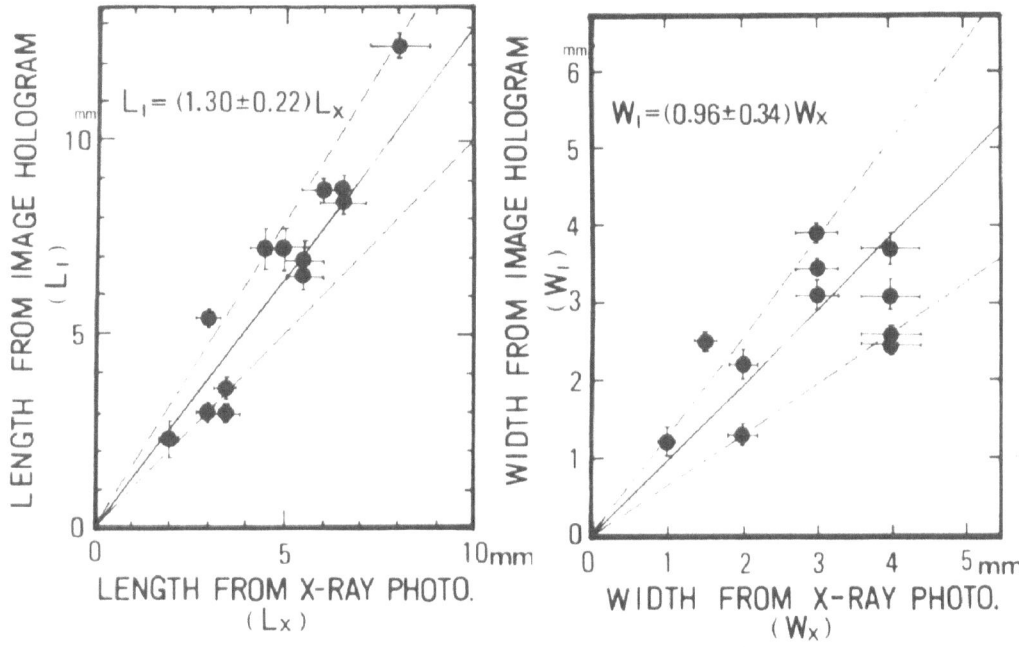

Fig. 9. Comparison of the measured size of flaws by image holography with the size of same flaws obtained by the X-ray technique.

After the completion of the ultrasonic testing by image holography, the test piece is cut off to a thickness of 20 mm along the welding line, and is then photographed by X-rays. The thirteen flaws are pictured and their size are measured on the film. The maximum height and width of the thirteen flaws measured by the X-ray technique and by image holography are compared as shown in Fig. 9. The most probable values for the height and width are shown by solidlines, while the standard deviations are shown by dotted lines. The figure shows that image holography overestimates the height of flaws about 30 % compared to the X-ray technique, while both methods agree within 4 % for the evaluation of width.

SUMMARY

Shearwave focused image holography is developed here for application in nondestructive testing. The results predicted by the theoretical analysis are examined by experiments. Satisfactory results are obtained for sizing artificial flaws with heights ranging from 2 mm to 4 mm. The deviations of the obtained heights from the true values measured by physical means are 10 % for the 4.0 mm height flaws at 3.0 MHz and 5 % at 5.6 MHz.

The method is also applied to measure the size of flaws in the electron beam welded interface of 70 mm thick steel block and the measured size of flaws is compared to the results obtained by the X-ray technique. The difference between the two methods for sizing the height of flaws is 30 %, while the difference for the width evaluation is 4 %.

REFERENCES

1. K. Suzuki and B. P. Hildebrand, "Holographic Interferometry with Acoustic Waves", Acoustical Holography, vol.6 edited by N. Booth, Plenum Press, 1975.

2. H. D. Collins, "Acoustical Interferometry Using Electronically Simulated Variable Reference and Multipul Path Techniques", Acoustical Holography, vol. 6, edited by N. Booth, Plenum Press, 1975.

3. B. P. Hildebrand and B. B. Brenden, An Introduction to Acoustical Holography, Plenum Press, 1972.

4. M. D. Fox, W. F. Ronson, J. R. Griffin, R. H. Petley, "Acoustic Holographic Interferometry," Acoustical Holography, vol. 5, edited by P. E. Green, Plenum Press, 1974.

5. W. S. Gan, "Fringe Localization in Acoustical Holographic Interferometry", Acoustical Holography, vol. 6., edited by N. Booth, Plenum Press, 1975.

ULTRASONIC CHARACTERIZATION OF DEFECTS

A. E. Holt and W. E. Lawrie

Babcock & Wilcox

Lynchburg, Virginia

INTRODUCTION

Since the introduction of commercially available, practical ultrasonic holographic systems several years ago, the majority of the activity has been focused on laboratory studies. Many of the holographically reconstructed images are of machined holes or objects of simple shape. The principal use of acoustical holography at Babcock and Wilcox has been to provide better characterization of natural defects detected using code-accepted NDE techniques. During the past several years, approximately 3 dozen natural defects have been examined by acoustical holography. Of these about 1/3 have been destructively analyzed. Normally acoustical holography provides dimensions of the reflecting surfaces that are accurate within one to two wavelengths of the true dimensions. In most cases, dimensions determined by holography slightly exceed the true values.

This paper describes the acoustical holographic evaluation of two natural defects and compares the holographic results with both ultrasonic size estimates and the dimensions determined destructively. In later sections of this paper, the relative merits and disadvantages of both conventional ultrasonic techniques and acoustical holography are discussed. In particular, the limitations inherent in defect sizing by conventional techniques are examined and the advantages of acoustical holography are indicated.

DATA ACQUISITION AND ANALYSIS

The results described in this paper are based on naturally occurring defects in thick section welds. Since the introduction of acoustical holography, this technique has been used to provide more definitive data on defects that were found by conventional techniques, either radiography or ultrasonics. The particular defects to be described were detected by ultrasonics. Additional measurements were then made using ultrasonic techniques, after which holographic data was obtained.

Ultrasonic Measurements

During routine ultrasonic examination in a vessel fabricated from 8" thick material, several defects were detected. Because of the nature of the material and of the defects, the defects presented an opportunity to compare conventional ultrasonics and acoustical holographic results to data obtained during the destructive evaluation.

The ultrasonic techniques that were employed were based on the use of a distance amplitude correction (DAC) curve and a 3 millimeter diameter flat-bottomed reference hole.

Two defects were examined in detail for later correlation with holographic results and with destructive analysis. One of the defects, at a depth of about 3-5/8 inches, was evaluated using nominal frequencies of 2.25 and 5 MHz. At 2.25 MHz the peak amplitude was 125% of DAC and at 5 MHz the peak amplitude was 80% of DAC. At 2.25 MHz, the length between the -6dB points was 0.44 inch.

The second defect appeared to be quite small so that a scanning measurement was not made. However, data was taken using five transducers. Two of the transducers were of the same frequency and size and the results obtained were very similar. Consequently, data is given for only four transducers. The data obtained from this second defect is tabulated in Table 1.

The data in Table 1 illustrates two important points relative to conventional ultrasonic testing. The first point is the significant difference between the nominal and actual frequencies in many "good" search units. The importance of this difference will be discussed at a later time. The second significant point is that the amplitude varies over a 5 to 1 range even when calibration was performed on the same reference hole for all transducers. Obviously a question arises. Which amplitude is truly indicative of the defect size?

Table 1

Ultrasonic Data for Second Flaw

Transducer	Nominal Frequency	Actual Frequency	Diameter	Relative Response	Relative Response
A	2.25 MHz	1.8 MHz	0.75"	75%	0
B	2.0	2.0	1.00	70	−0.6
C	4.0	3.0	1.00	25	−9.5
D	5.0	3.5	0.75	16	−13.4

It has long been recognized that the assumed linear relation-ship between area and amplitude exists only under special condi-tions. The defect must be smooth, of regular shape and oriented so that the ultrasonic energy approaches at normal incidence. If these conditions are not met, then the ultrasonic echo amplitude will depend upon more than just the area of the defect.

A very general way of discussing the dependence of echo ampli-tude upon defect characteristics is that the amplitude depends upon the interaction of the beam profile or radiation pattern of the search unit and also upon the reflected radiation pattern of the defect. Since the early ultrasonic measurements of the second flaw indicated that the flaw was relatively small, and because data had been obtained at a number of frequencies, an attempt was made to determine the defect area by taking into account the radia-tion pattern.

The first step was to compute the radiation pattern of each of the transducers from the measured frequency and the diameter. Tabu-lated values of k corresponding to known decreases from peak ampli-tudes were used to compute the semi-angle at which the decrease occurred, using the equation

$$\sin \theta/2 = k\lambda/D \qquad\qquad\qquad (1)$$

These curves are shown in Figure 1 for the four transducers. It is noted that radiation patterns for the highest frequency transducer (D) is broader than for the next highest transducer (C). This is caused by the smaller diameter of transducer D.

Since the dimensions of the defect are not known it becomes difficult to compute the radiation pattern at each of the transdu-cer frequencies. However, if we assume that the angles of incidence and reflection are small, then equation 1 may be rewritten in the form

$$D\theta = 2k\lambda \qquad\qquad\qquad (2)$$

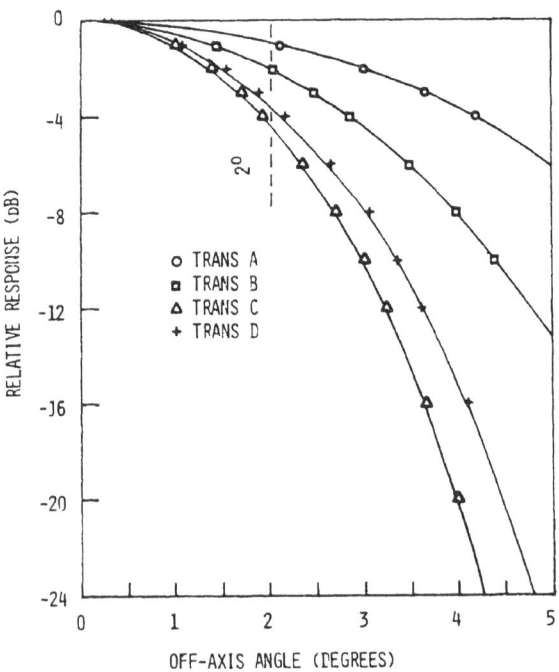

Figure 1.   Directivity characteristics for four transducers
operated in pulse–echo mode.

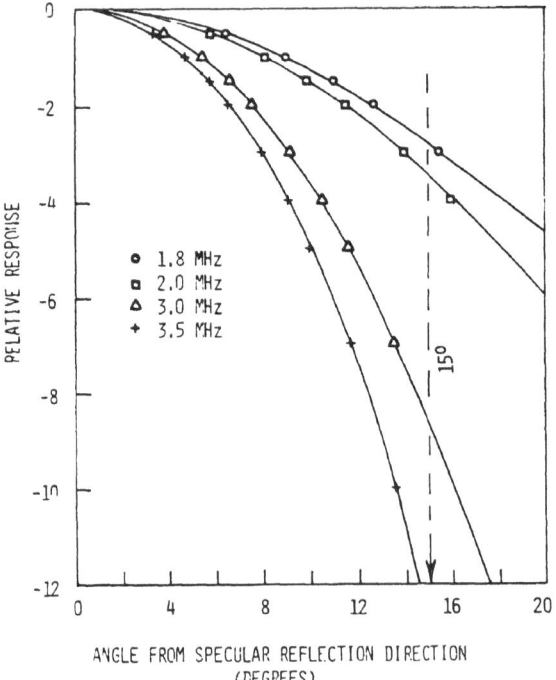

Figure 2.   Approximate directives for reflected beam from 1/4 inch
diameter deflect at transducer center frequencies.

This revision shows that for small defects, if we halve the defect dimension we simply double the width of the radiation pattern. As we shall see this is sufficient data for our purpose.

The radiation patterns of the defect were calculated using the actual frequencies of the transducers and the diameter of 1/4 inch for the defect. The four radiation patterns are shown in Figure 2. The radiation pattern shown in Figure 1 for the search unit is actually the square of the normal beam profile. In a pulse echo system, the beam profile determines the pressure distribution at each point in front of the transducer during transmission, and also determines the sensitivity to radiation from each direction when receiving. The radiation pattern of the defect is effective only during re-radiation or reflection.

An additional approximation is now necessary based on the lack of data. If a longitudinal wave search unit is scanned over a planar defect which is not parallel to the scanned surface, peak response will occur at different search unit positions for different search unit profiles. Maximum response will occur when a line joining the transducer exit point to the defect will be slightly off axis for both the transducer and defect radiation pattern. If the transducer's beam profile is much narrower than that of the defect, then the connecting line will make a smaller angle to the transducer beam axis than to the defect beam axis. This is illustrated in Figure 3. Data on the location of the transducers at the point of peak response is not available. Consequently, further calculations are based on the assumption that all transducers were at the same location for peak response.

The data in Table 1 and Figures 1 and 2 now allow us to determine the direction that the defect lies, relative to the transducer beam axis, and the product of the orientation and the size of the defect. By trial and error it is found that an angle of 2° from the transducer's beam axis and 15° from the defect reflected beam axis produces responses for each of the transducers that are quite close to the measured response. The comparison is shown in Table 2. The approximations that were made in the computations, and the normal spread of measured data probably account for the small differences between the normalized computed response and measured response. The computed size and orientation of the defect relative to the transducer is shown in Figure 4. It should be noted that if the defect was 1/8 inch in diameter, the normal to the defect would make an angle of 15° with the line to the transducer or the reflected signal to the transducer would be 30° off-axis.

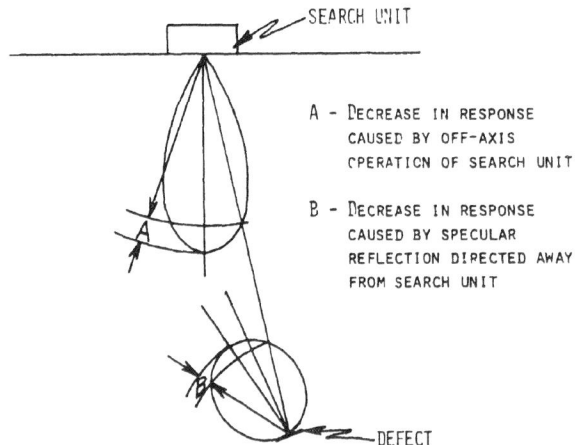

Figure 3.   Location of search unit for peak response from defect
            not parallel to surface.

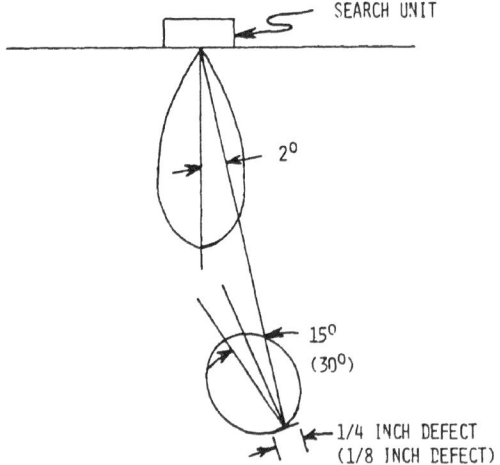

Figure 4.   Computed direction of specular reflection for 1/4 inch
            defect. If the defect was 1/8 inch diameter, the
            direction would be approximately 30°.

Table 2

Comparison of Measured and Computed Responses

Transducer	Rel. Trans. Reponse	Rel. Defect Response	Total Response	Normalized Response	Measured Response
A	−0.9dB	−2.8dB	−3.7dB	0	0
B	−1.9	−3.5	−5.4	−1.7	−0.6
C	−4.4	−8.8	−13.2	−9.5	−9.5
D	−3.6	−13.0	−16.6	−12.9	−13.4

If we assume that amplitude is proportional to area, then we may use the data of Table 2 to correct the amplitude response and thus permit us to calculate defect area relative to the hole area. With transducer A, the response is down by 3.7dB or down by 35% because of defect misorientation. In fact the response was 75% of the DAC curve at the same total metal path. Thus the diameter of the defect, if it were circular, must be $(75/65)^{1/2}$ times the diameter of the calibration hole. This corresponds to a defect diameter of 3.22 millimeters for a 3 millimeter reference hole. If we used the amplitude uncorrected for off-axis operation we would compute the diameter of the defect to be 2.6 millimeters.

Acoustical Holography Measurements

After completion of the conventional ultrasonic measurements, the two defects were examined using a Holosonics Model 200 scanning acoustical holographic system. Holograms were obtained using a frequency of 3 MHz for the first defect and frequencies of 3 and 5 MHz for the second defect. A reconstructed image of the first defect is shown in Figure 5. Measurements of the defect shown in Figure 5 and use of a computed magnification factor, showed that this defect was actually 0.8 inch long and 0.008 inch wide at the maximum width.

Holograms were obtained for the second defect at frequencies of both 3 and 5 MHz. The reconstructed images obtained from the holograms are shown in Figure 6. Measurements of the apparent dimensions from the reconstructed images and use of appropriate magnification factors for the two frequencies gave defect dimensions of 0.33 inch long and 0.18 inch wide at 3 MHz and 0.37 inch long and 0.124 inch wide at 5 MHz. There is obviously much closer agreement between the defect dimensions determined by acoustical holography at the two frequencies than there is from the relative amplitudes using conventional ultrasonics.

Results of Destructive Analysis

The dimensions of both defects discussed in preceeding sections were directly measured by destructive analysis. A section of the

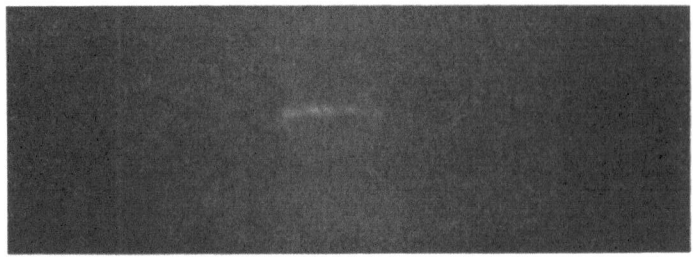

Figure 5 Image of First Defect Reconstructed from 3 MH₃
        Hologram

a) Image Reconstructed from 3 MH$_z$ Hologram

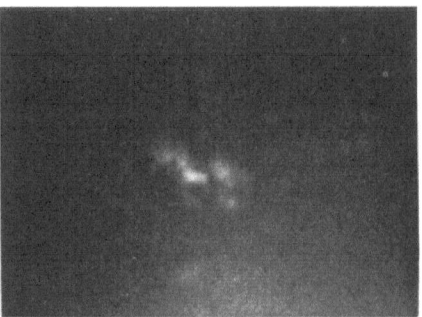

b) Image Reconstructed from 5 MH$_z$ Hologram

Figure 6 Reconstructed Images of Second Defect

weld containing the first defect, approximately 4 inches long, 4 inches deep and 2-1/2 inches wide, was removed for destructive evaluation. After additional radiographic examination to determine the location of the defect, the original sample was reduced in size. Thin layers were then removed until the defect was found. Additional layers of less than 1 mil in thickness was removed. As each new surface was exposed, the dimensions of the defect were obtained. The exposed section of the defect is shown at two different elevations in Figure 7. The maximum dimensions of the defect measured during this destructive analysis are 0.75 inch long and 0.06 inch wide.

The second defect did not undergo such extensive evaluations during removal. Air-arc removal was performed in the shop to re- move layers of approximately 1/8 inch in thickness. The top of the defect was exposed during one pass so that detailed knowledge of the reflecting surface is not available. The defect was basi- cally planer, and had a shape approximately like that shown in Figure 8. The maximum dimensions for this defect, determined du- ring removal were 0.25 inch long and 0.117 inch wide.

Comparison of Test Results

The defect sizes determined by conventional ultrasonics, by modified ultrasonics, by acoustical holography and by destructive evaluation are shown in Table 3.

Table 3

Comparison of Defect Dimension

	Defect 1	Defect 2
Conventional U.T. (Scanning or area amplitude)	0.44 long* (scanning)	0.104 diameter (area amplitude)
Correction for Beam Profiles	–	0.129 diameter
Acoustical Holography 3 MHz	0.80 x 0.08	0.33 x 0.176
5 MHz	–	0.37 x 0.124
Destructive Evaluation	0.76 x 0.06	0.25 x 0.117

*All dimensions are in inches

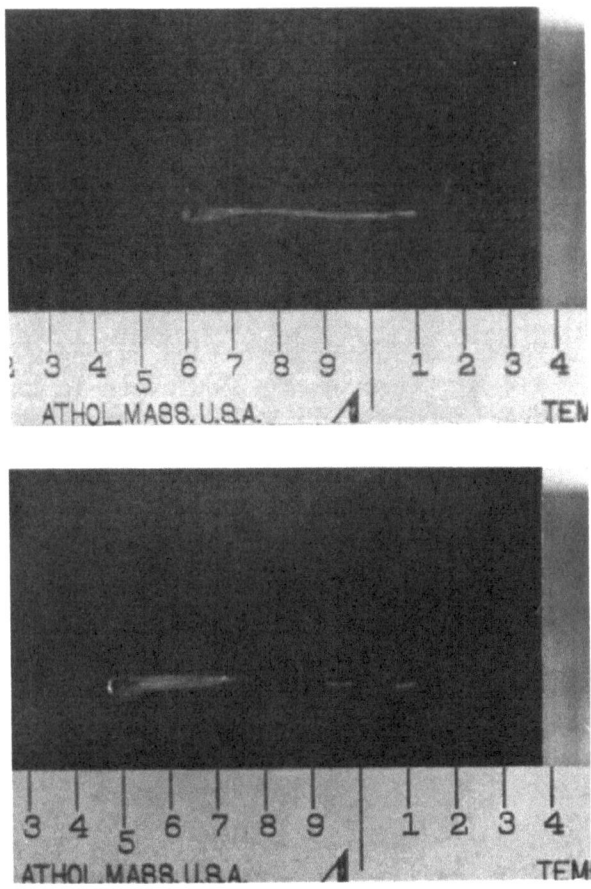

Figure 7   Photographs of First Defect as Exposed at two levels
          showing full extent of Defect

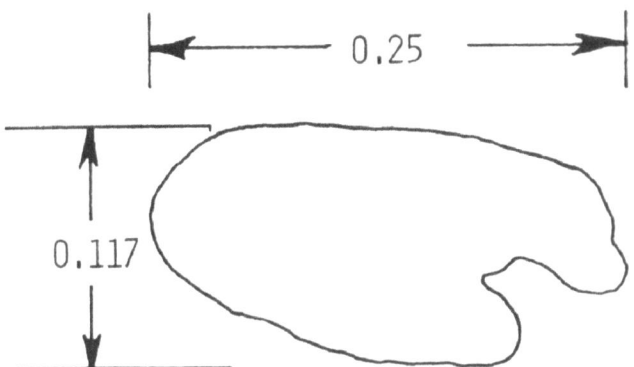

Figure 8   Size and Shape of Second Defect

On the first defect, the length determined by conventional ultrasonics was 0.44 inch long. Acoustical holography at 3 MHz showed the defect to be 0.8 inch long which was only slightly longer than the actual dimension of 0.76 inch. On the second defect, comparison of the maximum amplitude obtained with any transducer with the amplitude obtained from the reference hole indicated that the defect was 0.104 inch diameter. Even after correction for the effects of beam profiles the computed size was still only 0.129 inch diameter. Acoustical holography produced dimensions that were approximately 40% too long and 30% too wide, compared to the actual dimension of 0.25 and 0.117 inch. Part of the difference between the holographically determined dimensions and the actual dimensions may lie in the difficulty with which the actual dimensions of this second defect were obtained.

CONCLUSIONS

In both defects examined, the acoustical holographic dimensions were closer to the true dimensions than were those determined by other techniques. In addition, measurements made holographically at two frequencies were in much better agreement than were measurements made at different frequencies using conventional ultrasonic techniques. Changes in transducer frequency or size, and variations in defect shape, orientation and reflecting surface roughness will alter the radiation patterns of the search units and defects. The response of conventional ultrasonic systems then will vary within a broad range because of the interaction of the two beam profiles. The attempt to characterize a natural defect by a single number (e.g., 75% of DAC curve) must obviously result in ambiguities. In comparison the greater amount of signal data used to generate a hologram and, later, a reconstructed image, has inherent capabilities of providing more accurate characterization. Not only does acoustical holography provide better defect sizing, but the presentation of an image provides knowledge of defect shape and orientation which is much more reliable than that obtained by conventional ultrasonic techniques.

ACOUSTIC APERTURE DIFFRACTION IN A TRANSVERSELY MOVING MEDIUM

G. Flesher and S. Elliott

Department of Electrical Engineering
  and Computer Science
University of California
Santa Barbara, CA 93106
U.S.A.

ABSTRACT

The theory of diffraction from a screen in a static medium is well known. The diffraction pattern is altered, however, if there is relative motion between the diffraction plane, the observation plane or the medium between planes. The case of relative transverse motion of the medium is investigated theoretically by generalizing the wave equation to include this motion, assuming a uniform, constant velocity. An equivalent Rayleigh-Sommerfeld integral is developed for this case. It is used to obtain a closed-form solution for the diffraction pattern in the Fraunhofer region for medium velocities less than the speed of sound. This solution is used to predict the amplitude distributions for several diffracting screen configurations. The results of this theory have application in performance characterization for a number of high-scan-rate acoustic imaging systems.

INTRODUCTION

Many acoustic systems, including holographic systems and imaging systems, involve motion of the transmitter, the receiver, the object or the transmission medium. The effects of this relative motion on the resolution of these systems has, until recently, been neglected because the velocities involved were slow compared to the phase velocity of sound in the medium, c_o, as for example in mechanically-scanned systems. A current research emphasis, however, is on electronically-scanned, real-time systems for

medical diagnostics or nondestructive testing, where the scan
velocity of the transmitter or the receiver or both may be a
significant fraction of c_o. Hildebrand[1] and Coello-Vera, et al.[2],
have examined the distortion and loss of resolution produced by
these motions at the recording plane of a hologram. When a radiating aperture many wavelengths wide is scanned, it is incorrect
to suppose that the radiation pattern remains unchanged in shape
as it scans through the medium. Actually, the spatial frequency
waves which give rise to the radiation pattern are not superposed
in the usual manner when there is relative motion between the
aperture and the medium. In this paper, scalar diffraction theory
is generalized to account for relative motion, in the scanning
direction, between a radiating (or receiving) acoustic aperture
and the medium. The effects of this relative motion are illustrated
by several familiar examples of diffracting apertures, and the
results of the analysis are applied to predict the performance of
and to suggest improvements for several high-scan-rate acoustic
imaging systems.

APERTURE RADIATION IN A MOVING MEDIUM

The near- and far-field radiation patterns for an ultrasonic
transducer with known amplitude and phase distribution can be
calculated from classical scalar diffraction theory if the medium
is homogeneous and isotropic[3,4]. To solve the problem of a radiating
aperture moving through a stationary medium, we can transfer to a
moving coordinate system in which the aperture is stationary and
the medium moves. The technique of finding radiation from a
stationary aperture is applied using the waves propagating in the
moving medium. In a stationary medium, the wave equation for
sound waves is

$$(\partial_{x_1}^2 + \partial_{y_1}^2 + \partial_{z_1}^2)\rho = c_o^{-2} \partial_{t_1}^2 \rho \tag{1}$$

This equation is the linearized wave equation for the excess or
signal mass density field ρ which itself is a small variation
about the average density ρ_o. The velocity of plane waves in this
nonmoving medium is c_o. Consider a coordinate system (x,y,z,t)
coincident with (x_1,y_1,z_1,t_1) at $t=0$ and moving in the negative x_1-
direction with velocity u. The effects of special relativity can
be neglected at the low velocities considered here. The coordinate
transformations are

$$[x,y,z,t] = [x_1 + ut, y_1, z_1, t_1] \tag{2}$$

In the moving coordinates, the wave equation (1) transforms to

$$(\partial_x^2 + \partial_y^2 + \partial_z^2)\rho = c_o^{-2}(\partial_t + u\,\partial_x)^2\rho \tag{3}$$

Changing our relative viewpoint, this equation can be called the wave equation for the case of a medium moving through the coordinate system in the positive x-direction with a velocity u. (Turbulence and stream gradients are not present.) The only change in the wave equation from one coordinate system to the other is the modification of the time derivative term to include a spatial derivative. This gives rise to what we may call the Doppler change in wavelength, as seen below. Since the new wave equation has constant coefficients and is linear, we seek the usual exponential solutions of the form

$$\rho = \rho_o \exp[j(\omega t - \beta_x x - \beta_y y - \beta_z z)] \tag{4}$$

Substituting this equation in (3), we find the conditional equation which must be satisfied if solutions are to exist to be,

$$\beta_x^2 + \beta_y^2 + \beta_z^2 - \frac{1}{c_o^2}(\omega - \beta_x u)^2 = 0 \tag{5}$$

This equation can be viewed as the required condition on one of the wave-vector components when all of the others are chosen. For example, β_z is determined by the other wave and medium properties as

$$\beta_z = \pm\,[(\omega - \beta_x u)^2 c_o^{-2} - \beta_x^2 - \beta_y^2]^{1/2} \tag{6}$$

The positive sign is chosen for waves that travel in the positive z-direction.

These solutions describe plane waves of infinite extent. In the usual application, the physical structure is of finite size and the useful wave fields travel a finite distance. Fortunately, many useful solutions can be formed by the superposition of the plane-wave solutions via the Fourier integral. Consider the following form as a candidate for a solution of the wave equation

$$\rho(x,y,z,t) = \frac{1}{4\pi^2}\int_{-\infty}^{\infty}\int_{-\infty}^{\infty} \bar{\rho}(\beta_x,\beta_y)\,\exp[j(\omega t - \beta_x x - \beta_y y - \beta_z z)]\,d\beta_x\,d\beta_y \tag{7}$$

Assuming well-behaved solutions, ρ may be tried as a solution in the wave equation (3), and changing the order of differentiation and integration, this type of solution is found to be valid if and only if β_z obeys equation (6). The function $\bar{\rho}(\beta_x,\beta_y)$ is considered to be a weighting function for the continuous distribution of

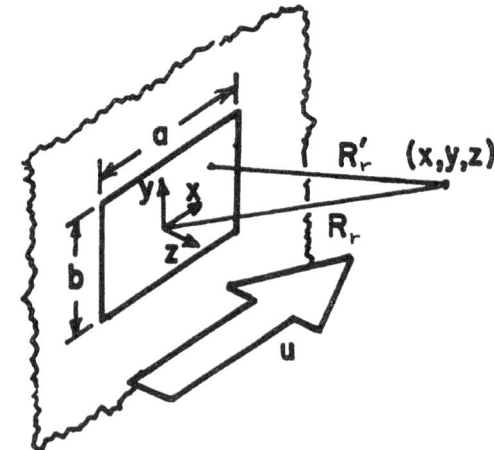

Fig. 1 An excited aperture radiating into a moving medium.

waves integrated in (7). Physical meaning for this function can
be found when we consider diffraction from an aperture radiating
into a half space as illustrated in Fig. 1. An opaque plane, z=0,
has an aperture which can have any shape (here illustrated as
rectangular) and which admits radiated waves into the half plane
z>0. As a boundary condition, assume that the value of ρ is known
for z=0, that is, at any point (x,y,0) on the opaque wall and in
the aperture, the value is $\rho(x,y,0,t)$. Setting z=0 in (7), the
resulting equation is recognized as one equation of a two-dimensional
Fourier transform pair. Therefore, we can find $\bar{\rho}$ from the other
member of that pair.

$$\bar{\rho}(\beta_x,\beta_y) = \int_{-\infty}^{\infty} \int_{-\infty}^{\infty} \rho(x,y,0) \ \exp[j(\beta_x x + \beta_y y)] \ dx \ dy \qquad (8)$$

(The exponential time function will be assumed but not written in
the following steady-state calculations.) Thus, the weighting
function $\bar{\rho}$ is the Fourier transform of ρ at z=0, and is a known function when $\rho(x,y,0)$ is known. Once $\bar{\rho}$ is known, it can be used in
(7) to find the desired density distribution ρ for any z. An
alternative integral to equation (7) can be found when it is
recognized that equation (7) is an inverse Fourier transform of
the product of two transform functions, and a convolution solution
is therefore possible. That is,

$$\rho(x,y,z) = F^{-1}[\bar{\rho}(\beta_x,\beta_y) \ \exp(-j\beta_z z)] = \rho(x,y,0) * F^{-1}[\exp(-j\beta_z z)] \qquad (9)$$

where * is the convolution operator. The inverse transform of the

exponential term is the impulse response of the space and is found
with some integration to be

$$F^{-1}(e^{-j\beta_z z}) = \frac{z}{2\pi\delta R^2} \left(\frac{j\omega}{c_o\delta} + \frac{1}{R}\right) \exp\left(\frac{j\omega u}{c_o^2\delta^2} x - \frac{j\omega R}{c_o\delta}\right) \quad (10)$$

where

$$\delta = (1 - u^2/c_o^2)^{1/2} \quad \text{and} \quad R^2 = x^2/\delta^2 + y^2 + z^2 \quad (11)$$

We will consider only distances far enough from the aperture where
the $1/R$ term is negligible. The convolution solution is then

$$\rho(x,y,z) = \frac{jz}{\lambda_o\delta^2} \int_{-\infty}^{\infty}\int_{-\infty}^{\infty} \frac{\rho(x',y',0)}{R'^2} \exp[-j\beta_o\frac{R'}{\delta} + j\frac{\beta_o u}{\delta^2 c_o}(x-x')]dx'dy' \quad (12)$$

where

$$\beta_o = \frac{\omega}{c_o} = \frac{2\pi}{\lambda_o} \quad \text{and} \quad R'^2 \equiv \frac{(x-x')^2}{\delta^2} + (y-y')^2 + z^2 \quad (13)$$

When the velocity of the medium is zero, the solution of (12) is
exactly the Rayleigh-Sommerfeld equation for radiation from an
aperture[4]. Thus (13) can be said to be the generalized Rayleigh-
Sommerfeld equation for radiation from an aperture into a moving
medium. The factor δ appears in several places in (12). Since δ
becomes zero as the velocity of the medium approaches the sonic
velocity c_o, care must be taken near this limit. Physically,
there are two features of this generalized equation which give a
preview of how the radiated field pattern will change when the
medium in front of the aperture is moving. First, there is a
phase advance term depending directly on $(x-x')$ and the velocity u
in (12). This will lead to a different value of x where the bulk
of the contributions from the integral add in phase. Second, the
x-coordinate of the solution always has the factor δ^{-2} multiplying
it. Consequently, under some conditions, we expect that the
solutions will show that the radiation patterns are contracted or
expanded in the x-direction when compared to those in a stationary
medium.

In application of the Rayleigh-Sommerfeld type of solution,
it is usually assumed with good approximation that the boundary
value $\rho(x,y,0)$ is a known function in the aperture and zero everywhere else on the opaque plane $z=0$. Sometimes it is convenient to
consider the radiation from a one-dimensional aperture or a slit
which extends to infinity in the y-direction. In this case a
cylindrical symmetry results since the solutions are identical in

each plane of y=constant. The solutions corresponding to (7) and
(8) are

$$\rho(x,z) = \frac{1}{2\pi} \int_{\infty}^{\infty} \bar{\rho}(\beta_x) \exp[-j(\beta_x x + \beta_z z)] \, d\beta_x$$

(14)

$$\bar{\rho}(\beta_x) = \int_{\infty}^{\infty} \rho(x,0) \exp(j\beta_x x) \, dx$$

When ρ no longer depends on y, the y' integral in the general
solution (12) can be performed utilizing an excellent asymptotic
value for the integral. The one-dimensional aperture solution
corresponding to (12) is

$$\rho(x,z) = \left(\frac{j}{\lambda_o}\right)^{1/2} \frac{z}{\delta^{3/2}} \int_{-\infty}^{\infty} \frac{\rho(x',0)}{R''^{3/2}} \exp[-j\beta_o\frac{R''}{\delta} + j\frac{\beta_o u}{\delta^2 c_o}(x-x')] \, dx' \quad (15)$$

where

$$R''^2 = \frac{(x-x')^2}{\delta^2} + z^2$$

(16)

The solution in (15) is the generalized Rayleigh-Sommerfeld equation
for radiation from an infinite slit aperture into a moving medium.
It has the same basic properties that equation (12) does for the
two-dimensional aperture. Equations (12) and (15) are exact and
their approximate integration for several examples will show the
net effect of the moving medium on the radiation patterns.

FRESNEL AND FRAUNHOFER APPROXIMATIONS

To get a physical feeling for the results of equations (12)
and (15), it is desirable to get closed-form solutions of the
integrals at the cost of certain approximations and restrictions.
The technique used here follows the usual one, with a modification
dictated by the fact that significant parts of our solutions may
be "downstream," that is, far off axis, so that the usual paraxial
approximations cannot be used. Referring to equation (13) we may
factor the term R out of this expression for R'. When the point
of observation is quite distant from the aperture, R is by far the
largest part of R', and the remaining factor can be approximated
by the first two terms of the binomial expansion of the R' in the
exponent. The critical approximation is in the phase factor of
the exponent, for it must be accurate to, say, less than one-tenth
of π/2 radians. Since the binomial expansion is an alternating

series, the error is less than the first term of the series that
is dropped. The approximation is thus valid for

$$R^3 > \frac{5}{\delta\lambda_o}[xx' + yy']^2 \quad \text{at maximum x' and y'} \tag{17}$$

When this is true, the resulting solution, called the Fresnel-zone
solution, is given by

$$\rho(x,y,z) = \frac{jz}{\lambda\delta^2 R^2} \exp[-j\frac{\beta_o}{\delta}\left(R - \frac{u}{c_o}\frac{x}{\delta}\right)]$$

$$\cdot \int_{\infty}^{\infty} \int_{\infty}^{\infty} \rho(x',y',0) \exp\{-j\frac{\beta_o}{2\delta R}[\left(\frac{x'}{\delta}\right)^2 + y'^2]\}$$

$$\cdot \exp\{-j\frac{\beta_o}{\delta}[\left(\frac{x}{R\delta^2} - \frac{u}{\delta c_o}\right)x' + \frac{y}{R}y']\} \, dx'dy' \tag{18}$$

For a variety of diffracting-screen transmission functions that
have linear or quadratic phase variations, the integrand can be
simply rearranged to form a Fresnel integral, which is tabulated.
However, one more approximation will be used to get the Fraunhofer
approximation. If R is large enough, the quadratic phase factor
in equation (18) is nearly zero and may be neglected for

$$R > \frac{20}{\lambda_o\delta}\left[\left(\frac{x'}{\delta}\right)^2 + y'^2\right] \tag{19}$$

With this approximation, the remaining integral in equation (18)
can be written explicitly as the Fraunhofer-zone solution

$$\rho(x,y,z) = \frac{jz}{\lambda_o\delta^2 R^2} \exp\left[-j\frac{\beta_o}{\delta}\left(R - \frac{u}{c_o}\frac{x}{\delta}\right)\right] \bar{\rho}(\beta'_x,\beta'_y), \tag{20}$$

where $\bar{\rho}$ is the Fourier transform of equation (8) evaluated at the
particular spatial frequencies

$$\beta'_x \equiv \frac{\beta_o x}{R\delta^3} - \frac{\beta_o}{\delta^2}\frac{u}{c_o} \quad \text{and} \quad \beta'_y \equiv \frac{\beta_o y}{R\delta} \tag{21}$$

The approximation requirements in (17) and (19) are sufficient,
but probably stronger than necessary for good accuracy. The
Fresnel and Fraunhofer approximations differ from the stationary

medium by the introduction of u, δ, and R, but become the stationary
solution when u=0. The examples below will demonstrate the effect
of medium velocity on the radiated patterns.

The Fresnel and Fraunhofer approximations for the one-dimensional aperture are found in the same way using, respectively,
equations (17) and (19) with y' removed. The Fresnel-zone solution
for equation (15) is

$$\rho(x,z) = \left(\frac{j}{\lambda_o}\right)^{1/2} \frac{z}{(\delta R_1)^{3/2}} \exp\left[-j\frac{\beta_o}{\delta}\left(R_1 - \frac{u}{c_o}\frac{x}{\delta}\right)\right]$$

$$\cdot \int_{-\infty}^{\infty} \rho(x,0) \exp[-j\frac{\beta_o}{2\delta R_1}\left(\frac{x'}{\delta}\right)^2] \exp[-j\frac{\beta_o}{\delta^2}\left(\frac{x}{R_1\delta} - \frac{u}{c_o}\right)x'] \ dx', \quad (22)$$

where R_1 is the same as R, except that the y-dependence is dropped.
The Fraunhofer-zone solution is

$$\rho(x,z) = \left(\frac{j}{\lambda_o}\right)^{1/2} \frac{z}{(R_1\delta)^{3/2}} \exp[-j\frac{\beta_o}{\delta}\left(R_1 - \frac{u}{c_o}\frac{x}{\delta}\right)] \ \bar{\rho}(\beta_x'). \quad (23)$$

ONE-DIMENSIONAL APERTURES

To illustrate the effect of a rapidly-moving medium, let us
examine the familiar case of Fraunhofer diffraction from a slit.
First, consider the insonification of the slit by a normally-
incident wave of unit amplitude, as illustrated in Fig. 2a. The
Fourier transform of this excitation is[4]

$$\bar{\rho}(\beta_x) = a \ \frac{\sin \frac{a}{2} \beta_x}{\frac{a}{2} \beta_x} \quad (24)$$

and using (23), the density distribution is given by

$$\rho(x,z) = \left(\frac{j}{\lambda_o}\right)^{1/2} \frac{az}{(\delta R_1)^{3/2}} \exp[-j\frac{\beta_o}{\delta}\left(R_1 - \frac{u}{c_o}\frac{x}{\delta}\right)] \ \cdot \ \frac{\sin \frac{a}{2} \frac{\beta_o}{\delta^2}\left(\frac{x}{\delta R_1} - \frac{u}{c_o}\right)}{\frac{a}{2} \frac{\beta_o}{\delta^2}\left(\frac{x}{\delta R_1} - \frac{u}{c_o}\right)} \quad (25)$$

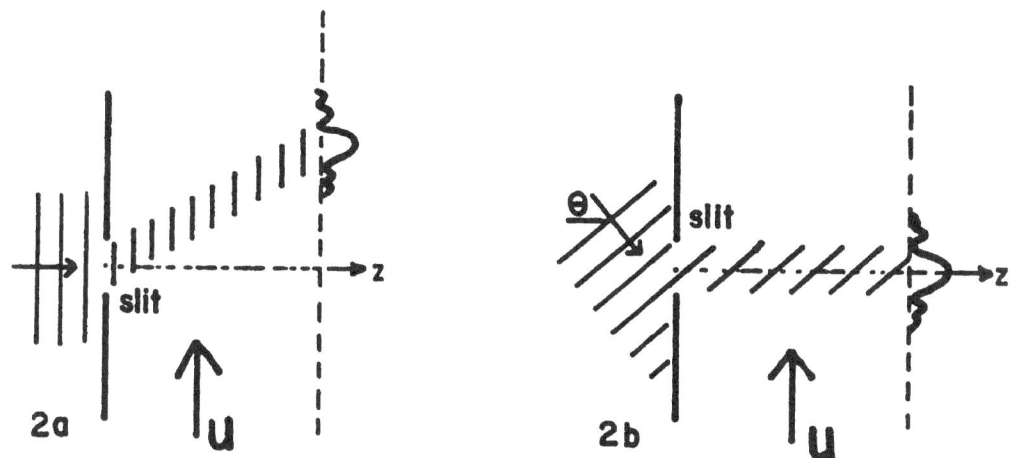

Fig. 2 a) A slit, excited by a normal wave, radiates into a moving
           medium. b) A slit, excited by a wave at a particular
           angle, radiates on the z-axis through the moving media.

A study of this equation shows that the maximum of the familiar
$(\sin x)/x$ pattern is moved downstream by approximately an amount
$z(u/c_0)$. This shift is physically reasonable when we think of a
rowboat being rowed straight across a moving river. The oarsman
finds himself landing downstream by a distance equal to the ratio
of the water velocity to the boat velocity times the width of the
river.

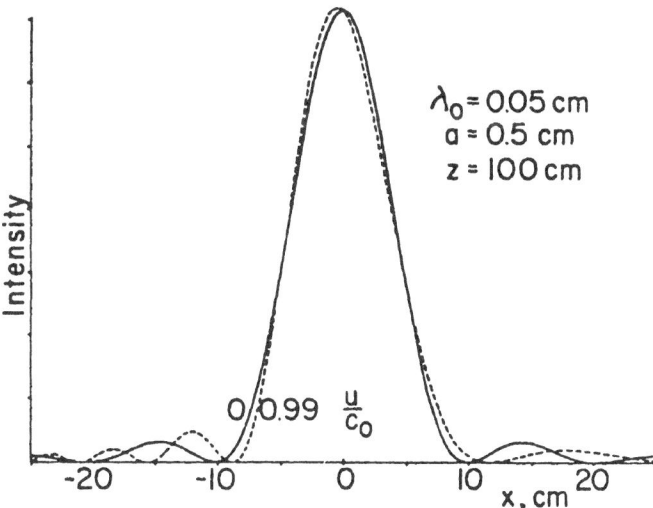

Fig. 3 The Fraunhofer-zone pattern of a slit radiating into a medium
         moving by with velocity u. The patterns are superimposed.

The width of the main lobe of the pattern is altered very little by the medium velocity and by the downstream transition. Figure 3 is a superposition of diffraction patterns (represented by the modulus squared of the density) for two values of medium velocity. Each pattern is shifted by an amount $z(u/c_o)$ for comparison. For these curves, the wavelength of sound in the nonmoving medium is taken to be 0.5 mm, the slit width is 0.5 cm and z = 100 cm. Notice that the side lobes become squeezed together on the upstream side and stretched out on the downstream side of the maximum.

A physical argument for these results is as follows. The Fourier optics method of analysis presented above assumes that the density pattern at each plane is the superposition of an infinite number of plane waves at different angles with weighted amplitudes. A plane wave traveling normal to the medium velocity is unaffected by that velocity. However, it can be shown that a plane wave traveling upstream at an angle ϕ to the normal will have its wavelength altered to

$$\lambda = \lambda_o (1 - \frac{u}{c_o} \sin \phi) \tag{26}$$

In the slit example above, most of the contribution to the diffraction pattern is from plane waves with propagation vectors at small angles ϕ; since the wavelengths are changed very little, the main lobe of the diffraction pattern is altered very little, except for a shift in position on the observation plane.

An interesting phenomenon occurs, however, when we insonify the slit with a plane wave incident upstream at an angle θ, as shown in Fig. 2b. Then the spatial frequency β_x' in equations (20), (21) and (23) is shifted by an amount $\beta_o \sin \theta$, so that

$$\bar{\rho}(\beta_x' + \beta_o \sin \theta) = \bar{\rho} \left(\frac{\beta_o x}{R_1 \delta^3} - \frac{\beta_o}{\delta^2} \frac{u}{c_o} + \beta_o \sin \theta \right) \tag{27}$$

Thus, in order to eliminate the downstream shift of the diffraction pattern, we can select the angle θ such that

$$\sin \theta = \frac{1}{\delta^2} \frac{u}{c_o} \tag{28}$$

The factors $1/\delta$, $1/\delta^2$, $1/\delta^3$ and $(1/\delta^2)(u/c_o)$ are plotted in Fig. 4 as a function of u/c_o up to 1, and a quick inspection shows that real values of θ are possible up to $u/c_o = 0.6$. From equation (23), the diffraction pattern for this case is

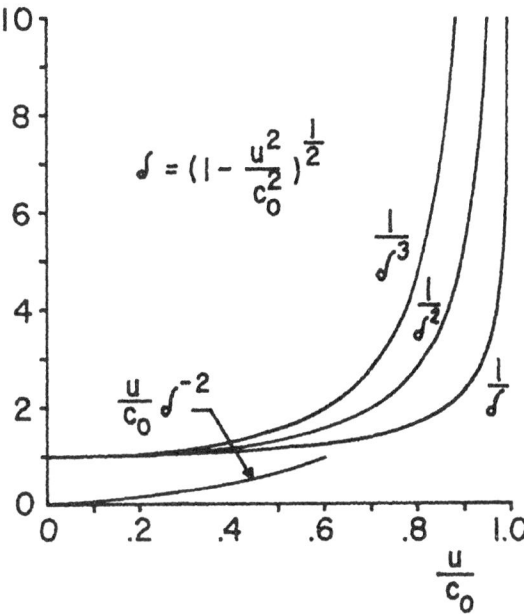

Fig. 4   Curves showing how various factors involving δ depend on u/c_o.

$$\rho(x,z) = \left(\frac{j}{\lambda_o}\right)^{1/2} \frac{az}{(\delta R_1)^{3/2}} \exp\left[-j\frac{\beta_o}{\delta}\left(R_1 - \frac{u}{c_o}\frac{x}{\delta}\right)\right] \frac{\sin\left(\frac{a\beta_o x}{2\delta^3 R_1}\right)}{\left(\frac{a\beta_o x}{2\delta^3 R_1}\right)} \quad (29)$$

A further simplification can be made if we note that the pattern is now centerd on the z-axis, thus we are interested in the density for small values of x/δ compared to z. Then R_1 can be approximated by z (or by z + $(1/2z)(x/\delta)^2$ in the exponent) and we can write

$$\rho(x,z) = \left(\frac{j}{\lambda_o z}\right)^{1/2} \frac{a}{\delta^{3/2}} \exp\left\{-j\frac{\beta_o}{\delta}\left[z+\frac{1}{2z}\left(\frac{x}{\delta}\right)^2 - \frac{u}{c_o}\frac{x}{\delta}\right]\right\} \frac{\sin\left(\frac{a\beta_o x}{2\delta^3 z}\right)}{\left(\frac{a\beta_o x}{2\delta^3 z}\right)} \quad (30)$$

Plots representative of equation (29) are shown in Fig. 5.

Notice that now the slit diffraction pattern is compressed by a factor of $1/\delta^3$. Physically, this is because plane-wave components contributing the most to construct the diffraction pattern are traveling into the water velocity vector. In fact, as θ approaches

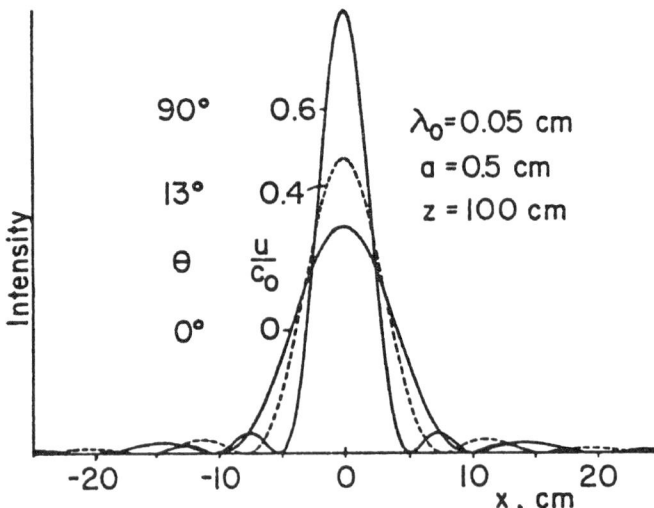

Fig. 5   The Fraunhofer-zone pattern of a slit radiating into a
        medium moving by with velocity u. The slit is excited with
        a wave at angle θ just sufficient to put the peak intensity
        on the z-axis.

90°, the wavelength of the highest amplitude plane-wave component
is reduced by a factor of

$$\lambda = \lambda_o \left[1 - \left(\frac{u}{c_o} \right)^2 \right] \Bigg|_{\frac{u}{c_o} = .6} = 0.64 \, \lambda_o \tag{31}$$

Therefore, the pattern looks as if it were being produced by
insonification of the slit at about half the wavelength actually
used, and thus the main lobe of the pattern is about half as wide
as in the case of normal incidence.

DIFFRACTION FROM A CYLINDRICAL LENS

To consider the analysis of Fresnel diffraction from a cylindrical lens, we shall assume a thin lens of focal length f, stopped
by a slit aperture of width a, as shown in Fig. 6. The approximate
transmission function for the stopped lens is given by[4]

$$t(x') = \text{rect}\left(\frac{2x'}{a} \right) \, \exp(j\gamma) \, \exp\left(j\frac{\beta_o}{2f} \, x'^2 \right) \tag{32}$$

where

$$\text{rect}(V) = \begin{cases} 1, & |V| \leq 1 \\ 0, & |V| > 1 \end{cases} \tag{33}$$

and γ is a constant phase factor, determined by the lens thickness and material, which we will assume to be zero. Consider, first, the case of a normally incident plane wave of unit amplitude. The density distribution at the output of the lens is then equal to $t(x')$, and assuming the Fresnel approximation, equation (32) may be substituted into equation (22) to yield

$$\rho(x,z) = \left(\frac{j}{\lambda_o}\right)^{1/2} \frac{z}{(\delta R_1)^{3/2}} \exp[-j\frac{\beta_o}{\delta}\left(R_1 - \frac{u}{c_o}\frac{x}{\delta}\right)]$$

$$\cdot \int_{-a/2}^{a/2} \exp[-j\frac{\beta_o}{2} x'^2\left(\frac{1}{\delta^3 R_1} - \frac{1}{f}\right)] \exp[-j\frac{\beta_o}{\delta^2}\left(\frac{x}{R_1\delta} - \frac{u}{c_o}\right)x'] \; dx' \tag{34}$$

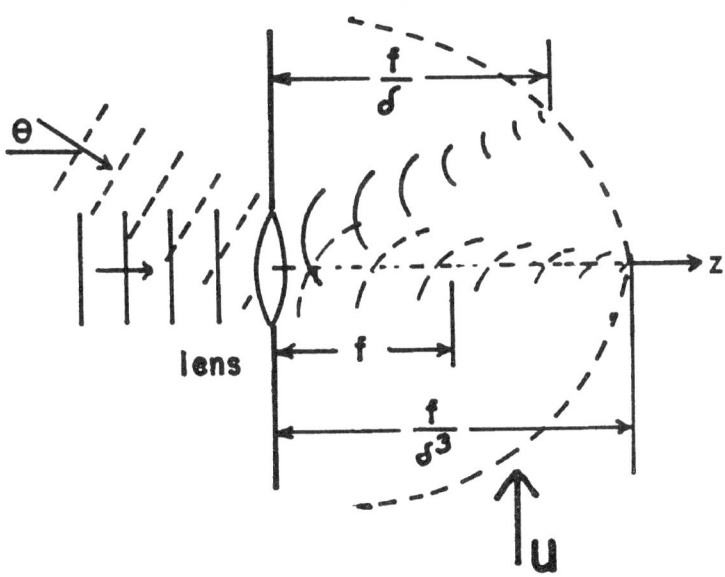

Fig. 6   The moving medium can change the effective focal point of an acoustic lens.

To observe a focal line, we must look on a surface defined by

$$R_1 = \frac{f}{\delta^3} = \sqrt{\left(\frac{x}{\delta}\right)^2 + z^2} \tag{35}$$

which is an elliptical cylinder with eccentricity u/c_o, as illustrated in cross-section by the broken line in Fig. 6. On this surface, the first exponential in the integrand of equation (34) is unity, resulting in equation (25) once again, with the constraint that this equation defines the density distribution on the surface of the elliptical cylinder only.

Suppose now that the insonifying plane wave is incident at an angle θ defined by equation (28) so that the diffraction pattern is centered on the z-axis. Then for u/c_o up to 0.6, most of the energy near the focal line will be concentrated at small values of x/δ so we can approximate R_1 by z in equations (34) and (35). It appears that the focal plane has then shifted to $z = f/\delta^3$, as indicated in Fig. 6. Making this substitution for z in equation (28), we find

$$\rho(x, \frac{f}{\delta^3}) = \left(\frac{j}{\lambda_o f}\right)^{1/2} a \exp[-j\frac{\beta_o}{\delta}\left(\frac{f}{\delta^3} - \frac{u}{c_o}\frac{x}{\delta}\right)] \frac{\sin\frac{a\beta_o}{2}\frac{x}{f}}{\frac{a\beta_o}{2}\frac{x}{f}} \tag{36}$$

This result shows that when viewed in the new focal plane, the focal pattern is not compressed by the moving medium with respect to that seen in the stationary medium. The effect of the medium velocity is to modify the phase of the wave distribution, and most importantly, to increase the distance of the focal plane by a factor of $1/\delta^3$. For a given focal distance and aperture dimension, this can be interpreted to mean that the resolution is <u>improved</u> by a factor of $1/\delta^3$ for the moving medium case, because a lower f/number lens is used to get focusing at the same z-axis plane. Physically, this "increase" in resolution is due to the shorter wavelength of plane-wave components traveling into the medium velocity.

Although the specific examples of a slit aperture and a cylindrical lens were used to illustrate the effects of a moving medium on the diffraction from a one-dimensional aperture, these effects are applicable to any one-dimensional diffracting screen within the limits of the approximations for Fresnel and Fraunhofer diffraction. We now consider the more general case of two-dimensional diffracting screens.

TWO-DIMENSIONAL APERTURES

The diffraction patterns formed by insonification of two-dimensional apertures are more complex than those of the one-dimensional case because of the anisotropy caused by the motion of the medium in one direction. Equation (18) is used to describe the density distribution in the Fresnel region, and equation (20) is used for the Fraunhofer region. A study of these equations reveals that the diffraction pattern is modified in the y-direction as well as in the x-direction by a medium velocity in the x-direction only. To illustrate these effects, the diffraction pattern for a rectangular aperture and the focal pattern for a lens with rectangular pupil function are presented.

The transmission function of the aperture shown in Fig. 1 is given by

$$t(x',y') = \text{rect}\left(\frac{2x'}{a}\right) \, \text{rect}\left(\frac{2a'}{b}\right) \tag{37}$$

For a normally-incident plane wave of unit amplitude, the two-dimensional Fourier transform of equation (37) is used in equation (18) to yield the density distribution in the Fraunhofer region

$$\rho(x,y,z) = \frac{jab}{\lambda_o} \frac{z}{(\delta R)^2} \exp[-j\frac{\beta_o}{\delta}\left(R - \frac{u}{c_o}\frac{x}{\delta}\right)]$$

$$\cdot \frac{\sin \frac{a\beta_o}{2\delta^2}\left(\frac{x}{\delta R} - \frac{u}{c_o}\right)}{\frac{a\beta_o}{2\delta^2}\left(\frac{x}{\delta R} - \frac{u}{c_o}\right)} \cdot \frac{\sin \frac{b\beta_o y}{2\delta R}}{\frac{b\beta_o y}{2\delta R}} \tag{38}$$

Fig. 7 shows plots of the modulus squared of this density for both the x-and y-directions for $u/c_o = 0$ and 0.9. Note that no change occurs in the y-direction for this case.

If the insonifying plane wave is incident at the angle θ defined by equation (28) with respect to the y,z-plane, the shift in the x-direction of the diffraction pattern is eliminated and the pattern is again compressed in the x-direction, as illustrated in Fig. 8. It is interesting to note that the pattern in the y-direction also suffers a compression, although not as much as in the x-direction.

As a final example, let us consider diffraction from a lens with a rectangular aperture. The nonsymmetry of the solutions so far suggests that we construct the lens with two different focal length cylindrical lenses with axes in the x- and y-directions, as

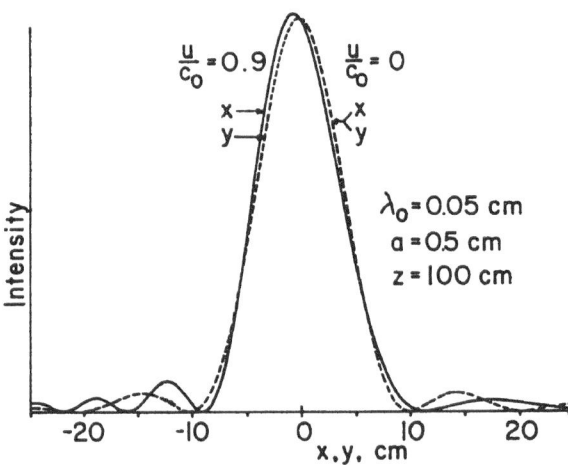

Fig. 7 The Fraunhofer-zone patterns of a square aperture radiating
        into a medium moving by with velocity u. The patterns are
        superimposed.

shown in Fig. 9. The transmission function for this lens system
is given by

$$t(x',y') = \text{rect}\left(\frac{2x'}{a}\right) \ \text{rect}\left(\frac{2y'}{b}\right) \ \exp[j\frac{\beta_0}{2}\left(\frac{x'^2}{f_1} + \frac{y'^2}{f_2}\right)] \tag{39}$$

Then, in the case of the normally-incident, unit-amplitude plane
wave, the distribution in the Fresnel zone is found from equation
(18) to be

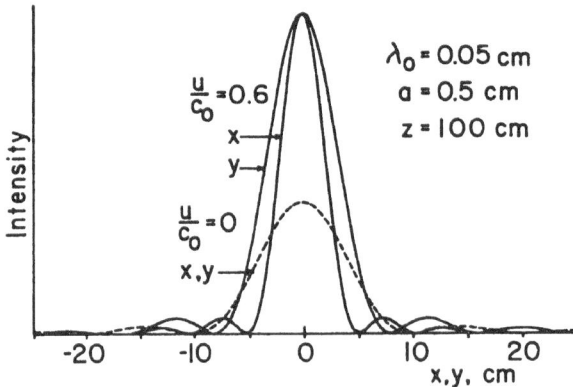

Fig. 8 The Fraunhofer-zone patterns of a square aperture radiating
        into a medium moving by with velocity u. The aperture is
        excited with a wave at angle θ just sufficient to put the
        peak intensity on the z-axis.

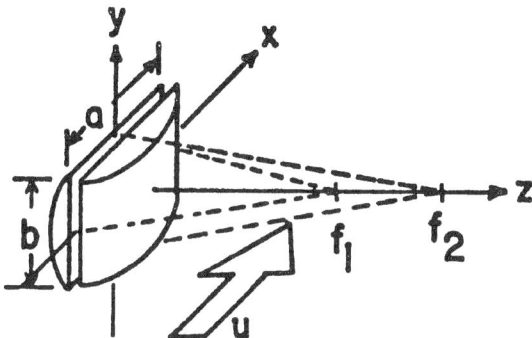

Fig. 9 Crossed cylindrical lenses used in a rectangular aperture
radiating into a medium moving by with velocity u.

$$\rho(x,y,z) = \frac{j}{2\pi} \frac{\beta_o z}{(\delta R)^2} \exp[-j\frac{\beta_o}{\delta}\left(R - \frac{u}{c_o}\frac{x}{\delta}\right)]$$

$$\cdot \int_{-a/2}^{a/2} \int_{-b/2}^{b/2} \exp\{-j\frac{\beta_o}{2}[x'^2\left(\frac{1}{\delta^3 R} - \frac{1}{f_1}\right) + y'^2\left(\frac{1}{\delta R} - \frac{1}{f_2}\right)]\}$$

$$\cdot \exp\{j\frac{\beta_o}{\delta}[\left(\frac{x}{\delta^2 R} - \frac{1}{\delta}\frac{u}{c_o}\right)x' + \frac{y}{R}y']\} \, dx'dy' \qquad (40)$$

If $f_1 = f_2 \equiv f$, as in the case of a spherical lens, there exists
no surface where the diffraction pattern is focused in both the x-
and y-directions. When $R = f/\delta^3$, the pattern is focused in the x-
direction, but not in the y-direction. When $R = f/\delta$, the converse
is true.

The system of Fig. 9 allows us to choose the focal lengths f_1
and f_2 such that

$$\frac{f_1}{\delta^3} = \frac{f_2}{\delta} = R \equiv [\left(\frac{x}{\delta}\right)^2 + y^2 + z^2]^{1/2} \qquad (41)$$

That is, our solution is valid only on this surface of focus
defined by equation (41), which is an oblate spheroid of eccentricity
u/c_o. On this surface, the first exponential in the integrand of
equation (40) is unity, so that the density distribution is given

again by equation (38). If the lenses are insonified at the angle θ
of equation (21), the shift in x-position is eliminated and R may
be approximated by z, as before, to yield

$$
\rho(x,y,z) = \frac{j}{2\pi} \frac{ab\beta_o}{\delta^2 z} \exp\left[-j\frac{\beta_o}{\delta}\left(z - \frac{u}{c_o}\frac{x}{\delta}\right)\right] \cdot
$$

$$
\cdot \frac{\sin\dfrac{a\beta_o x}{2f_1}}{\dfrac{a\beta_o x}{2f_1}} \cdot \frac{\sin\dfrac{b\beta_o y}{2f_2}}{\dfrac{b\beta_o y}{2f_2}} \tag{42}
$$

For this special case, the lenses produce a focal spot in both
directions at a single plane z, determined by the velocity of the
medium. If it is desired to have a focal spot that is symmetric
in x and y, then the ratio of aperture to focal length is made the
same in the two directions: $a/f_1 = b/f_2$. This also is the same
as choosing $a = b\delta^2$.

APPLICATION TO HIGH-SCAN-RATE ACOUSTICAL SYSTEMS

Most acoustic imaging systems that scan an object for image
information move the probing fields on the receiver pattern rather
than the object. This allows the possibility of real-time image
presentation if electronic scanning speeds are used. The scanning
speed is the product of the number of image lines in a frame with
the length of a line and the frame rate. For example, with a 200-
line raster, a 10-cm line length, and a frame rate of 30 per
second, the scanning speed is 600 m/s, or 40% of the speed of
sound in water. (With electonic scanning, it is possible to
conceive supersonic scannning speeds, a regime we have not considered.)

These scanning speeds are achieved with an electronically-
scanned, synthetic aperture lens. Such a lens can be synthesized,
for example, by a phase-and-amplitude-controlled transducer array,
as in the system of J.F. Havlice, et al.[5] or by a scanning, Gabor
zone-plate pattern as in the proposed system of Wang and Wade[6].
Thus, while nothing actually moves, the excitation pattern creating
the focal point moves at the scanning speed. To make the results
presented above applicable to these systems, we must re-transform
the above expressions, using equation (2), back to the coordinates
(x_1, y_1, z_1) in which the medium is stationary. This transformation
does not alter the conclusions of the above analysis. The scanned
focal point lags behind the excitation pattern and focuses at
different distances for the x- and y-directions unless the corrective measures described above are taken. If the receiver does not

scan with the source, there is a Doppler shift in the frequency
developed (corresponding to the Doppler change in wavelength of
the untransformed system) which can be found explicitly. If
normal incidence excitation is used, the side lobes will stretch
on one side and contract on the other, as shown above. This
effect seems to be evident in Fig. 7 of reference (5). If the
synthetic aperture directs the radiation pattern towards the
scanning direction, then better resolution is achievable. In
short, if scan speeds are an appreciable fraction of the sonic
speed, then the optimum system design can be improved using the
principles described above.

OTHER SCAN-RATE RELATED PROBLEMS

Rather simple solutions were obtained to the basic problem
because we were able to find coordinates where a steady-state,
single frequency solution was available. If the system is pulsed,
then many frequencies are present, and a transient analysis must
be made. The transit times of various frequency components from
the aperture to the observation screen may be different, thus the
straightforward superposition of plane waves used in this analysis
may not be applicable in the pulsed case. Transient effects also
occur in the starting or stopping times of the raster scan in
preparation for the flyback cycle. The assumption of continuous
motion is not always valid. Phased array systems are necessarily
made of discrete transducer elements separately controlled. The
effective aperture or lens does not travel perfectly smoothly, but
proceeds in a modulated fashion as control is passed from one
group of elements to the next. This transfer may take place in a
relatively smooth way, or in a discrete jump. In any case, one
has to account for the effect of the modulation of the scanning
excitation on the plane-wave components put into the medium.

A final problem not dealt with in this paper is the case of
medium velocities equal to or faster than the phase velocity of
the waves in a still medium. This case is ignored because it adds
much complexity to the analysis and to the physical understanding,
and because it is not frequently encountered in acoustic imaging
systems.

CONCLUSIONS

Development of a generalized Rayleigh-Sommerfeld equation
which includes the effects of a moving medium has allowed the
calculation of the radiation patterns of apertures radiating into
a moving medium. The converse problem of a moving aperture radiating into a stationary medium is also solved and is the usual

practical arrangement. The results clearly show that these effects must be taken into account in the design of equipment and that better resolution can be achieved by turning the results to the equipment's advantage.

ACKNOWLEDGEMENTS

The authors wish to thank Prof. Glen Wade, Prof. Jorge Fontana, Agustin Coello-Vera, Larry Schlussler and Kung-Yen Su for constructive criticism and discussion and Chuck Iverson for the meticulous preparation of this manuscript.

REFERENCES

1. B.P. Hildebrand, "Effect of High Scanning Velocities on the Holographic Image," Journal of the Opt. Soc. Amer., Vol. 60, No. 9, pp. 1166-1168 (1970).

2. A. Coello-Vera, L. Schlussler, G. Flesher, G. Wade, and J. Fontana, "Motion Limitations in Acoustic Holography Using an Electronic Reference" to be published in the Proc. of the Symposium on Optical, Electro-Optical, Laser and Photographic Technology, San Diego, CA (1976).

3. M. Born and E. Wolf, "Principles of Optics," Pergamon Press, N.Y. (1968).

4. J. Goodman, "Introduction to Fourier Optics," McGraw-Hill, N.Y. (1968).

5. J.F. Havlice, G.S. Kino, J.S. Kofol and C.F. Quate, "An Electronically Focused Acoustic Imaging Device," Acoustical Holography, Vol. 5, Plenum Press, N.Y., pp. 317-333 (1973).

6. K. Wang and G. Wade, "A Scanning Focused-Beam System for Real-Time Diagnostic Imaging," Acoustical Holography, Vol. 6, Plenum Press, N.Y., pp. 213-228 (1975).

PARTICIPANTS

I.P. Adachi
Minolta
3105 Lomita Boulevard
Torrance, Ca.   90505

R. Addison
American Optical
P.O. Box I
Southbridge, Mi.   01550

P. Alais
Laboratoire d'Acoustique Physique
  Informatique
Equipe de Recherche Associee AU
  CNRS
Universite Pierre et Marie Curie
Paris   France

G.A. Alphonse
RCA Laboratories
Princeton, N.J.   08540

B.H. Anderson
University of Michigan
2319 E.E.
525 E. University
Ann Arbor, Mi.   48109

E. Arkans
Kendall
411 Lake Zurich
Barrington, Il.   60010

A. Arnold
40 South Clay
Hinsdale, Il.   60521

H. Ashley
Bio-Dynamics
9115 Hagan Road
Indianapolis, In.   46250

G. Baum
Albert Einstein School of
  Medicine
1300 Morris Park Avenue
Bronx, N.Y.   10463

M. Birnboim
Rensselaer Polytech
Troy, N.Y.   12181

M.A. Blizard
Office of Naval Research
800 North Quincy Street
Arlington, Va.   22217

N. Booth
U.S. Navy
4643 El Cerrito Drive
San Diego, Ca.   92115

G. Brandenburger
Washington University
School of Medicine
Biomedical Computer Lab
700 South Euclid
St. Louis, Mo.   63110

B.B. Brenden
Holosonics, Inc.
2950 George Washington Way
Richland, Wa.   99352

631

E. Bridoux
Laboratoire d'Opto-Acousto-
Electronique
Centre Universitaire de
Valenciennes
59326 Valenciennes Cedex France

C. Bruneel
Laboratoire d'Opto-Acousto-
Electronique
Centre Universitaire de
Valenciennes
59326 Valenciennes Cedex France

C.B. Burckhardt
Hoffman-Laroche
Basel   Switzerland

R. Burr
Pratt & Whitney Aircraft
Aircraft Road Bldg. 140
Middletown, Ct. 06457

D. Burrows
40 South Clay
Hinsdale, Il. 60521

H. Busey
Picker Corporation
12 Clintonville Road
Northford, Ct. 06472

C.L. Buxbaum
Picker Corporation
12 Clintonville Road
Northford, Ct. 06472

E.N. Carlsen
Loma Linda University
Medical Center
Loma Linda, Ca. 92354

P.G. Cath
Ohio Nuclear
6000 Cochran Road
Solon, Oh. 44139

H. Chang
Electrical Engineering Dept.
University of Houston
Houston, Tx. 77004

M.A. Chaszeyka
Office of Naval Research
536 South Clark Street
Chicago, Il. 60605

D.A. Christensen
University of Utah
1400 East 2nd South
Salt Lake City, Ut. 84112

T.H. Christensen
N.O.R.D.A.
NSTL Bay
St. Louis, Ms. 39520

J.R. Cleveland
G.E. Medical Systems
3000 West Grandview
Waukesha, Wi. 53186

A. Coello-Vera
Electrical Engineering &
 Computer Science Department
University of California
Santa Barbara, Ca. 93106

H.D. Collins
Holosonics, Inc.
2950 George Washington Way
Richland, Wa. 99352

D. Cosgrove
Royal Marsden Hospital
Donous Road
Sutton, Surrey United Kingdom

H. Dardy
Naval Research Lab
Code 8133
Washington, D.C. 20375

P. Das
Rensselaer Polytechnic Institute
Troy, N.Y. 12181

W.T. Davids
Varian Associates
611 Hansen Way
Palo Alto, Ca.   94303

J. Deschamps
Thomson-CSF, TDI
B.P. 54   38120
Saint-Egreve   France

J.D. deVeer
American Optical Corporation
14 Mechanic Street
Southbridge, Ma.   01550

D.E. Dick
Electrical Engineering Dept.
University of Colorado
Boulder, Co.   80907

M.L. Dick
Electrical Engineering Dept.
University of Colorado
Boulder, Co.   80907

J.F. Dreyer
Dreyer Laboratories
9854 Zig Zag Road
Cincinnati, Oh.   45242

F.J. Eberhardt
U.S. Naval Academy
Electrical Engineering Dept.
Annapolis, Md.   21402

J.A. Edward
General Electric Company
Syracuse, N.Y.   13224

R.C. Eggleton
Indiana Univ. School of Medicine
 and ICFAR
1219 West Michigan Street
Indianapolis, In.   46202

S. Elliott
University of California
Electrical Enginerring and
 Computer Science Department
Santa Barbara, Ca.   93106

K.R. Erikson
Rohe Scientific Corporation
2158 South Hathaway
Santa Ana, Ca.   92705

A. Escarous
75 Saint Alphonses Street
Apt. 2005
Boston, Ma.   02120

K-F Etzold
RCA Laboratories
David Sarnoff Research Center
Princeton, N.J.   08540

R.C. Fairchild
University of Michigan
2319 Electrical Engineering
525 East University Avenue
Ann Arbor, Mi.   48109

J.B. Far
Western Geophysical Company
P.O. Box 2469
Houston, Tx.   77001

D. Feinstein
University of Pennsylvania
4116 Spruce Street
Philadelphia, Pa.   19104

W.R. Fenner
Aerospace Corporation
2350 El Segundo Boulevard
El Segundo, Ca.   90245

M. Fink
Laboratoire d'Acoustique Physique
 Informatique
Equipe de Recherche Associee AU
 CNRS
Universite Pierre et Marie Curie
Paris   France

G. Flesher
Electrical Engineering and
 Computer Science Department
University of California
Santa Barbara, Ca.   93106

C.M. Fortunko
Rockwell International
Science Center
1049 Camino Dos Rios
Thousand Oaks, Ca. 91360

K. Frank
Case Western Reserve University
2065 Adelbert Road
Cleveland, Oh. 44106

D. Frieda
University of California
Center for Health Sciences
Radiological Sciences Department
Los Angeles, Ca. 90024

W.S. Gan
Department of Physics
Nanyang University
Upper Jurong Road
Singapore 22,
Republic of Singapore

D.A. Gaubatz
Electrical Engineering Dept.
The Catholic Univ. of America
Washington, D.C. 20064

R.S. Gawlik
RCA Laboratories
2 Washington Road
Princeton, N.J. 08540

J.C. Geck
General Electric Company
1701 Blackhawk Trail
Waukesha, Wi. 53186

W.G. Gerhard
Dornier System GmbH
Postfach 1360
Friedrichshafen
Fed. Rep. of Germany D-7990

J.H. Gieske
Sandia Laboratories - 9352
Albuerquerque, N.M. 87115

C.L. Giles
University of Arizona
Optical Sciences
Tuscon, Az. 85721

G.A. Gilmour
Westinghouse Electric Corp.
Defense & Elec. Systems Center
P.O. Box 1693
Baltimore, Md. 21203

G.H. Glover
General Electric Company
Corporate Research & Development
P.O. Box 8
Schenectady, N.Y. 12301

L.C. Granger
Bendix Electrodynamics Division
15825 Roxford Street
Sylmar, Ca. 91342

P.S. Green
Stanford Research Institute
333 Ravenswood Avenue
Menlo Park, Ca. 94025

P.L. Greene
Bendix Electrodynamics Division
15852 Roxford Street
Sylmar, Ca. 91342

M. Greenfield
University of California
Radiological Sciences Dept.
Center for the Health Sciences
Los Angeles, Ca. 90024

J.F. Greenleaf
Mayo Foundation
Biophysical Sciences Unit
Dept. of Physiology & Biophysics
Rochester, Mn. 55901

R.P. Gribble
Holosonics, Inc.
2950 George Washington Way
Richland, Wa. 99352

A. Hanafy
Hewlett Packard
174 Wyman Street
Waltham, Ma. 02154

C.R. Hansen
Biophysical Sciences Unit
Dept. of Physiology & Biophysics
Mayo Foundation
Rochester, Mn. 55901

M.E. Haran
Bureau of Radiological Health
5600 Fishers Lane
Rockville, Md. 20852

J. Hart
Hewlett Packard
175 Wyman Street
Waltham, Ma. 02154

J.F. Havlice
Stanford Research Institute
333 Ravenswood Avenue
Menlo Park, Ca. 94025

G. Heidbreder
University of California
Electrical Engineering and
 Computer Science Department
Santa Barbara, Ca. 93106

J.S. Heyman
NASA-Langley Research Center
Mail Stop 499
Hampton, Va. 23665

B.P. Hildebrand
Battelle Pacific Northwest Labs
Richland, Wa. 99352

R.E. Hileman
Unirad Corporation
4785 Oakland Street
Denver, Co. 80239

B. Ho
Michigan State University
Electrical Engineering and
 Systems Science Department
E. Lansing, Mi. 49923

A.E. Holt
Babcock & Wilcox Company
Research & Development Division
P.O. Box 1260
Lynchburg, Va. 24505

K.E. Huff
Eastman Kodak Company
Research Lab Building 81
Rochester, N.Y. 14650

D.E. Huffern
Battelle Pacigic Northwest Labs
Richland, Wa. 99352

R.G. Hughes
NRL
Washington, D.C. 20375

L.H. Hussman
Loyola University Medical Center
111 South Northwest Highway
Palatine, Il. 60067

O. Ikeda
Tokyo Institute of Technology
4259 Nagatsuda
Midori-Ku, Yokohama   Japan

J. Ishii
Tokyo Institute of Technology
Faculty of Science & Engineering
4259 Nagatsuda
Midori-Ku, Yokohama 227   Japan

C. Jaffe
Yale New Haven Hospital
789 Howard Avenue
New Haven, Ct. 06504

C. Jagatich
Ohio Nuclear
6000 Cochran Road
Solon, Oh. 44139

R. Jaszczak
Searle
333 East Howard
Des Plaines, Il. 60018

B. Johnson
Columbia Gas System Service Corp.
1600 Dublin Road
Columbus, Oh. 43215

S.A. Johnson
Mayo Foundation
Biophysical Sciences Unit
Dept. of Physiology & Biophysics
Rochester, Mn. 55901

H.W. Jones
Acoustic Group
University of Calgary
Calgary, Alberta
T2N 1N4 Canada

J.P. Jones
Case Western Reserve University
School of Medicine
Cleveland, Oh. 44106

R.K. Jurgen
IEEE
345 East 47th Street
New York, N.Y. 10017

R.E. Jutila
Mayo Clinic - U.W.
17 North Bassett
Madison, Wi. 53703

H.B. Karplus
Argonne National Laboratory
Argonne, Il. 60439

F. Kearly
University of California
Radiological Sciences Department
Center for the Health Sciences
Los Angeles, Ca. 90024

P.N. Keating
Bendix Research Labs
Southfield, Mi. 48076

E. Kelly-Fry
Indiana University School of
 Medicine & ICFAR
1219 West Michigan Street
Indianapolis, In. 46202

J.C. Kennedy
The Boeing Company
P.O. Bo 3707
Seattle, Wa. 98124

L.W. Kessler
Sonoscan, Inc.
720 Foster Avenue
Bensenville, Il. 60106

N.B. Kindig
University of Colorado
Electrical Engineering Department
Boulder, Co. 80907

G. Kino
Stanford University
Stanford, Ca. 94305

R.J. Kleehammer
Thomson-CSF
Electron Tubes
750 Bloomfield Avenue
Clifton, N.J. 07015

J. Kleeper
Washington University
Physics Department
St. Louis, Mo. 63130

A. Korpel
Zenith Radio Corporation
6001 West Dickens
Chicago, Il. 60639

R.J. Kostelnicek
Exxon Product Research
P.O. Box 2189
Houston, Tx. 77001

S.T. Kowel
Syracuse University
111 Link Hall
Syracuse, N.Y. 13210

R.N. Kubik
Babcock & Wilcox
P.O. Box 1260
Lynchburg, Va. 24505

R. Kuc
Columbia University
1312 Mudd
New York, N.Y. 10027

W.E. Lawrie
Babcock & Wilcox
Research & Development Division
P.O. Box 1260
Lynchburg, Va. 24505

W.M. Lechelt
Holosonics, Inc.
2950 George Washington Eay
Richland, Wa. 99352

B.T. Lee
Bendix Electrodynamics Division
15852 Roxford Street
Sylmar, Ca. 91342

C.H. Lee
Electrical Engineering Dept.
Syracuse University
Syracuse, N.Y. 13210

C. LeMay
EMI, Ltd.
Trevor Road
Hayes Middlesex, England
United Kingdom

A. Lent
State Univ. of New York/Buffalo
Computer Science Department
Amherst, N.Y. 14226

W. Leung
Stanford University
Stanford, Ca. 94305

G.K. Lewis
Searle
2000 Nuclear Drive
Des Plaines, Il. 60018

K. Liang
Mayo Clinic
Rochester, Mn. 55901

M. Linzer
National Bureau of Standards
Division 313.01
Washington, D.C. 20234

M. Lo
University of California
Electrical Engineering and
 Computer Science Department
Santa Barbara, Ca. 93106

M. Luetkemeyer
KAE
POB 448545
28 Bremen
44 West Germany

J.F. McDonald
Rensselaer Polytechnic Institute
Troy, N.Y. 12181

R.E. McKeighen
Searle
200 Nuclear Drive
Des Plaines, Il. 60018

F.D. McLeod
Colorado State University
Physiology Department, CRHL
Fort Collins, Co. 80521

J.A. Majde
Office of Naval Research
536 South Clark Street
Chicago, Il. 60605

W.R. Mayberry
Peabody Testing
1118 Chess Drive
Foster City, Ca. 94404

J.D. Meindl
Stanford University
Electronics Laboratory
Stanford, Ca. 94305

R.D. Melen
Stanford University
Stanford, Ca. 94305

M.V. Menon
Office of Naval Research
536 South Clark Street
Chicago, Il. 60605

A.F. Metherell
University of California - Irvine
Radiology Department
101 City Drive South
Orange, Ca. 92668

R. Mezrich
RCA Laboratories
David Sarnoff Research Center
Princeton, N.J. 08540

Y. Michiguchi
Atomic Energy Research Lab
Hitachi Ltd.
Ozenji, Tamaku, Kangawa Japan

J.G. Miller
Washington University
Physics Department
St. Louis, Mo. 63130

R.B. Moyer
Carpenter Technology Corporation
101 West Bern Street
Reading, Pa. 19603

R.K. Mueller
University of Minnesota
9707 Manning Avenue North
Stillwater, Mn. 55082

W.F. Mullen
Stanford Research Institute
333 Ravenswood Avenue
Menlo Park, Ca. 94025

Y. Nakamura
Faculty of Science & Engineering
Tokyo Institute of Technology
Nagatsuda, Midori-ku, Yokohama
Japan 227

C. Nianios
University of Pennsylvania
4039 Chestnut Street
Philadelphia, Pa. 19104

A.K. Nigam
NYIT
Science & Technology Res. Ctr.
8000 North Ocean Drive
Dania, Fl. 33004

M. Nonaka
Faculty of Science & Engineering
Tokyo Institute of Technology
Nagatsuda, Midori-ku, Yokohama
227 Japan

B. Nongaillard
Laboratoire d'Opto-Acousto-
 Electronique
Centre Universitaire de
 Valenciennes
59326 Valenciennes Cedex France

W.D. O'Brien
University of Illinois
Electrical Engineering Dept.
Urbana, Il. 61801

H. Ohshima
Tokyo Institute of Technology
4259 Nagatsuda
Midori-Ku, Yokohama-Shi   Japam

C.P. Olinger
University of Cincinnati
Cincinnati, Oh.   45267

M. Olinger
Michigan State University
Electrical Engineering and
 Systems Science Department
E. Lansing, Mi.   48823

D. Perozek
Hewlett Packard
175 Wyman Street
Waltham, Ma.   02154

D.R. Phillips
University of North Carolina
School of Public Health
Chapel Hill, N.C.   27514

J.P. Powers
Naval Postgraduate School
Electrical Engineering Dept.
Monterey, Ca.   93940

R. Priemer
University of Illinois -
 Chicago Circle
Information Engineering Dept.
Chicago, Il.   60680

E.W. Purnell
Case Western Reserve University
School of Medicine
Cleveland, Oh.   44106

M. Ragozzino
University of Chicago
1515 East 54th
Chicago, Il.   60615

B. Rajagopalan
Mayo Clinic
Rochester, Mn.   55901

W.A. Riley
Bowman Gray School of Medicine
Neurology Department
Winston-Salem, N.C.   27103

D. Robinson
Columbia Gas System
1600 Dublin Road
Columbus, Oh.   43215

H.A.F. Rocha
General Electric Company
Corporate Researc & Development
P.O. Box 8
Schenectady, N.Y.   12345

D.S. Rodbell
RPI-CBME
RPI Campus
Troy, N.Y.   12181

S. Rosell
Luthernal Medical Center
514 49th Street
New York, N.Y.   10019

R. Rosenfeld
Eastman Kodak
Kodak Park   Building 59
Rochester, N.Y.   14650

J.M. Rouvaen
Laboratoire d'Opto-Acousto
 Electronique
Centre Universitaire de
 Valenciennes
59326 Valenciennes Cedex   France

R. Ryan
Federal Dept. of Transportation
Transportation Systems Center
Kendall Square
Cambridge, Ma.   02142

W.F. Samayoa
Mayo Foundation
Biophysical Science Unit
Physiology & Biophysics Dept.
Rochester, Mn.   55901

S.M. Sanzgiri
University of Illinoid
Information Engineering Dept.
Box 4348
Chicago, Il. 60680

K. Sasaki
Faculty of Science & Engineering
Tokyo Institute of Technology
Nagatsuda, Midori-Ku, Yokohama
227   Japan

T. Sato
Tokyo Institute of Technology
4259 Nagatsuda
Midori-Ku, Yokohama-Shi   Japan

T. Sawatari
Bendix Research Labs
20800 10½ Mile Road
Southfield, Mi. 48075

V. Schmitz
I.Z.F.P.
66 Saarbrucken University
Saarbrucken   Germany

M. Shah
University of Illinois
4240 Clarendon   # 63
Chicago, Il. 60613

H. Shaw
Stanford University
Stanford, Ca. 94305

H. Shen
University of Houston
Electrical Engineering Dept.

J.D. Shott
Stanford University
Stanford, Ca. 94305

M. Siegel
Michigan State University
Electrical Engineering and
 Systems Science Department
E. Lansing, Mi. 48823

G.P. Singh
Drexel University
32nd & Chestnut Street
Philadelphia, Pa. 19104

J.M. Smith
AMMRC
Arsenal Street   Building 313
Watertown, Ma. 02172

L.B. Smith
Beckman Instruments, Inc.
3900 River Road
Schiller Park, Il. 60176

J. Souquet
Stanford University
Stanford, Ca. 94305

J. Spalek
Holosonics, Inc.
2950 George Washington Way
Richland, Wa. 99352

P. Spiegler
University of California
Radiological Sciences Dept.
Center for the Health Sciences
Los Angeles, Ca. 90024

J.C. Stamm
University of Illinois -
 Chicago Circle
Information Engineering Dept.
Chicago, Il. 60680

R. Stern
University of California
Radiological Sciences Dept.
Center for the Health Sciences
Los Angeles, Ca. 90024

K. Su
University of California
Electrical Engineering and
 Computer Science Department
Santa Barbara, Ca. 93106

T. Sunada
Faculty of Science & Engineering
Tokyo Institute of Technology
Nagatsuda, Midori-Ku, Yokohama
227   Japan

V. Suryanarayana
University of Cincinnati
Stroke Clinic
Cincinnati, Oh.   45221

J.L. Sutton
Naval Undersea Center
Code   6513
San Diego, Ca.   92117

K. Suzuki
Atomic Energy Research Lab
Hitachi Ltd.
Ozenji, Tamaku, Kanagawa   Japan

J.C. Taenzer
Stanford Research Institute
333 Ravenswood Avenue
Menlo Park, Ca.   94025

F. Takahashi
Atomic Energy Research Lab
Hitachi Ltd.
Ozenji, Tamaku, Kawasaki   Japan

M. Tanaka
Biophysical Sciences Unit
Physiology & Biophysics Dept.
Mayo Foundation
Rochester, Mn.   55901

R.H. Tancrell
Raytheon Research
28 Seyon Street
Waltham, Ma.   02154

F. Thurstone
Duke University
Biomedical Engineering Dept.
Durham, N.C.   27706

H.F. Tiersten
Rensselaer Polytechnic Institute
Troy, N.Y.   12181

Toffer
United Nuclear Industries
P.O. Box   490
Richland, Wa.   99352

R. Torguet
Laboratoire d'Opto-Acousto-
 Electronique
Centre Universitaire de
 Valenciennes
59326 Valenciennes Cedex
France

M.F. Tse
Rensselaer Polytechnic Institute
Troy, N.Y.   12181

H.W. Tyrer
Becton Dickinson Res. Ctr.
P.O. Box   12016
Research Triangle Park, N.C.
27709

M. Vestrheim
University of Bergen
Department of Physics
5014 Bergen-U   Norway

D. Vilkomerson
RCA Laboratories
David Sarnoff Research Center
Princeton, N.J.   08540

F.S. Vinson
Indiana University School of
 Medicine and ICFAR
1219 West Michigan Street
Indianapolis, In.   46202

S. Wadaka
Faculty of Science & Engineering
Tokyo Institute of Technology
Nagatsuda, Midori-Ku, Yokohama
227   Japan

G. Wade
University of California
Electrical Engineering and
  Computer Science Department
Santa Barbara, Ca. 93106

K. Wang
University of Houston
Electrical Engineering Dept.

S. Wang
IBM Research
Yorktown Heights, N.Y.   10598

T.M. Waugh
Stanford University
Ginzton Laboratory
Stanford, Ca. 94305

F. Whitaker
G.E. Medical Systems
730 East North Street
Waukesha, Wi. 53186

F.B. Whitehead
Searle Radiographics
Des Plaines, Il. 60018

R. Whitman
Zenith Radio Corporation
6001 West Dickens
Chicago, Il. 60639

C.J. Williams
Acoustic Group
University of Calgary
Calgary, Alberta
T2N 1N4 Canada

I.M. Wilson
General Electric Co., Ltd.
Hirst Research Center
East Lane Wembley
Middlesex England HA9 7PP

D. Winslow
Stanford University
Stanford, Ca. 94305

R.L. Woolley
Brigham Young University
Provo, Ut. 84601

D.J. Wootton
EMI Ltd.
Trevor Road
Hayes Middlesex England U.K.

J.W. Young
AMETEK
Straza Division
790 Greenfield Drive
El Cajon, Ca. 92022

D.E. Yuhas
Sonoscan, Inc.
720 Foster Avenue
Bensenville, Il. 60106

J. Zagzebski
University of Wisconsin
Madison, Wi. 53709

M. Zambuto
New Jersey Institute of
  Technology
Newark, N.J. 07102

J.K. Zieniuk
Institute of Fundamental
  Technological Research
Polish Academy of Sciences
00-049, Warsaw,   Poland

G. Zilinskas
Bendix Electrodynamics Division
15825 Roxford Street
Sylmar, Ca. 91342

L. Zitelli
Stanford University
Stanford, Ca. 94305

R. Zuleeg
McDonnell Douglas Astronautics
5301 Bolsa Avenue
Huntington Beach, Ca. 92647